AF732493

Regenerative Approaches in Dentistry

Sepanta Hosseinpour • Laurence J. Walsh
Keyvan Moharamzadeh
Editors

Regenerative Approaches in Dentistry

An Evidence-Based Perspective

Springer

Editors
Sepanta Hosseinpour
School of Dentistry
University of Queensland
Herston, QLD, Australia

Laurence J. Walsh
School of Dentistry
University of Queensland
Herston, QLD, Australia

Keyvan Moharamzadeh
Hamdan Bin Mohammed College of
Dental Medicine (HBMCDM)
Mohammed Bin Rashid University of
Medicine and Health Sciences (MBRU)
Dubai, United Arab Emirates

ISBN 978-3-030-59808-2 ISBN 978-3-030-59809-9 (eBook)
https://doi.org/10.1007/978-3-030-59809-9

© Springer Nature Switzerland AG 2021
This work is subject to copyright. All rights are reserved by the Publisher, whether the whole or part of the material is concerned, specifically the rights of translation, reprinting, reuse of illustrations, recitation, broadcasting, reproduction on microfilms or in any other physical way, and transmission or information storage and retrieval, electronic adaptation, computer software, or by similar or dissimilar methodology now known or hereafter developed.
The use of general descriptive names, registered names, trademarks, service marks, etc. in this publication does not imply, even in the absence of a specific statement, that such names are exempt from the relevant protective laws and regulations and therefore free for general use.
The publisher, the authors, and the editors are safe to assume that the advice and information in this book are believed to be true and accurate at the date of publication. Neither the publisher nor the authors or the editors give a warranty, expressed or implied, with respect to the material contained herein or for any errors or omissions that may have been made. The publisher remains neutral with regard to jurisdictional claims in published maps and institutional affiliations.

This Springer imprint is published by the registered company Springer Nature Switzerland AG
The registered company address is: Gewerbestrasse 11, 6330 Cham, Switzerland

Preface

This book has been written to address the current need for a comprehensive reference on regenerative approaches in dentistry with a special focus on current clinical applications and future potentials. This topic has been evolved significantly in recent decades, and there is a plethora of papers (experimental and clinical studies) in various branches of dental science. This book aims to collect and compare what has been done to provide evidence-based information for clinicians and scientists in this emerging field of dentistry.

The book is intended to educate the readers about various therapeutic modalities used in the reconstruction of hard and soft tissues in the maxillofacial region. Different potential laboratory and clinical applications of engineered oral and dental tissue equivalents are discussed in the relevant chapters.

It is hoped that this book will be useful to students in dentistry (at both undergraduate and postgraduate levels), as well as scientists and researchers in the field of biomedical sciences in dentistry, and clinicians.

Queensland, Brisbane, Australia — Sepanta Hosseinpour
Queensland, Brisbane, Australia — Laurence J. Walsh
Dubai, United Arab Emirates — Keyvan Moharamzadeh

Acknowledgments

This book could not be completed without the valuable contributions of the authors and their wider research groups and collaborators.

We would like to thank the Springer team, specifically Ms. Deepika Devan, Ms. Alison Wolf, and We would like to thank the Springer team, specifically Ms. Deepika Devan, Ms. Alison Wolf, and Mr. Manohar Vignesh, for their administrative support during the publication process, for their administrative support during the publication process.

We also thank our colleagues and families for their help and support during the preparation of this book.

Contents

Editors and Contributors

About the Editors

Sepanta Hosseinpour is an Associate Lecturer at the School of Dentistry, the University of Queensland, Australia. He received his doctorate in dental surgery (DDS) and Master of Public Health (MPH) in 2017, and he pursued academic research in tissue regeneration as Research Associate for two years after graduation and his PhD at the University of Queensland. At the moment, he has 34 peer-reviewed journal articles in dentistry some of which were published in high-quality journals.

Laurence J. Walsh is an Emeritus Professor at the University of Queensland School of Dentistry where has been on the academic staff for over 36 years. He served for 20 years as research group leader for Advanced Materials and Technologies at the University of Queensland School of Dentistry. He is also the program director for the special needs dentistry Doctor of Clinical Dentistry. He has worked in medical and dental research for over 35 years, covering advanced aspects of material science, biomedical optics, lasers, and nanomaterials. Professor Walsh holds patents in six families of advanced technologies for dental diagnosis and treatment. He has led research into new technologies, from bench studies through to clinical trials and cost-benefit studies. His over 900 publications include over 300 refereed journal papers, over 90 published protocols and guidelines for professional associations, over 190 published technical reports and literature reviews, 5 sole author books, and a further 30 book chapters or edited books. In January 2018, he was appointed an Officer of the Order of Australia (AO) in recognition of his distinguished service to dentistry, and to dental education, as an academic and author. He is only the fourth dentist in the history of Australia to receive this national honor.

Keyvan Moharamzadeh is a Professor of Endodontics at Hamdan Bin Mohammed College of Dental Medicine (HBMCDM), Mohammed Bin Rashid University of Medicine and Health Sciences (MBRU), Dubai, United Arab Emirates (UAE). He was previously a Senior Clinical Lecturer at the University of Sheffield and Honorary Consultant in Restorative Dentistry at Charles Clifford Dental Hospital, UK. He is a specialist in endodontics, periodontics, prosthodontics, and restorative dentistry and was the program director for postgraduate DClinDent periodontology and prosthodontics courses at the University of Sheffield. His clinical work has included both private dental practice at Harley Street Dental Group in London and hospital-based treatment of patients referred by General Dental Practitioners for advanced restorative rehabilitation. He has authored a clinical reference book on *Diseases and Conditions in Dentistry* and coedited a textbook on *Biomaterials for Oral and Dental Tissue Engineering*. Professor Moharamzadeh is a Fellow of the Royal College of Surgeons of England and the UK Higher Education Academy. He has held an honorary research position in Marquette University in the United States, has extensively published in the literature, and has given many presentations and invited lectures in the national and international conferences.

Contributors

Aya Alali, BDS, MPhil School of Dentistry, The University of Queensland, Herston, QLD, Australia

Amelia Carr, BEng (Med) (Hons) School of Dentistry, The University of Queensland, Herston, QLD, Australia

Antonio Celentano, DDS, LDS, GCUT, PhD Melbourne Dental School, The University of Melbourne, Carlton, VIC, Australia

Ning Cheng, BDS, DDS, PhD Department of Otolaryngology-Head and Neck Surgery, University of California San Francisco, San Francisco, CA, USA

Behzad Cheshmi, DDS, MSc Private Practitioner, Professional Member of Maxillofacial Surgery and Implantology Research Foundation, Boroujerd, Iran

Yoo Jung Chung, DDS University of California San Francisco School of Dentistry, San Francisco, CA, USA

Omid Dianat, DDS, MS, MDS Endodontic Division, University of Maryland, School of Dentistry, Baltimore, MD, USA

Hamid Reza Fallahi, DDS, MSc Dental Research Center, Research Institute of Dental Sciences, Shahid Beheshti University of Medical Science, Tehran, Iran

Oral and Maxillofacial Surgeon, Founder and Director of Maxillofacial Surgery and Implantology Research Foundation, Ahvaz, Iran

Camile S. Farah, BDSc, MDSc (OralMed OralPath) Australian Centre for Oral Oncology Research and Education, Nedlands, WA, Australia

Australian Clinical Labs, Subiaco, WA, Australia

Simon Fox, PhD Systems Biology and Genomics, Harry Perkins Institute of Medical Research, Nedlands, WA, Australia

Alexis Gaudin, DDS, PhD Faculty of Dental Surgery, Department of Endodontics and Restorative Dentistry, University of Nantes, Nantes, France

Shohreh Ghasemi, DDS, MSc Adjunct Assistant Professor of Oral and Maxillofacial Surgery of DCG, Georgia, USA

Fernando Guastaldi, PhD Skeletal Biology Research Center, Massachusetts General Hospital and Harvard School of Dental Medicine, Boston, MA, USA

Yan He, PhD, MDS (Ortho.), DDS Skeletal Biology Research Center, Massachusetts General Hospital and Harvard School of Dental Medicine, Boston, MA, USA

Institute of Stem Cells and Tissue Engineering, Wenzhou Medical University, Wenzhou, China

Rita Hitching, MSc Palo Alto Veterans' Institute for Research (PAVIR), VA Palo Alto Health Care System, Palo Alto, CA, USA

Sepanta Hosseinpour, DDS, MPH, PhD candidate School of Dentistry, The University of Queensland, Herston, QLD, Australia

Seied Omid Keyhan, DDS, MSc Oral and Maxillofacial Surgeon, Founder and Director of Maxillofacial Surgery and Implantology Research Foundation, Isfahan, Iran

Cagdas Caglar Lacin, DDS, PhD Department of Periodontology, Faculty of Dentistry, Istanbul Kent University, Istanbul, Turkey

Chang Lei, BDS, MDS, PhD Australian Institute for Bioengineering and Nanotechnology, The University of Queensland, Brisbane, QLD, Australia

Tulio Fernandez-Medina, DDS, DClin Dent, PhD School of Dentistry, The University of Queensland, Herston, QLD, Australia

Keyvan Moharamzadeh, BSc, DDS, PhD, FHEA, FDSRCS Hamdan Bin Mohammed College of Dental Medicine (HBMCDM), Mohammed Bin Rashid University of Medicine and Health Sciences (MBRU), Dubai, United Arab Emirates

Ashwin Nanda, BDS, MDS (Prosthodontics), MPhil School of Dentistry, The University of Queensland, Herston, QLD, Australia

Necla Asli Kocak Oztug, DDS, PhD Department of Periodontology, Faculty of Dentistry, Istanbul University, Istanbul, Turkey

School of Dentistry, The University of Queensland, Herston, QLD, Australia

Giuseppe Pantaleo, PhD Oral Surgery, Department of Medicine and Surgery, University of Salerno, Fisciano, Italy

Avina Paranjpe, BDS, MS, MSD, PhD Department of Endodontics, University of Washington, School of Dentistry, Seattle, WA, USA

Ove A. Peters, DMD, MS, PhD School of Dentistry, The University of Queensland, Herston, QLD, Australia

Maryam Rezai Rad, DDS, PhD Research Institute of Dental Sciences, Dental school, Shahid Beheshti University of Medical Sciences, Tehran, Iran

Srinivas Sulugodu Ramachandra, BDS, MDS School of Dentistry, The University of Queensland, Herston, QLD, Australia

Elham Shadmehr, DDS, MS UCSF Endodontic Division, University of California San Francisco School of Dentistry, San Francisco, CA, USA

Kate Shearston, BA/BSc (Hons), PhD UWA Dental School, University of Western Australia, Nedlands, WA, Australia

Naisana Seyedasli, PhD Faculty of Medicine and Health, University of Sydney, and Centre for Cancer Research, Westmead Institute for Medical Research, Westmead, NSW, Australia

Laurence J. Walsh, AO BDSc (Hons), PhD, DDSc, GCEd School of Dentistry, The University of Queensland, Herston, QLD, Australia

Juan Wen, BDS, MDS (Orthodontics) Department of Orthodontics, Nanjing Stomatological Hospital, Medical School of Nanjing University, Nanjing, Jiang Su, China

Munira Xaymardan, BDS, MPhil (Oral Surgery), PhD Faculty of Medicine and Health, Oral Biosciences, Sydney Dental School, University of Sydney, Westmead, NSW, Australia

Chun Xu, BDS, MDS (Oral Surgery), PhD School of Dentistry, The University of Queensland, Herston, QLD, Australia

Shaomian Yao, PhD Department of Comparative Biomedical Sciences, School of Veterinary Medicine, Louisiana State University, LA, USA

Qingsong Ye, DDS, MOrth, PhD, FICD Skeletal Biology Research Center, Massachusetts General Hospital and Harvard School of Dental Medicine, Boston, MA, USA

Center of Regenerative Medicine, Wuhan University, Wuhan, China

Institute of Stem Cells and Tissue Engineering, Wenzhou Medical University, Wenzhou, China

The Paradigm of Regenerative Dentistry and Its Future Perspectives

Laurence J. Walsh and Sepanta Hosseinpour

1 Regenerative Approaches for Treatment

In the human body, there is a continuous renewal of tissues such as bone marrow, epithelia, bone, and connective tissue. The potential for regeneration after injury varies greatly, from tissues that have excellent capabilities for complete regeneration, such as the liver, through to sensory cells that give the special senses of hearing and vision, which do not regenerate when injured.

When considering the response to inflammatory diseases, trauma, or malignancy, the preferred outcome is always true regeneration, rather than repair with scarring. How much any one tissue can undergo repair, or truly regenerate, is influenced by the type and number of cells that are present, particularly stem cells, which can differentiate to replace missing tissues. In some sites in the oral cavity, the number of cells with a high regenerative capacity is limited because of the small volume of the tissue, e.g., in the dental pulp, as the pulp chamber reduces in size with age to a volume of tens of microliters.

Enamel is a unique tissue since the forming cells, the ameloblasts, are no longer present by the time the tooth is erupting into the oral cavity. This makes dental enamel a unique tissue in that it sits at the boundary of hard and soft tissues and must rely entirely on chemical repair by remineralization from ions in the saliva.

In the case of bone, the size and shape of defects created by trauma could be beyond the repair potential of the body. Such "critical-sized" bone defects will not heal spontaneously, however, in such situations, therapeutic intervention can help the cells to "generate again." In order to achieve this, an inductive material and scaffold can be used to improve the homing of endogenous bone-forming cells and their subsequent differentiation. Alternatively, stem cells could be transplanted into the treatment site [1]. Such therapeutic approaches exploit the principles of tissue engineering to restore the function and structure of a specific tissue or organ [2].

Regenerative dentistry is the branch of regenerative medicine that focuses on regeneration of oral and dental tissues. It most often follows the conceptual triad of tissue engineering, by including cells, scaffolds, and bioactive molecules [3]. The cells could come from any number of stem cell sources that have been identified, including embryonic stem cells, adult stem cells, and induced pluripotent stem cells. The choice of cell type is aligned with the treatment objectives. Scaffolds can be designed to carry appropriate cells, and to deliver signaling molecules to orchestrate tissue healing [4]. In addition, scaffolds can be used as a carrier or support for cul-

L. J. Walsh (✉) · S. Hosseinpour
School of Dentistry, The University of Queensland, Herston, QLD, Australia
e-mail: l.walsh@uq.edu.au; Sp.hosseinpour@uqconnect.edu.au

© Springer Nature Switzerland AG 2021
S. Hosseinpour et al. (eds.), *Regenerative Approaches in Dentistry*,
https://doi.org/10.1007/978-3-030-59809-9_1

turing cells ex vivo before the differentiated tissue is transplanted surgically into the defect site. Bioactive molecules, such as growth factors, genes, and drugs, can be released by the scaffold, or delivered independently [5].

It is important that the scaffold mimics the characteristics of the target tissue in terms of biological activity, mechanical integrity, and functionality [6, 7]. To achieve this, the optimum design for regenerative treatments will vary, based on the target tissue. For instance, the requirements of a scaffold for pulpal tissue regeneration are totally different from the one used for alveolar bone augmentation for dental implant placement. The features must meet the regenerative demands as defined from the target site, and then be optimized to achieve the best outcomes.

2 Therapeutic Targets in the Field of Dentistry

2.1 The Whole Tooth

In humans, tooth development for the primary (deciduous) dentition starts approximately at 6–8 weeks in utero. The tooth forming organ, the tooth germ or dental follicle, is comprised of an outer layer of cells from ectoderm (oral epithelium) and an inner layer of neural crest ectomesenchyme. In the early stages, odontogenic inducing signals pass from the epithelial cells to the mesenchymal cells, which condense to become the dental papilla. This sequence of tooth formation, which spans the stages from a tooth germ to a completely formed tooth, involves a complex orchestration of cellular activity, with cascades of cytokines and enzymes that tightly regulate cell arrangement, proliferation, differentiation, and secretion over a period of several years [8, 9].

As the tooth forms, a range of stem cells participate, including those in the outermost regions of the dental follicle where the tooth is forming (dental follicle stem cells), through to stem cells that remain after tooth formation is complete (e.g., dental pulp stem cells, periodontal ligament stem cells, stem cells from the apical region of the dental papilla, and stem cells in the gingival epithelium). All of these stem cell populations can be harvested and used in regenerative dentistry. They have properties that are broadly similar to bone marrow-derived mesenchymal cells.

Once a tooth has formed and then erupted, its durability and longevity in the oral cavity can be influenced by many processes including dental caries, periodontal diseases, accelerated tooth wear, and traumatic injuries. In oral rehabilitation, common methods for replacing a tooth that is missing or has been lost include a fixed or removable prosthesis, or a single tooth dental implant that supports a full dental crown [10]. However, by applying the evolving knowledge of tooth development and stem cell biology, the concept of bioengineered teeth has emerged. Although tooth-like structures have been created successfully in animal models [11–13], the complexity of any one tooth presents a major challenge, because of variables such as tooth type, size, color, and occlusal anatomy. These distinct morphological and functional characteristics provide a major challenge that causes more complexity in design than replacement of other structures such as bone. In chapter "Tooth Bioengineering and Whole Tooth Regeneration," the most recent findings regarding this topic will be discussed.

2.2 Individual Dental Structures

Dental caries is a global public health problem, and it remains one of the most prevalent microbial diseases around the world [14]. Dental caries leads to demineralization and proteolytic destruction of the crowns and exposed root surfaces of teeth, with the enamel and dentin being cavitated by the action of an aciduric, acidogenic, polymicrobial dental plaque biofilm. If untreated, the invasion can reach the dental pulp, causing irreversible pulpitis and finally necrosis of that tissue.

The current approach to treatment of a cavitated tooth is the removal of infected dental tissues, followed by restoration of missing tooth structure with synthetic dental materials [15]. However, this method can best be considered a surgical approach involving excision of the affected area followed by a small extent of repair at the level of the dental pulp. The strategy is not ideal, however once a tooth surface cavitates, its complete regeneration is not possible. Consequently, promoting healing of affected dentin and reversal of incipient lesions of dental caries that have not yet cavitated are important goals of preventive and minimal intervention dentistry, so that the tooth is retained as a vital, biologically functional unit. A range of novel approaches have been used to achieve reversal and arrest of early carious lesions where the tooth surface is still intact.

2.2.1 Enamel

Enamel is the hardest tissue in the body due to its unique structural properties, its high content of biological apatite minerals (such as hydroxyapatite), and the structural arrangement of enamel rods and prisms which resists the application of external forces [16]. The ameloblasts which form enamel are highly specialized epithelial cells, and only exist during tooth development. They are lost once the tooth crown has fully formed, which is some months before the tooth erupts into the oral cavity [17].

Although enamel should be able to withstand the intense forces of mastication during the whole lifespan, because of the ubiquitous presence of a dental plaque biofilm, it is vulnerable to cycles of demineralization. These acid attacks could be from organic acids produced by a dysbiotic dental plaque biofilm that has formed in a low pH environment, or from inorganic or organic acids found in foods and drinks, or in regurgitated gastric acid contents, as occurs in dental erosion. The stronger the acid and the better it chelates calcium ions, the more rapid crystallites of apatite mineral on the enamel surface will dissolve into the saliva.

Initial lesions of dental caries that are still contained in the outer half of the enamel are referred to as white spot lesions because of their unique appearance. When there are sufficient bioavailable calcium and phosphate ions, these lesions can remineralize. The likelihood of this occurring is increased when low levels of fluoride ions are present. If cycles of remineralization exceed periods of demineralization during biofilm acid production, the surface will be maintained because of this chemical regeneration process.

On the other hand, if cycles of mineral loss dominate, the surface will lose so much mineral that its integrity will be compromised. It will eventually collapse, and the resulting cavity that forms then provides a protected site for the microbial biofilm to continue its destruction of the tooth, now being located in a more protected site. Many dental preventive strategies have been developed to tip the balance to favor remineralization over demineralization, and thus break the cycle of progressive mineral loss. Regrettably, once a cavity has formed, the options for arresting the process are far fewer, and focus mostly on silver fluoride used in combination with ammonia or with stannous fluoride. These topical treatments can prevent the destruction caused by dental caries from progressing further, but they cannot repair the missing tooth structure [18]. They also cause dramatic discoloration of the tooth.

Given these limitations, there is interest in the concept of true regeneration of tooth structure. For white spot enamel lesions and incipient lesions on root surfaces, providing the correct stoichiometric ratios of calcium, phosphate and fluoride ions (5:3:1) can arrest lesions and cause subsurface regeneration of mineral, back to normal levels. Casein phosphopeptide-amorphous calcium phosphate provides such a ratio of ions, releasing these under acidic conditions to drive remineralization, but stabilizing these same ions under alkaline pH conditions.

Once a cavity has formed on the crown or root surface of a tooth, an operative or restorative approach has been the mainstay of treatment for many decades. Using techniques that conserve tooth structure and also encourage healing of the inner affected dentin is now commonplace. A number of biomimetic dental materials (such as glass ionomer cements) can have powerful influences on the healing of dentin, and some dental materials that are used in deep cavities as liners (such as alkaline bioceramic cements) are highly antimicrobial. Both material types lack the physical properties (such as high compressive strength) required for large restorations in high stress-bearing areas (such as occlusal surfaces of permanent molar teeth), hence both are typically used as a base or foundation, and then overlaid with tooth-colored materials (such as resin composites, ormocers, and ceramics) that are inert, from a purely chemical and biological perspective.

There have been attempts to synthesize human dental enamel or enamel-like materials by chemical reactions using solutions that could be applied topically onto cavities in teeth. Recently, a number of chemical strategies for forming biological apatites or enamel-like materials in vitro have been studied. Some positive results have been described for fluorapatite/phosphoric acid pastes [19] and for amelogenin-induced hydroxyapatite [20] (see chapter "Tooth Bioengineering and Whole Tooth Regeneration"). A major challenge with such approaches is that the material formed on the tooth is relatively thin and lacks the complex reinforced prismatic microstructure of natural dental enamel. This makes treating large or extensive lesions a major challenge.

2.2.2 The Dentine-Pulp Complex

The close interaction between dental pulp tissue and surrounding hard tissue (dentine) creates a functional unit known as the "dentine-pulp complex." This complex arises embryonically from ectomesenchymal cells [21]. The odontoblastic cellular layer differentiates in the "bell" stage of tooth development, and it secretes dentine as a distinct extracellular matrix (primary dentine). Once the tooth has formed fully, during the remainder of the lifespan, odontoblasts continue their synthetic function, but at a greatly reduced pace, forming secondary dentine.

In response to environmental stimuli such as bacterial invasion and trauma, odontoblasts react and form reparative dentine, to attempt to wall off the dental pulp from the noxious stimulus [22]. Despite these efforts, the rate or amount of destruction is often far greater than the ability of the odontoblasts to lay down a sufficient amount of protective dentine. Once the defensive line of the dentine has been breached, the pulpal soft tissues are exposed to chemical and microbial assaults from oral microorganisms, and this leads to an inflammatory response within the pulp tissue, and finally to its necrosis.

Removal of the inflamed/necrotic tissues, debridement of the pulp chamber and root canal system, and obturation of the space with root filling materials is the usual method of anterograde endodontic treatment. Rather than removing the pulpal soft tissue, an option that exists in roots with a partially open apex where there is good blood flow, is regenerative endodontics. With this method, the objective is dentine-pulp regeneration, and preservation of the vitality of the tooth. Several groups have been working on this concept, and it has been used in clinical practice in recent years [23–25]. Chapters "Dentin Pulp Complex Regeneration" and "Clinical Approach to Regenerative Endodontics" will explain regenerative endodontics in detail.

2.3 Periodontium

Periodontal diseases and occlusal trauma can adversely affect the supporting apparatus for teeth, known as the periodontium. This encompasses the periodontal ligament (PDL), cementum on the root surfaces of teeth, the alveolar bone, and the gingiva. In a healthy PDL, collagenous fibers from the cementum extend to the alveolar bone, and anchoring the root of the tooth into its socket. Plaque accumulation from poor oral hygiene elicits inflammation, and in susceptible individuals, the host inflammatory response causes destruction of the attachment and the alveolar bone, in bursts of varying duration. The loss of attachment and bone can be so severe that the tooth is lost as a result.

Regeneration of the periodontium is a challenging task. Not only is this a complex tissue with multiple elements, the local environment has a high level of microorganisms and the patient has already shown an inappropriate host immune reaction to those microorganisms. Current treatment modalities for periodontitis include debridement of teeth, and similar approaches are also used for dental implants. Regenerative surgical procedures are used for advanced cases [26, 27], with guided tissue regeneration (GTR) considered the current "gold standard" for treatment [28]. In this procedure, a barrier membrane is placed at the site against the root surface or implant surface, to support soft tissue regeneration (on its outer surface) and bone regeneration (on its inner surface). The membrane excludes invasion of the bony defect by rapidly migrating epithelial cells or fibroblasts. This topic is covered in chapter "Regenerative Approaches in Periodontics."

2.4 Bone Defects

In oral and maxillofacial surgery, the treatment of bone defects caused by trauma, infections, or malignancies is a significant challenge. Bone has a considerable capability for self-renewal [29], and as a tissue it must not only provide structural support and protection, but also serve several endocrine and hemopoietic functions [30]. While healthy normal bone can respond to strong biomechanical forces as well as daily microdamage, around 5–10% of bone fractures and most critical-sized bone defects are caused by trauma, pathology, or congenital malformations, do not heal fully despite timely clinical interventions [31].

In clinical practice, the standard therapeutic modality for treating large bony defects is bone grafting, using autogenous or allogenic bone grafts. These grafts provide osteoconduction and induction at the same time. Over 900,000 surgical procedures involving bone grafts are undertaken each year in the United States, and globally bone grafting surgery accounts for annual healthcare costs of some $30 billion [32], with bone being only second to blood as the most commonly transplanted tissue [32]. When harvesting bone autografts, a significant issue is morbidity of the donor site, which is a second surgical site in the same patient [33]. In chapters "Regenerative Approaches in Periodontics" and "Regenerative Approaches in Oral and Maxillofacial Surgery," current approaches for bone augmentation in dental settings will be described.

2.5 Oral Pathology

The application of stem cells for the treatment of oral lesions and conditions is a relatively new concept. Examples include submucosal injection of mesenchymal stem cells to promote healing of oral ulcers [34]. Immunomodulatory properties of stem cells may have a therapeutic benefit for oral vesiculobullous lesions [35, 36]. Moreover, stem cells could also be a delivery vehicle for therapeutic agents, for example, in the setting of treating malignant lesions [37]. This topic is discussed in the chapter "Regenerative Approaches in Oral Medicine."

3 Dental Tissue-Derived Stem Cells

Bone marrow mesenchymal stem cell (BMSC) are the best known and characterized cells used in tissue engineering [38]; however, harvesting them from bone is invasive, and there are issues with age-related differentiative potency [39, 40], which impede their usefulness. As a result, there has been interest in alternative tissue origins of stem cells, such as skeletal muscles [41], dental tissues [28, 42–44], and adipose tissues [45–47].

Dental tissues could be a convenient and accessible source of stem cells. Teeth are extracted due to various reasons (including as part of orthodontic treatment), and deciduous teeth exfoliate naturally when the erupting permanent teeth cause their roots to resorb. Dental stem cells (DSCs) have demonstrated multipotential differentiation, being able to differentiate and proliferate to form osteogenic, odontogenic, neurogenic, and adipogenic cell types [48, 49]. DSCs are highly effective at forming odontogenic structures, but can also form osseous tissues [50]. The precise differences between BMSCs and among the different types of DSCs remain to be explained fully. DSCs are derived from the neural crest (ectomesenchymal origin) [51], and so are superior for regenerating a wide variety of tissues [52], compared to BMMSCs that originate from mesoderm [53]. DSCs and their regenerative potential will be discussed in the chapter "Dental Tissues Originated Stem Cells for Tissue Regeneration."

4 Conclusions and Future Direction

Although many elements of current clinical dental practice rely on restorative materials and prostheses to replace lost tissue loss, there is growing interest in using regenerative approaches, and emerging horizons for bioengineered dentistry in the future. The concept of regenerating injured tissues in the oral cavity rather than replacing them with inert materials is appealing, however, many challenges remain to be addressed.

While stem cell-based therapies have demonstrated great potential for bone regeneration, it is important to better understand and control the local microenvironment. The microenvironment of the recipient site regulates the levels of endogenous cytokines, and the behavior of implanted cells, and thereby affects treatment outcomes. Understanding the cross talk between stem cells, biomaterials, and the host (including immune-modulatory effects) is a key requirement.

When using scaffolds and other biomaterials, it is necessary to optimize their structure and design, to meet both the mechanical and biological properties needed at the specific recipient site. As well, these materials must have the correct rate of degradation, be nontoxic to the host and be cost-effective for clinical use.

Given the growth and progress in regenerative dentistry, this area will grow in importance for both clinicians and researchers. Therefore, it is important that undergraduate and postgraduate dental students have a working knowledge of this exciting field and maintain a watching brief on new developments as these transition from concept through translation to reach clinical practice.

References

1. Mason C, Dunnill P. A brief definition of regenerative medicine. Regen Med. 2008;3(1):1–5.
2. Groll J, Boland T, Blunk T, Burdick JA, Cho D-W, Dalton PD, et al. Biofabrication: reappraising the definition of an evolving field. Biofabrication. 2016;8(1):013001. https://doi.org/10.1088/1758-5090/8/1/013001.
3. Bartold PM, Gronthos S, Ivanovski S, Fisher A, Hutmacher DW. Tissue engineered periodontal products. J Periodontal Res. 2016;51(1):1–15. https://doi.org/10.1111/jre.12275.
4. Hosseinpour S, Ahsaie MG, Rad MR, Taghi Baghani M, Motamedian SR, Khojasteh A. Application of selected scaffolds for bone tissue engineering: a systematic review. Oral Maxillofac Surg. 2017;21(2):109–29.
5. Brittberg M, Lindahl A, Nilsson A, Ohlsson C, Isaksson O, Peterson L. Treatment of deep cartilage defects in the knee with autologous chondrocyte transplantation. N Engl J Med. 1994;331(14):889–95.
6. Goldberg M, Smith AJ. Cells and extracellular matrices of dentin and pulp: a biological basis for repair and tissue engineering. Crit Rev Oral Biol Med. 2004;15(1):13–27.
7. Du C, Moradian-Oldak J. Tooth regeneration: challenges and opportunities for biomedical material research. Biomed Mater. 2006;1(1):R10.
8. Thesleff I. Epithelial-mesenchymal signalling regulating tooth morphogenesis. J Cell Sci. 2003;116(9):1647–8.
9. Kassai Y, Munne P, Hotta Y, Penttilä E, Kavanagh K, Ohbayashi N, et al. Regulation of mammalian tooth cusp patterning by ectodin. Science. 2005;309(5743):2067–70.
10. AFFAIRS ACOS. Titanium applications in dentistry. J Dent Res. 2003;134(3):347–9.
11. Yamamoto H, Kim E-J, Cho S-W, Jung H-S. Analysis of tooth formation by reaggregated dental mesenchyme from mouse embryo. Microscopy. 2003;52(6):559–66.
12. Mina M, Kollar E. The induction of odontogenesis in non-dental mesenchyme combined with early murine mandibular arch epithelium. Arch Oral Biol. 1987;32(2):123–7.
13. Cate AT. The role of epithelium in the development, structure and function of the tissues of tooth support. Oral Dis. 1996;2(1):55–62.
14. Balakrishnan M, Simmonds RS, Tagg JR. Dental caries is a preventable infectious disease. Aust Dent J. 2000;45(4):235–45. https://doi.org/10.1111/j.1834-7819.2000.tb00257.x.
15. Setzer F, Kim S. Comparison of long-term survival of implants and endodontically treated teeth. J Dent Res. 2014;93(1):19–26.
16. Stack MV. Organic constituents of enamel. J Dent Res. 1954;48(3):297–306.
17. Roberts-Clark D, Smith A. Angiogenic growth factors in human dentine matrix. Arch Oral Biol. 2000;45(11):1013–6.
18. Featherstone J. The continuum of dental caries—evidence for a dynamic disease process. J Dent Res. 2004;83(Suppl 1):39–42.
19. Wang X, Xia C, Zhang Z, Deng X, Wei S, Zheng G, et al. Direct growth of human enamel-like calcium phosphate microstructures on human tooth. J Nonosci Nanotechnol. 2009;9(2):1361–4.

20. Fan Y, Sun Z, Moradian-Oldak J. Controlled remineralization of enamel in the presence of amelogenin and fluoride. Biomaterials. 2009;30(4):478–83.
21. Mao JJ, Prockop DJ. Stem cells in the face: tooth regeneration and beyond. Cell Stem Cell. 2012;11(3):291–301.
22. Smith AJ, Cassidy N, Perry H, Begue-Kirn C, Ruch J-V, Lesot H. Reactionary dentinogenesis. Int J Dev Biol. 2003;39(1):273–80.
23. Mooney DJ, Powell C, Piana J, Rutherford B. Engineering dental pulp-like tissue in vitro. Biotechnol Prog. 1996;12(6):865–8.
24. Bohl KS, Shon J, Rutherford B, Mooney DJ. Role of synthetic extracellular matrix in development of engineered dental pulp. J Biomater Sci Polym Ed. 1998;9(7):749–64.
25. Prescott RS, Alsanea R, Fayad MI, Johnson BR, Wenckus CS, Hao J, et al. In vivo generation of dental pulp-like tissue by using dental pulp stem cells, a collagen scaffold, and dentin matrix protein 1 after subcutaneous transplantation in mice. J Endod. 2008;34(4):421–6.
26. Deas DE, Mealey BL. Response of chronic and aggressive periodontitis to treatment. Periodontol 2000. 2010;53(1):154–66.
27. Chen F-M, Zhang J, Zhang M, An Y, Chen F, Wu Z-F. A review on endogenous regenerative technology in periodontal regenerative medicine. Biomaterials. 2010;31(31):7892–927.
28. Bartold PM, Shi S, Gronthos S. Stem cells and periodontal regeneration. Periodontol 2000. 2006;40(1):164–72.
29. Hadjidakis DJ, Androulakis II. Bone remodeling. Ann NY Acad Sci. 2006;1092(1):385–96.
30. Doherty AH, Ghalambor CK, Donahue SW. Evolutionary physiology of bone: bone metabolism in changing environments. Physiology. 2015;30(1):17–29.
31. Bayer E, Gottardi R, Fedorchak M, Little S. The scope and sequence of growth factor delivery for vascularized bone tissue regeneration. J Control Release. 2015;219:129–40.
32. Elmore JC, Larsen C, CUS N. Markets for musculoskeletal tissue engineering and cell transplantation products. Market and technology reports. New York: Medtech Insights; 2010.
33. Gazdag AR, Lane JM, Glaser D, Forster RA. Alternatives to autogenous bone graft: efficacy and indications. J Am Acad Orthop Surg. 1995;3(1):1–8.
34. El-Menoufy H, Aly LA, Aziz MT, Atta HM, Roshdy NK, Rashed LA, et al. The role of bone marrow-derived mesenchymal stem cells in treating formocresol induced oral ulcers in dogs. J Oral Pathol Med. 2010;39(4):281–9. https://doi.org/10.1111/j.1600-0714.2009.00819.x.
35. Vanikar A, Modi P, Patel R, Kanodia K, Shah V, Trivedi V et al., editors. Hematopoietic stem cell transplantation in autoimmune diseases: the Ahmedabad experience. Transplantation proceedings. Elsevier; 2007.
36. Kanwar AJ, De D. Pemphigus in India. Indian J Dermatol Venereol Leprol. 2011;77(4):439.
37. Studeny M, Marini FC, Dembinski JL, Zompetta C, Cabreira-Hansen M, Bekele BN, et al. Mesenchymal stem cells: potential precursors for tumor stroma and targeted-delivery vehicles for anticancer agents. J Natl Cancer Inst. 2004;96(21):1593–603.
38. Charbord P. Bone marrow mesenchymal stem cells: historical overview and concepts. Hum Gene Ther. 2010;21(9):1045–56.
39. Zhou S, Greenberger JS, Epperly MW, Goff JP, Adler C, LeBoff MS, et al. Age-related intrinsic changes in human bone-marrow-derived mesenchymal stem cells and their differentiation to osteoblasts. Aging Cell. 2008;7(3):335–43.
40. Zomorodian E, Baghaban EM. Mesenchymal stem cells as a potent cell source for bone regeneration. Stem Cells Int. 2012;2012:1.
41. Bosch P, Musgrave DS, Lee JY, Cummins J, Shuler F, Ghivizzani SC, et al. Osteoprogenitor cells within skeletal muscle. J Orthop Res. 2000;18(6):933–44.
42. Morad G, Kheiri L, Khojasteh A. Dental pulp stem cells for in vivo bone regeneration: a systematic review of literature. Arch Oral Biol. 2013;58(12):1818–27.
43. Khojasteh A, Nazeman P, Rad MR. Dental stem cells in oral, maxillofacial and craniofacial regeneration. Dental stem cells. New York: Springer; 2016. p. 143–65.
44. Rezai-Rad M, Bova JF, Orooji M, Pepping J, Qureshi A, Del Piero F, et al. Evaluation of bone regeneration potential of dental follicle stem cells for treatment of craniofacial defects. Cytotherapy. 2015;17(11):1572–81.
45. Gimble J, Rad MR, Yao S. Adipose tissue–derived stem cells and their regeneration potential. Stem Cells Craniofacial Dev Regen. 2013;241–58.
46. Zuk PA, Zhu M, Mizuno H, Huang J, Futrell JW, Katz AJ, et al. Multilineage cells from human adipose tissue: implications for cell-based therapies. Tissue Eng. 2001;7(2):211–28.
47. Salehi-Nik N, Rezai Rad M, Kheiri L, Nazeman P, Nadjmi N, Khojasteh A. Buccal fat pad as a potential source of stem cells for bone regeneration: a literature review. Stem Cells Int. 2017;2017:1.
48. Marei MK, El Backly RM. Dental mesenchymal stem cell-based translational regenerative dentistry: from artificial to biological replacement. Front Bioeng Biotechnol. 2018;6:49.
49. Zhai Q, Dong Z, Wang W, Li B, Jin Y. Dental stem cell and dental tissue regeneration. Front Med. 2018;13:1–8.
50. Huang G-J, Gronthos S, Shi S. Mesenchymal stem cells derived from dental tissues vs. those from other sources: their biology and role in regenerative medicine. J Dent Res. 2009;88(9):792–806.

51. Ibarretxe G, Crende O, Aurrekoetxea M, García-Murga V, Etxaniz J, Unda F. Neural crest stem cells from dental tissues: a new hope for dental and neural regeneration. Stem Cells Int. 2012;2012:1.
52. Yu J, Wang Y, Deng Z, Tang L, Li Y, Shi J, et al. Odontogenic capability: bone marrow stromal stem cells versus dental pulp stem cells. Biol Cell. 2007;99(8):465–74.
53. Sheng G. The developmental basis of mesenchymal stem/stromal cells (MSCs). BMC Dev Biol. 2015;15(1):44.

Dental Tissues Originated Stem Cells for Tissue Regeneration

Maryam Rezai Rad, Sepanta Hosseinpour, Qingsong Ye, and Shaomian Yao

1 Introduction

Stem cell-based therapy as a major field in regenerative medicine has attracted scientists in the field of tissue engineering [1]. Stem cells possess a remarkable capability for proliferation and for differentiation into various cell types, and such capabilities can be useful for regenerative therapies. Nowadays, although multipotent mesenchymal stem cells derived from bone marrow (BMMSC) are among the best-known and best-characterized cells for tissue engineering [2], the invasive procedure needed to isolate these from bone marrow, and variations in their potency according to the age of the donor [3, 4] impede their usefulness. Therefore, several alternative tissue origins for stem cells have been explored including skeletal muscle [5], dental tissues [6–9], and adipose tissue [10–12].

In clinical dentistry, teeth are extracted due to various reasons, such as impacted third molars for orthodontic reasons. Collecting these extracted teeth does not require additional procedures. In addition, deciduous human teeth naturally exfoliate. Hence, harvesting MSCs from dental tissues of such exfoliated teeth is convenient. Dental stem cells (DSCs) have demonstrated the capability to differentiate into various cell lineages, including osteogenic, odontogenic, neurogenic, and adipogenic pathways [13, 14].

Although because of their origin DSCs seem to be more efficient for developing odontogenic structures than other osseous tissues when compared with BMMSCs [15], the precise differences between DSCs and BMMSCs, and among each of the different types of DSCs remain unclear. These cells are derived from the neural crest (i.e., they have an ectomesenchymal origin) [16], which gives them a superior capability for regenerating a wide variety of tissues [17], when compared to BMMSCs that originate from mesoderm [18]. Adding to this greater usefulness, DSCs can be obtained without any major ethical concerns [19, 20].

The first report of MSCs in the dental pulp was in 1985 by Yamamura [21, 22]. To date, five main types of dental tissue-derived stem cells have been

M. Rezai Rad
Research Institute of Dental Sciences, Dental school
Shahid Beheshti University of Medical Sciences,
Tehran, Iran

S. Hosseinpour
School of Dentistry, The University of Queensland,
Brisbane, QLD, Australia

Q. Ye
Skeletal Biology Research Center, Massachusetts
General Hospital and Harvard School of Dental
Medicine, Boston, MA, USA

Institute of Stem Cells and Tissue Engineering,
Wenzhou Medical University, Wenzhou, China

S. Yao (✉)
Department of Comparative Biomedical Sciences,
School of Veterinary Medicine, Louisiana State
University, LA, USA
e-mail: shaomia@lsu.edu

© Springer Nature Switzerland AG 2021

S. Hosseinpour et al. (eds.), *Regenerative Approaches in Dentistry*,
https://doi.org/10.1007/978-3-030-59809-9_2

reported. The first type was termed as "postnatal dental pulp stem cell" (DPSC) by Gronthos et al. [23] in 2000. These can regenerate a "dentine–pulp complex like" tissue in vitro. Subsequently, in 2003, Miura et al. reported the isolation and characterization of stem cells from exfoliated deciduous teeth which had proliferated from remnants of living DPSCs. These are known as stem cells from human exfoliated deciduous teeth (SHED) [24]. Morsczeck et al. [25] and Kemoun et al. [26] reported undifferentiated mesenchymal progenitor cells from the human dental follicle, in the connective tissue sac surrounding developing teeth. They named these dental follicle stem cells (DFSCs). The dental follicle forms the tooth and the supporting structures, and the follicle develops into the periodontal ligament when tooth is erupting. Stem cells have also been reported as being present in the periodontal ligament [27]. In 2006, Sonoyama et al. described unique undifferentiated stem cells from the dental apical papilla of human immature permanent teeth (SCAP) with the ability to generate osteoblasts, odontoblasts, and adipocytes in vitro [28, 29].

Figure 1 schematically shows the potential sources of these various dental tissue-derived stem cells. Although these DSCs have been investigated in many studies [13, 14, 30, 31], only limited work has been done to characterize their biological properties and compared their in vivo applications. In this chapter, we have compiled information regarding the isolation, characterization, and potential applications of DSCs in tissue regeneration and tissue engineering.

2 Stem Cells Obtained from Human Permanent Teeth or Exfoliated Deciduous Teeth

As mentioned earlier, the first isolated DSCs were dental pulp stem cells (DPSCs). DPSCs were isolated initially from permanent third molar teeth. They demonstrated colony formation and a high proliferation rate, as well as the ability to form calcified nodules [23]. DPSCs derived from impacted third molars at the stage

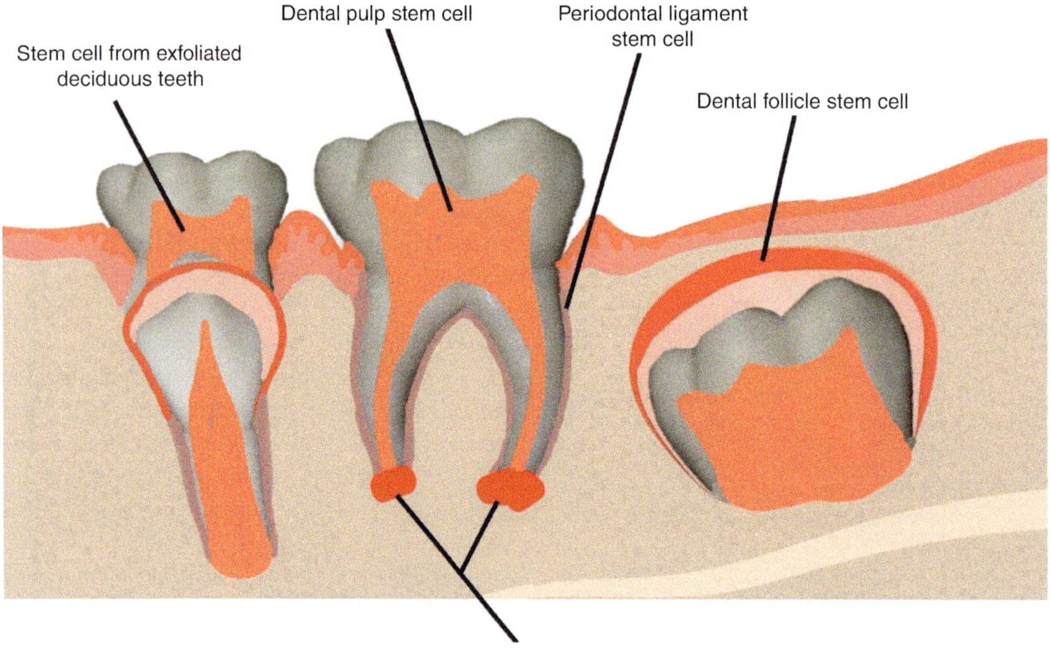

Fig. 1 Schematic view of various sources of stem cells from dental tissues

of root development can differentiate into active migratory odontoblast-like cells, which are able to produce a three-dimensional dentine-like mineralized structure [32]. The stem cell properties of DPSCs vary according to the donor's age when the specimens are obtained (Nakamura et al. 2009); i.e., the cells show a reduction in stem cell features as donor age is increased.

The transition of deciduous teeth in the primary dentition to the permanent dentition is a dynamic and distinctive process, in which the developing permanent teeth buds gradually resorb the roots of the overlying deciduous teeth [33]. As mentioned above, a unique population of DSCs can be harvested from the remaining living pulpal cells of exfoliated teeth, and cultivated in the laboratory. These cells, known as SHED, provide an easily accessible source of stem cells that can be preserved for future applications [24]. These cells differ from regular DPSCs because of their greater proliferation rate and enhanced population doubling rate, and their capability to form sphere-like cell clusters [34]. SHED express the minimum essential markers for stem cells defined by the International Society for Cellular Therapy criteria [15, 35]. They also express some embryonic stem cell markers [24], as listed in Table 1.

While SHED have been isolated using similar methods to those used to isolate DPSCs, there are two major differences between SHED and DPSCs. Firstly, the source of SHED is the dental pulp tissue of deciduous teeth, rather than permanent teeth. Secondly, the isolated cells do not grow as proliferative cells, but instead grow as clustered fibroblast-like cells [36]. It has been shown that 69.8% of SHED are in the S and G2

Table 1 Sources, surface markers, multipotentiality, and in vivo applications of dental stem cells[a]

Stem cell type	Source	Surface markers for characterization		In vitro multipotentiality	Regenerative in vivo application
		Positive	Negative		
Dental pulp stem cell (DPSC)	**Pulpal tissue of crown/root of** 1. Immature teeth that their pulp is exposed and accessible through the root 2. Extracted permanent or deciduous teeth; dental pulp extraction is accomplished through the dental crown by cutting the cementum–enamel junction using dental instruments 3. Carious teeth that its pulp exposed and removed during endodontic treatments	ALP, CD 9, CD 10, CD 13, CD 29, CD 44, CD 59, CD 73, CD 90, CD 105, CD 146, CD 166, CD 271, DSPP, DMP1, OPN, BSP, BBX, nestin, Oct4, STRO-1	CD 11b, CD 14, CD 19, CD 24, CD 31, CD 34, CD 45, CD 117, CD 133, HLA-DR	1. Odontogenic 2. Osteogenic 3. Chondrogenic 4. Neurogenic 5. Adipogenic 6. Myogenic 7. Melanogenic 8. Endothelial differentiation 9. Hepatogenic	• Bone regeneration • Dentin-pulp complex regeneration, regenerative endodontics, and dentin regeneration • Periodontal regeneration • Cornea regeneration • Angiogenic regeneration and blood vessel • Neural reconstruction • Hair follicle repair
Stem cell from human exfoliated deciduous teeth (SHED)	Pulpal tissue of crown/root of exfoliated deciduous teeth	CD 13, CD 29, CD 31, CD 44, CD 73, CD 90, CD 105, CD 146, CD 166, nestin, Oct4, STRO-1	CD 11b, CD 14, CD 19, CD 45, CD 34, CD 43, CD 45	1. Odontogenic 2. Osteogenic 3. Chondrogenic 4. Neurogenic 5. Adipogenic 6. Myogenic	• Dentin–pulp complex and regenerative endodontics • Cornea regeneration • Hepatocytes regeneration

(continued)

Table 1 (continued)

Stem cell type	Source	Surface markers for characterization Positive	Negative	In vitro multipotentiality	Regenerative in vivo application
Stem cell from apical papilla (SCAP)	Apical papilla of immature root in unerupted teeth, especially impacted third molars	ALP, CD 13, **CD 24**[b], CD 29, CD 44, CD 53, CD 59, CD 61, CD 73, CD 90, CD 105, CD 106, CD 146, CD 166, nestin, STRO-1	CD 14, CD 18, CD 34, CD 45, CD 117, CD 150	1. Odontogenic 2. Osteogenic 3. Chondrogenic 4. Neurogenic 5. Adipogenic 6. Angiogenic 7. Hepatogenic	• Dentin–pulp complex regeneration • Pulpal tissue regeneration • Neural regeneration • Angiogenic regeneration
Dental follicle stem cell (DFSC)	Dental follicle tissue which surrounded the crown of unerupted teeth	CD 9, CD 10, CD 13, CD 29, CD 44, CD 53, CD 59, CD 73, CD 90, CD 105, CD 106, CD 146, CD 166, CD 271, nestin, notch-1, STRO-1	CD 31, CD 34, CD 45, CD 133, HLA-DR	1. Osteogenic 2. Cementogenic 3. Chondrogenic 4. Neurogenic 5. Adipogenic 6. Hepatogenic	• Bone regeneration • Periodontal tissue regeneration
Periodontal ligament stem cell (PDLSC)	Soft connective tissue which surrounded the root surface of the tooth between alveolar bone proper and cementum	ALP, CD 10, CD 13, CD 29, CD 44, CD 49c, CD 59, CD 73, CD 90, CD 97, CD 105, CD 106, CD 146, CD 166, SSEA-3 and 4, STRO-1, HLA-A, HLA-B, HLA-C, Oct4, Nanog, Scleraxis, Sox2, STRO-1	CD 11b, CD 14, CD 34, CD 40, CD 45, CD 80, CD 86, CD 106, HLA-DR	1. Osteogenic 2. Cementogenic 3. Chondrogenic 4. Neurogenic 5. Adipogenic 6. Pancreatic islet cell 7. Endothelial differentiation	• Periodontal regeneration • Angiogenic regeneration and blood vessel regeneration • Bone regeneration • Tendon and cartilage regeneration

Abbreviations: *ALP* alkaline phosphatase, *BBX* bobby sox homolog, BSP, bone sialoprotein, *CD* cluster differentiation, *DMP* dentin matrix protein, *DSPP* dentin sialophosphoprotein, *HLA-DR* human leukocyte antigen D related, *Oct* octamer-binding transcription factor, *OPN* osteopontin, *PDL* periodontal ligament, *STRO* stromal precursor antigen

[a]References were quoted in specific paragraphs in the text

[b]Specific marker in comparison to other dental stem cells

stages of the cell cycle; however, 56% of DPSCs are in those cell cycle phases. This explains the different proliferative capacity of these cells [37]. In the ensuing sections, the isolation, characterization, and various applications of DPSCs and SHED will be discussed.

2.1 Isolation and Characterization

There are many approaches to the collection and isolation of DPSCs. A key consideration is to maintain the quality and safety of the isolated cells. Generally, enzymatic digestion of related tissues (such as the dental pulp of an adult tooth or the remaining dental pulp of an exfoliated tooth) or an outgrowth of a tissue explant [38] are the major approaches that have been used for DPSCs and SHED. Enzymatic digestion usually involves placing the source tissues into an appropriate mixture of enzymes such as collagenase I and dispase for 30–60 min at 37 °C, to break the tissue down in order to obtain single-cell suspensions. The tissue remnants are filtered through a sieve or strainer. This is followed by colony-based cultivation of the stem cells, or by cell sorting using magnetic or fluorescent markers [39].

On the other hand, in the outgrowth method, the harvested tissue is diced into 1–2 mm pieces and placed into culture plates, to allow outgrowth of cells from the tissue pieces [40]. Comparing these two methods of isolation, several studies have shown that DPSCs isolated by enzymatic digestion have a higher proliferation rate, higher expression of stromal precursor antigen (STRO-1) and CD 34, and higher rates of osteogenic or odontogenic differentiation [41–43].

After the isolation of DPSCs and SHED, the next step is to cultivate the cells to expand their number to reach the amount required for cell-based therapy. In addition, the pathway of differentiation can be adjusted by the choice of parameters used in the culture system [39, 44, 45]. Various culture systems have been used including serum-free versus serum-rich culture media, sphere-forming culture, and co-culture systems. Most commonly, a 10–20% concentration of fetal bovine serum (FBS) is included in the culture media for both isolation and expansion of DPSCs and SHED, to accelerate cell adhesion during the first stage of culture. Because of the risk of contamination of serum by bovine pathogens (and the presence of bacterial endotoxin) and an altered possibility of malignant transformation of the stem cells in a serum-containing culture system, the use of a chemically defined serum-free culture system has been recommended [46–49]. In this regard, a serum-free medium (Table 2) has been reported to be successful for obtaining DPSCs [50]. Cells from the adherent population appeared to be more engaged in the odontoblastic lineage than the adherent cells. The neural stem cells that are derived from various sources can proliferate in sphere-forming culture systems [51, 52], which are also recommended for differentiation of DSCs [53, 54].

Table 2 Serum-free medium for DPSC culture

Medium composition	Concentration	Source
Dulbecco's modified Eagle medium (DMEM)/Ham's F12		
Glucose	33 mM	
HEPES (pH 7.2)	5 mM	
N_2 supplement		Life Technologies, Invitrogen
Human EGF	10 ng/ml	
Human bFGF	5 ng/mL	Peprotech
Heparin solution[a]	0.2%	Peprotech
Streptomycin-penicillin	5 μg/mL–5 UI/mL	StemCell Technologies

[a]Final concentration in the medium is Heparin 5 μg/ml

Phenotypic markers expressed by DPSCs are summarized in Table 1. Expression of these markers relates to the unique capability of these cells [28, 55]. For example, STRO-1 positive cells show greater odontogenic or osteogenic differentiation, while CD34 and CD117 positive DPSCs show a greater ability for cell renewal and for mineralization [56]. Low expression of Class II HLA-DR surface antigen indicates that these cells may be immunologically privileged, which reduces concerns around antigenicity in the setting of tissue matching between the donor and the recipient. If the cells truly are not immunogenic then this raises the possibility of creating banks for DPSCs. This aspect is discussed in the last section of this chapter.

The phenotype of SHED differs from that of DPSCs. Pluripotency markers such as Pou5f1, Sox2, Oct3/Oct4, and Nanog are expressed more strongly in SHED than in DPSCs [57]. Nestin as a neuroepithelial marker is expressed less in SHED cells compared to DPSCs [58], which explains the reduced ability of SHED cells in comparison to DPSCs for neuronal regeneration [58]. However, due to the greater proliferative capability of SHED cells, they form sphere-like clusters in a neurogenic culture medium [24].

DPSCs have shown odonto/osteogenic, neurogenic, and adipogenic differentiation in preclinical studies [23, 59, 60] (as listed in Table 1). More recently, in vitro investigations have revealed osteogenic, chondrogenic, and myogenic differentiation of DPSCs [61–63]. Similar to DPSCs, SHED cells have also demonstrated the ability to undergo odonto/osteogenic, chondrogenic, neurogenic, adipogenic, and myogenic differentiation [13, 14, 30, 31] (Table 1).

2.2 Regenerative Applications

The concept of harvesting and banking dental tissue MSCs has opened up a new window for stem cell-based regenerative treatments. As already mentioned, DPSCs and SHED can differentiate effectively to various cell lineages. In this section, the regenerative capacity of these cells will be discussed further.

2.2.1 Dentine–Pulp Complex Regeneration

Dental pulp is a specialized connective tissue, which consists of odontoblasts, endothelial cells, neurons, fibroblasts, and other cells. The peculiar internal anatomy of teeth and the unique blood flow may influence the survival of stem cells [64]. When the dental pulp is infected by bacterial pathogens, it is hard to remove or inactivate these pathogens through antibiotics alone. For a number of clinical reasons, extirpation of the whole pulp may need to be undertaken [65].

For regeneration, dental pulp stem cells can be expanded in the laboratory. When seeded onto hydroxyapatite/tricalcium phosphate (HA/TCP), the cells have demonstrated formation of a dentine-pulp-like complex in mice [23]. In addition, SHED implanted into mice can generate odontoblast-like cells and dentine-like mineralized nodules. However, SHED cells do not seem to be able to form a complete dentine-pulp-like complex like that seen with transplantation of DPSCs [24]. In contrast, DPSCs can form mineralized nodules, which are covered by a layer of dentine-like reparative tissue in vivo [66]. DPSCs that have been seeded onto calcium phosphate [67], hexafluoropropanol silk [68], and polylactic acid [69] scaffolds have demonstrated dentine–pulp complex formation in animal models.

In addition to dentine–pulp complex regeneration, DPSCs have also been used in periodontal regeneration [31]. DPSCs transplanted into immunocompromised mice have been shown to differentiate into collagen forming cells, with the capacity to form a cementum-like tissue [70]. The expression of particular phenotypic markers such as SCD-1, STRO-1, CD44, and CD146 on DPSCs and SHED cells sensibly linked to their role in periodontal regeneration [71, 72]. In addition, both SHED and DPSCs are capable of bone regeneration [31]. This aspect will be discussed further in the next section.

2.2.2 Bone Regeneration

Bone formation, including the aggregation of osteoprogenitor cells, is similar in some ways to tooth bud development, but without the epithelial invagination aspect. Intramembranous and endochondral bone formation are the two main processes of bone regeneration. In endochondral regeneration, the aggregated MSCs first undergo chondrogenesis followed by ossification of the cartilage into bone tissue [73]. In intramembranous bone formation, MSCs first become osteoprogenitor cells and then further differentiate into osteoblasts, that create extracellular matrix containing collagen fibrils and other bone components, and this is called osteoid.

Bone tissue possesses the intrinsic capability of regenerating itself through adulthood [74]. However, when bone defects or fractures are larger than the regenerative capacity of the bone, other interventions are required to rehabilitate the defect site in the bone tissue [74]. One alternative is using stem cell-based therapy with stem cells of dental origin [75]. In vitro studies have demonstrated the osteogenic and chondrogenic multipotency of both DPSCs and SHED [75–77]. An in vivo investigation showed that DPSCs can regenerate osteoblasts and endotheliocytes that eventually formed bone in immunocompromised rats [78].

The application of SHED in combination with a hydroxyapatite (HA)/tricalcium phosphate scaffold has been shown to regenerate calvarial bone defects in animal models [79]. These studies have also shown that a large amount of bone and bone marrow-like structures are formed in the regenerated mineralized matrix. Other investigations have reported that the application of DPSCs in conjunction with fibroin/collagen scaffolds can significantly enhance bone formation after 4–8 months in rats [80–82]. The application of DPSCs loaded onto a HA-based hydrogel showed superior bone healing in rat calvarial defects compared to the scaffold only and to untreated control groups [83]. In addition, DPSCs cultured for 13 days on poly(lactide-*co*-glycolide) scaffolds caused significant bone regeneration in the same types of defects [84]. Other studies have reported the successful application of DPSCs and SHED in combination with HA/TCP in rats and mice in the treatment of calvarial defects [85].

The research team of d'Aquino et al. demonstrated that administration of DPSCs loaded onto a collagen sponge resulted in higher mineralization after 1 month and complete bone regeneration with a large amount of cortical bone formed after 3 months, in comparison to a scaffold only group [78]. Moreover, an increased level of clinical attachment was found in the gingiva overlying the defect site after transplantation of DPSCs loaded onto a collagen scaffold [86].

Several studies have investigated the use of DPSCs or SHED for bone regeneration in maxillary or mandibular bone defects [87–91]. Alkaisi et al. created a mandibular defect (from the first premolar to mental foramen) in rabbits, and assessed the formation of bone after applying SHED without using a scaffold [87]. Radiographic evaluations revealed that in SHED transplanted animals, partial bone defects were bridged after only 2. In addition, the regenerated bone in this group showed a bony ridge with higher radiopacity than insights treated with a scaffold only after 4 weeks. Finally, after 6 weeks, clear corticalization and strong radiodensity were observed in the defect gap in sites treated with SHED group. Histomorphometric analysis showed that the regenerated bone was significantly higher in the SHED group compared with the scaffold control.

Paino et al. created tissue-engineered woven bone tissue using DPSCs in vitro, and then transplanted the manufactured bone tissue into the mandibular defects in rats [88]. Their findings confirmed the remodeling capacity of the implanted tissue, and its integration into the surrounding normal bone with the passage of time. Lamellar bone tissue with vital osteocytes and vascularized Haversian canals were found after bone remodeling.

SHED have been administered intravenously in order to treat osteoporosis in animal models of this condition [79, 92]. Ma et al. have found that this type of administration route for SHED can ameliorate problems of low bone mineral density, and can improve the radiopacity of the trabecular bone structures [79]. Moreover, Liu et al. showed that systemic administration of SHED was able to enhance bone volume, and promote trabecular number, thickness, and density [92]. In general, SHED transplantation has been found to improve cortical bone parameters, including bone area, thickness, and cortical bone fraction [93]. Bone that has been regenerated using SHED has been found to express higher levels of Runx2, alkaline phosphatase, and osteocalcin than controls [94]. At the same time, systemic application of SHED markedly downregulates genes associated with bone resorption such as RANKL and C-terminal telopeptide, and upregulates osteoprotegrin (OPG) [79, 92, 95]. This is a powerful demonstration of the immunomodulatory effects of SHED.

According to a recent systematic review by Leyendecker et al., transplantation of DPSCs into cranial, maxillary, and mandibular bone defects gives superior regenerative outcomes compared to controls [93]. However, in one study Annibali et al. found no difference in bone regeneration between DPSCs and control groups [96], while in another study by Behnia et al. there was no difference after the application of SHED in combination with a collagen scaffold compared to a scaffold only for the treatment of mandibular defects in dogs [97].

The type of scaffold material that is used as the carrier for DSCs plays an important role in bone repair. For example, Zhang et al. did not find ectopic bone regeneration in DPSCs loaded on HA/TCP [98]. However, Kuo et al. demonstrated that DPSCs in combination with alpha-calcium sulfate hemihydrate/amorphous calcium phosphate could efficiently promote bone formation in comparison to DPSCs + calcium sulfate dihydrate or DPSCs + calcium sulfate dihydrate/TCP [90].

2.2.3 Neural Regeneration

Neural regeneration is a challenging objective for two major reasons. The first is the lack of neural progenitor cells, while the second reason is that the local microenvironment may impede regenerative processes [99]. DPSCs express markers of neural progenitors such as nestin and Pax6, due to the neural crest origin of these cells [100, 101].

Human DPSCs are able to form spheroids under serum-free neuronal stimulating culture conditions [102]. In addition, even without neural induction, DPSCs express neural markers such as CDH2, TUBB3, and NFM [103].

Transplantation of DPSCs and SHED into defect sites in the central nervous system (CNS) has been shown to enhance neural recovery [56, 101, 104]. DPSCs can coordinate axonal regrowth by secreting CXCR-4 and stimulating the SDF-1/CXCL12 axis that induces neuroplasticity [105]. It has been reported that the implantation of DPSCs into the hippocampus of immunocompromised mice can accelerate cell recruitment and proliferation, and the maturation of endogenous existing neural cells [118]. In fact, taken together all these findings suggest that DPSCs may have an application as a modulator and stimulator in neural recovery in the CNS [143].

In spinal cord trauma and cerebral ischemia models, DPSCs can significantly enhance neurological dysfunctions [57, 161]. Furthermore, in the case of a peripheral nerve injury, DPSCs can mediate neural tissue engineering with artificial nerve conduits, to regenerate myelinated neural fibers [122]. Luo et al. demonstrated that transplantation of a heparin–poloxamer hydrogel and DPSCs with fibroblast growth factor could significantly regenerate neurons in spinal cord injuries, and functionally repair nerve injury defects in rats after 28 days [101].

Despite several studies which have demonstrated that both DPSCs and SHED cells can (1) differentiate into functional neural cells which were voltage-sensitive in vitro, (2) express neural markers, and (3) migrate into the CNS in animal models [24, 105, 106], many in vivo investigations have reported that DPSCs and SHED are unable to differentiate into functional neurons that the sites of injuries to nerves [102]. The neural regenerative capacity of DPSCs and SHED seems to occur because of their production of neurotrophic products [107]. Furthermore, these cells impede axon growth inhibitor signals and improve the microenvironment for regeneration [108]. Further studies are necessary to clarify the neural regeneration capability of DPSCs and SHED.

2.2.4 Other Regenerative Applications

Muscle Regeneration

Arminan et al. reported the differentiation of DPSCs into cardiomyocytes in rats [109], while Yang et al. documented that DPSCs can differentiate into dystrophin-producing muscle cells in a mouse model of cardiac muscle injury [110]. Based on the observation that DPSCs infused into cardiac defect sites express dystrophin and myosin [111], it has been proposed that DPSCs can potentially be applied for muscle regeneration [110].

Corneal Regeneration

Some experimental evidence indicates that DPSCs are more similar to epithelial stem cells than BMMSCs [112, 113]. In fact, DPSCs can differentiate into keratinocytes and can express keratinocyte markers [114]. Gomez et al. reported that eye transparency in a rabbit corneal defect model was improved by the implantation of a sheet of tissue-engineered DPSCs into the defect sites [115]. In another study, DPSCs delivered by soft contact lenses in a clinical application were shown to promote corneal epithelial regeneration [116]. DPSCs transferred from the contact lenses to the corneal surface expressed the keratinocyte markers cytokeratin 3 and 12. Moreover, DPSCs can impede conjunctival cells from growing into the center of the cornea [117].

Cartilage Regeneration

As mentioned earlier, DPSCs can develop into dentine, cartilage, and bone tissues [118]. Yu et al. showed cartilage formation after 14 days after the implantation of pellets containing DPSCs into the renal capsule of rats [119]. In addition, Morito et al. reported similar results after the subcutaneous implantation of DPSCs in combination with FGF in immunocompromised mice [120].

Hair Follicle and Blood Vessel Regeneration

DPSCs have been transplanted to the surgically compromised hair follicles, where they have been shown to cause the formation of new head bolts and the regeneration of hair fibers [121].

DPSCs can promote angiogenesis and vasculogenesis in sites where there is peripheral nerve injury [122]. DPSCs can differentiate into endotheliocytes in rat models [78].

Endocrine Regenerative Potential

Previous studies have documented the possibility of using DPSCs and SHED in the treatment of diabetes mellitus using a regenerative approach, due to their multipotent capabilities, since they can differentiate into insulin-producing cells [13]. DPSCs are capable of differentiating into pancreatic cells and also into insulin-producing islet-like cells [123, 124]. Kanafi et al. implanted islet-like cell aggregates derived from DPSCs or SHED cells into diabetic mice, and found that the SHED group was superior to the group treated with DPSCs in terms of maintaining normal blood glucose levels [50].

SHED have also been studied for the treatment of injuries to the kidney, where they influence the proliferation of tubular epithelial cells [125], through a paracrine effect that induce cell migration and facilitates the repair of acute kidney damage.

Regenerative Therapy of Various Systemic Disease

Previous investigations demonstrated the positive impact of DPSCs and SHED when used in the treatment of various systemic diseases [126]. The enormous capacity of these DSCs makes them an attractive source for regenerative therapies in medicine. Both DPSCs and SHED can differentiate into hepatic cells [127, 128], which makes them promising in the treatment of liver cirrhosis [129].

As already discussed, due to their myogenic multipotency, DPSCs and SHED cells can be of value in the treatment of muscle injuries, including to the myocardium. Both cell types have been used experimentally to treat myocardial infarction [111] and Duchene muscular dystrophy [130, 131]. Moreover, the immunomodulatory and anti-inflammatory effects exerted by these cells have been applied successfully in the treatment of a range of inflammatory and autoimmune diseases including systemic lupus erythematosus [132], rheumatoid arthritis [133], autoimmune encephalomyelitis [134], Alzheimer's disease, and Parkinson's disease [135, 136].

There have been few systematic comparisons between BMMSCs and DSCs in terms of their immunophenotype, gene expression profile, and regenerative potentials [34]. Collectively, in vitro investigations show that DPSCs share a similar pattern of gene expression with BMMSCs [15]. In addition, signaling pathways of odontoblastic differentiation of DPSCs are similar to the pathways whereby bone marrow-derived stem cells take on osteoblastic features [15].

Shi et al. evaluated gene expression in DPSCs and BMMSCs, and showed that more than 4000 known human genes were similar between these cells [137]. However, they have found that collagen type 18, insulin-like growth factor-2, and cyclin-dependent kinase 6 were much more highly expressed in DPSCs, while insulin-like growth factor binding protein-7 and collagen types I and II were expressed more in BMMSCs [137].

Yamada et al. characterized and compared DPSCs and BMMSCs using a cDNA microarray system, including 12814 genes and a clustering algorithm [138]. They demonstrated that after osteoinduction DPSCs expressed alkaline phosphatase, DSPP, and DMP-1 at levels that were higher than BMMSCs. However, in the clustering assessment, it became apparent that both cells share similar gene regulation pathways for signaling, cell metabolism, and communication [138].

Despite DPSCs and BMMSCs having many similarities in regulating roles, in signaling factors, and in their expression profile, these two types of stem cells are very different in their proliferative capacity and their differentiation potential, and this has led to distinct patterns of use for tissue regeneration in preclinical studies [15]. For instance, the chondrogenic potential of DPSC, and the adipogenic capacity of both SCAP and DPSCs are weaker than those of BMMSCs. Conversely, the neurogenic capabilities of DSCs are far more potent than those of BMMSCs. In addition, SHED and PDLSCs have a much higher growth potential compared to BMMSCs [139]. These differences may be due to the neural crest origin of DSCs.

3 Stem Cells from the Apical Papilla

The apical part of dental papilla is a cell-rich zone containing stem cells (Fig. 1). The apical papilla can be harvested from an immature extracted tooth, and used for isolation of SCAP [140]. The main difference between SCAP and DPSCs is that SCAP is the precursor of the radicular pulp. The characteristic differences between these cells are listed in Table 1. In general, SCAP derives from the developing dental tissues, which consists of early progenitor cells that are distinctive from the cells of mature tissues (DPSCs) [141].

3.1 Isolation and Characterization

After harvesting the apical papilla of a developing root, the tissue is diced into smaller pieces and subjected to enzymatic digestion with collagenase and dispase [140] as already described for the isolation of DPSCs and SHED.

Cultivated SCAP possess low immunogenicity, as seen by lymphocyte assays in the laboratory [142]. Flow-cytometry analysis shows that SCAP express typical cell markers including CD73, 90, and 105 (Table 1) [143]. In addition, the perivascular location of SCAP reflected by their expression of STRO-1 and CD146, which gradually fades with extended passaging of the cells [144–146]. Although expressed at a relatively low amount CD24 seems to be exclusively positive in SCAP compared to other DSCs and other MSCs [147, 148]. Expression of CD24 expression reduces to zero after the 10th passage [149, 150]. If CD 24 is a specific marker for determining the "stemness" of SCAP, these findings imply that loss of stemness in SCAP occurs after the 10th passage.

3.2 Regenerative Applications

SCAP can differentiate into various cell lineages [149], which make these cells an attractive source for tissue engineering.

3.2.1 Pulp Regeneration and Angiogenesis

Regenerative endodontic therapy includes the process of regenerating the dentine–pulp complex [151]. SCAP are one of the most promising stem cell sources for such therapy due to their known odontogenic potency and their expression of dentine-related differentiation markers such as dentine sialphosphoprotein (DSPP) [152].

Nowadays, with the application of scaffolds and the inclusion of growth factors such as vascular endothelial growth factor (VEGF), and fibroblast growth factor (FGF), there is increasing optimism regarding the possibility of regeneration of dentin-pulp complex [153–155]. Cell homing therapies have determined that chemotactic factors for SCAP include SDF-1, TGF-β, and granulocyte-colony stimulating factor (G-CSF). These not only can improve the migration of SCAP, but they also can promote their differentiation [156].

Sonoyama et al. reported that human SCAP can develop into a functional root in an animal model [29]. De novo dentine–pulp complex regeneration by SCAP begins with the migration of these cells onto the dentine surface, followed by odontoblastic differentiation and the expression of phenotypic markers of dentinogenesis [157].

Hikens et al. demonstrated that SCAP are able to express a number of markers of angiogenesis including VEGF), thrombospondin-1, angiopoietin-1, endostatin, matrix metalloproteinases, and FGF [153]. Revascularization is a critical requirement for pulpal tissue engineering [158]. SCAP cells appear to be a good candidate for promoting revascularization because of their original niche in the perivascular location [159].

Moreover, in repairing dentine–pulp complex, the administration of scaffolds can be challenging, as some of the biomaterials that are used (e.g., HA and TCP) are osteoinductive, and thus there is a risk of generalized calcification occurring in the pulp space [157]. Thus, the properties of the scaffold need to be adjusted to suit the requirements of pulp tissue regeneration. Amirkia et al. used a three-dimensional silk fibroin as a natural scaffold, and found that this improved the attachment of SCAP and their differentiation [160]. In addition, chitosan-based scaffolds seem to be an appropriate carrier for SCAP as they improve their odontogenic potential [161].

3.2.2 Neural Regeneration

The neurogenic potential of SCAP has been exploited for neural tissue engineering [158]. SCAP originated from the neural crest, and they express high level of the neural marker nestin [152]. In addition, SCAP can drive a neuroprotective mechanism, by decreasing inflammation and inducing differentiation of oligodendrocytes [155]. SCAP also express various markers (genes) for neurogenesis (Table 1). In vitro and in vivo investigations show SCAP can participate in neurite outgrowth and in axonal induction [158].

Under neurogenic induction, SCAP start to mimic spindle-shaped neurocytes, with long cellular process [152, 162, 163]. In a study by De Berdt et al., the entire apical papilla was transplanted into an area of artificial spinal cord damage in an animal model. The apical papilla functioned as a scaffold for SCAP to regenerate neural tissue in the original niche of the cells [164]. Interestingly, this study revealed that hypoxic conditions could stimulate SCAP to express neural-specific genes and to secrete growth factors [164].

3.2.3 Bone Regeneration

The osteogenic potency of SCAP has been confirmed by studies which have shown differentiation of these cells into osteoblasts, as determined by alizarin red staining of calcium deposition and the expression of osteogenic markers, including bone sialoprotein, alkaline phosphatase, gamma-carboxyglutamate protein, runt-related transcription factor-2 (RUNX2), and bone morphogenetic proteins (BMPs) [32]. The osteogenic differentiation capability of SCAP is comparable to that of BMMSCs [148].

The cultivation of SCAP in an osteogenic medium results in the formation of osteoblast-like cells that are able to produce mineralized nodules [165]. Nada et al. observed that SCAP isolated from different teeth showed different differentiation patterns. For instance, SCAP isolated from third molars tended to produce diffuse mineralization, whereas SCAP isolated from premolar teeth showed a localized pattern of calcific deposits [165]. To date, no in vivo investigations have been conducted to evaluate the bone regeneration capacity of SCAP.

3.2.4 Other Regenerative Applications

The chondrogenic potential of SCAP has been assessed using Alcian Blue staining of the chondrocytes that have formed in vitro [166], but to date no study has been carried out to evaluate the molecular evidence for the chondrogenic potential of SCAP, such as the expression of chondrogenic genes in SCAP under chondrogenic induction. Thus far, no in vivo study of cartilage formation by SCAP has been reported.

Several studies have demonstrated the adipogenic capacity of SCAP [165]. SCAP can express lipoprotein and lipase, suggesting the adipogenic capability of these cells [152]. However, in comparison to BMMSCs, their adipogenic potential is low [148, 153, 167].

4 Dental Follicle Stem Cells

The dental follicle is a loose connective tissue sac surrounding the tooth bud. It plays an important role in tooth development and eruption. The dental follicle is involved in tooth eruption by controlling osseous remodeling through the timely production of various secreted mediators [168]. Stem cells have been harvested from dental follicle of different species in various developmental stages [169, 170]. DFSCs are multipotent, and can differentiate to form periodontium, bone, and cementum [169, 171].

4.1 Isolation and Characterization

Human extracted third molar teeth are the major tissue source that has been used to isolate human DFSCs. The method for DFSCs isolation is similar to that described for other DSCs.

In culture, DFSCs have a typical fibroblast-like morphology. They express CD9, CD10, CD13, notch-1, and nestin, but they do not express CD31, CD34, CD45, and HLA-DR [172, 173] (Table 1). DFSCs also express cementum attachment protein and cementum protein-23 (CP-23), which are two putative cementoblast markers [15]. DFSCs also express STRO-1 and BMP receptors in vivo [26].

4.2 Regenerative Applications

DFSCs can differentiate into osteoblasts, cementoblasts [30], and adipocytes in the appropriate inducing culture media [170]. Although experimental investigations have revealed mineralized tissue formation by DFSCs, DPSCs have a greater capacity for all hard tissue formation. Yagyuu et al. reported hard tissue formation of DFSC at preclinical studies [176], while Yokoi et al. demonstrated that DFSCs are capable of regenerating soft tissue and PDL in vivo [174].

4.2.1 Bone Regeneration

DFSCs have the capacity to create calcification, as seen both in cell culture [25] and in animal [175] studies. Several investigations have reported the osteogenic differentiation of DFSCs in appropriate osteogenic medium [176–179]. In vivo bone formation by DFSCs has been demonstrated in critical size bone defects in rat calvaria [180]. Moreover, in vitro investigations have indicated that BMP-6 and BMP-9 promote the osteogenic differentiation of DFSCs [176].

Rezai Rad et al. showed that a temperature of 37–40 °C was optimal for inducing osteogenesis of DFSCs in vitro [181]. Honda et al. reported the results of two different animal studies to evaluate the bone regenerative capacity of DFSCs [180, 182]. The findings of both studies suggest that DFSCs supported bone regeneration, although it was not clear whether the transplanted stem cells had in fact differentiated into osteoblasts.

4.2.2 Periodontal Regeneration

DFSCs are capable of generating osteoblasts, cementoblasts, and PDL [14]. Honda et al. isolated DFSCs from the third molar teeth of 6-month-old pigs [180]. DFCSs were seeded in the bottom of a tube and DPSCs and the enamel organ epithelium (which originated from same pigs) were added in order to produce a recombination which mimicked the tooth primordia. The whole mixture was then transplanted into the omentum of immunocompromised rats (Fig. 2). A thick layer of dentine with viable odontoblasts and a layer of cementum-like tissue were found at 24 weeks post-surgery. Moreover, collagen fibers were present in a pattern that resembled the PDL, and they were attached to the cementum-like layer.

Another histological evaluation also confirmed the possibility of whole periodontium regeneration via expanded DFSCs [182]. A further investigation reported that DFSCs could differentiate into PDL cells, and could regenerate PDL-like structures including cementum-like tissues [183]. Other researchers have indicated that DFSCs used in combination with a treated dentine matrix could regenerate root-like tissues with a dentine–pulp complex [184].

5 Stem Cells from the Periodontal Ligament

Although early studies provided some evidence to support the differentiation capability of periodontal ligament (PDL) cells, such as the ability to differentiate into cementum-forming cells and osteoblasts [185, 186], Seo et al., conclusively identified a population of MSCs within the periodontal ligament that can express stem cell markers and that have the capability to differentiate into various cell lines [27].

5.1 Isolation and Characterization

To isolate PDLSCs, extracted teeth with an intact PDL are immersed into a digestion solution consisting of collagenase and trypsin [187]. The resultant cells from the enzymatic digestion can then be cultured in various media such as serum-

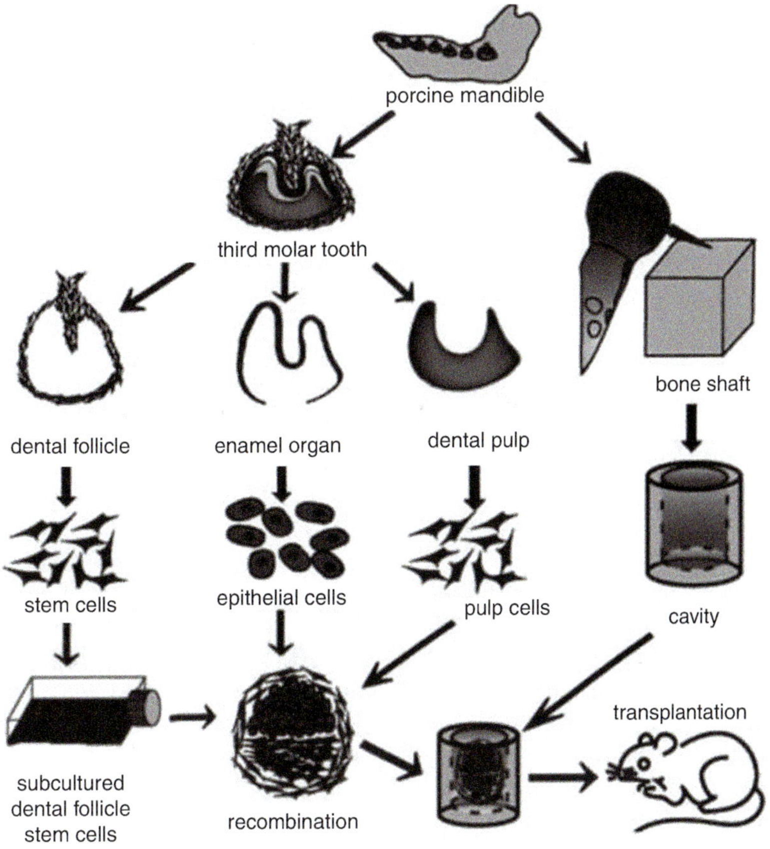

Fig. 2 Schematic diagram of procedure used to regenerate engineered dental root analogue in Honda et al. study [180]

containing media and neurosphere-forming medium, according to the intended therapeutic purpose [53].

PDLSCs possess low immunogenicity, and they can modulate behavior of peripheral blood mononuclear cells by secreting TGF-β and HGF [188]. PDLSCs isolated from inflamed periodontium can significantly reduce the activity and proliferation of T lymphocytes [189]. They express high levels of interleukin (IL)-10, and IL-17 in comparison to PDLSCs derived from healthy tissues [190].

PDLSCs express MSC-related markers and tendon specific transcription factor (scleraxis) (Table 1). Scleraxis expression is significantly higher in PDLSCs than in DPSCs and in BMMSCs [15]. PDLSCs do not express hematopoietic markers such as CD14, CD19, CD34, and CD45 [15], or markers associated with hematopoietic cells including CD40, CD80, and CD86 [188].

5.2 Regenerative Applications

The current literature report that PDLSCs can differentiate into cementum-like structures, and along osteogenic, adipogenic, and chondrogenic cell lineages (Table 1).

5.2.1 Periodontal Regeneration

PDLSCs were firstly applied in animal models and used to reconstruct cementum-PDL-like structures [27]. Subsequently, several studies investigated the capability of PDLSCs for periodontal regeneration [191–193]. Ninomiya et al. implanted a HA scaffold loaded with PDLSCs into the dorsal muscle of rats, and demonstrated bone-like tissue formation [194]. PDLSCs seeded on HA/TCP successfully formed PDL and cementum-like tissues surrounding the scaffold [195, 196]. Complete PDL regeneration was achieved, with formation of Sharpey's fibers between the newly formed cementum and the fibers of the PDL [197].

Likewise, when transplanting PDLSCs that were treated with recombinant human plasminogen activator inhibitor-1, and then seeded on HA/TCP scaffold, into the dorsal region of immunodeficient mice, cementum-like tissue surrounded by PDL-like tissue was observed after 10 weeks [198].

Using stem cell all sheets derived from PDLSCs for tissue engineering has been attempted. HA/TCP wrapped with PDLSCs cell sheets has been shown to generate PDL/cementum-like structures in rats and mice [199, 200]. Moreover, adding platelet-rich fibrin (PRF) to a HA/TCP scaffold wrapped with PDLSCs sheets was shown to regenerate not only cementum and PDL, but also blood vessels [201].

Transplantation of human PDLSCs sheets into infra-bony mandibular defects in dogs showed new cementum regeneration, with the formation of collagen and nerve fibers around the roots of the teeth after 8 weeks [202]. When PDLSCs and gelatin sponge scaffolds were grafted onto fenestration defects in rats, complete healing of bone, cementum, and PDL was seen after only 3 weeks [203]. These findings suggested that there is considerable potential for regeneration of the periodontium using PDLSCs, and that this approach is likely to be successful when used in the clinic to manage real bony defects [204].

However, a significant point is that the regenerative capability of PDLSCs is affected considerably by the presence of inflammation and by the age of the PDL tissue from which the cells are harvested [200, 204, 205]. Gao et al. compared various PDLSCs from donors of different ages [200]. They found that PDLSCs from younger donors had greater cementum/PDL formation potential than those from older donors. Other studies have reported that PDLSCs derived from donors with periodontitis have a significantly lower capacity to form bone than PDLSCs derived from healthy sites or healthy donors [205].

5.2.2 Bone Regeneration

The osteogenic potential of PDLSCs has been shown in several in vitro studies, which have reported the formation of mineralized nodules [27, 206–208]. Although PDLSCs typically form cementum-like structures [15], PDLSCs implanted into periodontal defects of animal models appear to accelerate the regeneration of trabecular bone next to PDL-like structures, thus demonstrating their capacity for alveolar bone regeneration [27, 209].

Transplantation of human PDLSCs encapsulated in a RGD-modified alginate has been shown to enhance bone formation in critical-sized calvarial defects in rats [210]. Several studies have observed a lower bone regenerative potential of PDLSCs compared to BMMSCs. By way of comparison, BMMSCs were reported to be more effective for alveolar bone repair in canine models [211]. Another study confirmed the lower osteogenic capability of PDLSCs compared to BMMSCs [212]. The reason for this may be the presence of more end-differentiated cells in the PDLSCs population [213]. However, there are a few contrary reports in the literature suggesting similar or even better osteogenic potentials of PDLSCs compared to BMMSCs [214], and other DSCs [80, 215]. This point needs further research to resolve it.

5.2.3 Tendon and Cartilage Regeneration

Gronthos et al. showed that PDLSCs can express a tendon-specific marker (scleraxis) in vitro [206]. A combination of human PDLSCs with an RGD-coupled alginate could form a tendon-like tissue in mice. When used for tendon regeneration, compared with BMMSCs, PDLSCs gave a more highly organized tissue with a greater amount of collagen fibers [216].

Moshavernia et al. showed that cartilage healing could be achieved by applying encapsulated PDLSCs in an alginate hydrogel [216]. Moreover, several animal studies have successfully used PDLSCs for cartilage tissue engineering. Ectopic cartilage formation was observed at PDLSC transplantation sites [102, 217, 218]. It is well-known that cartilage has a very restricted ability for self-renewal and regeneration [219]. In this regard, the chondrogenic potential of PDLSCs is noteworthy, and it is likely that they will be of interest for cartilage repair [102].

Table 3 Licensed dental tissue-derived stem cells bank all around the world[a]

Name	Website	Country
BioEDEN	http://www.bioeden.com/	United States
Store-A-Tooth	http://www.store-atooth.com/	
StemSave	http://www.stemsave.com/	
Three brackets (Hiroshima University)	http://www.teethbank.jp/	Japan
Teeth Bank Co.	http://www.teethbank.jp/	
Advanced Center for Tissue Engineering	http://www.acte-group.com/	
MoBaTann: Tooth biobank	http://www.uib.no/en/rg/biomaterial/64723/mobatann-tooth-biobank	Norway
The Norwegian Tooth Bank	http://www.fhi.no/morogbarn	
Stemade Biotech Pvt.	http://www.stemade.com/	India

[a]All information gathered from Chalisserry et al. and Liu et al. systematic reviews

5.2.4 Other Regenerative Applications

Recently, it has been shown that PDLSCs have both neurogenic and angiogenic differentiation potentials [53]. PDL-derived spheres are able to differentiate into mesodermal and neural cells. MSCs originated from the PDL can form Schwann cells by inducing the Erk1 signaling pathway [220]. Furthermore, these cells can regenerate retinal ganglion-like cells via functional synapses and respond to calcium [221]. Cen et al. demonstrated that human PDLSCs ameliorate ganglion cells and axonal regeneration when used in the retina of animals with a traumatized optic nerve [222].

Another novel application of PDLSCs is its possible use for cardiogenic differentiation. PDLSCs express some cardiac cell markers, such as sarcomeric actin and cardiac troponin T [223].

6 Banking of DSCs

According to the diversity of DSCs and their beneficial aspects, the potential uses of stem cell-based treatments using DSCs in both dentistry and medicine are significant. The administration of a patient's own DSCs during therapy may not often be practical, since they may have more conditions requiring treatment in their later years, but have higher numbers of stem cells in their tissues when they are in their childhood years. Banking of DSCs is one way of easily maintaining a suitable supply of DSCs to meet the needs of patients later in their life. Such an approach can pave a new road for progress in healthcare by maintaining a promising source of autologous cells for personalized regenerative treatments.

Given these considerations, the ability to harvest and safely preserve DSCs becomes more important. Nowadays, DSCs can be cryopreserved for a long period of time [224–226]. In a number of developed countries, licensed tooth banks have been founded (Table 3) [30, 34, 227]. When such banks have been established there are a number of ethical controversies, as well as social, and legal issues that need to be considered. Consistent and well-documented laboratory procedures are needed to evaluate and preserve the cells. There is also a need for appropriate regulations or legislation regarding stem cell banking.

7 Limitations

Although stem cell-based therapeutic approaches have shown much promise with an appealing path to their use in tissue regeneration and functional repair, multiple factors need to be optimized. This will require thorough clinical investigations as well as cell culture studies to enhance methods to grow up and maintain a large quantity of cells. One must bear in mind that the availability of the dental tissues over a lifetime will alter, and the time of tooth extraction may not match the time when the patient needs therapy with dental stem cells. Banking of dental stem cells may partially address this issue, but more work is needed to optimize preservation methods, and make such banks less expensive and easier to use for clinical applications. Cell banks must address quality and safety issues such as the stability of the cell phenotype over time, and the possible risks of contamination with endotoxins or with pathogens [30, 228].

8 Conclusions and Future Direction

DSCs as one of the more versatile MSCs have been used widely utilized in preclinical studies for tissue engineering purposes. DSCs exert a range of immunomodulatory activities, and these have not yet been characterized fully. DSCs have a low immunogenicity and may also have immunosuppressive actions. This makes dental tissues a promising source for stem cells for the repair of bone, dental pulp, periodontium, nerves, and other tissues. The various recognized DSCs not only have the potential to differentiate into different cell lines and undergo self-renewal, but their collection could be done as part of normal dental treatment, harvesting them from extracted teeth. Isolation of DSCs usually does not require additional surgical procedures, as exfoliated and extracted teeth are often discarded as medical waste. The collection of DSCs does not pose any major ethical concerns, unlike the use of embryonic stem cells. Nonetheless, in future, it is important to optimize DSC cryopreservation protocols, and address issues including donor-related diversity, and the influence of cell culture conditions [34, 229, 230]. Because of their potential use in a range of cell-based therapies, the ability to expand stem cells of dental origin while also maintaining their original stemness properties is critical. Thus, development of safe and efficient cell expansion strategies should be a focus for research in the future.

References

1. Volarevic V, Markovic BS, Gazdic M, Volarevic A, Jovicic N, Arsenijevic N, et al. Ethical and safety issues of stem cell-based therapy. Int J Med Sci. 2018;15(1):36.
2. Charbord P. Bone marrow mesenchymal stem cells: historical overview and concepts. Hum Gene Ther. 2010;21(9):1045–56.
3. Zhou S, Greenberger JS, Epperly MW, Goff JP, Adler C, LeBoff MS, et al. Age-related intrinsic changes in human bone-marrow-derived mesenchymal stem cells and their differentiation to osteoblasts. Aging Cell. 2008;7(3):335–43.
4. Zomorodian E, Baghaban EM. Mesenchymal stem cells as a potent cell source for bone regeneration. Stem Cells Int. 2012;2012:980353.
5. Bosch P, Musgrave DS, Lee JY, Cummins J, Shuler F, Ghivizzani SC, et al. Osteoprogenitor cells within skeletal muscle. J Orthop Res. 2000;18(6):933–44.
6. Morad G, Kheiri L, Khojasteh A. Dental pulp stem cells for in vivo bone regeneration: a systematic review of literature. Arch Oral Biol. 2013;58(12):1818–27.
7. Bartold PM, Shi S, Gronthos S. Stem cells and periodontal regeneration. Periodontol 2000. 2006;40(1):164–72.
8. Khojasteh A, Nazeman P, Rad MR. Dental stem cells in oral, maxillofacial and craniofacial regeneration. Dental stem cells. New York: Springer; 2016. p. 143–65.
9. Rezai-Rad M, Bova JF, Orooji M, Pepping J, Qureshi A, Del Piero F, et al. Evaluation of bone regeneration potential of dental follicle stem cells for treatment of craniofacial defects. Cytotherapy. 2015;17(11):1572–81.
10. Gimble J, Rad MR, Yao S. Adipose tissue–derived stem cells and their regeneration potential. Stem Cells Craniofacial Dev Regen. 2013;241–58.
11. Zuk PA, Zhu M, Mizuno H, Huang J, Futrell JW, Katz AJ, et al. Multilineage cells from human adipose tissue: implications for cell-based therapies. Tissue Eng. 2001;7(2):211–28.
12. Salehi-Nik N, Rezai Rad M, Kheiri L, Nazeman P, Nadjmi N, Khojasteh A. Buccal fat pad as a potential source of stem cells for bone regeneration: a literature review. Stem Cells Int. 2017;2017:8354640.
13. Marei MK, El Backly RM. Dental mesenchymal stem cell-based translational regenerative dentistry: from artificial to biological replacement. Front Bioeng Biotechnol. 2018;6:49.
14. Zhai Q, Dong Z, Wang W, Li B, Jin Y. Dental stem cell and dental tissue regeneration. Front Med. 2018;13:1–8.
15. Huang G-J, Gronthos S, Shi S. Mesenchymal stem cells derived from dental tissues vs. those from other sources: their biology and role in regenerative medicine. J Dent Res. 2009;88(9):792–806.

16. Ibarretxe G, Crende O, Aurrekoetxea M, García-Murga V, Etxaniz J, Unda F. Neural crest stem cells from dental tissues: a new hope for dental and neural regeneration. Stem Cells Int. 2012;2012:103503.
17. Yu J, Wang Y, Deng Z, Tang L, Li Y, Shi J, et al. Odontogenic capability: bone marrow stromal stem cells versus dental pulp stem cells. Biol Cell. 2007;99(8):465–74.
18. Sheng G. The developmental basis of mesenchymal stem/stromal cells (MSCs). BMC Dev Biol. 2015;15(1):44.
19. Rezai Rad M. Characteristics of dental follicle stem cells and their potential application for treatment of craniofacial defects. 2014. doi: etd-07052014-002034.
20. Jaquery A. Implantation of dental pulp stem cells in a biodegradable scaffold for dental pulp tissue engineering; 2007.
21. Ferro F, Spelat R, D'Aurizio F, Puppato E, Pandolfi M, Beltrami AP, et al. Dental pulp stem cells differentiation reveals new insights in Oct4A dynamics. PLoS One. 2012;7(7):e41774.
22. Rodríguez-Lozano FJ, Bueno C, Insausti CL, Meseguer L, Ramirez M, Blanquer M, et al. Mesenchymal stem cells derived from dental tissues. Int Endod J. 2011;44(9):800–6.
23. Gronthos S, Mankani M, Brahim J, Robey PG, Shi S. Postnatal human dental pulp stem cells (DPSCs) in vitro and in vivo. Proc Natl Acad Sci USA. 2000;97(25):13625–30.
24. Miura M, Gronthos S, Zhao M, Lu B, Fisher LW, Robey PG, et al. SHED: stem cells from human exfoliated deciduous teeth. Proc Natl Acad Sci USA. 2003;100(10):5807–12.
25. Morsczeck C, Götz W, Schierholz J, Zeilhofer F, Kühn U, Möhl C, et al. Isolation of precursor cells (PCs) from human dental follicle of wisdom teeth. Matrix Biol. 2005;24(2):155–65.
26. Kémoun P, Laurencin-Dalicieux S, Rue J, Farges J-C, Gennero I, Conte-Auriol F, et al. Human dental follicle cells acquire cementoblast features under stimulation by BMP-2/-7 and enamel matrix derivatives (EMD) in vitro. Cell Tissue Res. 2007;329(2):283–94.
27. Seo B-M, Miura M, Gronthos S, Bartold PM, Batouli S, Brahim J, et al. Investigation of multipotent postnatal stem cells from human periodontal ligament. Lancet. 2004;364(9429):149–55.
28. Sonoyama W, Liu Y, Fang D, Yamaza T, Seo B-M, Zhang C, et al. Mesenchymal stem cell-mediated functional tooth regeneration in swine. PLoS One. 2006;1(1):e79.
29. Sonoyama W, Liu Y, Yamaza T, Tuan RS, Wang S, Shi S, et al. Characterization of the apical papilla and its residing stem cells from human immature permanent teeth: a pilot study. J Endod. 2008;34(2):166–71.
30. Chalisserry EP, Nam SY, Park SH, Anil S. Therapeutic potential of dental stem cells. J Tissue Eng. 2017;8:2041731417702531.
31. Hollands P, Aboyeji D, Orcharton M. Dental pulp stem cells in regenerative medicine. Br Dent J. 2018;224(9):747.
32. Bakopoulou A, Leyhausen G, Volk J, Tsiftsoglou A, Garefis P, Koidis P, et al. Comparative analysis of in vitro osteo/odontogenic differentiation potential of human dental pulp stem cells (DPSCs) and stem cells from the apical papilla (SCAP). Arch Oral Biol. 2011;56(7):709–21.
33. Parner E, Heidmann JM, Kjaer I, Væth M, Poulsen S. Biological interpretation of the correlation of emergence times of permanent teeth. J Dent Res. 2002;81(7):451–4.
34. Liu J, Yu F, Sun Y, Jiang B, Zhang W, Yang J, et al. Concise reviews: characteristics and potential applications of human dental tissue-derived mesenchymal stem cells. Stem Cells. 2015;33(3):627–38.
35. Pivoriūnas A, Surovas A, Borutinskaitė V, Matuzevičius D, Treigytė G, Savickienė J, et al. Proteomic analysis of stromal cells derived from the dental pulp of human exfoliated deciduous teeth. Stem Cells Dev. 2009;19(7):1081–93.
36. Jamal M, Chogle S, Goodis H, Karam SM. Dental stem cells and their potential role in regenerative medicine. J Med Sci. 2011;4(2):53–61.
37. Suchánek J, Visek B, Soukup T, El-Din Mohamed S, Ivancakova R, Mokry J, et al. Stem cells from human exfoliated deciduous teeth-isolation, long term cultivation and phenotypical analysis. Acta Med (Hradec Kralove). 2010;53(2):93–9.
38. Karamzadeh R, Eslaminejad MB. Dental-related stem cells and their potential in regenerative medicine. InTech: Regenerative Medicine and Tissue Engineering; 2013.
39. Yan M, Yu Y, Zhang G, Tang C, Yu J. A journey from dental pulp stem cells to a bio-tooth. Stem Cell Rev Rep. 2011;7(1):161–71.
40. Kerkis I, Kerkis A, Dozortsev D, Stukart-Parsons GC, Massironi SMG, Pereira LV, et al. Isolation and characterization of a population of immature dental pulp stem cells expressing OCT-4 and other embryonic stem cell markers. Cells Tissues Organs. 2006;184(3–4):105–16.
41. Huang GT-J, Sonoyama W, Chen J, Park SH. In vitro characterization of human dental pulp cells: various isolation methods and culturing environments. Cell Tissue Res. 2006;324(2):225.
42. Bakopoulou A, Leyhausen G, Volk J, Tsiftsoglou A, Garefis P, Koidis P, et al. Assessment of the impact of two different isolation methods on the osteo/odontogenic differentiation potential of human dental stem cells derived from deciduous teeth. Calcif Tissue Int. 2011;88(2):130–41.
43. Karamzadeh R, Eslaminejad MB, Aflatoonian R. Isolation, characterization and comparative differentiation of human dental pulp stem cells derived from permanent teeth by using two different methods. J Vis Exp. 2012;69:4372.
44. Morsczeck C, Völlner F, Saugspier M, Brandl C, Reichert TE, Driemel O, et al. Comparison of human dental follicle cells (DFCs) and stem cells from human exfoliated deciduous teeth (SHED) after neural differentiation in vitro. Clin Oral Investig. 2010;14(4):433–40.

45. Kim S-H, Kim Y-S, Lee S-Y, Kim K-H, Lee Y-M, Kim W-K, et al. Gene expression profile in mesenchymal stem cells derived from dental tissues and bone marrow. J Periodontal Implant Sci. 2011;41(4): 192–200.
46. Tseng P-Y, Chen C-J, Sheu C-C, Yu C-W, Huang Y-S. Spontaneous differentiation of adult rat marrow stromal cells in a long-term culture. J Vet Med Sci. 2007;69(2):95–102.
47. Torsvik A, Røsland GV, Svendsen A, Molven A, Immervoll H, McCormack E, et al. Spontaneous malignant transformation of human mesenchymal stem cells reflects cross-contamination: putting the research field on track–letter. Cancer Res. 2010;70(15):6393–6.
48. Gou S, Wang C, Liu T, Wu H, Xiong J, Zhou F, et al. Spontaneous differentiation of murine bone marrow-derived mesenchymal stem cells into adipocytes without malignant transformation after long-term culture. Cells Tissues Organs. 2010;191(3): 185–92.
49. Ren Z, Wang J, Zhu W, Guan Y, Zou C, Chen Z, et al. Spontaneous transformation of adult mesenchymal stem cells from cynomolgus macaques in vitro. Exp Cell Res. 2011;317(20):2950–7.
50. Kanafi MM, Rajeshwari YB, Gupta S, Dadheech N, Nair PD, Gupta PK, et al. Transplantation of islet-like cell clusters derived from human dental pulp stem cells restores normoglycemia in diabetic mice. Cytotherapy. 2013;15(10):1228–36. https://doi.org/10.1016/j.jcyt.2013.05.008.
51. Reynolds BA, Weiss S. Generation of neurons and astrocytes from isolated cells of the adult mammalian central nervous system. Science. 1992;255(5052):1707–10.
52. Kopen GC, Prockop DJ, Phinney DG. Marrow stromal cells migrate throughout forebrain and cerebellum, and they differentiate into astrocytes after injection into neonatal mouse brains. Proc Natl Acad Sci USA. 1999;96(19):10711–6.
53. Techawattanawisal W, Nakahama K, Komaki M, Abe M, Takagi Y, Morita I. Isolation of multipotent stem cells from adult rat periodontal ligament by neurosphere-forming culture system. Biochem Biophys Res Commun. 2007;357(4):917–23.
54. Sasaki R, Aoki S, Yamato M, Uchiyama H, Wada K, Okano T, et al. Neurosphere generation from dental pulp of adult rat incisor. Eur J Neurosci. 2008;27(3):538–48.
55. Espagnolle N, Guilloton F, Deschaseaux F, Gadelorge M, Sensébé L, Bourin P. CD 146 expression on mesenchymal stem cells is associated with their vascular smooth muscle commitment. J Cell Mol Med. 2014;18(1):104–14.
56. Yang K-L, Chen M-F, Liao C-H, Pang C-Y, Lin P-Y. A simple and efficient method for generating Nurr1-positive neuronal stem cells from human wisdom teeth (tNSC) and the potential of tNSC for stroke therapy. Cytotherapy. 2009;11(5): 606–17.
57. Govindasamy V, Abdullah AN, Ronald VS, Musa S, Aziz ZACA, Zain RB, et al. Inherent differential propensity of dental pulp stem cells derived from human deciduous and permanent teeth. J Endod. 2010;36(9):1504–15.
58. Dahlstrand J, Lardelli M, Lendahl U. Nestin mRNA expression correlates with the central nervous system progenitor cell state in many, but not all, regions of developing central nervous system. Dev Brain Res. 1995;84(1):109–29.
59. Gronthos S, Zannettino AC. A method to isolate and purify human bone marrow stromal stem cells. Mesenchymal stem cells. New York: Springer; 2008. p. 45–57.
60. Gronthos S, Brahim J, Li W, Fisher L, Cherman N, Boyde A, et al. Stem cell properties of human dental pulp stem cells. J Dent Res. 2002;81(8):531–5.
61. Laino G, d'Aquino R, Graziano A, Lanza V, Carinci F, Naro F, et al. A new population of human adult dental pulp stem cells: a useful source of living autologous fibrous bone tissue (LAB). J Bone Miner Res. 2005;20(8):1394–402.
62. d'Aquino R, Graziano A, Sampaolesi M, Laino G, Pirozzi G, De Rosa A, et al. Human postnatal dental pulp cells co-differentiate into osteoblasts and endotheliocytes: a pivotal synergy leading to adult bone tissue formation. Cell Death Differ. 2007;14(6): 1162.
63. Zhang W, Walboomers XF, Shi S, Fan M, Jansen JA. Multilineage differentiation potential of stem cells derived from human dental pulp after cryopreservation. Tissue Eng. 2006;12(10):2813–23.
64. Bhaskar S. Orban's oral histology and embryology, vol. 178. 11th ed. St Louis: CV Mosby; 1991.
65. Cipolleschi MG, Sbarba PD, Olivotto M. The role of hypoxia in the maintenance of hematopoietic stem cells. Blood. 1993;82(7):2031–7.
66. Batouli S, Miura M, Brahim J, Tsutsui T, Fisher L, Gronthos S, et al. Comparison of stem-cell-mediated osteogenesis and dentinogenesis. J Dent Res. 2003;82(12):976–81.
67. Takeda T, Tezuka Y, Horiuchi M, Hosono K, Iida K, Hatakeyama D, et al. Characterization of dental pulp stem cells of human tooth germs. J Dent Res. 2008;87(7):676–81.
68. Zhang W, Ahluwalia IP, Literman R, Kaplan DL, Yelick PC. Human dental pulp progenitor cell behavior on aqueous and hexafluoroisopropanol based silk scaffolds. J Biomed Mater Res Part A. 2011;97(4):414–22.
69. Wang J, Ma H, Jin X, Hu J, Liu X, Ni L, et al. The effect of scaffold architecture on odontogenic differentiation of human dental pulp stem cells. Biomaterials. 2011;32(31):7822–30.
70. Kinaia BM, Chogle SM, Kinaia AM, Goodis HE. Regenerative therapy: a periodontal-endodontic perspective. Dent Clin. 2012;56(3):537–47.
71. Du L, Yang P, Ge S. Stromal cell-derived factor-1 significantly induces proliferation, migration, and collagen type I expression in a human periodontal ligament stem

cell subpopulation. J Periodontol. 2012;83(3):379–88. https://doi.org/10.1902/jop.2011.110201.
72. Rettori E, De Laurentiis A, Zorrilla Zubilete M, Rettori V, Elverdin JC. Anti-inflammatory effect of the endocannabinoid anandamide in experimental periodontitis and stress in the rat. Neuroimmunomodulation. 2012;19(5):293–303. https://doi.org/10.1159/000339113.
73. Rodan GA. Introduction to bone biology. Bone. 1992;13(Suppl 1):S3–6.
74. Dimitriou R, Jones E, McGonagle D, Giannoudis PV. Bone regeneration: current concepts and future directions. BMC Med. 2011;9(1):66. https://doi.org/10.1186/1741-7015-9-66.
75. Graziano A, d'Aquino R, Laino G, Papaccio G. Dental pulp stem cells: a promising tool for bone regeneration. Stem Cell Rev. 2008;4(1):21–6. https://doi.org/10.1007/s12015-008-9013-5.
76. Yamada Y, Ito K, Nakamura S, Ueda M, Nagasaka T. Promising cell-based therapy for bone regeneration using stem cells from deciduous teeth, dental pulp, and bone marrow. Cell Transplant. 2011; 20(7):1003–13. https://doi.org/10.3727/096368910x539128.
77. Mori G, Brunetti G, Oranger A, Carbone C, Ballini A, Lo Muzio L, et al. Dental pulp stem cells: osteogenic differentiation and gene expression. Ann NY Acad Sci. 2011;1237:47–52. https://doi.org/10.1111/j.1749-6632.2011.06234.x.
78. d'Aquino R, Graziano A, Sampaolesi M, Laino G, Pirozzi G, De Rosa A, et al. Human postnatal dental pulp cells co-differentiate into osteoblasts and endotheliocytes: a pivotal synergy leading to adult bone tissue formation. Cell Death Differ. 2007;14(6):1162–71. https://doi.org/10.1038/sj.cdd.4402121.
79. Ma L, Makino Y, Yamaza H, Akiyama K, Hoshino Y, Song G, et al. Cryopreserved dental pulp tissues of exfoliated deciduous teeth is a feasible stem cell resource for regenerative medicine. PLoS One. 2012;7(12):e51777. https://doi.org/10.1371/journal.pone.0051777.
80. de Mendonca CA, Bueno DF, Martins MT, Kerkis I, Kerkis A, Fanganiello RD, et al. Reconstruction of large cranial defects in nonimmunosuppressed experimental design with human dental pulp stem cells. J Craniofac Surg. 2008;19(1):204–10. https://doi.org/10.1097/scs.0b013e31815c8a54.
81. Pisciotta A, Riccio M, Carnevale G, Beretti F, Gibellini L, Maraldi T, et al. Human serum promotes osteogenic differentiation of human dental pulp stem cells in vitro and in vivo. PLoS One. 2012;7(11):e50542. https://doi.org/10.1371/journal.pone.0050542.
82. Riccio M, Maraldi T, Pisciotta A, La Sala GB, Ferrari A, Bruzzesi G, et al. Fibroin scaffold repairs critical-size bone defects in vivo supported by human amniotic fluid and dental pulp stem cells. Tissue Eng Part A. 2012;18(9–10):1006–13. https://doi.org/10.1089/ten.TEA.2011.0542.
83. Petridis X, Diamanti E, Trigas G, Kalyvas D, Kitraki E. Bone regeneration in critical-size calvarial defects using human dental pulp cells in an extracellular matrix-based scaffold. J Craniomaxillofac Surg. 2015;43(4):483–90. https://doi.org/10.1016/j.jcms.2015.02.003.
84. Asutay F, Polat S, Gul M, Subasi C, Kahraman SA, Karaoz E. The effects of dental pulp stem cells on bone regeneration in rat calvarial defect model: micro-computed tomography and histomorphometric analysis. Arch Oral Biol. 2015;60(12):1729–35. https://doi.org/10.1016/j.archoralbio.2015.09.002.
85. Seo BM, Sonoyama W, Yamaza T, Coppe C, Kikuiri T, Akiyama K, et al. SHED repair critical-size calvarial defects in mice. Oral Dis. 2008;14(5): 428–34.
86. d'Aquino R, De Rosa A, Lanza V, Tirino V, Laino L, Graziano A, et al. Human mandible bone defect repair by the grafting of dental pulp stem/progenitor cells and collagen sponge biocomplexes. Eur Cell Mater. 2009;18:75–83.
87. Alkaisi A, Ismail AR, Mutum SS, Ahmad ZA, Masudi S, Abd Razak NH. Transplantation of human dental pulp stem cells: enhance bone consolidation in mandibular distraction osteogenesis. J Oral Maxillofac Surg. 2013;71(10):1758.e1-13. https://doi.org/10.1016/j.joms.2013.05.016.
88. Paino F, La Noce M, Giuliani A, De Rosa A, Mazzoni S, Laino L, et al. Human DPSCs fabricate vascularized woven bone tissue: a new tool in bone tissue engineering. Clin Sci (Lond). 2017;131(8):699–713. https://doi.org/10.1042/cs20170047.
89. Cao Y, Liu Z, Xie Y, Hu J, Wang H, Fan Z, et al. Adenovirus-mediated transfer of hepatocyte growth factor gene to human dental pulp stem cells under good manufacturing practice improves their potential for periodontal regeneration in swine. Stem Cell Res Ther. 2015;6:249. https://doi.org/10.1186/s13287-015-0244-5.
90. Kuo TF, Lee SY, Wu HD, Poma M, Wu YW, Yang JC. An in vivo swine study for xeno-grafts of calcium sulfate-based bone grafts with human dental pulp stem cells (hDPSCs). Mater Sci Eng C Mater Biol Appl. 2015;50:19–23. https://doi.org/10.1016/j.msec.2015.01.092.
91. Jahanbin A, Rashed R, Alamdari DH, Koohestanian N, Ezzati A, Kazemian M, et al. Success of maxillary alveolar defect repair in rats using osteoblast-differentiated human deciduous dental pulp stem cells. J Oral Maxillofac Surg. 2016;74(4):829.e1-9. https://doi.org/10.1016/j.joms.2015.11.033.
92. Liu Y, Wang L, Liu S, Liu D, Chen C, Xu X, et al. Transplantation of SHED prevents bone loss in the early phase of ovariectomy-induced osteoporosis. J Dent Res. 2014;93(11):1124–32. https://doi.org/10.1177/0022034514552675.
93. Leyendecker Junior A, Gomes Pinheiro CC, Lazzaretti Fernandes T, Franco BD. The use of human dental pulp stem cells for in vivo bone tissue engineering: a systematic review. J Tissue Eng. 2018;9:2041731417752766. https://doi.org/10.1177/2041731417752766.

94. Kukowska-Latallo JF, Bielinska AU, Johnson J, Spindler R, Tomalia DA, Baker JR. Efficient transfer of genetic material into mammalian cells using Starburst polyamidoamine dendrimers. Proc Natl Acad Sci USA. 1996;93(10):4897–902.
95. Ma L, Aijima R, Hoshino Y, Yamaza H, Tomoda E, Tanaka Y, et al. Transplantation of mesenchymal stem cells ameliorates secondary osteoporosis through interleukin-17-impaired functions of recipient bone marrow mesenchymal stem cells in MRL/lpr mice. Stem Cell Res Ther. 2015;6:104. https://doi.org/10.1186/s13287-015-0091-4.
96. Annibali S, Cicconetti A, Cristalli MP, Giordano G, Trisi P, Pilloni A, et al. A comparative morphometric analysis of biodegradable scaffolds as carriers for dental pulp and periosteal stem cells in a model of bone regeneration. J Craniofac Surg. 2013;24(3):866–71. https://doi.org/10.1097/SCS.0b013e31827ca530.
97. Behnia A, Haghighat A, Talebi A, Nourbakhsh N, Heidari F. Transplantation of stem cells from human exfoliated deciduous teeth for bone regeneration in the dog mandibular defect. World J Stem Cells. 2014;6(4):505–10. https://doi.org/10.4252/wjsc.v6.i4.505.
98. Zhang W, Walboomers XF, van Osch GJ, van den Dolder J, Jansen JA. Hard tissue formation in a porous HA/TCP ceramic scaffold loaded with stromal cells derived from dental pulp and bone marrow. Tissue Eng Part A. 2008;14(2):285–94. https://doi.org/10.1089/tea.2007.0146.
99. Horner PJ, Gage FH. Regenerating the damaged central nervous system. Nature. 2000;407(6807):963.
100. Xiao L, Tsutsui T. Human dental mesenchymal stem cells and neural regeneration. Hum Cell. 2013;26(3):91–6.
101. Luo L, He Y, Wang X, Key B, Lee BH, Li H, et al. Potential roles of dental pulp stem cells in neural regeneration and repair. Stem Cells Int. 2018;2018:1731289. https://doi.org/10.1155/2018/1731289.
102. Xiao L, Nasu M. From regenerative dentistry to regenerative medicine: progress, challenges, and potential applications of oral stem cells. Stem Cells Cloning. 2014;7:89.
103. Xiao L, Tsutsui T. Characterization of human dental pulp cells-derived spheroids in serum-free medium: stem cells in the core. J Cell Biochem. 2013;114(11):2624–36.
104. Yamagata M, Yamamoto A, Kako E, Kaneko N, Matsubara K, Sakai K, et al. Human dental pulp-derived stem cells protect against hypoxic-ischemic brain injury in neonatal mice. Stroke. 2013;44(2):551–4.
105. Arthur A, Rychkov G, Shi S, Koblar SA, Gronthos S. Adult human dental pulp stem cells differentiate toward functionally active neurons under appropriate environmental cues. Stem Cells. 2008;26(7):1787–95.
106. Wang J, Wei X, Ling J, Huang Y, Gong Q, Huo Y. Identification and characterization of side population cells from adult human dental pulp after ischemic culture. J Endod. 2012;38(11):1489–97.
107. Sakai K, Yamamoto A, Matsubara K, Nakamura S, Naruse M, Yamagata M, et al. Human dental pulp-derived stem cells promote locomotor recovery after complete transection of the rat spinal cord by multiple neuro-regenerative mechanisms. J Clin Invest. 2012;122(1):80–90.
108. Mead B, Logan A, Berry M, Leadbeater W, Scheven BA. Intravitreally transplanted dental pulp stem cells promote neuroprotection and axon regeneration of retinal ganglion cells after optic nerve injury. Invest Ophthalmol Vis Sci. 2013;54(12):7544–56.
109. Arminan A, Gandia C, Bartual M, Garcia-Verdugo JM, Lledo E, Mirabet V, et al. Cardiac differentiation is driven by NKX2.5 and GATA4 nuclear translocation in tissue-specific mesenchymal stem cells. Stem Cells Dev. 2009;18(6):907–18. https://doi.org/10.1089/scd.2008.0292.
110. Yang R, Chen M, Lee CH, Yoon R, Lal S, Mao JJ. Clones of ectopic stem cells in the regeneration of muscle defects in vivo. PLoS One. 2010;5(10):e13547. https://doi.org/10.1371/journal.pone.0013547.
111. Gandia C, Arminan A, Garcia-Verdugo JM, Lledo E, Ruiz A, Minana MD, et al. Human dental pulp stem cells improve left ventricular function, induce angiogenesis, and reduce infarct size in rats with acute myocardial infarction. Stem Cells. 2008;26(3):638–45. https://doi.org/10.1634/stemcells.2007-0484.
112. Garzon I, Martin-Piedra MA, Alaminos M. Human dental pulp stem cells. A promising epithelial-like cell source. Med Hypotheses. 2015;84(5):516–7. https://doi.org/10.1016/j.mehy.2015.02.020.
113. Karaoz E, Demircan PC, Saglam O, Aksoy A, Kaymaz F, Duruksu G. Human dental pulp stem cells demonstrate better neural and epithelial stem cell properties than bone marrow-derived mesenchymal stem cells. Histochem Cell Biol. 2011;136(4):455–73. https://doi.org/10.1007/s00418-011-0858-3.
114. Syed-Picard FN, Du Y, Lathrop KL, Mann MM, Funderburgh ML, Funderburgh JL. Dental pulp stem cells: a new cellular resource for corneal stromal regeneration. Stem Cells Transl Med. 2015;4(3):276–85. https://doi.org/10.5966/sctm.2014-0115.
115. Gomes JA, Geraldes Monteiro B, Melo GB, Smith RL, Cavenaghi Pereira da Silva M, Lizier NF, et al. Corneal reconstruction with tissue-engineered cell sheets composed of human immature dental pulp stem cells. Invest Ophthalmol Vis Sci. 2010;51(3):1408–14. https://doi.org/10.1167/iovs.09-4029.
116. Kushnerev E, Shawcross SG, Sothirachagan S, Carley F, Brahma A, Yates JM, et al. Regeneration of corneal epithelium with dental pulp stem cells using a contact lens delivery system. Invest Ophthalmol Vis Sci. 2016;57(13):5192–9. https://doi.org/10.1167/iovs.15-17953.
117. Yam GH, Peh GS, Singhal S, Goh BT, Mehta JS. Dental stem cells: a future asset of ocular cell

therapy. Expert Rev Mol Med. 2015;17:e20. https://doi.org/10.1017/erm.2015.16.
118. Nakashima M, Iohara K, Murakami M. Dental pulp stem cells and regeneration. Endodontic Topics. 2013;28(1):38–50. https://doi.org/10.1111/etp.12027.
119. Yu J, He H, Tang C, Zhang G, Li Y, Wang R, et al. Differentiation potential of STRO-1+ dental pulp stem cells changes during cell passaging. BMC Cell Biol. 2010;11:32. https://doi.org/10.1186/1471-2121-11-32.
120. Morito A, Kida Y, Suzuki K, Inoue K, Kuroda N, Gomi K, et al. Effects of basic fibroblast growth factor on the development of the stem cell properties of human dental pulp cells. Arch Histol Cytol. 2009;72(1):51–64.
121. Reynolds AJ, Jahoda CA. Cultured human and rat tooth papilla cells induce hair follicle regeneration and fiber growth. Differentiation. 2004;72(9–10):566–75. https://doi.org/10.1111/j.1432-0436.2004.07209010.x.
122. Sasaki R, Aoki S, Yamato M, Uchiyama H, Wada K, Okano T, et al. Tubulation with dental pulp cells promotes facial nerve regeneration in rats. Tissue Eng Part A. 2008;14(7):1141–7. https://doi.org/10.1089/ten.tea.2007.0157.
123. Govindasamy V, Ronald VS, Abdullah AN, Nathan KRG, Ab Aziz ZAC, Abdullah M, et al. Differentiation of dental pulp stem cells into islet-like aggregates. J Dent Res. 2011;90(5):646–52. https://doi.org/10.1177/0022034510396879.
124. Carnevale G, Riccio M, Pisciotta A, Beretti F, Maraldi T, Zavatti M, et al. In vitro differentiation into insulin-producing beta-cells of stem cells isolated from human amniotic fluid and dental pulp. Dig Liver Dis. 2013;45(8):669–76. https://doi.org/10.1016/j.dld.2013.02.007.
125. Hattori Y, Kim H, Tsuboi N, Yamamoto A, Akiyama S, Shi Y, et al. Therapeutic potential of stem cells from human exfoliated deciduous teeth in models of acute kidney injury. PLoS One. 2015;10(10):e0140121. https://doi.org/10.1371/journal.pone.0140121.
126. Botelho J, Cavacas MA, Machado V, Mendes JJ. Dental stem cells: recent progresses in tissue engineering and regenerative medicine. Ann Med. 2017;49(8):644–51. https://doi.org/10.1080/07853890.2017.1347705.
127. Ishkitiev N, Yaegaki K, Imai T, Tanaka T, Nakahara T, Ishikawa H, et al. High-purity hepatic lineage differentiated from dental pulp stem cells in serum-free medium. J Endod. 2012;38(4):475–80. https://doi.org/10.1016/j.joen.2011.12.011.
128. Okada M, Ishkitiev N, Yaegaki K, Imai T, Tanaka T, Fukuda M, et al. Hydrogen sulphide increases hepatic differentiation of human tooth pulp stem cells compared with human bone marrow stem cells. Int Endod J. 2014;47(12):1142–50. https://doi.org/10.1111/iej.12262.
129. Hirata M, Ishigami M, Matsushita Y, Ito T, Hattori H, Hibi H, et al. Multifaceted therapeutic benefits of factors derived from dental pulp stem cells for mouse liver fibrosis. Stem Cells Transl Med. 2016;5(10):1416–24. https://doi.org/10.5966/sctm.2015-0353.
130. Kerkis I, Ambrosio CE, Kerkis A, Martins DS, Zucconi E, Fonseca SA, et al. Early transplantation of human immature dental pulp stem cells from baby teeth to golden retriever muscular dystrophy (GRMD) dogs: local or systemic? J Transl Med. 2008;6:35. https://doi.org/10.1186/1479-5876-6-35.
131. Pisciotta A, Riccio M, Carnevale G, Lu A, De Biasi S, Gibellini L, et al. Stem cells isolated from human dental pulp and amniotic fluid improve skeletal muscle histopathology in mdx/SCID mice. Stem Cell Res Ther. 2015;6(1):156. https://doi.org/10.1186/s13287-015-0141-y.
132. Yamaza T, Kentaro A, Chen C, Liu Y, Shi Y, Gronthos S, et al. Immunomodulatory properties of stem cells from human exfoliated deciduous teeth. Stem Cell Res Ther. 2010;1(1):5. https://doi.org/10.1186/scrt5.
133. Ishikawa J, Takahashi N, Matsumoto T, Yoshioka Y, Yamamoto N, Nishikawa M, et al. Factors secreted from dental pulp stem cells show multifaceted benefits for treating experimental rheumatoid arthritis. Bone. 2016;83:210–9. https://doi.org/10.1016/j.bone.2015.11.012.
134. Mita T, Furukawa-Hibi Y, Takeuchi H, Hattori H, Yamada K, Hibi H, et al. Conditioned medium from the stem cells of human dental pulp improves cognitive function in a mouse model of Alzheimer's disease. Behav Brain Res. 2015;293:189–97. https://doi.org/10.1016/j.bbr.2015.07.043.
135. Gnanasegaran N, Govindasamy V, Simon C, Gan QF, Vincent-Chong VK, Mani V, et al. Effect of dental pulp stem cells in MPTP-induced old-aged mice model. Eur J Clin Investig. 2017;47(6):403–14. https://doi.org/10.1111/eci.12753.
136. Wang J, Wang X, Sun Z, Wang X, Yang H, Shi S, et al. Stem cells from human-exfoliated deciduous teeth can differentiate into dopaminergic neuron-like cells. Stem Cells Dev. 2010;19(9):1375–83. https://doi.org/10.1089/scd.2009.0258.
137. Shi S, Robey P, Gronthos S. Comparison of human dental pulp and bone marrow stromal stem cells by cDNA microarray analysis. Bone. 2001;29(6):532–9.
138. Yamada Y, Fujimoto A, Ito A, Yoshimi R, Ueda M. Cluster analysis and gene expression profiles: a cDNA microarray system-based comparison between human dental pulp stem cells (hDPSCs) and human mesenchymal stem cells (hMSCs) for tissue engineering cell therapy. Biomaterials. 2006;27(20):3766–81.
139. Liu H, Gronthos S, Shi S. Dental pulp stem cells. Methods in enzymology. London: Elsevier; 2006. p. 99–113.
140. Huang GT, Sonoyama W, Liu Y, Liu H, Wang S, Shi S. The hidden treasure in apical papilla: the potential role in pulp/dentin regeneration and bioroot engineering. J Endod. 2008;34(6):645–51. https://doi.org/10.1016/j.joen.2008.03.001.

141. Huang GT, Yamaza T, Shea LD, Djouad F, Kuhn NZ, Tuan RS, et al. Stem/progenitor cell-mediated de novo regeneration of dental pulp with newly deposited continuous layer of dentin in an in vivo model. Tissue Eng Part A. 2010;16(2):605–15. https://doi.org/10.1089/ten.TEA.2009.0518.
142. Ding G, Liu Y, An Y, Zhang C, Shi S, Wang W, et al. Suppression of T cell proliferation by root apical papilla stem cells in vitro. Cells Tissues Organs. 2010;191(5):357–64. https://doi.org/10.1159/000276589.
143. Hilkens P, Bronckaers A, Ratajczak J, Gervois P, Wolfs E, Lambrichts I. The angiogenic potential of DPSCs and SCAPs in an in vivo model of dental pulp regeneration. Stem Cells Int. 2017;2017:2582080. https://doi.org/10.1155/2017/2582080.
144. Schneider R, Holland GR, Chiego D Jr, Hu JC, Nor JE, Botero TM. White mineral trioxide aggregate induces migration and proliferation of stem cells from the apical papilla. J Endod. 2014;40(7):931–6. https://doi.org/10.1016/j.joen.2013.11.021.
145. Liu C, Xiong H, Chen K, Huang Y, Huang Y, Yin X. Long-term exposure to pro-inflammatory cytokines inhibits the osteogenic/dentinogenic differentiation of stem cells from the apical papilla. Int Endod J. 2016;49(10):950–9. https://doi.org/10.1111/iej.12551.
146. Bakopoulou A, Leyhausen G, Volk J, Koidis P, Geurtsen W. Effects of resinous monomers on the odontogenic differentiation and mineralization potential of highly proliferative and clonogenic cultured apical papilla stem cells. Dent Mater. 2012;28(3):327–39. https://doi.org/10.1016/j.dental.2012.01.002.
147. Ding G, Wang W, Liu Y, An Y, Zhang C, Shi S, et al. Effect of cryopreservation on biological and immunological properties of stem cells from apical papilla. J Cell Physiol. 2010;223(2):415–22. https://doi.org/10.1002/jcp.22050.
148. Dong R, Yao R, Du J, Wang S, Fan Z. Depletion of histone demethylase KDM2A enhanced the adipogenic and chondrogenic differentiation potentials of stem cells from apical papilla. Exp Cell Res. 2013;319(18):2874–82. https://doi.org/10.1016/j.yexcr.2013.07.008.
149. Zhang W, Zhang X, Ling J, Liu W, Zhang X, Ma J, et al. Proliferation and odontogenic differentiation of BMP2 genetransfected stem cells from human tooth apical papilla: an in vitro study. Int J Mol Med. 2014;34(4):1004–12. https://doi.org/10.3892/ijmm.2014.1862.
150. Zhang J, Wang Z, Jiang Y, Niu Z, Fu L, Luo Z, et al. Nuclear factor I-C promotes proliferation and differentiation of apical papilla-derived human stem cells in vitro. Exp Cell Res. 2015;332(2):259–66. https://doi.org/10.1016/j.yexcr.2015.01.020.
151. Murray PE, Garcia-Godoy F, Hargreaves KM. Regenerative endodontics: a review of current status and a call for action. J Endod. 2007;33(4):377–90.
152. Bakopoulou A, Leyhausen G, Volk J, Koidis P, Geurtsen W. Comparative characterization of STRO-1neg/CD146pos and STRO-1pos/CD146pos apical papilla stem cells enriched with flow cytometry. Arch Oral Biol. 2013;58(10):1556–68.
153. Hilkens P, Fanton Y, Martens W, Gervois P, Struys T, Politis C, et al. Pro-angiogenic impact of dental stem cells in vitro and in vivo. Stem Cell Res. 2014;12(3):778–90.
154. He L, Zhong J, Gong Q, Cheng B, Kim SG, Ling J, et al. Regenerative endodontics by cell homing. Dent Clin. 2017;61(1):143–59.
155. De Berdt P, Bottemanne P, Bianco J, Alhouayek M, Diogenes A, Llyod A, et al. Stem cells from human apical papilla decrease neuro-inflammation and stimulate oligodendrocyte progenitor differentiation via activin-A secretion. Cell Mol Life Sci. 2018:1–14.
156. Fayazi S, Takimoto K, Diogenes A. Comparative evaluation of chemotactic factor effect on migration and differentiation of stem cells of the apical papilla. J Endod. 2017;43(8):1288–93.
157. Huang GT-J, Yamaza T, Shea LD, Djouad F, Kuhn NZ, Tuan RS, et al. Stem/progenitor cell–mediated de novo regeneration of dental pulp with newly deposited continuous layer of dentin in an in vivo model. Tissue Eng Part A. 2009;16(2):605–15.
158. de Almeida JFA, Chen P, Henry MA, Diogenes A. Stem cells of the apical papilla regulate trigeminal neurite outgrowth and targeting through a BDNF-dependent mechanism. Tissue Eng Part A. 2014;20(23–24):3089–100.
159. de Souza LEB, Malta TM, Kashima Haddad S, Covas DT. Mesenchymal stem cells and pericytes: to what extent are they related? Stem Cells Dev. 2016;25(24):1843–52.
160. Amirikia M, Shariatzadeh SMA, Jorsaraei SGA, Mehranjani MS. Impact of pre-incubation time of silk fibroin scaffolds in culture medium on cell proliferation and attachment. Tissue Cell. 2017;49(6):657–63.
161. Bellamy C, Shrestha S, Torneck C, Kishen A. Effects of a bioactive scaffold containing a sustained transforming growth factor-β1–releasing nanoparticle system on the migration and differentiation of stem cells from the apical papilla. J Endod. 2016;42(9):1385–92.
162. Bakopoulou A, Kritis A, Andreadis D, Papachristou E, Leyhausen G, Koidis P, et al. Angiogenic potential and secretome of human apical papilla mesenchymal stem cells in various stress microenvironments. Stem Cells Dev. 2015;24(21):2496–512.
163. Yuan C, Wang P, Zhu L, Dissanayaka WL, Green DW, Tong EH, et al. Coculture of stem cells from apical papilla and human umbilical vein endothelial cell under hypoxia increases the formation of three-dimensional vessel-like structures in vitro. Tissue Eng Part A. 2014;21(5–6):1163–72.
164. De Berdt P, Vanacker J, Ucakar B, Elens L, Diogenes A, Leprince J, et al. Dental apical papilla

as therapy for spinal cord injury. J Dent Res. 2015;94(11):1575–81.
165. Nada OA, El Backly RM. Stem cells from the apical papilla (SCAP) as a tool for endogenous tissue regeneration. Front Bioeng Biotech. 2018;6:103.
166. Cao Y, Xia D, Qi S, Du J, Ma P, Wang S, et al. Epiregulin can promote proliferation of stem cells from the dental apical papilla via MEK/Erk and JNK signalling pathways. Cell Prolif. 2013;46(4):447–56.
167. Lin LM, Kim SG, Martin G, Kahler B. Continued root maturation despite persistent apical periodontitis of immature permanent teeth after failed regenerative endodontic therapy. Aust Endod J. 2018;44(3):292–9.
168. Wise G, King G. Mechanisms of tooth eruption and orthodontic tooth movement. J Dent Res. 2008;87(5):414–34.
169. Honda MJ, Imaizumi M, Tsuchiya S, Morsczeck C. Dental follicle stem cells and tissue engineering. J Oral Sci. 2010;52(4):541–52.
170. Yao S, Pan F, Prpic V, Wise G. Differentiation of stem cells in the dental follicle. J Dent Res. 2008;87(8):767–71.
171. Nanci A, editor. Ten Cate's oral histology: development, structure and function. St. Louis: MOSBY E; 2008.
172. Lindroos B, Mäenpää K, Ylikomi T, Oja H, Suuronen R, Miettinen S. Characterisation of human dental stem cells and buccal mucosa fibroblasts. Biochem Biophys Res Commun. 2008;368(2):329–35.
173. Yagyuu T, Ikeda E, Ohgushi H, Tadokoro M, Hirose M, Maeda M, et al. Hard tissue-forming potential of stem/progenitor cells in human dental follicle and dental papilla. Arch Oral Biol. 2010;55(1):68–76.
174. Yokoi T, Saito M, Kiyono T, Iseki S, Kosaka K, Nishida E, et al. Establishment of immortalized dental follicle cells for generating periodontal ligament in vivo. Cell Tissue Res. 2007;327(2):301–11.
175. Handa K, Saito M, Tsunoda A, Yamauchi M, Hattori S, Sato S, et al. Progenitor cells from dental follicle are able to form cementum matrix in vivo. Connect Tissue Res. 2002;43(2–3):406–8.
176. Ramamoorthi M, Bakkar M, Jordan J, Tran SD. Osteogenic potential of dental mesenchymal stem cells in preclinical studies: a systematic review using modified ARRIVE and CONSORT guidelines. Stem Cells Int. 2015;2015:378368.
177. Xu L-L, Liu H-C, Wang D-S, Ling-Ling E, Xu L, Jin Z-L, et al. Effects of BMP-2 and dexamethasone on osteogenic differentiation of rat dental follicle progenitor cells seeded on three-dimensional β-TCP. Biomed Mater. 2009;4(6):065010.
178. Tsuchiya S, Ohshima S, Yamakoshi Y, Simmer JP, Honda MJ. Osteogenic differentiation capacity of porcine dental follicle progenitor cells. Connect Tissue Res. 2010;51(3):197–207.
179. Chen X, Zhang T, Shi J, Xu P, Gu Z, Sandham A, et al. Notch1 signaling regulates the proliferation and self-renewal of human dental follicle cells by modulating the G1/S phase transition and telomerase activity. PLoS One. 2013;8(7):e69967.
180. Honda MJ, Imaizumi M, Suzuki H, Ohshima S, Tsuchiya S, Satomura K. Stem cells isolated from human dental follicles have osteogenic potential. Oral Surg Oral Med Oral Pathol Oral Radiol Endod. 2011;111(6):700–8.
181. Rezai Rad M, Wise GE, Brooks H, Flanagan MB, Yao S. Activation of proliferation and differentiation of dental follicle stem cells (DFSCs) by heat stress. Cell Prolif. 2013;46(1):58–66.
182. Honda MJ, Tsuchiya S, Shinohara Y, Shinmura Y, Sumita Y. Recent advances in engineering of tooth and tooth structures using postnatal dental cells. Jpn Dent Sci Rev. 2010;46(1):54–66.
183. Handa K, Saito M, Yamauchi M, Kiyono T, Sato S, Teranaka T, et al. Cementum matrix formation in vivo by cultured dental follicle cells. Bone. 2002;31(5):606–11.
184. Guo W, Chen L, Gong K, Ding B, Duan Y, Jin Y. Heterogeneous dental follicle cells and the regeneration of complex periodontal tissues. Tissue Eng Part A. 2012;18(5–6):459–70.
185. Isaka J, Ohazama A, Kobayashi M, Nagashima C, Takiguchi T, Kawasaki H, et al. Participation of periodontal ligament cells with regeneration of alveolar bone. J Periodontol. 2001;72(3):314–23. https://doi.org/10.1902/jop.2001.72.3.314.
186. McCulloch CA, Bordin S. Role of fibroblast subpopulations in periodontal physiology and pathology. J Periodontal Res. 1991;26(3 Pt 1):144–54.
187. Kaneda T, Miyauchi M, Takekoshi T, Kitagawa S, Kitagawa M, Shiba H, et al. Characteristics of periodontal ligament subpopulations obtained by sequential enzymatic digestion of rat molar periodontal ligament. Bone. 2006;38(3):420–6. https://doi.org/10.1016/j.bone.2005.08.021.
188. Wada N, Menicanin D, Shi S, Bartold PM, Gronthos S. Immunomodulatory properties of human periodontal ligament stem cells. J Cell Physiol. 2009;219(3):667–76. https://doi.org/10.1002/jcp.21710.
189. Ding G, Liu Y, Wang W, Wei F, Liu D, Fan Z, et al. Allogeneic periodontal ligament stem cell therapy for periodontitis in swine. Stem Cells. 2010;28(10):1829–38. https://doi.org/10.1002/stem.512.
190. Liu D, Xu J, Liu O, Fan Z, Liu Y, Wang F, et al. Mesenchymal stem cells derived from inflamed periodontal ligaments exhibit impaired immunomodulation. J Clin Periodontol. 2012;39(12):1174–82. https://doi.org/10.1111/jcpe.12009.
191. Tomokiyo A, Yoshida S, Hamano S, Hasegawa D, Sugii H, Maeda H. Detection, characterization, and clinical application of mesenchymal stem cells in periodontal ligament tissue. Stem Cells Int. 2018;2018:5450768.

192. Liu Z, Yin X, Ye Q, He W, Ge M, Zhou X, et al. Periodontal regeneration with stem cells-seeded collagen-hydroxyapatite scaffold. J Biomater Appl. 2016;31(1):121–31. https://doi.org/10.1177/0885328216637978.
193. Sun X, Xu C, Wu G, Ye Q, Wang C. Poly(lactic-co-glycolic acid): applications and future prospects for periodontal tissue regeneration. Polymers. 2017;9(6):189.
194. Ninomiya T, Hiraga T, Hosoya A, Ohnuma K, Ito Y, Takahashi M, et al. Enhanced bone-forming activity of side population cells in the periodontal ligament. Cell Transplant. 2014;23(6):691–701.
195. Park JC, Kim JM, Jung IH, Kim JC, Choi SH, Cho KS, et al. Isolation and characterization of human periodontal ligament (PDL) stem cells (PDLSCs) from the inflamed PDL tissue: in vitro and in vivo evaluations. J Clin Periodontol. 2011;38(8):721–31.
196. Wang H, Li J, Zhang X, Ning T, Ma D, Ge Y, et al. Priming integrin alpha 5 promotes the osteogenic differentiation of human periodontal ligament stem cells due to cytoskeleton and cell cycle changes. J Proteome. 2018;179:122–30.
197. Feng F, Akiyama K, Liu Y, Yamaza T, Wang TM, Chen JH, et al. Utility of PDL progenitors for in vivo tissue regeneration: a report of 3 cases. Oral Dis. 2010;16(1):20–8.
198. Jin H, Choung H-W, Lim K-T, Jin B, Jin C, Chung J-H, et al. Recombinant human plasminogen activator inhibitor-1 promotes cementogenic differentiation of human periodontal ligament stem cells. Tissue Eng Part A. 2015;21(23–24):2817–28.
199. Zhang J, An Y, Gao L-N, Zhang Y-J, Jin Y, Chen F-M. The effect of aging on the pluripotential capacity and regenerative potential of human periodontal ligament stem cells. Biomaterials. 2012;33(29):6974–86.
200. Gao L-N, An Y, Lei M, Li B, Yang H, Lu H, et al. The effect of the coumarin-like derivative osthole on the osteogenic properties of human periodontal ligament and jaw bone marrow mesenchymal stem cell sheets. Biomaterials. 2013;34(38):9937–51.
201. Wang Z-S, Feng Z-H, Wu G-F, Bai S-Z, Dong Y, Chen F-M, et al. The use of platelet-rich fibrin combined with periodontal ligament and jaw bone mesenchymal stem cell sheets for periodontal tissue engineering. Sci Rep. 2016;6:28126.
202. Tsumanuma Y, Iwata T, Washio K, Yoshida T, Yamada A, Takagi R, et al. Comparison of different tissue-derived stem cell sheets for periodontal regeneration in a canine 1-wall defect model. Biomaterials. 2011;32(25):5819–25.
203. Han J, Menicanin D, Marino V, Ge S, Mrozik K, Gronthos S, et al. Assessment of the regenerative potential of allogeneic periodontal ligament stem cells in a rodent periodontal defect model. J Periodontal Res. 2014;49(3):333–45.
204. Pereira L, Rubini M, Silva J, Oliveira D, Silva I, Poças-Fonseca M, et al. Comparison of stem cell properties of cells isolated from normal and inflamed dental pulps. Int Endod J. 2012;45(12):1080–90.
205. Tang HN, Xia Y, Yu Y, Wu RX, Gao LN, Chen FM. Stem cells derived from "inflamed" and healthy periodontal ligament tissues and their sheet functionalities: a patient-matched comparison. J Clin Periodontol. 2016;43(1):72–84.
206. Gronthos S, Mrozik K, Shi S, Bartold P. Ovine periodontal ligament stem cells: isolation, characterization, and differentiation potential. Calcif Tissue Int. 2006;79(5):310–7.
207. Fujita T, Iwata T, Shiba H, Igarashi A, Hirata R, Takeda K, et al. Identification of marker genes distinguishing human periodontal ligament cells from human mesenchymal stem cells and human gingival fibroblasts. J Periodontal Res. 2007;42(3):283–6. https://doi.org/10.1111/j.1600-0765.2006.00944.x.
208. Lekic P, Rojas J, Birek C, Tenenbaum H, McCulloch CA. Phenotypic comparison of periodontal ligament cells in vivo and in vitro. J Periodontal Res. 2001;36(2):71–9.
209. Lekic PC, Rajshankar D, Chen H, Tenenbaum H, McCulloch CA. Transplantation of labeled periodontal ligament cells promotes regeneration of alveolar bone. Anat Rec. 2001;262(2):193–202.
210. Moshaverinia A, Xu X, Chen C, Akiyama K, Snead ML, Shi S. Dental mesenchymal stem cells encapsulated in an alginate hydrogel co-delivery microencapsulation system for cartilage regeneration. Acta Biomater. 2013;9(12):9343–50. https://doi.org/10.1016/j.actbio.2013.07.023.
211. Kim SH, Kim KH, Seo BM, Koo KT, Kim TI, Seol YJ, et al. Alveolar bone regeneration by transplantation of periodontal ligament stem cells and bone marrow stem cells in a canine peri-implant defect model: a pilot study. J Periodontol. 2009;80(11):1815–23. https://doi.org/10.1902/jop.2009.090249.
212. Xu J, Wang W, Kapila Y, Lotz J, Kapila S. Multiple differentiation capacity of STRO-1+/CD146+ PDL mesenchymal progenitor cells. Stem Cells Dev. 2009;18(3):487–96. https://doi.org/10.1089/scd.2008.0113.
213. Gay IC, Chen S, MacDougall M. Isolation and characterization of multipotent human periodontal ligament stem cells. Orthod Craniofac Res. 2007;10(3):149–60. https://doi.org/10.1111/j.1601-6343.2007.00399.x.
214. Houshmand B, Behnia H, Khoshzaban A, Morad G, Behrouzi G, Dashti SG, et al. Osteoblastic differentiation of human stem cells derived from bone marrow and periodontal ligament under the effect of enamel matrix derivative and transforming growth factor-beta. Int J Oral Maxillofac Implants. 2013;28(6):e440–50. https://doi.org/10.11607/jomi.te24.
215. Chadipiralla K, Yochim JM, Bahuleyan B, Huang CY, Garcia-Godoy F, Murray PE, et al. Osteogenic differentiation of stem cells derived from human

periodontal ligaments and pulp of human exfoliated deciduous teeth. Cell Tissue Res. 2010;340(2):323–33. https://doi.org/10.1007/s00441-010-0953-0.
216. Moshaverinia A, Xu X, Chen C, Ansari S, Zadeh HH, Snead ML, et al. Application of stem cells derived from the periodontal ligament or gingival tissue sources for tendon tissue regeneration. Biomaterials. 2014;35(9):2642–50. https://doi.org/10.1016/j.biomaterials.2013.12.053.
217. Ebihara G, Sato M, Yamato M, Mitani G, Kutsuna T, Nagai T, et al. Cartilage repair in transplanted scaffold-free chondrocyte sheets using a minipig model. Biomaterials. 2012;33(15):3846–51. https://doi.org/10.1016/j.biomaterials.2012.01.056.
218. Yan H, Yu C. Repair of full-thickness cartilage defects with cells of different origin in a rabbit model. Arthroscopy. 2007;23(2):178–87. https://doi.org/10.1016/j.arthro.2006.09.005.
219. Tuan RS, Chen AF, Klatt BA. Cartilage regeneration. J Am Acad Orthop Surg. 2013;21(5):303–11. https://doi.org/10.5435/JAAOS-21-05-303.
220. Dapeng L, Xiaojie L, Ping G, Yan D, Gang S. Erk1/2 signalling is involved in the differentiation of periodontal ligament stem cells to Schwann cells in dog. Arch Oral Biol. 2014;59(5):487–91.
221. Ng TK, Yung JS, Choy KW, Cao D, Leung CK, Cheung HS, et al. Transdifferentiation of periodontal ligament-derived stem cells into retinal ganglion-like cells and its microRNA signature. Sci Rep. 2015;5:16429.
222. Cen LP, Ng TK, Liang JJ, Zhuang X, Yao X, Yam GH, et al. Human periodontal ligament-derived stem cells promote retinal ganglion cell survival and axon regeneration after optic nerve injury. Stem Cells. 2018;36(6):844–55. https://doi.org/10.1002/stem.2812.
223. Pelaez D, Torres ZA, Ng TK, Choy KW, Pang CP, Cheung HS. Cardiomyogenesis of periodontal ligament-derived stem cells by dynamic tensile strain. Cell Tissue Res. 2017;367(2):229–41.
224. Gioventù S, Andriolo G, Bonino F, Frasca S, Lazzari L, Montelatici E, et al. A novel method for banking dental pulp stem cells. Transfus Apher Sci. 2012;47(2):199–206.
225. Chen YK, Huang AHC, Chan AWS, Shieh TY, Lin LM. Human dental pulp stem cells derived from different cryopreservation methods of human dental pulp tissues of diseased teeth. J Oral Pathol Med. 2011;40(10):793–800.
226. Perry BC, Zhou D, Wu X, Yang F-C, Byers MA, Chu T-MG, et al. Collection, cryopreservation, and characterization of human dental pulp–derived mesenchymal stem cells for banking and clinical use. Tissue Eng Part C Methods. 2008;14(2):149–56.
227. Arora V, Arora P, Munshi AK. Banking stem cells from human exfoliated deciduous teeth (SHED): saving for the future. J Clin Pediatr Dent. 2009;33(4):289–94.
228. Almela T, Brook IM, Moharamzadeh K. The significance of cell-related challenges in the clinical application of tissue engineering. J Biomed Mater Res Part A. 2016;104(12):3157–63.
229. Hilkens P, Driesen RB, Wolfs E, Gervois P, Vangansewinkel T, Ratajczak J, et al. Cryopreservation and banking of dental stem cells. Adv Exp Med Biol. 2016;951:199–235. https://doi.org/10.1007/978-3-319-45457-3_17.
230. Hilkens P, Meschi N, Lambrechts P, Bronckaers A, Lambrichts I. Dental stem cells in pulp regeneration: near future or long road ahead? Stem Cells Dev. 2015;24(14):1610–22. https://doi.org/10.1089/scd.2014.0510.

Dentine–Pulp Complex Regeneration

Ove A. Peters, Avina Paranjpe, and Alexis Gaudin

1 Introduction

Therapeutic strategies in dentistry are traditionally subtractive rather than additive, for example, in case of dental caries, diseased tissue is removed and replaced with a restoration. In endodontics, an irreversibly diseased dental pulp is removed, and the canal space debrided and obturated with a root canal filling. The latter practice is well established and enjoys reasonably high clinical success rates; however, the procedure leaves the tooth with reduced structural strength [1].

The discovery of dental pulp stem cells two decades ago [2] opened the door for a regenerative approach to the diseased pulp. While the healing potential of the pulp had been recognized for a long time, attempts to predictably regenerate a functional dentin–pulp complex had been futile. In a best-case scenario, the pulp stayed vital, and a dentine bridge formed. This reparative dentine typically did not regenerate a fully functional dentine–pulp complex; this joint structure is characterized by a common embryologic pathway from the dental papilla and by a tight histologic interconnectivity [3].

Box: The Dentine–Pulp Complex

The view that dentine and pulp are embryologically, histologically, and functionally similar tissues have been held for many decades. An intact dentine–pulp complex serves important purposes:

- Synthesize and secrete dentine
- Maintaining tissue homeostasis
- Mediating reparative processes
- Screening for invading pathogens
- Supporting enamel in force dissipation

True regeneration of the dentine–pulp complex would mean "*restitutio ad integrum*," which must include its histological appearance, physiology, and mechanical properties. Current clinical approaches have been characterized as "guided pulpal repair," which likely offer clinical benefits without fulfilling the set criteria for

O. A. Peters (✉)
School of Dentistry, The University of Queensland, Herston, QLD, Australia

Department of Endodontics, Arthur A. Dugoni School of Dentistry, University of the Pacific, San Francisco, CA, USA
e-mail: o.peters@uq.edu.au

A. Paranjpe
Department of Endodontics, University of Washington, School of Dentistry, Seattle, WA, USA
e-mail: avina@uw.edu

A. Gaudin
Faculty of Dental Surgery, Department of Endodontics and Restorative Dentistry, University of Nantes, Nantes, France
e-mail: alexis.gaudin@univ-nantes.fr

© Springer Nature Switzerland AG 2021
S. Hosseinpour et al. (eds.), *Regenerative Approaches in Dentistry*,
https://doi.org/10.1007/978-3-030-59809-9_3

regeneration. Regeneration of the dentine–pulp complex, in contrast, has been defined as healing progression starting with inflammation, using immune signaling, and cellular interaction toward tissue restoration [4]. Considered by some as the "holy grail" of endodontic research, true regeneration of the dentine–pulp complex has yet to be achieved predictably (Fig. 1). This chapter will discuss the structure and physiology of the dentine–pulp complex and highlight pathways for regeneration along with associated limitations.

2 Dentine–Pulp Complex Biology

Human teeth share a similar structure with other vertebrates. However, there is considerable variation in their form and position. The tooth is composed of both mineralized tissues (enamel, dentine, and cementum) and non-mineralized tissue (dental pulp). Dentine and dental pulp can be considered as similar tissues, because of their close embryological, histological, and functional similarities. However, while this anatomical

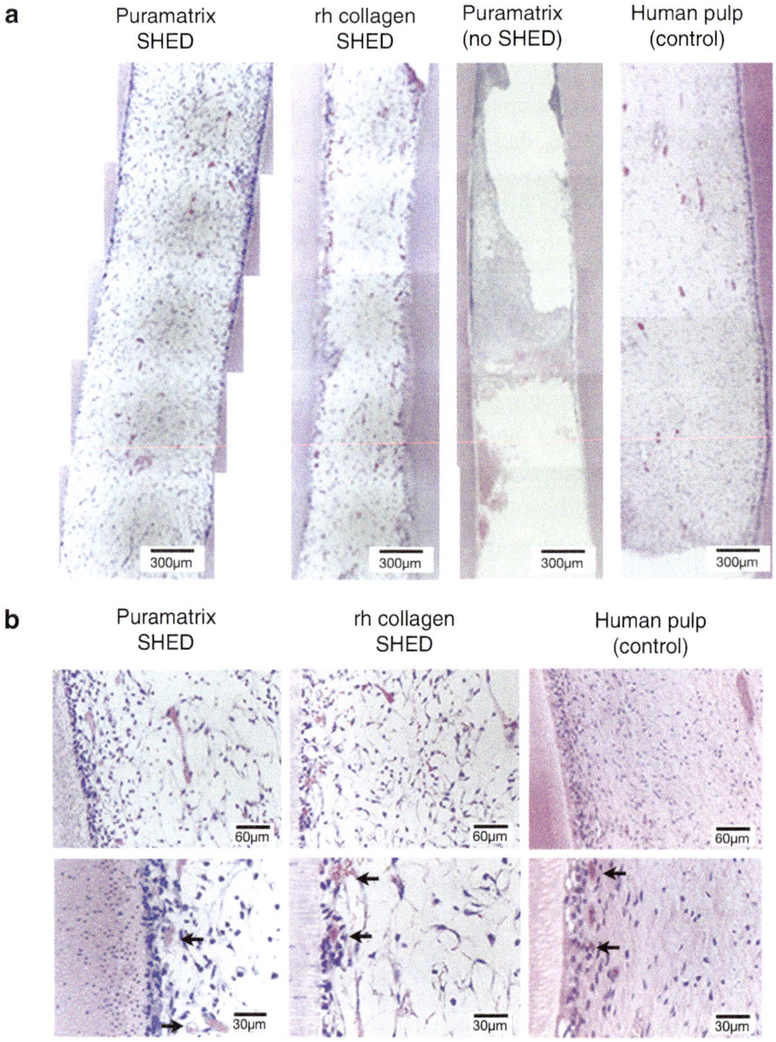

Fig. 1 Dental pulp tissue engineering with SHED injected into human root canals and transplanted into immunodeficient mice. (**a**) Low-magnification and (**b**) high-magnification images of tissues formed when SHED mixed with scaffolds (Puramatrix™, rhCollagen type I groups) were injected into full-length root canals of human premolars. A vascularized connective tissue occupied the full extension of the root canal. Cell densification and many blood vessels were observed along dentine walls. Scaffolds (Puramatrix™) injected into the root canals without cells were used as controls for SHED. Freshly extracted human premolars were used as tissue controls. Black arrows point to blood vessels close to the odontoblastic layer. Modified from Rosa et al., J Dent Res, 92(11):970–975, 2013, with permission

entity is therefore named the "dentine–pulp complex," it has no direct analogue from a biological point of view, and this terminology may indeed be an oversimplification.

2.1 Tooth Development

Tooth development or odontogenesis occurs during embryonic, fetal, neonatal, and childhood stages of development. Teeth form from embryonic cells, then grow, and erupt into the mouth. Odontogenesis is a complex process. The human dentition begins to form between the 6th and 8th week of prenatal development with primary teeth, whereas the permanent teeth begin to form in the 20th week. Biomineralization starts during the 14th week of gestation [5].

The tooth germ is organized into the enamel organ, the dental papilla, and the dental follicle. The cells of the tooth germ are derived from the ectoderm of the first pharyngeal arch, and the ectomesenchyme of the neural crest. The process of odontogenesis is regulated by epithelial–mesenchymal interactions. Although tooth formation occurs as one continuous process, odontogenesis is classically described by the succession of several stages, i.e., initiation, tooth germ morphogenesis, terminal cytodifferentiation, and matrix apposition, resulting successively in the following anatomical stages: dental lamina, bud stage, cap stage, early bell stage and late bell stage (with terminal differentiation of odontoblasts and ameloblasts), root formation, functional differentiation of cementoblasts, and dental eruption.

At the initial stage of tooth development, a basement membrane already separates the epithelium from the underlying ectomesenchyme. The epithelium thickens, at the origin of the dental lamina that will later become the dental bud. A more substantial and localized epithelial thickening corresponding to the outlines of future teeth. The tooth bud increases in size and then turn into dental caps. These are characterized by a concave shape of the epithelial tissue which partially envelops the proliferating underlying mesenchyme, and the future dental pulp [6].

During the cap stage, a new specific structure arises due to the epithelial outgrowth: the enamel organ. This enamel organ is composed of the outer enamel epithelium, inner enamel epithelium, stellate reticulum, and stratum intermedium. The cells of the inner dental epithelium will gradually lengthen and form the future ameloblasts. Moreover, a particular and transitory structure appears at the center of the enamel organ: the enamel knot. The enamel knot is an organizing center of the tissue, controlling the shape of the crown, and expressing molecules belonging to the different families of growth factors, such as FGF (fibroblast growth factor), BMP (bone morphogenetic protein), Hg (hedgehog), and Wnt (wingless). These growth factors are known for their essential role in embryogenesis and in the formation of organs [7].

The ectomesenchyme condenses in the concavity of the enamel organ and forms the dental papilla. This dental papilla forms the odontoblasts and the dental pulp. The cap stage evolves into the bell stage, during which the dental crowns acquire its final shape (i.e., morphodifferentiation), and the formation of cusp patterns is observed.

The outer enamel epithelium and the inner enamel epithelium join at the cervical loop or "zone of reflection." The growth of cervical loop cells into the deeper tissues forms Hertwig's epithelial root sheath. This determines the root shape, including root odontoblast differentiation. The cervical loop progresses in apical direction due to an increase in cell divisions. The growth of the germ is consequently amplified, and the pulp is individualized in relation to the peripheral layer of the odontoblasts [8].

In the late bell stage, tooth morphogenesis is followed by a phase of cell differentiation (i.e., histodifferentiation). These cells will differentiate in pre-ameloblasts and pre-odontoblasts in order to become polarized and secreting cells, to ultimately form enamel and dentine respectively.

The condensed ectomesenchyme located at the periphery of the enamel organ and dental papilla is the dental follicle, and gives rise to cementoblasts, osteoblasts, and fibroblasts.

Thus, the dental follicle is involved and responsible for the formation of the root and tooth eruption [9, 10].

Determining the processes that initiate tooth development led to a significant amount of research which has provided the basis for the current understanding of the processes involved in dentine and dental pulp repair (Fig. 2a).

2.2 Dentinogenesis

Details of the basic principles of odontoblast differentiation is particularly relevant when considering the tissue engineering of dentine–pulp complex.

2.2.1 Odontoblast Life Cycle

Odontoblasts are postmitotic and highly differentiated cells that originate from the cranial neural crest-derived cells of the dental papilla. These cells produce and regulate an organic matrix that will be secondarily mineralized, namely the dentine.

From Pre-odontoblasts to Polarizing Pre-secretory Odontoblasts

The differentiation of odontoblasts from neural crest cells is a drawn-out process involving a series of changes in the morphology, transcriptional profile, and expression of proteins secreted by cells in the odontoblast lineage regulated by the epithelium–dental mesenchyme interactions

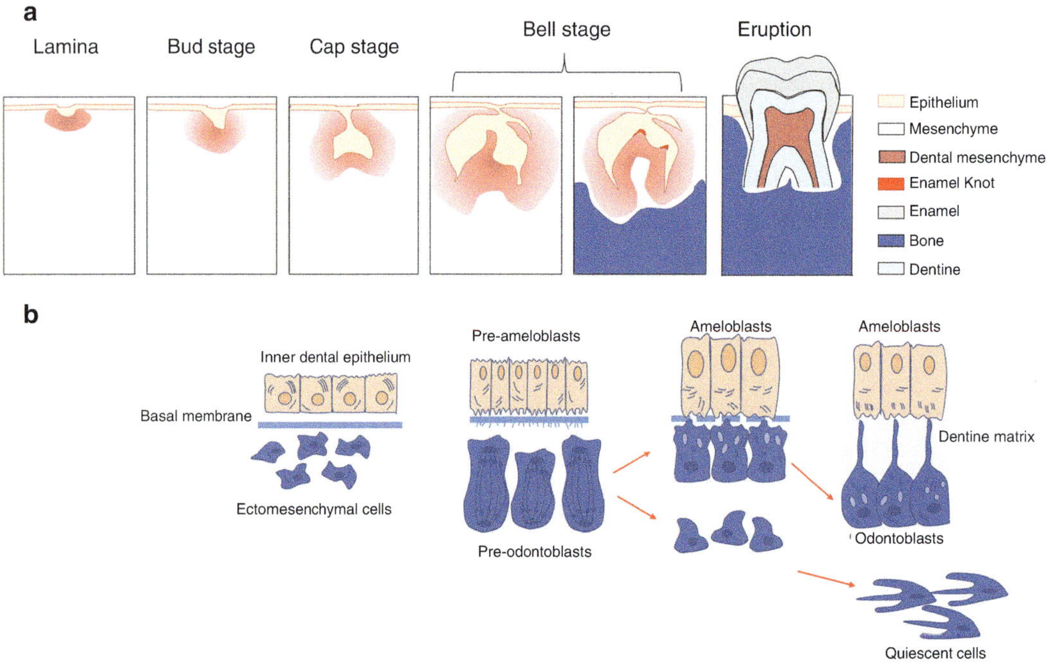

Fig. 2 Schematic drawings of the development of the dentine–pulp complex. (**a**) Stages of tooth development. The succession of the different anatomical stages: dental lamina, bud stage, cap stage, early bell, and late bell stages (terminal differentiation of odontoblasts and ameloblasts), root formation, functional differentiation of cementoblasts and dental eruption. The enamel knot appears before the terminal differentiation of cells and controls the shape of the crown due to FGF, BMPs, Hg, and Wnt. (**b**) Terminal events leading to odontoblast differentiation. Odontoblasts originate from the cranial neural crest-derived cells of the dental papilla. They differentiate from ectomesenchymal cells that are located near the basement membrane. Short, columnar-shaped pre-odontoblasts elongate and extend cellular processes toward the basement membrane where dental epithelium and ectomesenchyme interface. Secretory odontoblasts are fully differentiated polarized columnar cells containing numerous organelles in their supranuclear area. During the last mitosis, the daughter cells in contact with the basement membrane differentiate into odontoblasts while the other cells in the peripheral zone will join the cells of the Höhl layer

mediated by the basement membrane. Initiated at the tip of the cusp in the most peripheral layer of the cells of the dental papilla, differentiation continues according to a genetically predetermined temporo-spatial pattern. Inductive signals from internal epithelial cells involve members of the TGF-ß family (BMP-2, BMP-4, and TGF-ß1) [11], and other growth factors (IGF: insulin-like growth factor), which are partially sequestered in the basement membrane onto which the peripheral cells of the dental papilla align [12].

Functional competence of odontoblasts is achieved after a predetermined number of cell divisions and when cells express specific growth factor receptors. The fixed number of divisions allows these cells to reach the periphery of the dental pulp. During the last cell division cycle, only the most peripheral layer of cells underlying the basement membrane (pre-odontoblasts) responds to signals from the internal dental epithelium, to become completely differentiated into odontoblasts. The other cell resulting from the cell division cycle, "the daughter cell" that which is not in contact with the basement membrane, moves away from the previous one, to join the layer of Höhl [13]. This cell layer has been considered as a potential reservoir of cells, containing incompletely differentiated cells that could be involved in the healing process of reparative dentinogenesis [14].

Odontoblastic Differentiation and Terminal Polarization

Once the odontoblasts are differentiated, they undergo a terminal polarization. As the differentiation takes place in an apical direction, the cells change shape, going from round to cuboidal to an increasingly elongated appearance. At the subcellular level, the cells acquire a pronounced synthetic and secretory apparatus. The Golgi apparatus migrates from the basal part to a supranuclear region simultaneously with the development of cytoskeletal proteins, microtubules, odontoblastic cilium, actin microfilaments, and intermediate filaments containing vimentin and nestin. A distal junction complex appears with desmosome-like, communicating junctions (gap junctions). These junction complexes constitute a solid permeability membrane limited to molecules of low molecular weight. Finally, amino acids, fatty acids, sugars, and ions cross the space between endothelial cells and the basement membrane to be taken up into odontoblasts.

Although understanding of the molecular events preceding the terminal differentiation of odontoblasts has markedly improved, the final determinants of differentiation of odontoblasts remain to be characterized [12, 15–18]. The odontoblasts in their terminal cell division are initially positioned roughly parallel to the basal lamina, but after a short time their major axis becomes perpendicular, and they form a palisade-like structure. The terminal polarization leads to a partition into a cell body and a long process. The cell body houses all the organelles involved in the synthesis of the extracellular matrix: rough endoplasmic reticulum, Golgi apparatus, and immature and mature secretory vesicles, associated with equipment lysosomal (smooth endoplasmic reticulum, lysosomal vesicles, multivesicular structures).

The cell process, on the other hand, protruding a variable distance into the predentine, and adheres to the dentine walls of the tubules (odontoblast process). To accommodate these organelles and to prepare for the secretion of the components of the dentine matrix apically and unidirectionally, the nucleus moves to the opposite pole of the cell, in a position opposite to the internal dental epithelial cells. Nuclear repolarization is one of the important characteristics of the differentiation of terminal odontoblasts, and is a critical step both in the formation of primary tubular dentine and in the regeneration of dentine tissue (Fig. 2b) [18].

2.2.2 Role, Structure, and Composition of Dentine

Dentine is a calcified tissue that usually is covered on its coronal aspects by enamel, and on its radicular (root) aspect by cementum. It and houses the entire dental pulp. Dentine is less mineralized and less brittle than enamel. Dentine is also necessary for the support of enamel. Dentine rates at approximately "3" on the Mohs scale of mineral hardness.

As a mineralized connective tissue, dentine constitutes the major part of the tooth, and is composed by volume of 40–45% mineral (mainly hydroxyapatite and some noncrystalline amorphous calcium phosphate), and 30% organic material, of which 90% is collagen type I and the remaining 10% is ground substance. The latter includes dentine-specific proteins (SIBLINGs: small integrin-binding ligand N-linked glycoprotein). Dentine also contains 20–25% water [19].

Dentine is porous, and contains microscopic channels, called dentinal tubules, which radiate outward through the dentine from the pulp toward the exterior cementum or enamel border, with permanent connections to the odontoblastic layer located at the periphery of the pulp due to the penetration of cytoplasmic extensions of odontoblasts into the dentinal tubules.

Dentine is divided into two areas at the coronal level: the mantle dentine on the periphery, and the circumpulpal dentine near the dental pulp. The mantle dentine can be identified by the presence of various characteristics. The collagen fibers here are found perpendicular to the enamel–dentine junction.

At the root level, two layers are specifically observed: the mantle dentine is replaced by the hyaline layer of Hopewell-Smith on the periphery of dentine, with the granular layer of Tomes beneath this. These superficial layers are to be distinguished from circumpulpal dentine both by their composition and by their structure. The mantle dentine and the hyaline layer of Hopewell-Smith are atubular layers, unlike the granular layer of Tomes which contains fine canaliculi [20].

The innermost layer of dentine is laid down prior to mineralization, and is the predentine. This predentine is the initial dentine matrix. The presence of odontoblastic processes here allows the secretion of matrix components. Predentine can be 10–40 μm in width, depending on its rate of deposition [21, 22].

2.2.3 Types of Dentine

There are three types of dentine: primary, secondary, and tertiary. Primary dentine is the most prominent dentine in the tooth, and lies between the enamel and the pulp chamber. As soon as the odontoblasts are polarized and the junction complexes between the cell bodies delimit the basolateral and apical compartments, the transformation of predentine into dentine contributes to the formation of primary dentine. It is a tissue containing canaliculi with an S-curved path, and is produced by functional odontoblasts at a speed of 4 μm/day. This process ends with the function of the tooth in the arch.

Secondary dentine is formed when the tooth becomes functional. The S-shaped trajectory of the canaliculi becomes accentuated. In addition, as the number of odontoblasts increases compared to a smaller surface, the number of canaliculi gradually increases as one gains the innermost layers. In theory, this dentine is formed throughout life, without time limits, but its production in the elderly is, however, gradually reduced. Primary and secondary dentine are adjacent, and form in continuity. They are physiological types of dentine, and are made up of intercanalicular and pericanalicular dentine.

Tertiary dentine is formed as a reaction to external stimulation such as bacterial attack, trauma, and tooth wear. The pulp seeks to preserve its own vitality by synthesizing a scar tissue called tertiary dentine. This newly formed tissue will form a calcified barrier separated from physiological dentine by a more or less marked calciotraumatic line. Depending on the intensity of the stimulus and the nature of the lesions induced in the pulp, there are two types of tertiary dentine that may be formed: reactionary, where dentine is formed from a pre-existing odontoblast, or reparative dentine, wherein newly differentiated odontoblast-like cells are formed due to the death of the original odontoblasts.

The architecture and structure of tertiary dentine depend on the intensity and duration of the stimulus. Tertiary dentine is deposited rapidly, with a sparse and irregular tubular pattern and some cellular inclusions; and is referred to as "osteodentine." However, if the stimulus is less active, it is laid down less rapidly with a more regular tubular pattern with minimal cellular inclusions [23–25].

2.2.4 Dentinal Tubules

One of the characteristics of human dentine is the presence of dentinal tubules that occupy between 1% (in the superficial part of the dentine) and 30% (in the internal part of the dentine near the pulp) of the tissue volume. These tubules run through the dentine right from the amelodentinal junction (ADJ) or cementodentinal junction (CDJ), whether located at the level of the pulp chamber or the root canal. These tubules contain the dentinal fluid as well as the odontoblastic processes. The intra-pulpal pressure is approximately 10.3 mm Hg, so there is a pressure gradient between the pulp and the oral cavity [26].

The configuration of the tubules indicates the course of the odontoblasts during dentinogenesis. The tubules follow a sinusoidal S-shape from the outer surface of the dentine to the area near the pulp. Their number and diameter (1 μm in the dentine mass to 2.5 μm near the pulp) vary depending on their location. In a human premolar or molar tooth, the number of tubules is approximately 59,000 to 76,000 per mm^2 in the pulp area, and decreases by half toward the enamel area. A decrease in the density of the tubules is also found in root dentine compared to dentine in the cervical area [27]. These canaliculi represent a real channel of communication and diffusion between the "outside and inside" of the tooth. The tubular nature of dentine gives this hard tissue a degree of permeability, which can accelerate the carious process as well as accentuate the response of the pulp to dental restorative procedures.

To date, the composition of dentinal fluid is not known completely, but it is considered to be that of a tissue fluid containing serum proteins and immunoglobulins. The dentine fluid contained in the dentinal tubules of vital teeth is believed to be composed of a transudate from the pulp. Therefore, based on the analysis of the dentinal fluid, future research should focus on identifying biomarkers that would be relevant for pupal diagnosis [28–30].

2.3 Dental Pulp Histology

2.3.1 Structural Organization of the Dental Pulp

Dental pulp is a connective tissue that shares characteristics with other connective tissues in the body, but also has specific properties related to its anatomy and its environment. The dental pulp is located inside a rigid envelope of mineralized dentine to form the dentine–pulp complex, and this imposes constraints on its development, its physiology, and its response to different stimuli and injuries [25, 31]. The role of this specialized connective tissue is to enable the function of odontoblasts throughout the life of the tooth.

At the periphery of the pulp, odontoblasts have a characteristic palisading cellular arrangement, which circumscribes the outermost part of the pulp. The cell body of the odontoblast is found in the pulp, while long cytoplasmic processes (odontoblast processes), extend into the dentinal tubules. Odontoblasts are rather tall and columnar in the coronal pulp, shorten in the middle part of the tooth, and cuboidal and rather flat in the root portion of the tooth [32].

The axons of nerve fibers coming out of the Raschkow plexus pass between the odontoblasts, and constitute free nerve endings. Moreover, a network of terminal capillaries run through the odontoblastic layer, and many dendritic cells can be found close to the odontoblastic layer [33].

Below the odontoblastic layer is a relatively cell-free area. This area is known as the "Weil area." The main constituents of this area include the rich network of essentially nonmyelinated nerve fibers, blood capillaries, and fibroblast processes. This area is less visible histologically when odontoblasts effectively secrete dentine.

Beneath the cell-poor layer is a layer with relatively high cell density, located deeper in the pulp. This area is noticeable because it has a high fibroblast density. This area may represent a source of stem/progenitor cells that are able to differentiate into odontoblast-like upon damage to primary odontoblasts [31, 34].

The central portion of the connective tissue of the dental pulp contains fibroblasts, large blood vessels, and nerves. Undifferentiated mesenchymal cells and immune cells such as macrophages are frequently located in the perivascular region. The bundles of collagen fibers here are much more numerous in the root pulp than in the coronal pulp.

2.3.2 Pulp Cells

Odontoblasts
Structure

Odontoblasts are the most highly differentiated cells of the pulp. As secretory cells, odontoblasts have a rough endoplasmic reticulum, a Golgi complex, numerous mitochondria, and numerous vesicles. Quiescent odontoblasts are shorter and less polarized than active mineralizing cells. When cells are in the transition phase between active synthesis and quiescence, their organelles tend to show a perinuclear distribution [35]. Autophagic vacuoles can be observed in the cytoplasm inducing a reduction of the organelles. At the final stage of the cell's life cycle, these organelles are only located in the subnuclear region. The supranuclear region is devoid of organelles, except for large vacuoles filled with lipids [36].

Odontoblastic Process

The odontoblastic process is a direct extension of the cell body and occupies most of the space in the dentinal tubules. Its diameter is 3–4 μm at the pulpo–dentine junction, and narrows as it extends into the dentine tubules. The odontoblastic process has many lateral branches that can encounter the other branches of other odontoblasts. Unlike the main cell body, this process is practically devoid of major organelles required for synthetic activity. In contrast, the process shows a well-developed cytoskeleton, with many microfilaments with microtubules oriented parallel to its long axis. The extensions extend variably in the dentinal tubules, but do not seem to go beyond the internal third [37–39].

Synthesis and Secretion of Odontoblasts

The main function of odontoblasts is to synthesize and secrete the various components of the organic matrix of dentine, mainly collagen type I and proteoglycans. Odontoblasts are also capable of producing bioactive molecules (chemokines, growth factors such as TGF-ß, matrix metalloproteinases (MMP)), and they express molecules that are involved in defense and repair reactions [Toll like receptor (TLR)] [40].

Odontoblasts also synthesize various non-collagenous proteins associated with mineralization, including bone sialoprotein, dentine sialoprotein, phosphoryn (dentine phosphoprotein), dentine matrix protein 1 (DMP-1), osteocalcin, osteonectin, and osteopontin [41]. These molecules are secreted at the apical end of the odontoblast cell bodies as well as along the cytoplasmic extensions within the tubules in the area of predentine [42].

In addition to their synthetic activity, odontoblasts also participate in the degradation of organic components via the production of matrix metalloproteinases (MMP-8 (collagenase 2), MMP-2 and -9 (gelatinases), MMP-14 and MMP-20 (enamelysin), and stromelysin), which catalyze the degradation of matrix macromolecules such as collagens and proteoglycans [22, 30, 43, 44].

Fibroblasts

Although distributed almost everywhere in the dental pulp, fibroblasts are found in a high density in areas rich in cells of the coronal pulp. They come from multipotent mesenchymal stem cells (MSCs). Quiescent cells, which are commonly found in older pulp, are smaller than active cells, and tend to be spindle-shaped with fewer processes. The amount of rough endoplasmic reticulum in these cells is also lower. When quiescent cells are properly stimulated, their synthetic activity can be reactivated [45].

Fibroblasts play an important role in the production and remodeling of extracellular matrix (ECM). Fibroblasts are responsible for the synthesis and secretion of type I and type III collagens and non-collagenous extracellular matrix, such as proteoglycans and fibronectin. Pulp fibroblasts also produce certain non-collagenous proteins responsible for the mineralization of hard tissue, such as osteonectin, bone sialoprotein, and osteopontin [46]. In addition, pulp fibroblasts are involved in the breakdown of components of the extracellular matrix. These cells are also a source of MMP degrading matrix macromolecules such as collagens and proteoglycans [47, 48]. Fibroblasts are also able to secrete growth factors, such as members of the TGF-ß superfamily essential for remodeling pulp tissue [49].

Fibroblasts also have the capacity to produce pro-inflammatory cytokines, and to express adhesion molecules in response to molecular models associated with pathogens (PAMP) [50]. Fibroblasts express TLR2, TLR3, TLR4, and TLR5 [45, 51] and other cytosolic PRRs such as NOD-like receptors [52]. Finally, recent research has shown that fibroblasts are actively involved in the process of phagocytosis by producing complement C3b protein and opsonizing bacteria [53].

Immune Cells

Human dental pulp is a highly dynamic tissue equipped with a network of immunocompetent cells. These cells are believed to play a major role in the maintenance of tissue homeostasis. The average percentage of CD45 + immune cells (leukocytes) is approximately 1% of the total cell percentage. Among this population of CD45 + cells, granulocytes/neutrophils represent the largest cell population of CD45 + cells (50% ± 9%), followed by T lymphocytes (33% ± 11%), monocytes (9% ± 6%), and dendritic cells (5% ± 1%). Minor populations include natural killer cells (3% ± 1%) and B lymphocytes (2% ± 1%) [54].

Dental Pulp Stem Cells

Populations of mesenchymal stem cells with a phenotype close to bone marrow stromal cells have been described in dental pulp. These are known as DPSCs (Dental Pulp Stem Cells). These cells may be located in the perivascular zone and in the cell-rich zone adjacent to the odontoblastic layer. They have significant self-renewal and mineralization capabilities, like bone marrow cells. In addition, they behave like multipotent cells and are able, after adequate *in vitro* or *in vivo* induction, to engage in various differentiation programs, such as chrondrocytic, adipocytic, and neurogenic differentiation [55]. Given their potential, these cells have attracted considerable interest for therapy of pulpal involvement (see the following part).

2.3.3 Extracellular Matrix

The extracellular matrix (ECM) of the dental pulp consists of a loose and irregular network of collagen fibrils, glycosaminoglycans, and proteoglycans, acting either by stabilizing the network of fibrils or by participating as a stimulating or inhibiting mineralization agent.

Collagen

Collagen is the major organic component of dental pulp. The amount of collagen is greater at the root level than at the coronary level [56, 57]. Type I is predominant, and helps to establish the architecture of the pulp tissue. The relative proportion of type III collagen is also high (43% of the total collagen in the human dental pulp), which gives elastic properties to dental pulp. Type III collagen forms finer fibrils than those of type I collagen. Type III collagen is distributed in acellular areas, and in areas rich in cells [31, 57].

Glycosaminoglycans and Proteoglycans

Many proteoglycans are found in dental pulp, including chondroitin sulfate, dermatan sulfate, and hyaluronic acid [58].

Non-collagenous Proteins

Fibronectin is the main non-collagenous protein in dental pulp. It forms a cross-linked network of fibrils which increases near blood vessels. Fibronectin is also found in the odontoblastic layer, where it forms fibers in a "corkscrew" arrangement that passes from the pulp to the predentine parallel to the longitudinal axis of the odontoblasts. Fibronectin is believed to control the interactions between fully differentiated odontoblasts and the fibers of the extracellular matrix, and thereby to contribute to the maintenance of the specific morphology of these cells. Fibronectin is also involved in the terminal differentiation and polarization of odontoblasts [59].

2.3.4 Vascularization

As for other connective tissues, dental pulp vascularization provides nutrients, oxygen, and also the means of detection and elimination of toxic waste and materials. However, unlike other tissues, the dental pulp is highly vascularized (about 15% of its volume is occupied by the vessels) and it is located within and constrained by a rigid mineralized environment. This closed and inextensible environment should be very harmful to the pulp in the event of inflammation, however, there are different mechanisms at the vascular level which make the dental pulp unique and specific [60, 61].

In the dental pulp microcirculatory system, the widest vessels are the arterioles and venules. This microcirculation regulates the blood flow and also the lymphatic flow. Understanding pulp vascularization is one of the keys to unlocking pulp vitality and regenerative therapies [62].

The arterioles measure approximately 50 μm in diameter. They are composed of several layers of smooth muscle fibers that regulate vascular tone. The arterioles enter at the level of the apical foramen and go up towards the central region of the coronary pulp, ending in rich networks of capillaries at the periphery of the pulp. A parallel vascular network drains the blood flow through pulp venules, which exit the apical foramen. The blood flow is high, at around 40–50 mL/min/100 g of pulp tissue [63].

Pulpal blood flow is subject to complex control. It depends on local factors produced by cells, as well as molecules produced at a distance such as hormones or cytokines, and it is also under the influence of local nerve impulses. The sympathetic perivascular nerve fibers release noradrenaline and neuropeptide Y, which cause a reduction in pulpal blood flow. The perivascular nerve endings are either adrenergic (containing norepinephrine), or somatosensitive (containing substance P or peptides linked to the calcitonin gene). These nerve fibers participate in the regulation of pulp blood flow by controlling the tone of smooth muscles and the diameter of the vessels [64, 65].

2.3.5 Innervation

Dental pulp is one of the most richly innervated tissues in the body. There are approximately 2000 to 2500 axons at the apex of an adult premolar tooth. However, little information is received by the brain apart from that caused by inflammatory phenomena. The innervation comes from the trigeminal nerve. Nerve fibers from the teeth, the lower alveolar nerve, the posterior superior alveolar nerve, the medial superior anterior and anterior superior nerve, penetrate through the apical foramen, go up the root canal and divide when approaching the pulp periphery to the acellular Weil zone, forming the sub-odontoblastic nerve plexus or Raschkow plexus [66, 67].

In addition to nerve fibers, there are several receptors in the dental pulp, including nociceptors, thermoreceptors, mechanoreceptors, bacterial receptors, and receptors for cytokines and chemokines. The pulp contains transient potential receptors (TRPV1 and TRPV2) that capture noxious thermal and chemical information. Odontoblasts have certain TRPs. There also have receptors that are sensitive to chemical variations in the environment, such as ionic receptors of the acid-sensing ionic channel (ASIC) type. Mechanoreceptors of dental pulp (such as potassium channels TREK1, TREK2, TRAAK, ENaC, and ASIC 3) are present on pulp myelinated fibers, and have the capacity to detect tissue deformations via stretching of membranes [68, 69].

Among their many functions, odontoblasts are also involved in tooth sensitivity, such as seen in cervical dentinal hypersensitivity [70]. They do not have a gap junction-type synapse or connection with the nerve fibers, which suggests the absence of primary sensory function. However, studies have shown paracrine activity, which may explain a neuromodulatory role for odontoblasts. Ion channels, particularly mechanosensitive potassium channels, explain the excitability of odontoblasts. In addition, odontoblasts can attract nerve fibers and can express neurotrophic factors and receptors during development and after pulp inflammation develops [67].

2.4 Immune Responses in the Dentine–Pulp Complex

2.4.1 Odontoblasts as Immunocompetent Cells

Odontoblasts are the first cells of the dentine–pulp complex that are challenged with transdentinal bacterial attacks. In addition to their secretory activity, odontoblasts can also be considered as immunocompetent cells, for the following reasons: (1) Several pattern recognition receptors (PRR), which are cellular receptors capable of recognizing molecular patterns characteristic of pathogens, including TLR (Toll-like receptors) and NODs (nucleotide oligomerization domain receptors), are expressed by odontoblasts in humans and rodents [71–73]. (2) Odontoblasts are able to produce several chemokines or express chemokines receptors [40, 73–75]. Observations suggest that odontoblasts are more potent attractants than pulp fibroblasts [45]. (3) Odontoblasts can produce antimicrobial peptides, such as ß-defensins, that can directly kill bacteria [45, 76, 77]. As a result, the odontoblast is a multifunctional cell that is able to identify dangers to the pulp tissue, and can activate the immune response of the pulp [78]. A better understanding of the underlying mechanisms controlling odontoblasts could lead to the development of new therapeutic molecules or strategies in the treatment of pulpitis.

2.4.2 Response of Pulp Immune Cells to Tooth-Invading Pathogens

Bacteria are the major cause of inflammation and pulp infections. Carious lesions, trauma, wear or cracks cause bacteria and their degradation products to cross the enamel and make their way through the dentinal tubules in the direction of the dental pulp. The dental pulp is equipped to trigger an innate immune response, and then a specific adaptive antigen response. Innate immunity plays an important role, especially in superficial enamel [79, 80]. In this type of lesion, the pulp tissue is not in direct contact with the bacteria, and phagocytosis activity is therefore not possible. The transition from the innate to adaptive response probably appears in the case of irreversibly inflamed pulps, in the front of the carious lesion that is located less than 2 mm from the pulp [81].

As described above, the first line of defense is the palisade of odontoblasts. Changes in blood flow happen quickly during pulpal inflammation, with vasodilation and increased blood flow. These variations lead to an exudation of plasma proteins and the recruitment of leukocytes. Neutrophils are the first recruits, followed by monocytes, which differentiate into macrophages. Neutrophils and macrophages gradually infiltrate the pulp tissue as the carious lesion progresses. The complement system is activated by bacteria, and this causes the production of inflammatory mediators such as C3a and C5a, and the formation of a membrane attack complex (C5b-9) on the surface of the pathogens [82]. Complement also participates in the recruitment of leukocytes via the anaphylatoxins C3a and C5a [83].

Many inflammatory mediators are secreted (IL-1α, IL-1β, TNFα, IL-4, IL-6, IL-8, 1L-10) [84–86]. In addition to these secreted molecules, there is a release of molecules linked to the demineralization of dentine by bacterial acids, which may have a role in chemotaxis and the recruitment of cells from the systemic circulation [13, 87].

As the carious lesions progress, the number of immunocompetent cells in the dental pulp will increase. The deeper or greater the carious lesions extend, the more the infiltrate of immune cells increases, and the situation goes from a localized mode to an extension within the pulp tissue. Neutrophils and macrophages gradually infiltrate the pulp tissue as the carious lesion progresses. Macrophages are able to phagocytose bacteria and activate T cells, thereby triggering an adaptive response.

The dendritic cells (DC) initially present in the steady state are attracted by chemokines from odontoblasts at the site of infection, and these eventually leave the dental pulp and migrate to the lymph nodes, where they present the antigens to T lymphocytes. The T cells that have been activated by DC will differentiate into different effectors, and will secrete cytokines to recruit more immune cells along with B cells. Hence, there is a significant increase in the numbers of B cells in irreversible pulpitis in humans [88].

2.4.3 The Fine Balance and the Interplay between Inflammation and Regeneration in the Dentine–Pulp Complex

The pulp environment, the diversity of aggressive events, the diverse microflora at the level of the carious lesion, and the possibilities of supply of nutrients lead to a relative efficiency of the immune response within the pulp tissue. The inflammatory response initiated can progress to cell death and tissue destruction, or to scarring and/or tissue regeneration.

This balance, which can evolve in one direction or the other, is linked to the presence and concentration of inflammatory mediators. A carious lesion with slow shallow progression potentially leads to repair or regeneration phenomena which will no longer be possible if the aggression continues. Scarring and fibrosis in pulpitis prevent the return to homeostasis, since healing, regeneration, and reconstruction of diseased tissues are significantly hampered.

Extensive work over the past few decades has revealed that resolution of acute inflammation is crucial to avoid persistent chronic inflammation, and to support repair or regeneration. Macrophages are considered as the primary effector cells in regulating tissue repair. They participate in the evolution of the inflammatory response by shifting from an inflammatory phenotype (M1) to a regenerative phenotype (M2), as healing progresses. Recent work has revealed that macrophages in regenerating pulp tissue undergo a distinct transition from M1- to M2-dominant, suggesting that the M1-to-M2 transition of macrophages plays an important role in creating a microenvironment that is necessary for pulp tissue regeneration. In addition, conditioned medium from M2 macrophages enhances the odontogenic/osteogenic differentiation of human dental pulp stem cells [89]. These findings suggest strongly that deleterious inflammatory events could be reverted by creating a local environment that induces M2 phenotype polarization.

In order to avoid irreversible damage to the pulp tissue, the immune response must be controlled to eliminate pathogens without destroying the host tissue (the pulp). Although progress remains to be made on understanding the regulation of the immune response within the dental pulp, some molecules and therapeutic proposals have been identified in order to modulate the inflammatory response of the dental pulp.

Numerous studies have demonstrated the ability of mesenchymal stem cells, antioxidant molecules, low intensity lasers, and molecules of epigenetic regulation to modulate the inflammatory response [90]. Recently emphasis has been placed on immune modulatory roles of specialized pro-resolving mediators (SPM). These are secreted by neutrophils and macrophages, and regulate functions of the innate immune system (attenuate monocyte recruitment. They induce a pro-resolving M2 phenotype [91, 92], inhibit functions of dendritic cells [93, 94] and also modulate the adaptive immune system by decreasing memory B-cell responses (Fig. 3).

3 Stem Cell-Based Dentine–Pulp Complex Regeneration

3.1 Key Players for Dentine–Pulp Complex Regeneration

The regeneration of the dentine–pulp complex is a biological process whose goal is to replace *ad integrum* the damaged structures: the pulp and dentine. This procedure encompasses different clinical approaches. The first is to use the biological potential of the residual pulp tissue. In this situation, it is crucial to preserve the vitality of the pulp through so-called pulp vital pulp therapies (VPT). The direct application of a biomaterial to the vital pulp tissue aims to reduce the inflammation of the pulp, to stimulate the healing process, and to form a mineral tissue in order to create a physical barrier. However, current VPT only allows repair in contact with the biomaterial, and not true regeneration of the pulp tissue within the biomaterial. In this context, new proposals are emerging in order to regenerate the pulp volume destroyed by carious lesions or dental trauma.

The other approach aims to fully regenerate new vital tissue in an empty and disinfected canal space. This approach is complex due to the particular histological structure of the dental pulp and its functional relationship with the surrounding environment. Although earlier studies have demonstrated the role of the blood clot in the healing and repair of the pulp [95], the emergence of regenerative techniques in endodontics

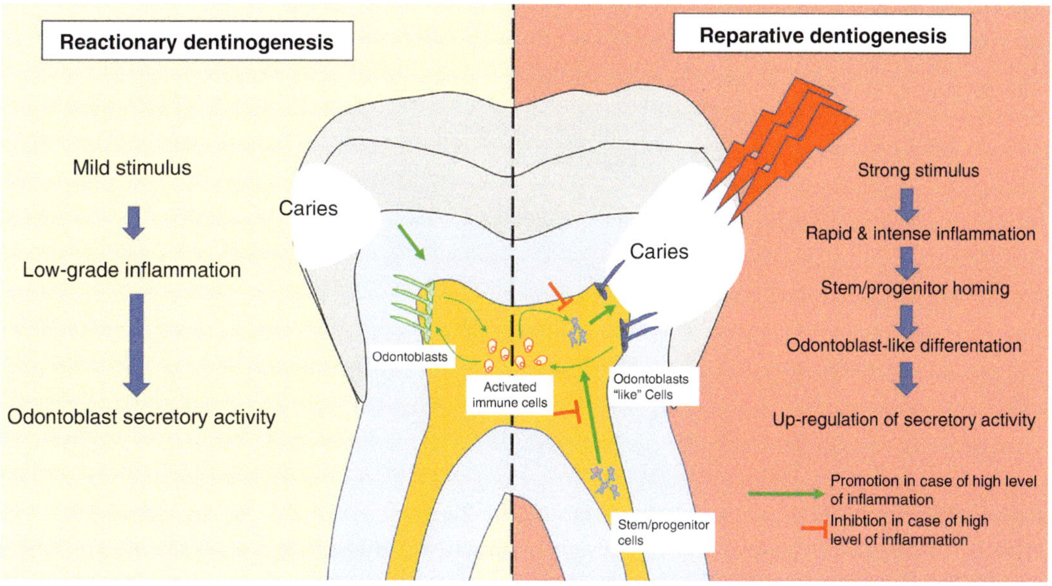

Fig. 3 The fine balance of the interplay between inflammation and regeneration interplay in the dentine–pulp complex. The complex is able to defend itself and to heal when challenged with bacterial aggression. According to the degree of stimulus, different types of tertiary dentine can be secreted. Reactionary dentine occurs following a mild stimulus and is secreted by odontoblasts originally present in the dental pulp. Reparative dentine is secreted by odontoblast-like cells following stem/progenitor cell differentiation, and occurs when the odontoblasts that were originally present have been destroyed

dates from the early 1960s [95] and then later on in the early 2000s with research from Banchs and Trope [96]. Clinical success is possible because the restoration of a blood supply in the canal space may be followed by reinnervation from the sensory axons from the apical region. The terms given to these phenomena are those of revascularization or revitalization. In order to mimic this process, different procedures rely on the creation of bleeding from the apical region into the canal space, filling the canal space and forming a blood clot. Numerous published case reports have shown resolution of apical periodontitis and radiographic signs of continuous root development and apical narrowing, and in some situations, positive responses to vitality tests [97, 98].

The explanation for these phenomena was advanced in 2011 when studies demonstrated that the influx of apical blood into the disinfected root canals was followed by a clinically significant transfer of mesenchymal stem cells into the canal system [99]. This was an important step in the field of regenerative endodontics, since it establishes that these procedures were based on the principles of tissue engineering, and the triad which includes the stem cells, scaffolds, and growth factors [100]. However, the newly formed tissues do not correspond to a functional dental pulp, but rather to a connective tissue associating the periodontal, bone, and cementum elements. Greater awareness of the role of stem cells, scaffolds, growth factors, and their properties has led researchers and clinicians to reanalyze the principles of tissue engineering, in order to develop more predictable and successful outcomes of regenerative procedures. This lead to the addition of another important factor to the existing triad, which includes adequate disinfection of the canal space (Fig. 4).

3.1.1 Pulp Stem/Progenitor Cells

Stem cells are found in all multicellular organisms. They are commonly defined as clonogenic cells with a capacity for multi-lineage differentiation and self-renewal. In accordance with the minimum criteria defined by the International Society for

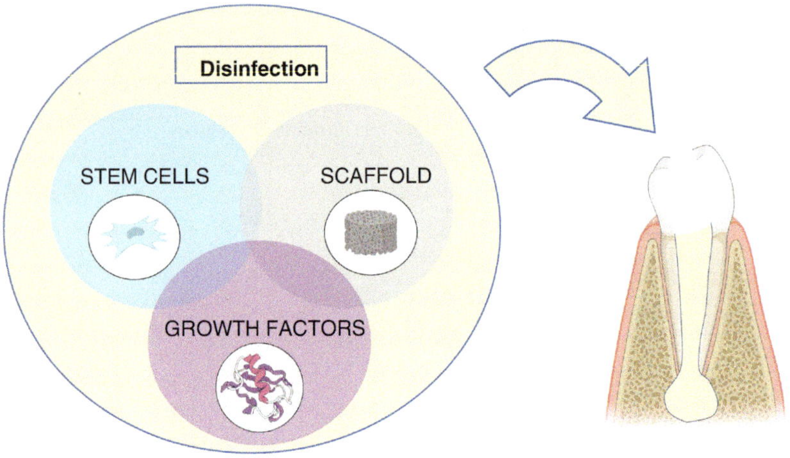

Fig. 4 A schematic showing that disinfection interacts with the interplay between stem cells, scaffolds, and growth factors (the classic triad of tissue engineering) for regenerative endodontics

Cellular Therapy (ISCR), stem cells have the properties of adhesion to plastic, the positive or negative expression of specific markers of surface antigen, and the formation of colonies *in vitro*.

Stem cells can come from the recipient patient themselves (autologous) or from a donor (allogenic origin). Stem cells can be divided into four main sources: adult mesenchymal stem cells, embryonic stem cells (ESC), neonatal stem cells from the umbilical cord, and induced pluripotent stem cells (iPSC) obtained from already existing cells. Although ESCs are of great interest, their practical application in tissue engineering and cell therapy is limited due to the high risk of tumorigenicity, as well as the ethical and legal issues associated with their embryonic origin. However, there is an increasing interest in research into the development of tissue regeneration therapies using identified postnatal stem cells from various tissues and organs of the human body.

Like any other organs or tissues in the human body, dental tissue is considered a potential niche for MSC. Numerous studies have shown that dental pulp constitutes an appreciable and easily accessible source of postnatal stem cells that can be harvested from extracted permanent teeth or deciduous teeth. There are different populations of pulp stem cells: dental pulp stem cells (DPSC), exfoliated decidual teeth stem cells (SHED), stem cells from the dental follicle (DFSC), and stem cells of the apical papilla (SCAP). They have all shown great potential for regeneration of dentine–pulp complex (Fig. 5).

Recently, techniques for sorting cell subpopulations have been discussed in order to obtain optimal phenotypes. Cell sorting does not seem to be a profitable option since there is no convincing evidence suggesting a considerable improvement in the regeneration of tissues resembling dentine or dental pulp [101].

Since the availability and quality of dental pulp tissue decline sharply with age, non-odontogenic stem cells have been studied as alternative sources. Stem cells from bone marrow and adipose tissue have shown particular promise because of their beneficial biological properties and their gene expression profile [102].

3.1.2 Scaffolds

Along with stem cells and bioactive molecules, scaffolds are one of the components of tissue engineering strategies. Functionally, the scaffolds ensure the seeding, adhesion, proliferation, and spatial distribution of cells. The aim is to promote the complete replacement of the extracellular scaffold and serve as a bioactive platform regulating cellular activities, intra-, and intercellular communication, so as to provide a physiological microenvironment.

Many materials have been proposed and assessed in scientific literature. A blood clot has been used as a physical scaffold in many cases [103]. However, many other scaffolds have been introduced into endodontics, including natural polymers (fibrin, collagen, chitosan, glycosaminoglycan, and dentine scaffold) or synthetic

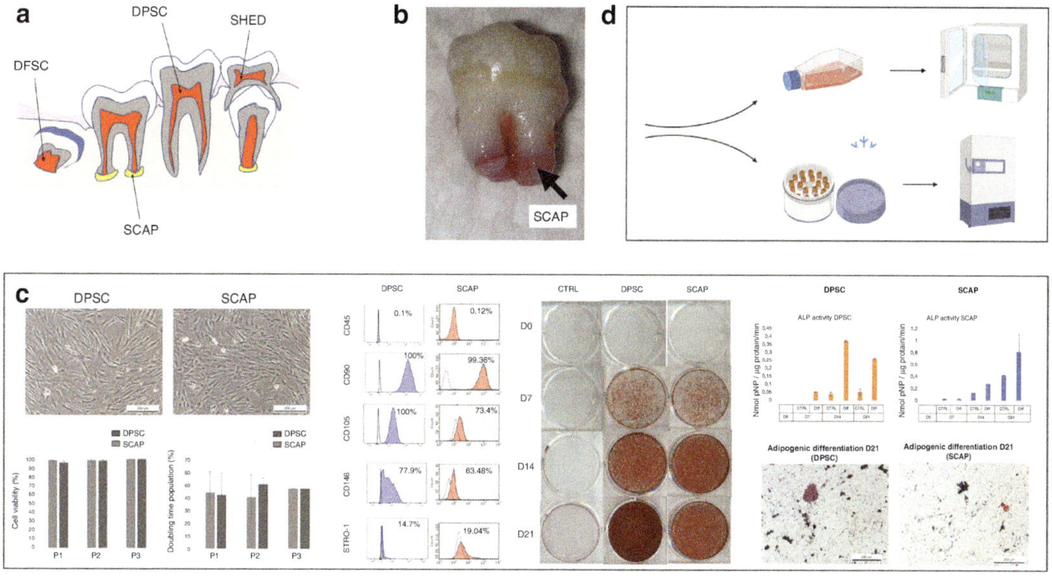

Fig. 5 Stem cells from dental origin for dentine–pulp complex regeneration. (**a**) Different mesenchymal stem cells from dental origin. Human dental mesenchymal stem cells can be harvested from healthy tooth-related pulp tissue (DPSC and SHED), dental follicle (DFPC), and apical papilla (SCAP). (**b**) DPSC and SCAP can be easily isolated from an extracted tooth. The apical papilla is a soft tissue loosely attached to the apices of immature permanent teeth and can be easily detached with a pair of tweezers. DPSC are stem cells present in the dental pulp. (**c**) A comparison between DPSC and SCAP. The *in vitro* aspects are seen with phase-contrast microscopy, cell surface-specific markers as analyzed by flow cytometry, alizarin red staining, alkaline phosphatase expression, and adipogenic differentiation. (**d**) Primary cultures of stem cells can be further cultured or cryopreserved

polymers (polyglycolic acid [PGA], poly-l-lactic acid [PLLA], polylactide-co-glycolic acid [PLGA]). There are also inorganic and composite materials that are synthesized in the form of porous scaffolds, nanofibrous materials, microparticles, or hydrogels.

The specifications for the ideal scaffold for tissue engineering in endodontics include several elements. The ideal scaffold must be biodegradable, and have similar viscoelastic properties and a similar rate of synthesis as the extracellular matrix. If the rate of degradation is too fast, dissolution of the material will occur before the formation of new tissue. Conversely, if the rate of degradation is too slow, the synthesis of extracellular matrix will be disturbed. The proper interactions between the cells and the scaffold are essential. The design of the scaffold (porosity, surface topography, charge, stiffness, viscosity) may modify cellular behavior and the expression profile of the cells (adhesion, migration, proliferation, synthesis of extracellular scaffold) [104]. The peculiarity of endodontic tissue engineering is that this scaffold must be injectable into a narrow space: the root canal and/or the pulp chamber. After injection, the scaffold must be able to quickly reach its initial stiffness values. Finally, the ideal scaffold must be sterilizable and free from the risk of contamination, and must allow the biomineralization that occurs in the natural state in the pulp.

Although natural polymers and synthetic materials have shown promising therapeutic effects when used for the regeneration of dentine and pulp, none of them encompass all the properties necessary for the regeneration of the cellular physiological microenvironment. Natural polymers are often affected by problems of purity and antigenicity. Synthetic polymers bypass the potential risks of immunoreactivity, but lack bioactivity and have poor degradability. The recent appearance of nanofibrous polymers mimicking the ECM has shown interesting results in terms of cell–ECM interactions [105].

Table 1 Classification, advantages and drawbacks of scaffolds used for tissue engineering in endodontics

Classification	Natural polymers, polysaccharides	Components of the extracellular matrix	Synthetic polymers	Bioceramics and derived substances			
Biomaterials	Alginates Chitosan Cellulose	Hyaluronic acid Collagen Gelatin Proteins/peptides Fibrin Self-assembling peptides	Polylactic acid (PLA) Polyglycolic acid (PGA)	Calcium phosphate			Bio-glass
				Hydroxyapatite	Biphasic calcium phosphate	Tricalcium phosphates	
Advantages	• Low cost • Ease to handle	• Dynamic environment similar to physiological extracellular matrix • Guide cellular behavior thanks to the signal molecules present. • Possibility of degradation and selective remodeling • Biological materials on a nano-scale • Allow controlled release of growth factors	• Low production cost • Low antigenicity • Reproducible • Possibility of adjusting the mechanical properties, the degradation rate • Biodegrad-able • Easy seeding of stem cells	• Low immunogenicity • Good resorption rate • Similar to mineralized tissues • Osteoconductive			
Drawbacks	• Inter batch variability • Poor mechanical properties	• Inter batch variability • Difficulty in sterilizing • Significant degradability • Volumetric retractions	• Do not have the biomolecules of the physiological extracellular matrix • Potential accumulation of acid degradation products	• Limited applications for tissue engineering in endodontics • Requires a viscous vector for injection • Tendency to cause mineralization of the pulp chamber			

Currently used biomaterials for dentine–pulp complex regeneration are summarized in Table 1.

3.1.3 Growth Factors/Bioactive Molecules

Growth factors/bioactive molecules are chemical signals or signaling molecules that control various cellular responses. Bioactive molecules are essential in order to obtain functional regeneration of the pulp. By inducing specific signaling pathways, growth factors modulate each key cellular event of pulpal homeostasis: cell proliferation, differentiation, and mineral synthesis.

Several growth factors have proven their importance in dental tissue engineering, such as the BMP superfamily (BMP-2, BMP-4, BMP-7)

and TGF-β1. These molecules appear promising for modulating cellular functions involved in the differentiation and mineralization of odontoblasts. Basic fibroblast growth factor (bFGF) and granulocyte colony-stimulating factor (G-CSF) are involved in the migration of cells to their original tissue (i.e., cell-homing), as well as in angiogenesis, and during certain neuronal and anti-apoptotic events. Vascular endothelial growth factor (VEGF) and platelet-derived growth factor (PDGF) are involved in the induction of blood vessel formation [106].

The effects of biomolecules depend on their in situ concentration and on cellular interactions. Interestingly, several biomolecules are sequestered in the dentine matrix, which, in the appropriate conditions, is able to modulate tissue responses. Microbial assaults (dental caries) or iatrogenic factors (chemical attacks and restoration procedures/materials) can lead to demineralization of dentine, which then releases these sequestered biomolecules within the dentine matrix [107, 108]. The released biomolecules may play a key role in several events, and in particular, they may influence the healing functions of the dentine–pulp complex [109]. However, the direct clinical application of these biomolecules is not feasible for a number of reasons. A bolus dose can cause toxic effects. Free biomolecules can have a limited half-life. Finally, it is difficult to maintain an appropriate concentration of biomolecules for the entire therapeutic period. To overcome these drawbacks, current research is focused on controlled biomolecule delivery systems, thus mimicking a biological response for engineering functional tissues.

3.1.4 Disinfection

The etiologies of pulp necrosis and preoperative infection are important in cases when regenerative endodontics is considered as a treatment option [110]. One of the main concerns during this procedure is an effective and efficient disinfection protocol that removes the bacteria completely and renders the disinfected root canal space conducive to repopulation by stem cells. This is especially important since recent research studies have demonstrated that the presence of remnant bacteria within the canal space during these procedures can have a negative effect on the outcomes of regenerative procedures [111]. One of the mechanisms suggested for this, is the activation of the immune system by the residual bacteria in the canal, which in turn, leads to the death of the SCAP [112].

Various intracanal medicaments, disinfectants, and disinfectant techniques have been used for this purpose. A triple antibiotic paste (consisting of metronidazole, minocycline, and ciprofloxacin) has been shown to be effective in disinfecting the root canal space in teeth with immature apices, without affecting the SCAP [113]. Other studies have advocated the use of calcium hydroxide instead of the triple antibiotic paste, due to their effects on the SCAP [114]. Irrespective of the type of medicament or irrigating solution used, it is imperative that adequate disinfection of the canal space be achieved, for a more predictable outcome.

3.2 Complete Dentine–Pulp Complex Regeneration

Transplanting human pulp stem cells into root canals has been shown to potentially regenerate pulp-like tissue, with odontoblasts, blood vessels, and innervation similar to a healthy human dental pulp [2, 115, 116]. Most protocols using stem cells also employ a scaffold in order to provide an adequate environment for the cells, allowing their proliferation, migration, and differentiation into a functional pulp. There are also alternative approaches using a self-assembled three-dimensional pulp tissue made directly from the pulp cells without the addition of an exogeneous scaffold [117] (Fig. 6).

A recent clinical trial in humans carried out in Japan by Nakashima and collaborators showed a complete regeneration of the pulp after a pulp stem cell transplant (DPSC) [118]. In this study, five patients with irreversible pulpitis were treated, and were followed up to 24 weeks after the DPSC transplant. DPSCs were isolated from another healthy tooth extracted with an original protocol based on the mobilization of DPSC

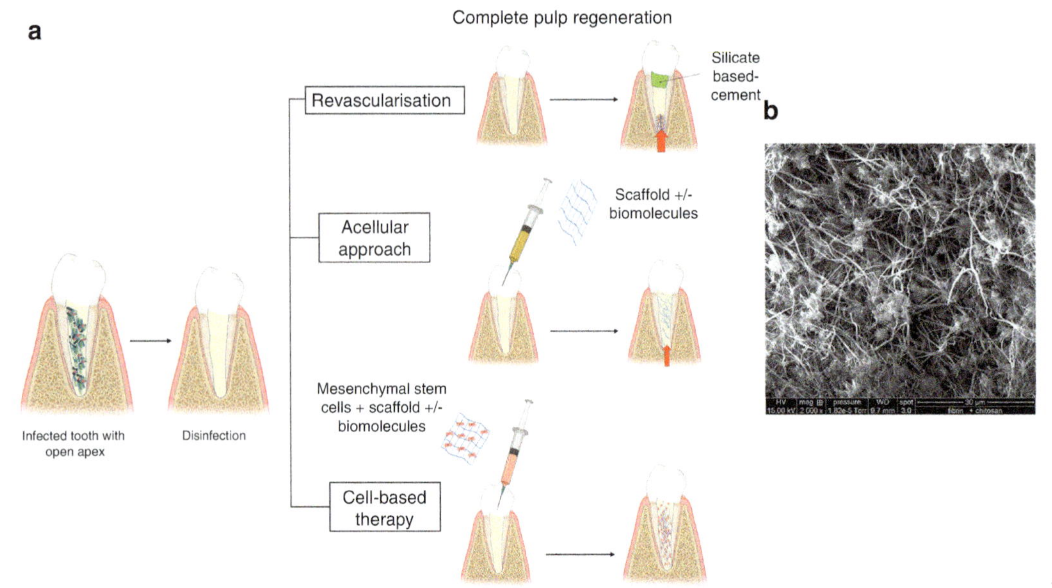

Fig. 6 Regenerative approaches in endodontics. (**a**) Different strategies may be considered for regenerative approach in endodontics. Revascularization consists of evoking apical bleeding through the apical foramina. The following blood clot will serve as a scaffold and a reservoir for growth factors for the stem/progenitor cells. The acellular approach consists of injecting scaffolds with or without biomolecules in order to recruit and to help differentiation of stem/progenitor cells. The cell-based therapy consists of injecting a combination of a scaffold and stem/progenitor cells able to differentiate into dentine-forming odontoblasts. (**b**) An example of a fibrin–chitosan hydrogel scaffold for dental pulp tissue engineering. Scanning electron microscopy shows rounded aggregates of chitosan trapped between and around the fibrin fibrils of the hydrogel (Courtesy of Prof. J.C. Farges)

using G-CSF. These cells were called MDPSC, that is, "mobilized" DPSC. The cells were grown in accordance with Good Manufacturing Practice (GMP). The quality of the stem cells at passages 9 or 10 was verified by karyotype analyzes. The MDPSCs were transplanted, with G-CSF, into collagen and placed into teeth that had undergone a pulpectomy. Clinical and laboratory evaluations did not show any toxicity or adverse effects. The electrical pulp sensitivity test at 4 weeks demonstrated a robust positive response. After 24 weeks, the intensity of the magnetic resonance imaging (MRI) signal of the regenerated tissue in the root canal was similar to that of the normal dental pulp of the untreated control teeth. Finally, CBCT scans revealed the formation of functional dentine in three of the five patients.

These promising results are first steps in the use of human tissue engineering for the regeneration of a complete pulp entity. However, the accessibility, the clinical reality, and the generalizability of this type of treatment are low. Many questions remain, and the protocols remain experimental. We do not know what the optimal disinfection protocol would be to guarantee the result without compromising the stem cell transplant. The long-term fate and long-term functionality of the pulp have not yet been explored. Technical feasibility, ensuring that the transplanted cells are free of infection risks, and the high cost of such procedures all constitute major obstacles.

3.3 Partial Dentine–Pulp Complex Regeneration

An *in vivo* model of pulpotomy in dogs involves the transplantation of subpopulations of sorted (CD146- and CD31-) autologous stem cells that have been cultured three-dimensionally and are then implanted in a collagen scaffold at a pulpotomy site. This approach results in complete regeneration of the pulp tissue, with vascular and neural processes, within 14 days. The transplanted cells express pro-angiogenic factors that cause a trophic action on endothelial cells. The regenerated pulp contains well-developed vascularization and innervation, as well as a tubular mineral structure along the dentine walls [119]. Similarly, partial pulp regeneration has been demonstrated in rat molars [120].

Conversely, another study carried out using a larger animal model (minipig) showed the absence of pulp regeneration after partial pulpotomy and implantation of stem cells in a scaffold of a self-assembling peptide (PuraMatrix™), with the appearance of mineralization at the canal openings [121]. A recent study using bone marrow-derived stem cells showed hard tissue bridge formation after pulpotomy in dogs [122].

Many factors can explain these differences in results (the animal model, the methods of cell isolation, the capacities of the stem cells themselves, the local environment, the maturity of the treated teeth, the extent of vascularization in situ, the operating protocol, and the scaffolds and biomaterials used). In addition, a failure of regeneration in favor of pulp mineralization at the level of the canal opening may also be linked to the inflammation generated in these models which could negatively influence neo-angiogenesis and promote the formation of mineralized tissue.

4 Acellular Material-Based Dentine–Pulp Complex Formation

4.1 Complete Dentine–Pulp Complex Regeneration

In order to overcome the drawbacks and limitations of approaches using cell transplantation, several so-called acellular approaches have been proposed. These acellular approaches are based on cell recruitment. In the case of total pulp regeneration, the target cells to be recruited are located in the periapical region. This is the most clinically developed approach, pioneered by Trope et al. [123]. This procedure entails disinfection of the canal space, followed by the generation of a blood clot in the root canal by intentionally instrumenting beyond the apical foramen into the peri-apex. The blood clot acts as a scaffold for the newly regenerated tissue, and also serves as a reservoir of natural biomolecules, allowing the recruitment of cells. A calcium silicate cement (MTA®, Biodentine®, Bioceramic cements) is placed to seal the coronal portion of the canal space (Fig. 6).

This approach does not allow the regeneration of structural and functional pulp tissue from a histological point of view. Applying the classic principles of tissue engineering would allow a better control of the cells and tissues involved, to achieve more predictable results. Studies on the use of biomolecules or even cocktails of molecules (bFGF, VEGF, PDGF and NGF, and BMP-7) promoting chemotaxis constitute an alternative strategy.

The use of optimized scaffolds for this type of application is crucial, and has been the subject of various studies. Proteins derived from dentine may induce chemotaxis and the formation of pulp-like tissue [124, 125]. The processes to promote their release are still under investigation. Recently, it has been demonstrated that the use of citric acid appears to be more advantageous than EDTA, which has traditionally been proposed clinically to promote cell homing in regenerative procedures [126].

The use of optimized scaffolds for pulp regeneration is under active investigation. The pulp extracellular matrix (ECM) is a natural source for producing scaffolds that mimic the chemical and mechanical properties of native tissue. The introduction of a predefined decellularized ECM with its spatial distribution of trophic factors seems to be a promising approach. Recently, scaffolds based on extracellular matrix have shown promising results in terms of recruitment of progenitor cells, promotion of constructive remodeling, and modulation of the host response. These matrices are expected to trigger a migration of progenitors and stem cells, in order to recolonize the decellularized matrix. Cells recruited from surrounding tissues migrate to the new network of extracellular matrix in a similar process to the natural pulp tissue [127, 128] (Fig. 6).

4.2 Partial Dentine–Pulp Complex Regeneration

The same principles may be used to recruit progenitor cells from the remaining dental pulp using homing or migration factors. This method is attractive compared to the stem cell approach, since the costs involved are less (no need for isolation, manipulation, and storage of stem cells). The risks of immune reactions, contamination, or transmission of pathogens or tumorigenicity associated with cell transplants are absent.

The most studied migration factors are fibroblast growth factor 2 (FGF-2) and stromal cell-derived factor-1α (SDF-1). The uncontrolled release of FGF-2 accelerates the formation of tertiary repair dentine in the residual pulp, while the controlled release of FGF-2 from a gelatin hydrogel induces the formation of osteodentine at the level of dentinal defects [129].

MMP-3 has also been used as a migration/homing factor. MMP-3 drives cell proliferation and migration, and it has an anti-apoptotic effect on endothelial cells *in vitro*. In addition, MMP-3 promotes angiogenesis and pulp healing in models of pulpotomy in rats as well as in models of reversible pulpitis in dogs [130, 131]. However, this attractive approach has a number of limitations. The results to date encompass only animal models without pre-existing pulpal inflammation, and therefore more long-term clinical studies with follow-up are needed before a human translation.

5 3D Printing for Endodontic Regeneration

5.1 Concept 3D Bioprinting

The use of 3D scaffolds has many challenges when considering tissue engineering of the dentine–pulp complex. In addition to providing mechanical features for cell support, an ideal scaffold also must create a bioactive and dynamic environment that regulates cellular functions and intracellular communication to replicate the native extracellular matrix (ECM) [101]. The use of traditional techniques for scaffolds (electrospinning, freeze-drying, gas foaming, melt molding, solvent casting and particulate leaching, and phase separation) do not allow exhaustive control of every parameter.

Recently, additive manufacturing techniques such 3D printing have been adopted for tissue engineering. The 3D bioprinting concept is based on the invention of stereolithography by Charles Hull in 1986 [132]. The goal of 3D bioprinting is to develop highly customized cell-laden scaffolds. 3D bioprinting enables the precise positioning of cells. It can precisely control the external and internal morphology of scaffolds, and can be used for high throughput production. The porous structure inside the 3D-printed scaffold facilitates the infiltration of nutrients and oxygen, thus providing the necessary conditions for normal metabolic activity of cells.

The three main bioprinting techniques that may be used for endodontic regeneration are inkjet bioprinting, laser-assisted bioprinting, and extrusion bioprinting. Inkjet printing is a widely understood printing technology derived from the conventional 2D desktop inkjet printers. Inkjet printers eject droplets of biomaterials through a nozzle by thermal energy or by a piezoelectric actuator. Thermal inkjet technology is simple, efficient, and economical. However, one of the disadvantages of this technology is the frequent clogging of nozzles by bio-ink. Gelation causes unequal sized drops and this disturbs the printing process. Another challenge involves the thermal and shear stresses involved in creating bio-ink drops that may affect cell viability [133].

Laser-assisted bioprinting (LAB) focuses laser pulses onto the donor slide, thus creating high pressure, to propel droplets of cell-laden hydrogel onto the collector slide [134]. LAB is a highly versatile method for fabricating heterogeneous tissue constructs with high cell densities, at high resolution (10–100 μm) and various sizes. LAB is an attractive approach for 3D tissue fabrication due to its automation, reproducibility, and high throughput. Attention must be paid to the selection of the biomaterial, as this should exhibit fast gelation kinetics (i.e., rapid cross-linking). The laser wavelength used must preserve the resolution and arrangement of cells and biomaterials in the 3D printed constructs [133]. However, major concerns are the long time that is required for fabrication and the problem of gravitational settling of cells in solution.

Extrusion or robotic dispensing bioprinters extrude biopolymers or cell-laden hydrogels through the nozzle by applying air pressure (pneumatic force) or using mechanical systems (such as pistons or screws) (Fig. 7a). More recent extrusion bioprinters are equipped with multiple printer heads, to allow simultaneous deposition of different bio-inks with minimal cross-contamination [135]. Furthermore, they offer advantages of better control over porosity, shape, and cell distribution in the printed construct [133]. Extrusion bioprinting is gaining popularity due to its high versatility and therefore it is being considered to fabricate scaffolds for tissue engineering (Fig. 7b).

5.2 Bio-Inks for 3D Bioprinting Applied for Endodontic Regeneration

Although advantages and applications depicted in other medical fields are becoming very popular, 3D bioprinting for endodontic regeneration is only just beginning, and few research articles have been published on this topic. The precise strategy needed for 3D bioprinting for endodontic applications remains elusive. Which technology of bioprinters to use? Which bio-ink would be the most suitable for endodontic applications? What kind of approach would be the most successful: to fabricate dental pulp, dental tissue constructs, the whole tooth with enamel, or the whole tooth with cement and ligament?

Cell-laden hydrogels, decellularized extracellular matrix, and cell aggregates are the main types of bio-inks used in bioprinting devices. However, these materials cannot represent fully the complex structure of the ECM of the dentine–pulp complex. There currently is a lack of bio-inks devoted to regeneration of dental tissues,

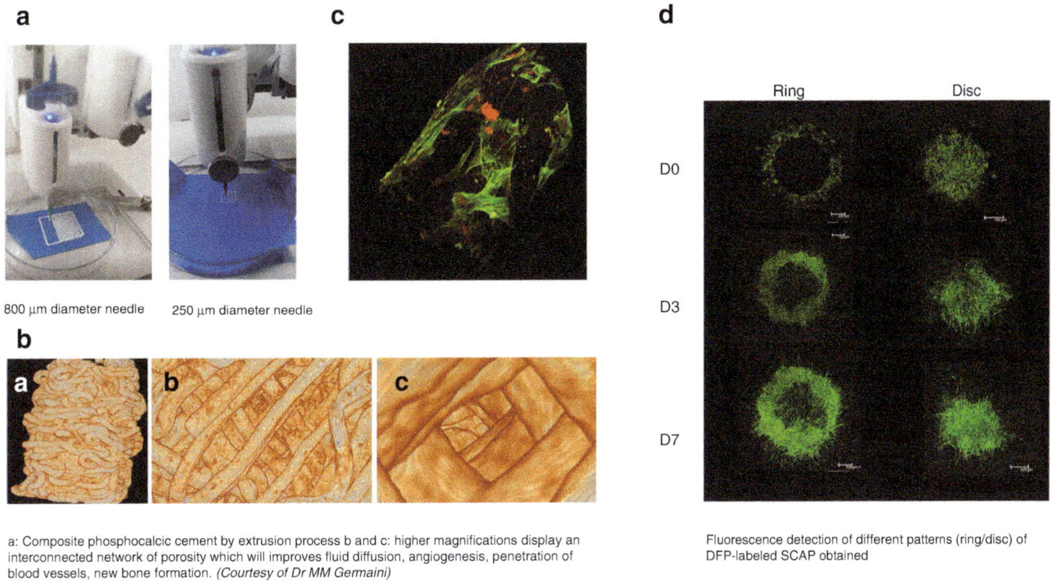

Fig. 7 3D Bioprinting in endodontics. (**a**) A pneumatic printing head (extrusion 3D bioprinter) printing a phosphocalcific paste. (**b**) 3D reconstruction as shown by microtomography of a cylindrical 3D printed implant. (**c**) Mesenchymal stem cells seeded onto a 3D implant. (Courtesy of Dr. M.M. Germaini). (**d**) Viability of SCAP after laser-assisted bioprinting *in vitro*, comparing baseline (Day 0) and days 3 and 7 after printing (Courtesy of Dr. O. Kérourédan)

and specifically those with odontogenic potential, which may be used in the rapidly evolving field of regenerative endodontics.

Recent work has explored the cytocompatibility and the odontogenic potential of different bio-inks (Fig. 7c, d). Yu et al. assessed the effects of 3D bioprinting an alginate/gelatin hydrogel (Alg-Gel) scaffold extract on proliferation and differentiation of DPSCs in comparison with a traditional Alg-Gel hydrogel scaffold. More cells were grown on and adhered to the 3D-printed Alg-Gel scaffolds than the Alg-Gel scaffolds. This could be because the extract of 3D-printed Alg-Gel scaffolds contained more calcium and phosphorus ions, indicating that hDPSCs have stronger proliferation and osteogenic/odontoblastic differentiation capacity when seeded onto the 3D printed scaffold [136]. In another recent study, Athirasala et al. engineered and characterized a novel dentine-derived ECM hybrid cell-laden hydrogel bio-ink, comprised of alginate and dentine matrix proteins. The viscosity, the printability, the survival rate, and the odontogenic potential of SCAPs may be modulated by fine-tuning the composition of the bio-ink. A concentration of 100 μg/ml of soluble dentine molecules significantly enhances odontogenic differentiation of encapsulated SCAPs [137].

Besides the assessment of suitable bio-inks and bioprinters, strategies for human therapeutics are also under consideration. One approach is to enable bioprinting of cell-loaded collagen-based bio-inks with suitable rheological, structural, and biological properties directly in situ into the root canal. This process has been assessed using human teeth *ex vivo*, with a hand-held drop-on-demand bioprinter. New blood vessel formation (vasculogenesis) was successfully demonstrated, and was comparable in quality and quantity to that seen with fibrin and collagen non-bioprintable hydrogel controls [138].

Another approach is to consider whole tooth regeneration. Different studies have successfully shown the possibility of producing tooth-shaped tissues and structures. However, there are limitations in controlling the size and shape of these teeth. It is difficult to culture tooth germs [139, 140]. In addition, scaffold-based approaches do not seem suitable for the regeneration of tooth-like composite tissues because they cannot place multiple types of cells in a pre-defined manner.

Han et al. [141] proposed a fibrin-based bio-ink to regenerate patient-specific shaped and tooth-like composite tissue utilizing 3D bioprinting technology. A dentine–pulp complex having patient-specific shape was produced by co-printing human DPSC-laden bio-inks with polycaprolactone, which is a thermoplastic material. Localized differentiation of DPSCs in the outer region of the three-dimensional cellular construct was successfully achieved, with localized mineralization [141]. This result demonstrates the possibility of producing patient-specific composite tissues for specific tooth engineering.

The use of 3D bioprinting to regenerate the dentine–pulp complex is very recent. The methods being used need much more study and refinement so that they become reproducible, feasible, safe, and cost-effective. Most of the data that have been generated have come from *in vivo* or *ex vivo* models, and these do not take in account the clinical context, such as the surrounding environment, the presence of bacteria and their by-products, the nature of dentine, and the effect of irrigant solutions on the remaining dentine. There have not been any randomized clinical trials conducted to evaluate the efficiency of these methods for regenerative endodontic procedures.

6 Future Directions

Regenerative procedures in the fields of medicine and dentistry have evolved considerably in the past two decades. The recognition and presence of multiple different stem cells in the oral cavity has changed the way that regenerative procedures are now being performed. These dental-derived MSCs have gained importance in recent years because they are easier to obtain than stem cells from other parts of the body (e.g., umbilical cord blood stem cells, and bone marrow stem cells). These dental stem cells are present in the oral cavity and can be obtained easily by routine dental procedures (such as extraction of third molars or premolars for orthodontic reasons).

Furthermore, previous research has demonstrated that these dental stem cells have the ability to differentiate into several lineages including osteogenic, odontogenic, chondrogenic, adipogenic, myogenic, and neurogenic [142–144]. This is one of the main reasons why these dental stem cells are now used in various tissue engineering procedures beyond the oral cavity in regenerative medicine, such as in the treatment of acute lung injury, neurotrauma, myocardial infarction, muscular dystrophy, autoimmune diseases, and for corneal reconstruction [145–150]. However, many studies have been conducted either *in vitro* or in using small animal models. The extrapolation of these results to a human model and the clinical applicability of these procedures is still unknown.

There are some aspects of these regenerative procedures that should be considered for greater attention, especially since these cells are being used more often. These include:

1. Disinfection: As discussed earlier, the presence of bacteria can affect the prognosis of any regenerative procedure. Hence, it is imperative that the effects of bacteria on the stem cells be analyzed.
2. Stem cell properties: Stem cells differ in their inherent properties. Some stem cells are more prone to apoptosis and immune cell mediated cell death compared to others. Hence, it is unknown whether all these different cell types can be used in the same way. More information on stem cell properties and their interaction with other cells is needed.
3. Interactions with the immune system: Different stem cells interact differently with the immune system. This is necessary information, especially when using them in regenerative medicine. Some stem cells have been shown to increase the levels of pro-inflammatory cytokines in certain situations, which could be detrimental when used for stem cell therapy [151].
4. Feasibility and cost effectiveness: Key questions include: Are these procedures feasible and predictable? Can the cells be obtained and preserved in a predictable manner? Is a steam cell based approach cost-effective?

Currently, stem cells are being used for whole tooth regeneration. Such an approach could potentially change dental treatment modalities in the future. In medicine, stem cells have been used to regenerate in the laboratory the bladder, trachea, neurogenic tissue, and cardiac tissue. The future looks promising, and what we learn today about these cells will definitely influence the future.

References

1. Gluskin AH, Peters CI, Peters OA. Minimally invasive endodontics: challenging prevailing paradigms. Br Dent J. 2014;216:347–53.
2. Gronthos S, Mankani M, Brahim J, Robey PG, Shi S. Postnatal human dental pulp stem cells (DPSCs) in vitro and in vivo. Proc Natl Acad Sci USA. 2000;97:13625–30.
3. Bergenholtz G, Mjör IA, Cotton WR, Hanks CT, Kim S, Torneck CD, et al. The biology of dentin and pulp. Consensus report. J Dent Res. 1985;64:631–3.
4. Shah D, Lynd T, Ho D, Chen J, Vines J, Jung H-D, et al. Pulp–dentin tissue healing response: a discussion of current biomedical approaches. J Clin Med. 2020;9:434.
5. Zheng L, Ehardt L, McAlpin B, About I, Kim D, Papagerakis S, et al. The tick tock of odontogenesis. Exp Cell Res. 2014;325:83–9.
6. Thesleff I. Current understanding of the process of tooth formation: transfer from the laboratory to the clinic. Aust Dent J. 2014;59:48–54.
7. Thesleff I, Keränen S, Jernvall J. Enamel knots as signaling centers linking tooth morphogenesis and odontoblast differentiation. Adv Dent Res. 2001;15:14–8.
8. Yoshiba N, Yoshiba K, Stoetzel C, Perrin-Schmitt F, Cam Y, Ruch JV, et al. Temporospatial gene expression and protein localization of matrix metalloproteinases and their inhibitors during mouse molar tooth development. Dev Dyn. 2003;228:105–12.
9. Harokpakis-Haojishengallis E. Physiologic root resorption in primary teeth: molecular and histological events. J Oral Sci. 2007;49:1–12.
10. Li R, Guo W, Yang B, Guo L, Sheng L, Chen G, et al. Human treated dentin matrix as a natural scaffold for complete human dentin tissue regeneration. Biomaterials. 2011;32:4525–38.
11. Niwa T, Yamakoshi Y, Yamazaki H, Karakida T, Chiba R. The dynamics of TGF-β in dental pulp, odontoblasts and dentin. Sci Rep. 2018:1–14.
12. Ruch JV, Lesot H, Begue-kirn C. Odontoblast differentiation. Int J Dev Biol. 1995;68:51–68.
13. Goldberg M, Smith AJ. Cells and extracellular matrices of dentin and pulp: a biological basis for repair

and tissue engineering. Crit Rev Oral Biol Med. 2004;15:13–27.
14. Simon S, Cooper P, Lumley P, Berdal A, Tomson P, Smith A. Understanding pulp biology for routine clinical practice. Endod Pract Today. 2009;3:171–84.
15. Maurin JC, Couble ML, Staquet MJ, Carrouel F, About I, Avila J, et al. Microtubule-associated protein 1b, a neuronal marker involved in odontoblast differentiation. J Endod. 2009;35:992–6.
16. Tziafas D, Kodonas K. Differentiation potential of dental papilla, dental pulp, and apical papilla progenitor cells. J Endod. 2010;36:781–9.
17. Mitsiadis TA, Feki A, Papaccio G, Caton J. Dental pulp stem cells, niches, and notch signaling in tooth injury. Adv Dent Res. 2011;23:275–9.
18. Chang B, Svoboda KKH, Liu X. Cell polarization: from epithelial cells to odontoblasts. Eur J Cell Biol. 2019;98:1–11.
19. Goldberg M, Kulkarmi A, Young M, Boskey A. Dentin: structure, composition and mineralization: the role of dentin ECM in dentin formation and mineralization. Front Biosci. 2012;3:711–35.
20. Ten Cate AR. An analysis of Tomes' granular layer. Anat Rec. 1972;172:137–47.
21. Bishop MA, Malhotra M, Yoshida S. Interodontoblastic collagen (von Korff fibers) and circumpulpal dentin formation: an ultrathin serial section study in the cat. Am J Anat. 1991;191:67–73.
22. Hall R, Septier D, Embery G, Goldberg M. Stromelysin-1 (MMP-3) in forming enamel and predentine in rat incisor-coordinated distribution with proteoglycans suggests a functional role. Histochem J. 1999;31:761–70.
23. Simon SRJ, Tomson PL, Berdal A. Regenerative endodontics: regeneration or repair? J Endod. 2014;40:4–9.
24. Tziafa C, Koliniotou-Koumpia E, Papadimitriou S, Tziafas D. Dentinogenic responses after direct pulp capping of miniature swine teeth with biodentine. J Endod. 2014;40:1967–71.
25. Simon SRJ, Berdal A, Cooper PR, Lumley PJ, Tomson PL, Smith AJ. Dentin–pulp complex regeneration: from lab to clinic. Adv Dent Res. 2011;23:340–5.
26. Heyeraas KJ, Berggreen E. Interstitial fluid pressure. Crit Rev Oral Biol Med. 1999;10:328–36.
27. Nanci A. Dentin–pulp complex. In: Nanci A, editor. Ten Cate's oral histology: development structure, and function. Saint-Louis: Elsevier; 2003. p. 192–239.
28. Bletsa A, Berggreen E, Fristad I, Tenstad O, Wiig H. Cytokine signalling in rat pulp interstitial fluid and transcapillary fluid exchange during lipopolysaccharide-induced acute inflammation. J Physiol. 2006;573:225–36.
29. Chidiac JJ, Al-Asmar B, Rifai K, Jabbur SJ, Saadé NE. Inflammatory mediators released following application of irritants on the rat injured incisors. The effect of treatment with anti-inflammatory drugs. Cytokine. 2009;46:194–200.
30. Zehnder M, Wegehaupt FJ, Attin T. A first study on the usefulness of matrix metalloproteinase 9 from dentinal fluid to indicate pulp inflammation. J Endod. 2011;37:17–20.
31. Goldberg M, Farges J-C, Lacerda-Pinheiro S, Six N, Decup F, Nadège J, et al. Inflammatory and immunological aspects of dental pulp repair. Pharmacol Res. 2008;58:137–47.
32. Arana-Chavez VE, Massa LF. Odontoblasts: the cells forming and maintaining dentine. Int J Biochem Cell Biol. 2004;36:1367–73.
33. Jontell M, Okiji T, Dahlgren U, Bergenholtz G, Nayebzadeh A, Stangel I, et al. Immune defense mechanisms of the dental pulp. Crit Rev Oral Biol Med. 1998;9:179–200.
34. Farges J-C, Carrouel F, Keller J-F, Baudouin C, Msika P, Bleicher F, et al. Cytokine production by human odontoblast-like cells upon toll-like receptor-2 engagement. Immunobiology. 2011;216:513–7.
35. Couve E. Ultrastructural changes during the life cycle of human odontoblasts. Arch Oral Biol. 1986;31:643–51.
36. Couve E, Osorio R, Schmachtenberg O. The amazing odontoblast: activity, autophagy, and aging. J Dent Res. 2013;92:765–72.
37. Holland GR. The odontoblast process: form and function. J Dent Res. 1985;64 Spec No:499–514.
38. Yoshiba K, Yoshiba N, Ejiri S, Iwaku M, Ozawa H. Odontoblast processes in human dentin revealed by fluorescence labeling and transmission electron microscopy. Histochem Cell Biol. 2002;118:205–12.
39. Magloire H, Couble M-L, Romeas A, Bleicher F. Odontoblast primary cilia: facts and hypotheses. Cell Biol Int. 2004;28:93–9.
40. Veerayutthwilai O, Byers MR, Pham T-TT, Darveau RP, Dale BA. Differential regulation of immune responses by odontoblasts. Oral Microbiol Immunol. 2007;22:5–13.
41. Butler WT, Ritchie H. The nature and functional significance of dentin extracellular matrix proteins. Int J Dev Biol. 1995;39:169–79.
42. Linde A. Dentin mineralization and the role of odontoblasts in calcium transport. Connect Tissue Res. 1995;33:163–70.
43. Shin S-J, Lee J-I, Baek S-H, Lim S-S. Tissue levels of matrix metalloproteinases in pulps and periapical lesions. J Endod. 2002;28:313–5.
44. Loreto C, Galanti C, Musumeci G, Rusu MC, Leonardi R. Immunohistochemical analysis of matrix metalloproteinase-13 in human caries dentin. Eur J Histochem 2014;58.
45. Staquet MM-J, Durand SH, Colomb E, Roméas A, Vincent C, Bleicher F, et al. Different roles of odontoblasts and fibroblasts in immunity. J Dent Res. 2008;87:256–61.
46. Garcia JMQ, Martins MD, Jaeger RG, Marques MM. Immunolocalization of bone extracellular matrix proteins (type I collagen, osteonectin and bone sialoprotein) in human dental pulp and cultured pulp cells. Int Endod J. 2003;36:404–10.
47. Sulkala M, Paakkonen V, Larmas M, Salo T, Tjaderhane L. Matrix metalloproteinase-13 (MMP-

13, collagenase-3) is highly expressed in human tooth pulp. Connect Tissue Res. 2004;45:231–7.
48. Jain A, Bahuguna R. Role of matrix metalloproteinases in dental caries, pulp and periapical inflammation: an overview. J Oral Biol Craniofacial Res. 2015;5:212–8.
49. Sloan AJ, Perry H, Matthews JB, Smith AJ. Transforming growth factor-beta isoform expression in mature human healthy and carious molar teeth. Histochem J. 2000;32:247–52.
50. Nakanishi T, Takegawa D, Hirao K, Takahashi K, Yumoto H, Matsuo T. Roles of dental pulp fibroblasts in the recognition of bacterium-related factors and subsequent development of pulpitis. Jpn Dent Sci Rev. 2011;47:161–6.
51. Keller J-F, Carrouel F, Colomb E, Durand SH, Baudouin C, Msika P, et al. Toll-like receptor 2 activation by lipoteichoic acid induces differential production of pro-inflammatory cytokines in human odontoblasts, dental pulp fibroblasts and immature dendritic cells. Immunobiology. 2009;215:53–9.
52. Hirao K, Yumoto H, Takahashi K, Mukai K, Nakanishi T, Matsuo T. Roles of TLR2, TLR4, NOD2, and NOD1 in pulp fibroblasts. J Dent Res. 2009;88:762–7.
53. Le Fournis C, Hadjichristou C, Jeanneau C, About I. Human pulp fibroblast implication in phagocytosis via complement activation. J Endod. 2019;45:584–90.
54. Gaudin A, Renard E, Hill M, Bouchet-delbos L, Farges J, Cuturi M. Phenotypic analysis of immunocompetent cells in healthy human dental pulp. J Endod. 2015;41:621–7.
55. Huang GT-J, Gronthos S, Shi S. Mesenchymal stem cells derived from dental tissues vs. those from other sources: their biology and role in regenerative medicine. J Dent Res. 2009;88:792–806.
56. van Amerongen JP, Lemmens IG, Tonino GJ. Immunofluorescent localization and extractability of fibronectin in human dental pulp. Arch Oral Biol. 1984;29:93–9.
57. van Amerongen JP, Lemmens IG, Tonino GJ. The concentration, extractability and characterization of collagen in human dental pulp. Arch Oral Biol. 1983;28:339–45.
58. Veis A, Goldberg M. Pulp extracellular matrix. In: Goldberg M, editor. The dental pulp: biology, pathology, and regenerative therapies. Berlin: Springer; 2014. p. 35–46.
59. Mizuno M, Banzai Y. Calcium ion release from calcium hydroxide stimulated fibronectin gene expression in dental pulp cells and the differentiation of dental pulp cells to mineralized tissue forming cells by fibronectin. Int Endod J. 2008;41:933–8.
60. Matthews B, Vongsavan N. Interactions between neural and hydrodynamic mechanisms in dentine and pulp. Arch Oral Biol. 1994;39(Suppl):87S–95S.
61. Iijima T, Zhang JQ. Three-dimensional wall structure and the innervation of dental pulp blood vessels. Microsc Res Tech. 2002;56:32–41.
62. Ikeda H, Suda H. Circulation of the pulp. In: Hargreaves KM, Goodis HE, Tay FR, editors. Seltzer and Bender's dental pulp. Chicago: Quintessence; 2012. p. 109–32.
63. Berggreen E, Bletsa A, Heyeraas KJ. Circulation in normal and inflamed dental pulp. Endod Top. 2007;17:2–11.
64. Kim S, Dörscher-Kim J. Hemodynamic regulation of the dental pulp in a low compliance environment. J Endod. 1989;15:404–8.
65. Olgart L. Neural control of pulpal blood flow. Crit Rev Oral Biol Med. 1996;7:159–71.
66. Nair PN. Neural elements in dental pulp and dentin. Oral Surg Oral Med Oral Pathol Oral Radiol Endod. 1995;80:710–9.
67. Magloire H, Maurin JC, Couble ML, Shibukawa Y, Tsumura M, Thivichon-Prince B, et al. Topical review. Dental pain and odontoblasts: facts and hypotheses. J Orofac Pain. 2010;24:335–49.
68. Ichikawa H, Sugimoto T. The co-expression of ASIC3 with calcitonin gene-related peptide and parvalbumin in the rat trigeminal ganglion. Brain Res. 2002;943:287–91.
69. Ichikawa H, Fukuda T, Terayama R, Yamaai T, Kuboki T, Sugimoto T. Immunohistochemical localization of gamma and beta subunits of epithelial Na+ channel in the rat molar tooth pulp. Brain Res. 2005;1065:138–41.
70. Chung G, Jung SJ, Oh SB. Cellular and molecular mechanisms of dental nociception. J Dent Res. 2013;92:948–55.
71. Mutoh N, Tani-Ishii N, Tsukinoki K, Chieda K, Watanabe K. Expression of toll-like receptor 2 and 4 in dental pulp. J Endod. 2007;33:1183–6.
72. He W, Yu Q, Zhou Z, Wang P. CpG oligonucleotides induce an immune response of odontoblasts through the TLR9, MyD88 and NF-kappaB pathways. Biochem Biophys Res Commun. 2010;399:274–8.
73. Durand SH, Flacher V, Roméas A, Carrouel F, Colomb E, Vincent C, et al. Lipoteichoic acid increases TLR and functional chemokine expression while reducing dentin formation in in vitro differentiated human odontoblasts. J Immunol. 2006;176:2880–7.
74. Levin LG, Rudd A, Bletsa A, Reisner H. Expression of IL-8 by cells of the odontoblast layer in vitro. Eur J Oral Sci. 1999;107:131–7.
75. Botero TM, Shelburne CE, Holland GR, Hanks CT, Nör JE. TLR4 mediates LPS-induced VEGF expression in odontoblasts. J Endod. 2006;32:951–5.
76. Dommisch H, Winter J, Willebrand C, Eberhard J, Jepsen S. Immune regulatory functions of human beta-defensin-2 in odontoblast-like cells. Int Endod J. 2007;40:300–7.
77. Farges J, Alliot-licht B, Baudouin C, Msika P, Bleicher F, Carrouel F. Odontoblast control of dental pulp inflammation triggered by cariogenic bacteria. Front Physiol. 2013;4:1–3.
78. Yumoto H, Hirao K, Hosokawa Y, Kuramoto H, Takegawa D, Nakanishi T, et al. The roles of odontoblasts in dental pulp innate immunity. Jpn Dent Sci Rev. 2018;54:105–17.

79. Izumi T, Kobayashi I, Okamura K, Matsuo K, Kiyoshima T, Ishibashi Y, et al. Responses of immunocompetent cells to cavity preparation in rat molars: an immunohistochemical study using OX6-monoclonal antibody. Arch Oral Biol. 1995;41:627–30.
80. Hahn C-L, Liewehr FR. Innate immune responses of the dental pulp to caries. J Endod. 2007;33: 643–51.
81. Cooper PRP, McLachlan JLJ, Simon S, Graham LW, Smith AJ. Mediators of inflammation and regeneration. Adv Dent Res. 2011;23:290–5.
82. Cruvinel W d M, Mesquita DJ, JAP A, TTT C, de Souza AWS, da Silva NP, et al. Immune system – part I Fundamentals of innate immunity with emphasis on molecular and cellular mechanisms of inflammatory response. Rev Bras Reumatol. 2010;50:434–61.
83. Hartmann K, Henz BM, Kruger-Krasagakes S, Kohl J, Burger R, Guhl S, et al. C3a and C5a stimulate chemotaxis of human mast cells. Blood. 1997;89: 2863–70.
84. Kokkas AB, Goulas A, Varsamidis K, Mirtsou V, Tziafas D. Irreversible but not reversible pulpitis is associated with up-regulation of tumour necrosis factor-alpha gene expression in human pulp. Int Endod J. 2007;40:198–203.
85. Kupper TS, Horowitz M, Birchall N, Mizutani H, Coleman D, McGuire J, et al. Hematopoietic, lymphopoietic, and proinflammatory cytokines produced by human and murine keratinocytes. Ann NY Acad Sci. 1988;548:262–70.
86. McLachlan JL, Smith AJ, Bujalska IJ, Cooper PR. Gene expression profiling of pulpal tissue reveals the molecular complexity of dental caries. Biochim Biophys Acta. 1741;2005:271–81.
87. Li D, Fu L, Zhang Y, Yu Q, Ma F, Wang Z, et al. The effects of LPS on adhesion and migration of human dental pulp stem cells in vitro. J Dent. 2014;2:1327–34.
88. Hahn C-L, Liewehr F. Update on the adaptive immune responses of the dental pulp. J Endod. 2007;33:773–81.
89. Park HCC, Quan H, Zhu T, Kim Y, Kim B, Yang HCC. The effects of M1 and M2 macrophages on odontogenic differentiation of human dental pulp cells. J Endod. 2017;43:596–601.
90. Cooper PR, Takahashi Y, Graham LW, Simon S, Imazato S, Smith AJ. Inflammation-regeneration interplay in the dentine–pulp complex. J Dent. 2010;38:687–97.
91. Maderna P, Godson C. Lipoxins: resolutionary road. Br J Pharmacol. 2009;158:947–59.
92. Serhan CN, Chiang N, Dalli J. The resolution code of acute inflammation: novel pro-resolving lipid mediators in resolution. Semin Immunol. 2015;27: 200–15.
93. Aliberti J, Hieny S, Reis Sousa C, Serhan CN, Sher A. Lipoxin-mediated inhibition of IL-12 production by DCs: a mechanism for regulation of microbial immunity. Nat Immunol. 2002;3:76–82.
94. Ramon S, Bancos S, Serhan CN, Phipps RP. Lipoxin A4 modulates adaptive immunity by decreasing memory B-cell responses via an ALX/FPR2-dependent mechanism. Eur J Immunol. 2014;44:357–69.
95. Ostby BN. The role of the blood clot in endodontic therapy. An experimental histologic study. Acta Odontol Scand. 1961;19:324–53.
96. Diogenes AR, Ruparel NB, Shiloh MU, Hargreaves KM. Regenerative endodontics. Am Assoc Endodontists. 2016;147:372–80.
97. Conde MCM, Chisini LA, Sarkis-Onofre R, Schuch HS, Nör JE, Demarco FF. A scoping review of root canal revascularization: relevant aspects for clinical success and tissue formation. Int Endod J. 2016;50:860–74.
98. Chan EKM, Desmeules M, Cielecki M, Dabbagh B, Ferraz dos Santos B. Longitudinal cohort study of regenerative endodontic treatment for immature necrotic permanent teeth. J Endod. 2017;43:395–400.
99. Lovelace TW, Henry MA, Hargreaves KM, Diogenes A. Evaluation of the delivery of mesenchymal stem cells into the root canal space of necrotic immature teeth after clinical regenerative endodontic procedure. J Endod. 2011;37:133–8.
100. Murray PE, Garcia-Godoy F, Hargreaves KM. Regenerative endodontics: a review of current status and a call for action. J Endod. 2007;33: 377–90.
101. Gong T, Heng BC, Lo ECM, Zhang C. Current advance and future prospects of tissue engineering approach to dentin/pulp regenerative therapy. Stem Cells Int. 2016;9204574.
102. Shi S, Robey PG, Gronthos S. Comparison of human dental pulp and bone marrow stromal stem cells by cDNA microarray analysis. Bone. 2001;29:532–9.
103. Wigler R, Kaufman AY, Lin S, Steinbock N, Hazan-Molina H, Torneck CD. Revascularization: a treatment for permanent teeth with necrotic pulp and incomplete root development. J Endod. 2013;39:319–26.
104. Moussa DG, Aparicio C. Present and future of tissue engineering scaffolds for dentin–pulp complex regeneration. J Tissue Eng Regen Med. 2019;13:58–75.
105. Guo B, Lei B, Li P, Ma PX. Functionalized scaffolds to enhance tissue regeneration. Regen Biomater. 2015;2:47–57.
106. Orti V, Collart-Dutilleul P-Y, Piglionico S, Cuisinier F, Panayotov I. Tissue engineering to clinical approaches pulp regeneration concepts for non-vital teeth: from tissue. Tissue Eng. 2018:1–75.
107. Zhao S, Sloan AJ, Murray PE, Lumley PJ, Smith AJ. Ultrastructural localisation of TGF-β exposure in dentine by chemical treatment. Histochem J. 2000;32:489–94.
108. Aubeux D, Beck L, Weiss P, Guicheux J, Enkel B, Pérez F, et al. Assessment and quantification of noncollagenic matrix proteins released from human dentin powder incorporated into a silated hydroxy-

propylmethylcellulose biomedical hydrogel. J Endod. 2016;42:1371–6.
109. Shrestha S, Kishen A. Bioactive molecule delivery systems for dentin–pulp tissue engineering. J Endod. 2017;43:733–44.
110. Almutairi W, Yassen GH, Aminoshariae A, Williams KA, Mickel A. Regenerative endodontics: a systematic analysis of the failed cases. J Endod. 2019;45:567–77.
111. Verma P, Nosrat A, Kim JRR, Price JBB, Wang P, Bair E, et al. Effect of residual bacteria on the outcome of pulp regeneration in vivo. J Dent Res. 2017;96:100–6.
112. Latham J, Fong H, Jewett A, Johnson JD, Paranjpe A. Disinfection efficacy of current regenerative endodontic protocols in simulated necrotic immature permanent teeth. J Endod. 2016;42:1218–25.
113. Montero-Miralles P, Martín-González J, Alonso-Ezpeleta O, Jiménez-Sánchez MC, Velasco-Ortega E, Segura-Egea JJ. Effectiveness and clinical implications of the use of topical antibiotics in regenerative endodontic procedures: a review. Int Endod J. 2018;51:981–8.
114. Ruparel NB, Teixeira FB, Ferraz CCR, Diogenes A. Direct effect of intracanal medicaments on survival of stem cells of the apical papilla. J Endod. 2012;38:1372–5.
115. Casagrande L, Cordeiro MM, Nör SA, Nör JE. Dental pulp stem cells in regenerative dentistry. Odontology. 2011;99:1–7.
116. Rosa V, Zhang Z, Grande RHM, Nör JE. Dental pulp tissue engineering in full-length human root canals. J Dent Res. 2013;92:970–5.
117. Syed-Picard FN, Ray HL, Kumta PN, Sfeir C. Scaffoldless tissue-engineered dental pulp cell constructs for endodontic therapy. J Dent Res. 2014;93:250–5.
118. Nakashima M, Iohara K, Murakami M, Nakamura H, Sato Y, Ariji Y, et al. Pulp regeneration by transplantation of dental pulp stem cells in pulpitis: a pilot clinical study. Stem Cell Res Ther. 2017;8:1–13.
119. Nakashima M, Iohara K. Regeneration of dental pulp by stem cells. Adv Dent Res. 2011;23:313–9.
120. Souron J-B, Petiet A, Decup F, Tran XV, Lesieur J, Poliard A, et al. Pulp cell tracking by radionuclide imaging for dental tissue engineering. Tissue Eng Part C. 2014;20:188–97.
121. Mangione F, EzEldeen M, Bardet C, Lesieur J, Bonneau M, Decup F, et al. Implanted dental pulp cells fail to induce regeneration in partial pulpotomies. J Dent Res. 2017;96:1406–13.
122. El-Zekrid MH, Mahmoud SH, Ali FA, Helal ME, Grawish ME. Healing capacity of autologous bone marrow–derived mesenchymal stem cells on partially pulpotomized dogs' teeth. J Endod. 2019;45:287–94.
123. Trope M. Treatment of the immature tooth with a non-vital pulp and apical periodontitis. Dent Clin N Am. 2010;54:313–24.
124. Galler KM, Widbiller M. Perspectives for cell-homing approaches to engineer dental pulp. J Endod. 2017;43:S40–5.
125. Widbiller M, Driesen RB, Eidt A, Lambrichts I, Hiller K, Buchalla W, et al. Cell homing for pulp tissue engineering with endogenous dentin matrix proteins. J Endod. 2018;44:956–62.e2.
126. Ivica A, Zehnder M, Mateos JM, Ghayor C, Weber FE. Biomimetic conditioning of human dentin using citric acid. J Endod. 2018;45:45–50.
127. Song JS, Takimoto K, Jeon M, Vadakekalam J, Ruparel NB, Diogenes A. Decellularized human dental pulp as a scaffold for regenerative endodontics. J Dent Res. 2017;22034517693606.
128. Alqahtani Q, Zaky SH, Patil A, Beniash E, Ray H, Sfeir C. Decellularized swine dental pulp tissue for regenerative root canal therapy. J Dent Res. 2018;97:1460.
129. Kitamura C, Nishihara T, Terashita M, Tabata Y, Washio A. Local regeneration of dentin–pulp complex using controlled release of FGF-2 and naturally derived sponge-like scaffolds. Int J Dent. 2012;2012:1.
130. Zheng L, Amano K, Iohara K, Ito M, Imabayashi K, Into T, et al. Matrix metalloproteinase-3 accelerates wound healing following dental pulp injury. Am J Pathol. 2009;175:1905–14.
131. Eba H, Murasawa Y, Iohara K, Isogai Z, Nakamura H, Nakamura H, et al. The anti-inflammatory effects of matrix metalloproteinase-3 on irreversible pulpitis of mature erupted teeth. PLoS One. 2012;7:e52523.
132. Hull C. Apparatus for production of three-dimensional objects by stereolithography. US Patent 4575330 A 1986.
133. Matai I, Kaur G, Seyedsalehi A, McClinton A, Laurencin CT. Progress in 3D bioprinting technology for tissue/organ regenerative engineering. Biomaterials. 2020;226:119536.
134. Vijayavenkataraman S, Fuh JYH, Lu WF. 3D printing and 3D bioprinting in pediatrics. Bioengineering. 2017;4:63.
135. Munaz A, Vadivelu RK, St. John J, Barton M, Kamble H, Nguyen NT. Three-dimensional printing of biological matters. J Sci Adv Mater Devices. 2016;1:1–17.
136. Yu H, Zhang X, Song W, Pan T, Wang H, Ning T, et al. Effects of 3-dimensional bioprinting alginate/gelatin hydrogel scaffold extract on proliferation and differentiation of human dental pulp stem cells. J Endod. 2019;45:706–15.
137. Athirasala A, Tahayeri A, Thrivikraman G, Franca CM, Monteiro N, Tran V, et al. A dentin-derived hydrogel bioink for 3D bioprinting of cell laden scaffolds for regenerative dentistry. Biofabrication 2018;10:0–23.
138. Duarte Campos DF, Zhang S, Kreimendahl F, Köpf M, Fischer H, Vogt M, et al. Hand-held bioprinting for de novo vascular formation applicable

to dental pulp regeneration. Connect Tissue Res. 2020;61:205–15.
139. Leyendecker Junior A, Gomes Pinheiro CC, Lazzaretti Fernandes T, Franco Bueno D, Junior AL, Cristina C, et al. The use of human dental pulp stem cells for in vivo bone tissue engineering: a systematic review. J Tissue Eng. 2018;9:2041731417752766.
140. Li L, Tang Q, Wang A, Chen YP. Regrowing a tooth: in vitro and in vivo approaches. Curr Opin Cell Biol. 2019;61:126–31.
141. Han J, Kim DS, Jang H, Kim HR, Kang HW. Bioprinting of three-dimensional dentin–pulp complex with local differentiation of human dental pulp stem cells. J Tissue Eng. 2019;10:2041731419845849.
142. Tecles O, Laurent P, Zygouritsas S, Burger A-S, Camps J, Dejou J, et al. Activation of human dental pulp progenitor/stem cells in response to odontoblast injury. Arch Oral Biol. 2005;50:103–8.
143. Huang GTJ, Sonoyama W, Liu Y, Liu H, Wang S, Shi S. The hidden treasure in apical papilla: the potential role in pulp/dentin regeneration and bioroot engineering. J Endod. 2008;34:645–51.
144. Sonoyama W, Liu Y, Yamaza T, Tuan RS, Wang S, Shi S, et al. Characterization of the apical papilla and its residing stem cells from human immature permanent teeth: a pilot study. J Endod. 2008;34:166–71.
145. Mao JJ, Giannobile WV, Helms JA, Hollister SJ, Krebsbach PH, Longaker MT, et al. Craniofacial tissue engineering by stem cells. J Dent Res. 2006;85:966–79.
146. Gandia C, Armiñan A, García-Verdugo JM, Lledó E, Ruiz A, Miñana MD, et al. Human dental pulp stem cells improve left ventricular function, induce angiogenesis, and reduce infarct size in rats with acute myocardial infarction. Stem Cells. 2008;26:638–45.
147. Iohara K, Zheng L, Wake H, Ito M, Nabekura J, Wakita H, et al. A novel stem cell source for vasculogenesis in ischemia: subfraction of side population cells from dental pulp. Stem Cells. 2008;26:2408–18.
148. Wakayama H, Hashimoto N, Matsushita Y, Matsubara K, Yamamoto N, Hasegawa Y, et al. Factors secreted from dental pulp stem cells show multifaceted benefits for treating acute lung injury in mice. Cytotherapy. 2015;17:1119–29.
149. Botelho J, Cavacas MA, Machado V, Mendes JJ. Dental stem cells: recent progresses in tissue engineering and regenerative medicine. Ann Med. 2017;49:644–51.
150. Armiñán A, Gandía C, García-Verdugo JM, Lledó E, Trigueros C, Ruiz-Saurí A, et al. Mesenchymal stem cells provide better results than hematopoietic precursors for the treatment of myocardial infarction. J Am Coll Cardiol. 2010;55:2244–53.
151. Whiting D, Chung WO, Johnson JD, Paranjpe A. Characterization of the cellular responses of dental mesenchymal stem cells to the immune system. J Endod. 2018;44:1126–31.

Clinical Approach to Regenerative Endodontics

Omid Dianat, Elham Shadmehr, and Yoo Jung Chung

1 Introduction

A developing young permanent tooth generally has an open apex until apical closure occurs approximately 3 years after its eruption [1]. A traumatic injury, rapidly progressing dental caries, or developmental anomalies may result in early pulpal necrosis, development of apical periodontitis and incomplete root formation [2]. Up to 50% of traumatized teeth may present with pulpal necrosis and exhibit signs and symptoms of pulpal and periapical disease [3]. Another common etiology is dental anomalies, which can be lead to pulpal necrosis in immature permanent teeth. Dens evaginatus and dens invaginatus are the most common anomalies resulting in pulpal necrosis in immature teeth [4].

Treatment of necrotic permanent teeth with open apices has proved to be a great challenge to clinicians. Due to the thin dentinal walls of the root canal, mechanical debridement cannot be performed with the conventional endodontic procedures [5]. The absence of an apical barrier can bring challenges in adequately sealing the apical third of the root canal space. In addition, these teeth are prone to fracture due to their structural weaknesses. Therefore, it is hard to achieve an optimal clinical outcome with conventional root canal procedures performed on pulpless immature teeth with a blunderbuss apex [6]. Permanent teeth with pulpal necrosis and incomplete roots have conventionally been treated by clinicians with an endodontic procedure called apexification [6–8]. In this approach, the formation of a calcified apical barrier is induced in an incompletely formed root in teeth with open apices and necrotic pulps [9], using long-term calcium hydroxide dressings. Another method is to create an artificial apical barrier in a single visit using a plug of mineral trioxide aggregate (MTA) [10, 11] or another bioceramic (BC) material [12].

The apexification approach uses calcium hydroxide as the intracanal medicament of choice, delivered in a powder or paste form, and left in the canal for a long period of time. Every 6 months, this calcium hydroxide dressing is replaced, and a radiograph is taken. This procedure is repeated until apical closure via the formation of a dentinal bridge is noted on the radiograph [5]. Dentinal bridge formation at apical foramen can also be confirmed clinically with an instrument such as an endodontic file or a paper point [5]. Once the apex is determined to

O. Dianat
Endodontic Division, University of Maryland, School of Dentistry, Baltimore, MD, USA

E. Shadmehr (✉)
UCSF Endodontic Division, University of California San Francisco School of Dentistry, San Francisco, CA, USA
e-mail: Elham.Shadmehr@ucsf.edu

Y. J. Chung
University of California San Francisco School of Dentistry, San Francisco, CA, USA

© Springer Nature Switzerland AG 2021
S. Hosseinpour et al. (eds.), *Regenerative Approaches in Dentistry*,
https://doi.org/10.1007/978-3-030-59809-9_4

be closed, the whole canal is then obturated with gutta percha [13]. While the success rate of calcium hydroxide apexification is approximately 95% [14], this treatment has several main drawbacks that do not support its clinical application anymore:

1. The average time to make an apical barrier is between 3 months and 2 years [5, 15–17].
2. Patient compliance is required since they need to make multiple visits to have calcium hydroxide replaced periodically over a long period of time, until apical closure occurs [5, 8].
3. It has been reported that long-term intracanal dressing of calcium hydroxide may put the tooth at a greater risk of fracture [18], especially horizontal root fractures at the cementoenamel junction [14]. This view has however been challenged in the recent literature.
4. The dentinal bridge formed at the apex after the placement of long-term calcium hydroxide dressings is very porous [5, 19, 20], and it does not provide strong apical closure.

Due to these drawbacks with apexification, an alternative approach was developed in which an artificial barrier was placed at the apex, instead of inducing one to form. This alternative method reduces the number of visits required. This procedure involves making an artificial barrier approximately 3–4 mm in length at the apex with a plug made of either MTA [11, 21–23] or a bioceramic material [12], and then filling the rest of the canal with gutta percha [5].

The success rate of MTA apexification is reported to be 94%, which is comparable to calcium hydroxide apexification. However, this procedure has its own shortcomings. Although it provides an artificial barrier, this treatment modality still leaves the dentinal walls thin, and it does not strengthen or improve the weak structural integrity of the immature root [8]. Also, the treated roots are still weak and thus prone to fracture [18].

Regenerative endodontic treatment (RET) was developed to overcome these shortcomings, and it has been found to have the potential to be an ideal treatment approach for treating immature, necrotic permanent teeth.

2 History and Emergence of Regenerative Endodontics

With the advent of a better understanding of regeneration and tissue engineering and its application to dentistry, regenerative endodontics has developed. This approach represents a paradigm shift in the treatment of immature permanent teeth with a necrotic dental pulp [24, 25]. RET is also sometimes referred to as "revitalization" and "revascularization." However, the term "revascularization" is no longer used, since the recruited tissue in the canal space is much more than just blood vessels.

The development of the concept of regenerative endodontics was facilitated by the publication of two significant case reports by Iwaya [26] and by Banch and Trope [27]. However, this treatment was historically introduced by Nygaard-Ostby utilizing an apical blood clot in the resolution of apical periodontitis and for pulp repair [28].

3 Goals of Regenerative Endodontics

Essentially, there are three goals for RET [29]: First, to treat apical periodontitis and successfully allow healing of damaged apical tissue, eliminating clinical symptoms and signs [25–27]. The second, and more important goal of RET is to regenerate the structure and function of dentin-pulp complex [30]. Though apexification methods using either long-term calcium hydroxide or a one-visit MTA plug resolve apical periodontitis, these two traditional approaches cannot restore the original anatomic structure and physiological function of the dentin-pulp complex. The third goal of RET is to regain vitality of the tooth.

With the use of the stem cell tissue engineering concept, RET seeks to regenerate vital pulp- and dentin-like tissues in the root canal space in order to accomplish the reconstitution of the neurovascular system, with tubular dentin with an odontoblastic layer, and a natural immune defense system. An increased root canal wall thickness and/or increased root length is not an essential goal of this treatment, according to the

American Association of Endodontists (AAE), but it is considered as desirable. However, the European Society of Endodontology (ESE) considers an increase of root thickness and length as one of the criteria for success with RET [31]. Through dentin-pulp tissue deposition, this novel procedure is expected to produce a stronger mature root [32, 33], and therefore, reduces the risk of fracture of the root.

4 Why Do We Need a Vital Pulp?

A vital dental pulp is essential since it provides the innate immune system which responds when the tissue is faced with bacterial invasion. Healthy pulp tissue has the potential to respond to bacterial insult in response that involves cells, neurons, and the microvasculature [34].

First, odontoblasts act as the first defensive mechanism in a healthy pulp. They detect microbes through toll-like receptors (TLRs) [34]. These receptors are triggered when they bind the lipoteichoic acid (LTA) of Gram-positive bacteria or the lipopolysaccharide (LPS) of Gram-negative bacteria. LTA and LPS trigger different types of TLRs; for example, LTA mainly binds to the TLR-2 receptor on odontoblasts, whereas LPS mainly binds to TLR-4. Receptor binding induces odontoblasts to produce various cytokines which start the inflammatory process, as part of the response to dentinal caries. For instance, odontoblasts secrete transforming growth factor-beta (TGF-β) during the initial stage of inflammation, which drives dentin mineralization, as a way of providing a physical barrier against the advancing caries front, through calcifications within the dentinal tubules. In addition, TGF-β and other chemokines secreted by odontoblasts activated through the TLR pathway, such as CCL2, CXCL12, and CXCL14, promote the recruitment of immature dendritic cells (iDC). These dendritic cells phagocytose the bacteria and subsequently present the bacterial antigens to T-lymphocytes, to trigger an adaptive immune response [35–37].

Odontoblasts also secrete a low level of interleukin-8 (IL-8), which activates neutrophils, and also promotes the recruitment of T-lymphocytes [34]. Cytotoxic T-lymphocytes (CD8) release perforin and lytic granule enzymes that can damage bacterial cell membranes and can lyse bacterial cells, and also can trigger apoptosis [38]. Cytotoxic T-lymphocytes also produce cytokines such as IFN-γ, which inhibits microbial replication, while causing macrophage activation at the same time.

In addition, odontoblasts express beta-defensin-2 (BD-2). This is a bactericidal agent for bacteria (especially *Streptococcus mutans* and *Lactobacillus casei*), and it binds to the membrane of the microbes and increases bacterial cell permeability [39]. BD-2 also plays a pivotal role in activating the immune system. It is a chemoattractant for natural killer cells, T-lymphocytes, and immature DCs [34]. BD-2 is made by neutrophils, as well as by odontoblasts.

Odontoblasts challenged with LTA or LPS also secrete vascular endothelial growth factor (VEGF), which stimulates angiogenesis and increases vascular permeability, to bring more immune cells to the site of the injury, for better defense by neutrophils and other parts of the innate immune system against bacterial invasion.

A vital dental pulp also has sensory neuropeptides, such as substance P (SP), calcitonin gene-related peptide (CGRP), and neurokinin A (NKA). Following pulpal injury from microbes, the concentration of these neuropeptides increases significantly, causing vasodilation, increased vascular permeability, and increased pulpal blood flow [34].

Overall, these cellular, neuronal, and vascular responses of odontoblasts in a healthy pulp to the threat posed by bacterial invasion increase the interstitial tissue pressure. The increased intrapulpal pressure drives an increase in the outward flow of dentinal fluid. It also results in the stimulation of A-delta fibers and ultimately causes pain, which alerts the patient to the alarming situation confronted by the vital dental pulp as it faces the threat of bacterial invasion [34].

Interestingly, the sensibility, immunity, and even vitality of immature teeth with necrotic pulps might be restored after RET.

5 Concepts of Regenerative Endodontics

In general, as mentioned earlier in chapter "Dentin Pulp Complex Regeneration," there are three essential factors required for tissue engineering: cells, signaling molecules, and scaffolds [40, 41]. This triad of tissue engineering must be present to develop a biologically functional tissue.

First, stem cells that have the capacity to regenerate the appropriate tissue need to be recruited or formed at the target site. Second, appropriate signaling molecules (e.g., growth factors) must be delivered to or induced in the target tissue. The purpose of having the signaling molecules is to guide the stem cells to differentiate into their final structure. The goal of RET is to develop a dentin-pulp complex. For stem cells to regenerate dentin, they must encounter specific signaling molecules such as bone morphogenic proteins [42], transforming growth factor-β [43], and fibroblast growth factors [44].

Third, the stem cells need to be recruited to or placed within the scaffold. A scaffolds is highly porous and may be composed of natural materials (such as collagen) or synthetic polymers (such as polylactic acid). Scaffolds facilitate three-dimensional tissue formation by providing a structural template for cell migration and cell attachment [40, 41]. With this concept of stem cell recruitment and differentiation using signaling molecules and scaffolds, two different approaches of RET have been developed: cell-based and cell homing.

5.1 Cell-Based Approach

The approach of using exogenous stem cells and transferring them to the site through a scaffold is called "cell-based regeneration." This technique is able to yield pulp/dentin-like tissues. In a cell-based animal study, Huang and his colleagues extracted and cultivated DPSCs and SCAP from human third molar teeth, and then injected them into a poly (lactide-co-glycolide) (PLG) scaffold. The scaffold was then inserted into a root fragment that had MTA on one end and was open on the other. This tooth fragment with the scaffold and the odontoblast-lineage cells (DPSC and SCAP) were transplanted into immunocompromised mice subcutaneously. As a result, the root canal space was completely filled with pulp-like tissues with a continuous layer of a dentin-like structure adjacent to the existing dentinal wall. In other words, from a scaffold with growth factors, SCAP and DPSCs, pulp/dentin-like structures were generated in an empty root canal space.

Thus, this cell-based animal study proved, for the first time, that de novo regeneration of pulp-like tissue was in fact possible. As this study shows, the cell-based approach can regenerate the dentin-pulp complex in the canal space without any competition between odontoblast- and non-odontoblast lineage progenitors, due to the direct insertion of DPSC and SCAP. Therefore, with the cell-based technique, the regenerated tissues are well-vascularized and are consist of pulp-like and dentin-like structures in the root canal space [30].

Although this technique exhibits a great success in de novo regeneration of dentin-pulp complex in the intracanal space, there are some drawbacks to this approach. First, the mineralized dentin tissues that are formed with this technique do not actually resemble the original dentin structure. Unlike natural dentin, which has a tubular structure, the newly regenerated mineralized tissues show a disorganized alignment and entrapment of odontoblast-like cells, which is more similar to tertiary and reparative dentin [30]. Additionally, the amount of deposited tissue cannot be controlled.

Moreover, this approach is highly technique sensitive. Currently, it is not practical to apply this approach clinically, due to its variability, poor cost efficiency, the need for stem cell banks or other ways to store the stem cells, and the absence of a safe and reliable protocol that has been proven through multiple clinical trials [24].

Recently, a systematic review and a meta-analysis were conducted on stem/progenitor cell-mediated pulpal tissue regeneration. While transplantation of stem/progenitor cells may represent a promising future approach for pulp regeneration [45], only one clinical trial has been conducted, despite several animal experiments [46].

In a randomized controlled clinical trial, Xuan and his colleagues implanted human deciduous pulp stem cells (hDPSCs) into immature, permanent teeth with necrosis. They compared the results to apexification, using various parameters including radiovisiography (RVG), cone-beam computed tomography (CBCT), continuous-wave Doppler flowmetry, and electric pulp sensibility testing. Unlike the apexification approach, hDPSC implantation was found to regenerate pulp tissue that contained blood vessels and sensory nerves. CBCT imaging showed that hDPSC implantation resulted in root elongation, increased dentin thickness, and a reduction in the width of the apical foramen. Laser Doppler flowmetry showed that the cell-based approach caused an increase in blood vessel formation and thus pulpal blood flow. Furthermore, their histological studies revealed that the regenerated tissue contained normal dental pulp structures, including an odontoblast layer, connective tissue, and blood vessels. No adverse events were observed after 24 months. In sum, this study demonstrates that a cell-based approach involving implanting stem cells can regenerate the entire pulp and can support root development. These positive results suggest that the cell-based approach may be a viable, promising treatment option for immature necrotic permanent teeth in the future [46].

5.2 Cell Homing Approach

The cell homing concept for dental pulp and/or dentine regeneration was first introduced in 2010 [47]. In modern clinical endodontic practice, despite a number of preclinical research studies on cell-based approaches, the clinical procedures for RET use the cell homing approach. This consists of two distinct cellular processes: recruitment and differentiation. There are four goals that cell homing RET is expected to achieve:

1. Promotion of intracanal angiogenesis
2. Migration of stem cells
3. Regeneration of the pulp; and
4. Dentin mineralization [24]

The cell homing regenerative approach revitalizes and recreates the damaged dentin-pulp tissue of an immature, necrotic permanent tooth, by inducing an intentional bleeding with a file that is used to instrument the area apical to the apex [25]. Nygaard-Ostby was the first to hypothesize that the induction of bleeding by intentional laceration of the periapical tissues produced new blood vessel formation and promoted further apex development [28]. This intentional laceration of the apical papillary tissue provokes intracanal bleeding, and this recruits stem cells from the periradicular tissues into the root canal [48]. The cell homing approach uses one of three kinds of scaffolds to support stem cell proliferation and growth factor storage: an ordinary blood clot [48], platelet-rich plasma (PRP) [49–52] or platelet-rich fibrin (PRF) [52–57].

With the appropriate signaling molecules and scaffolds, this cell homing technique aims to restore the pulp vitality and close the apex with dentin-pulp complex formation. The stem cells are induced to migrate into the disinfected intracanal space by growth factors. Of these stem cells, those that originate from the residual pulp and the apical papilla (DPSCs and SCAP) differentiate into odontoblast-like cells. Then, these cells deposit dentin-pulp tissues using scaffolds as their structural matrices. As a result of dentin mineralization on the canal walls and the root apex, the dentinal walls become thicker, and the root becomes longer. Thus, the immature necrotic tooth that had once had an open apex and thin dentinal walls now become structurally stronger [25, 27, 58].

5.2.1 Problems with Cell Homing

While RET using the cell homing approach is currently practiced by clinicians today, the approach still has some issues, as follows.

First, the stem cells are recruited by the intentional creation of bleeding at the apical tissue [59]. Therefore, there are some degrees of uncertainty regarding the number and type of cells that are recruited by this method [24]. Further, the success rates of inducing bleeding are inconsistent, which makes it even harder to predict the number of recruited cells [60, 61].

Second, different sources of stem cells compete with one another in the cell homing regenerative approach. In other words, the intentional laceration of apical papilla induces bleeding, and as a result, stem cells, including SCAPs, DPSCs, PDLSCs, and other MSCs migrated into the canal. Further, released growth factors from the dentin are not specific to the induction of odontoblast-lineage stem cells only. In effect, many other sources of stem cells migrate, proliferate, and differentiate into their corresponding final structures. The ideal goal of RET is to generate pulp-dentin complex. However, it is difficult with the current technology to stimulate only the odontoblast precursor stem cells. Thus, the final tissues that are generated in the root canal space with the cell homing approach may not be limited to pulp-dentin complex, but instead they may also contain bone-like or cementum-like tissues as a major component [24].

Third, it is difficult clinically to identify the presence of viable apical papilla. In order to successfully generate dentin tissues, recruitment of dentin progenitor cells is required [30]. The main dentin progenitor cells are SCAPs. These are stem cells that are committed to the odontoblast lineage, and they ultimately differentiate to form dentin. SCAPs are located at the apex of a growing tooth. The apical papilla is only present during root development. Immature necrotic teeth usually have accompanying periapical lesions. Due to the radiolucent appearance of periapical lesions, it is hard to differentiate a viable apical papilla from a periapical lesion with current periapical radiography. Therefore, it is uncertain whether viable apical papilla tissues are present at the apex, which is a requirement for SCAPs to be induced to migrate to the canal space for pulp/dentin tissue generation.

Fourth, a number of studies of cell homing RET have concluded from their histological analysis of the regenerated structure that the generated tissues in the canal from this cell homing approach were mainly cementum or bone-like, rather than a normal pulp/dentin complex [24, 62, 63]. In other words, with the cell homing approach, PDLSCs and cementum/bone-like tissue progenitors are more likely to be the migrating cells than odontoblasts and DPSCs. In addition, it is impossible with today's periapical radiography and CBCT technologies to determine whether or not the regenerated tissue in the canal space in the increased and widened root is truly a dentin-pulp complex, which is the ideal goal of RET [25].

Hence, one can conclude that unfortunately, the modern RET technique does not reliably regenerate odontoblasts and the pulp-dentin complex from migrated stem cells from the periapical region. This issue has generated questions and controversy regarding whether RET in this form is a true regenerative process. Some have called the technique "endodontic repair," arguing that it is a reparative process, rather than a regenerative process, since most of the generated tissue is cementum/bone-like [64]. Even when dentin is regenerated, it may resemble tertiary dentin rather than the original primary dentin [24].

However, the primary goal of RET of immature permanent teeth with a necrotic pulp is to eliminate clinical signs and symptoms and to achieve healing of apical periodontitis. Although it is not ideal, wound healing with repair by a tissue that is different from the original pulp/dentin structure may not be considered a clinical treatment failure.

6 Common Challenges in Regenerative Endodontics

Regenerative endodontics has three main challenges:

1. Microbial control (bacterial reduction): adequate microbial control is required for recruitment, viability, and proliferation of stem cells.

2. Adequate growth factor release and stem cell recruitment: specific growth factors are required to be released in order for stem cells to differentiate into odontoblasts.
3. Proper scaffolds for regenerating the dentin-pulp complex.

In order to meet these three goals, a number of clinical procedures have been carried out using different irrigants and intracanal medicaments [65, 66].

6.1 Microbial Control, Growth Factors, and Stem Cell

RET exploits tissue engineering with stem cells with the aim of regenerating the pulp-dentin complex in the root canal space [66]. Bacteria are the major cause of the development of periradicular lesions. Since bacteria in the root canal space will affect the viability of stem cells, it is crucial to perform adequate disinfection, to ensure microbial control before the recruitment of stem cells into the root canal space. However, disinfectants are cytotoxic not only to bacteria, but also to stem cells [66]. In other words, it is important to use a biocompatible disinfection strategy that is bactericidal, while having minimal effects on the viability and proliferative capacity of stem cells [66]. In order to resolve this Goldilocks issue, it is necessary to discover the most appropriate clinical protocol for disinfection for RET [66].

There are three different techniques to perform microbial control in the root canal space: (1) instrumentation, (2) irrigation, and (3) intracanal medication. There have been many studies testing different concentrations of various irrigation solutions, intracanal medications, and techniques to accomplish adequate microbial control [65, 66].

6.1.1 Instrumentation

Mechanical instrumentation is beneficial in the eradication of bacterial biofilms attached to the dentinal walls that are resistant to irrigants and intracanal medicaments. However, it is generally not recommended for RET since instrumentation makes the dentinal wall of the immature permanent teeth even thinner. Therefore, bacterial reduction in RET is mainly accomplished with copious irrigation and the use of intracanal medication. However, since the nature of the biofilm is resistant to irrigation and medication, minimal instrumentation is recommended [67, 68].

Biofilms are microbial, sessile communities of cells that are irreversibly attached and interfaced with one another [69]. They demonstrate altered growth rates and gene transcription. These layers of microbes in the biofilm cannot be removed with irrigation alone. They need to be mechanically disturbed to make them more susceptible to chemical disinfectant irrigant solution [67, 68]. The XP Endo Finisher (XPF; FKG Dentaire SA, La Chaux-de-Fonds, Switzerland) can be used for a minimal instrumentation technique [70]. It disturbs the biofilm in the dentinal wall and therefore promotes the penetration of irrigation fluids [71].

6.1.2 Irrigation in Regenerative Endodontics

Intracanal disinfection with irrigation is crucial in RET because of the need for disinfection via bactericidal and bacteriostatic effects [66]. However, a significant variability of clinical protocols is seen in published cases of RET as this aspect has not been standardized, probably because little is known regarding the effects of various disinfectant irrigants on the viability and differentiation capacity of stem cells [40]. Various disinfection protocols have been introduced, including application of NaOCl (in concentrations ranging from 1 to 6%) [72–74], chlorhexidine (0.12–2%) [55, 75, 76], EDTA (17%) [77–79], and combinations of different irrigants. After a careful review of case reports and in vitro studies, the AAE has recommended 1.5% sodium hypochlorite and 17% EDTA. However, they did not recommend chlorhexidine [29]. These irrigation solutions will be discussed in depth in the following section.

Sodium Hypochlorite

Most published cases of RET have used NaOCl as the primary irrigant solution [40]. From earlier studies, NaOCl is known to have antimicrobial efficacy at concentrations of 1% or greater [80–82]. However, some variability is found in the concentration of NaOCl used, ranging from 1% to as much as 6% [72–74]. Disinfectants may affect the survival and differentiation of stem cells due to their cytotoxic effects [66]. Therefore, it is crucial to find an appropriate NaOCl concentration to be used in RET that is bactericidal, while having minimal effects on stem cells.

Using human extracted teeth, Martin and his colleagues compared various concentrations of NaOCl (0.5, 1.5, 3, and 6%) in terms of their effects on the survival of SCAPs and on the expression of dentin sialophosphoprotein (DSPP) (a marker for dentin formation). In their study, SCAPs were inserted in a hydrogel solution, and placed inside the standardized root canals of extracted human teeth. These root canals were previously irrigated with NaOCl followed by final irrigation with either 17% EDTA or saline. The greatest SCAP survival and DSPP expression were observed in the 1.5% NaOCl and 17% EDTA group. Thus, they concluded that the detrimental effects of NaOCl on the survival and differentiation of SCAPs could be minimized or avoided by using 1.5% NaOCl followed by 17% EDTA [83] (Fig. 1).

Therefore, NaOCl should be used at 1.5% for irrigation in order to promote the survival and proliferation of stem cells [83]. However, this recommendation poses challenges to clinicians since a weaker concentration of the irrigant may be less effective in microbial control [84]. An in vitro study by Siqueira et al. demonstrated that regular exchange and using a higher volume of NaOCl may compensate for using a lower concentration [82]. Thus, frequent replacement of NaOCl and using a copious amount of irrigant is recommended to compensate for using 1.5% concentration, in order to balance the microbial efficacy and cytotoxicity of the irrigant.

Chlorhexidine

Chlorhexidine has also been applied in some clinical protocols for RET due to its antimicrobial activity, and its intrinsic ability to be substantive and thus to be retained and gradually released over a period of time [40, 85]. While chlorhexidine has been used in various concentrations, ranging from 0.12% to 2% [55, 75, 76], its effects on the survival and differentiation of odontoblast-lineage stem cells are questionable.

In order to test the effect of chlorhexidine on the survival of stem cells from the apical papilla, Trevino and his colleagues applied four different irrigation protocols [(1) 17%

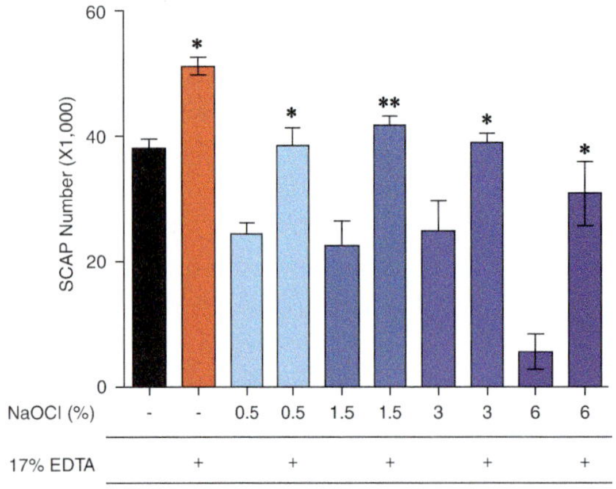

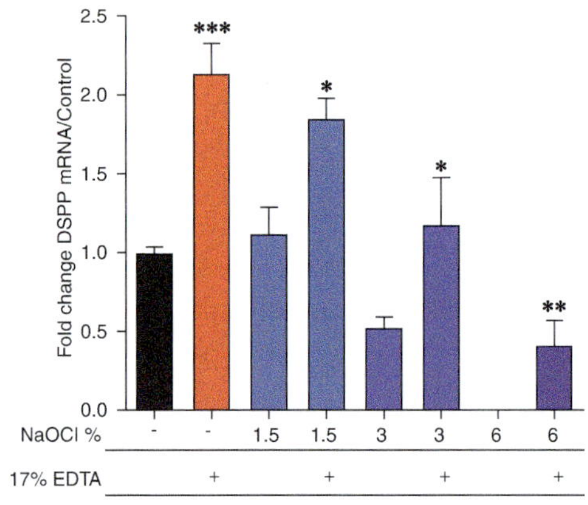

Fig. 1 Effects of NaOCl and EDTA on SCAP. A significant reduction in SCAP survival and DSPP expression was found in the group treated with 6% NaOCl. A final irrigation with 17% EDTA increased SCAP survival and DSPP expression. The groups treated with lower concentrations of NaOCl had comparable cell survival, but the greatest overall cell survival and DSPP expression were observed in the 1.5% NaOCl and 17% EDTA group. Based on reference [83]

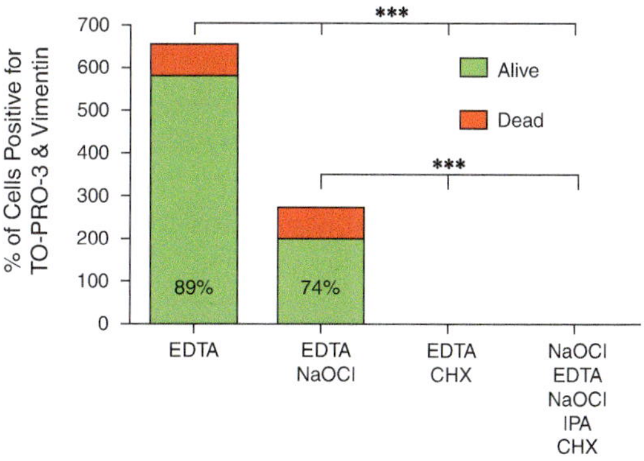

Fig. 2 Effects of chlorhexidine on SCAP. Irrigation with 17% EDTA gave the greatest cell survival, followed by a 6% NaOCl/17% EDTA/6% NaOCl irrigation protocol [66]. The groups that included 2% chlorhexidine in their protocols had no viable cells

EDTA; (2) 6% NaOCl/17% EDTA/6% NaOCl; (3) 17% EDTA/2% chlorhexidine; and (4) 6% NaOCl/17% EDTA/6% NaOCl/isopropyl alcohol/2% CHX] (Fig. 2). They seeded a mixture of SCAPs from immature human third molars and platelet-rich plasma into standardized human root segments that were irrigated with one of the four solutions. Irrigation with 17% EDTA gave the best cell survival, whereas in all protocols that included 2% chlorhexidine there were no viable cells remaining [66]. Therefore, their study suggests that chlorhexidine is not considered a suitable irrigant solution for RET due to its cytotoxicity to stem cells and substantivity. The use of chlorhexidine is not recommended in RET by the AAE [29].

EDTA

In RET, EDTA has is used as an irrigant two purposes, namely smear layer removal, and growth factor release. During the first visit, EDTA is often used as an irrigant because of its ability to clean the inorganic smear layer from the canal walls, and thus allow the medication to penetrate better into the dentinal tubules in the time period between appointments [86–88]. In addition, it is crucial to use an irrigant that has the greatest potential to release growth factors from the dentin into the root canal space in the second visit. As discussed above, 17% EDTA does not have deleterious effects on the survival and differentiation of stem cells [66, 83]. EDTA is used in the second visit as an irrigant to remove the intracanal medicament placed in the first visit, and more importantly, to release growth factors from the dentin.

Growth factors are trapped or become embedded in the dentin matrix during dentinogenesis [89]. They can be reactivated and released later in life by demineralization. Their effects on the migration, proliferation, and differentiation of pulpal stem cells could be beneficial for RET. After disinfection of the root canal with copious irrigation, bleeding is provoked, as a trigger for stem cells to migrate in [48]. For these stem cells to regenerate dentin-pulp tissues, they need to proliferate and differentiate, and this is where growth factors become important [89]. Therefore, using EDTA irrigation to demineralize dentin to release these bound growth factors contributes to successful RET.

A number of studies have shown that EDTA as an irrigant solution can in fact release growth factors from the dentinal walls of the canal. Comparing EDTA to NaOCl, citric acid, and saline, Zhao and his colleagues reported that treatment with EDTA gave the greatest release of transforming growth factor-beta1 (TGF-β1) from dentin [90]. TGF-β1 is an important growth factor sequestered in the dentin matrix, and it stimulates tertiary dentinogenesis.

In order to develop a standardized protocol for growth factor release from human dentin, Galler and her colleagues assessed the effect of irrigation with EDTA on growth factor release and compared it to citric acid, citrate buffer, and citric acid in phosphate buffer [89]. With EDTA treatment, TGF-β1, fibroblast growth factor 2 (FGF-2), and vascular endothelial growth factor (VEGF) were detected. TGF-β1stimulates the chemotaxis [91] and differentiation of stem cells [92, 93], whereas FGF-2 and VEGF increase cell proliferation and angiogenesis, respectively [94, 95]. They reported that EDTA gave the greatest amount of released TGF-β1 [89]. They also compared irrigation with chlorhexidine to irrigation with sodium hypochlorite, before EDTA conditioning. Using chlorhexidine before final irrigation with EDTA gave more TGF-β1 release, whereas sodium hypochlorite decreased this release. From these studies, the AAE has recommended that the use of sodium hypochlorite should be limited to the first visit of RET, due to its deleterious effect on growth factor release [29].

6.1.3 Intracanal Medicaments

Intracanal medicament placement is another method used to achieve microbial control in a root canal space in RET. A wide range of intracanal medicament protocols has been proposed in RET, with an antibiotic combination paste emerging as the primary agent of choice for the intracanal medicament, followed by calcium hydroxide [40]. Commonly used intracanal medicaments are calcium hydroxide [72, 96], triple antibiotic paste (a combination of ciprofloxacin, metronidazole, and minocycline) [72, 75], modified triple antibiotic paste (a combination of metronidazole, ciprofloxacin, and cefaclor) [76, 79], and double antibiotic paste (a combination of metronidazole and ciprofloxacin) [26, 97].

As already discussed for irrigants, intracanal medicaments also need to have the least detrimental effects on stem cells, in order to regenerate dentin-pulp complex. In an attempt to assess the effects of different intracanal medicaments at different concentrations on stem cell survival, Ruparel and his colleagues used SCAP culture assays, to test the effects of calcium hydroxide and four different antibiotic preparations (triple antibiotic paste (TAP), double antibiotic paste (DAP), modified TAP, and amoxicillin with clavulanic acid (Augmentin™)). All four antibiotic preparations considerably reduced SCAP survival in a concentration-dependent fashion. When 1 mg/mL or greater of antibiotic paste was used, some 50% of SCAPs became non-viable. On the other hand, dressing with calcium hydroxide did not decrease SCAP survival at any concentration. Thus, calcium hydroxide or low concentrations of antibiotics should be used as intracanal medicaments for RET [65] (Fig. 3).

Antibiotics should be used at low concentrations as intracanal medicaments, in order to prevent detrimental effects on SCAP survival. However, in order to be an effective intracanal medicament, the antibiotic concentration also needs to be high enough to be effective on bacteria that are in the quiescent biofilm state. In order to evaluate the effect of various antibiotics used in RET on a *Enterococcus faecalis* biofilm grown on dentin, Tagelsir and her colleagues treated the biofilm with different antimicrobial agents, including calcium hydroxide, 500 mg/mL of DAP, 1 mg/mL of DAP, 0.1 mg/mL of DAP, 1.5% sodium hypochlorite, and 2% chlorhexidine solution. Sodium hypochlorite and 500 mg/mL of DAP caused complete inactivation of the bacteria, while comparable but incomplete antimicrobial effects were observed with 2% chlorhexidine, calcium hydroxide, and 1 mg/mL of DAP. These results demonstrate that

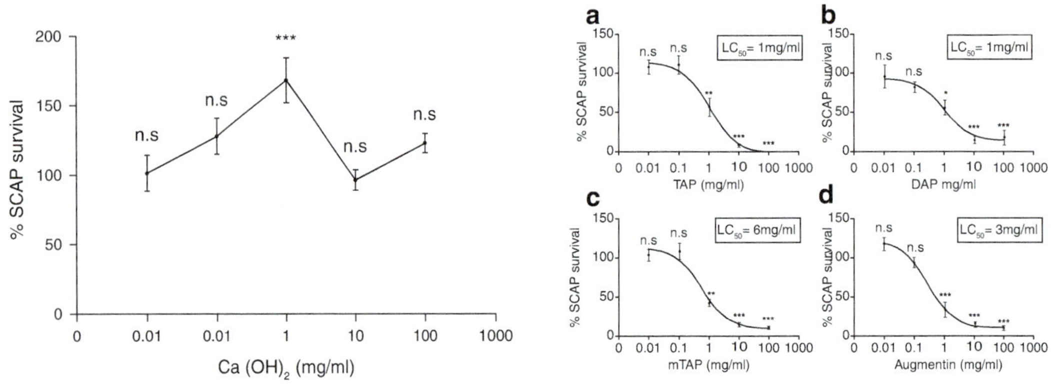

Fig. 3 Percentage of SCAP cells surviving following the exposure to different concentrations of calcium hydroxide, TAP, DAP, mTAP, and Augmentin. Based on Ruparel et al. [65]

if an antibiotic paste is to be used as an intracanal medicament, at least 1 mg/mL is required [98], but caution is needed as this level might be cytotoxic to stem cells.

The antimicrobial dressing also influences the extent of growth factor release from the dentin matrix. Galler and her colleagues tested different intracanal medicaments used with a final irrigation with 10% EDTA and compared their effects on TGF-β1 release [89]. The studied intracanal microbial medicaments included 1% chlorhexidine gel, Ledermix™ (a corticosteroid-antibiotic paste containing demeclocycline), TAP, water-based calcium hydroxide, and oil-based calcium hydroxide. All the tested dressings interfered with TGF-β1 release, except for water-based calcium hydroxide. Thus, their results suggest that calcium hydroxide is a better choice of intracanal medicament in RET. Based on these studies, the AAE recommended calcium hydroxide as the intracanal medicament of choice in RET since it does not produce negative effects on stem cell viability or growth factor release [29].

6.2 Scaffolds

In regenerative endodontic procedures, a blood clot acts as a scaffold for stem cells. The blood clot is formed by intentionally lacerating the periapical tissue and thereby making blood enter the root canal space. However, the induction of intracanal bleeding is not always successful [60, 61]. This brings a significant challenge to clinicians when treating a permanent immature tooth with apical periodontitis. The clinician should refrain from using vasoconstrictor-containing local anesthetics during the final visit, so as to minimize challenges associated with achieving adequate bleeding into the tooth. Mepivacaine without epinephrine is commonly recommended for the second appointment of RET.

However, if the clinician faces problems inducing bleeding into the root canal, there are several ways to overcome this problem. First, the failure may be because the periapical tissue has been severely destroyed [25]. In this situation, Kim and his colleagues advise postponing the procedure until the periapical tissue has recovered.

A second method when bleeding induction is not successful is to withdraw blood from the patient by venipuncture, and use this to prepare platelet-rich plasma (PRP) [50, 51], platelet-rich fibrin (PRF) [56, 57], or a platelet pellet (PP) [99, 100]. PRP, PRF, and PP are concentrated sources of platelets [101]. Since PRP has a five times greater concentration of platelets than blood [102, 103], it should bring a greater concentration of growth factors for RET [104, 105]. PRF has a similar concentration of platelets as PRP, but its polymerization only contains endogenous components [106]. Therefore, it provides a better fibrin network for growth factor reservation and cell migration. PP has better adhesive properties due to its gel consistency [107], and it has been used for this reason in regenerative periodontics [99, 100]. Using autologous platelet concentrates, PRP, PRF, and PP can yield similar clinical and radiographic outcomes for RET as an induced blood clot. However, this method does not have a high acceptance rate in younger patients since it requires drawing intravenous blood [101].

Another method that is used to overcome a failure of bleeding provocation is the use of the XP Endo Finisher (FKG Dentaire SA, La Chaux-de-Fonds, Switzerland). This instrument is normally used for cleaning the root canal walls and for irrigation activation [108]. Having high flexibility with no taper, it simply touches the canal walls and removes debris and smear layer without damaging dentin. When the induction of periapical bleeding has failed, the XP Finisher can be pushed past the apex in order to provoke bleeding by further lacerating the apical tissue. However, the risk of instrument separation should be taken into consideration, and hence this instrument should be used gently with minimum apical pressure.

After successful induction of bleeding into the root canal system, Emdogain™ is sometimes injected as a scaffold. Emdogain is an enamel matrix derivative (EMD), which is a protein extract comprising amelogenins [109]. It provides an excellent scaffold for RET since it has high levels of growth factors such as TGF- β1 and BMP-2 [110, 111]. It has also been reported to reduce the size of periapical lesions and to promote root elongation and increased root thickness [112].

7 AAE Guideline on Irrigation and Intracanal Medicament Protocol

As discussed previously, successful regenerative endodontic treatment is based on root canal space disinfection, preservation and recruitment of undifferentiated mesenchymal stem cells, the release of required growth factors, a stable scaffold, an adequate coronal barrier, and a suitable long-term restoration. There are multiple methods and protocols available for achieving a successful outcome in RET, with over 200 published cases of RET in the literature. Interestingly, these published cases show a tremendous heterogeneity in the clinical protocols that have been used. With this high variability in protocols, it was crucial for a specific, evidence-based, and standardized guideline for microbial control in RET to be created. Using the evidence from recently published studies, the American Association of Endodontists (AAE) announced its "Clinical Considerations for a Regenerative Procedure" as a standardized clinical protocol for endodontic regeneration therapy [29]. However, the AAE further recommended that this guideline needs to be considered as only one possible available source and that given the rapid rate of growth in new information in regenerative technologies, clinicians should keep themselves updated and actively review newly published evidence.

First Visit

- The first step is obtaining the informed consent and reviewing the risks and alternative treatments (including the no treatment option) with the patient or his/her guardians.
- Local anesthesia should be administered, followed by isolation of the tooth with a dental dam, and access cavity preparation.
- Copious and gentle irrigation of the root canal system with 20 mL of sodium hypochlorite (NaOCl) is undertaken using an irrigation system that minimizes the risk of extrusion risk of irrigants into the periradicular space (e.g., closed-end or side-vented needles, or EndoVac™). Lower concentrations of NaOCl are recommended [1.5% NaOCl (20 mL/canal, 5 min)]. The next step is irrigation with normal saline or 17% EDTA (20 mL/canal, 5 min), with the needle positioned 1 mm short of the root end, to minimize cytotoxicity to the stem cells that are present in the apical tissues.
- Canals should then be dried with sterile paper points.
- A low concentration of an antibiotic paste or calcium hydroxide should be placed. If TAP is used: (1) Sealing of the pulp chamber with a dentin bonding agent should be considered [to reduce the risk of discoloration of the root] and (2) 1:1:1 ciprofloxacin: metronidazole: minocycline to a final concentration of 1–5 mg/ml should be mixed. Minocycline in TAP can be associated with tooth staining or discoloration. DAP without minocycline or a modified TAP with substitution of minocycline with another antibiotic (e.g., clindamycin; amoxicillin; cefaclor) are other possible alternative intracanal medicaments. Clinicians should be aware that several studies have been published using higher concentrations of TAP/DAP, but a recommendation to use a higher concentration cannot be made at this time due to limited published evidence.
- The intracanal medicament should be delivered into the root canal system via a syringe.
- If TAP is used, it must remain below the cementoenamel junction (CEJ) in order to reduce the risk of crown discoloration.
- The access cavity should be sealed with 3–4 mm of a suitable temporary restorative material such as Cavit™, IRM™, or a glass ionomer cement.
- The patient should be dismissed and reappointed to be seen again after 1–4 weeks.

Second Visit (1–4 Weeks After the First Visit)

- The effect of the initial treatment should be assessed. If signs or symptoms of persistent infection are still present, additional treatment, including replacing the antimicrobial mixture or using alternative antimicrobial agents should be considered.
- In this session, local anesthetic agents without a vasoconstrictor are to be used, such as 3% mepivacaine. Once again, the tooth is isolated with a dental dam.
- The interim restoration is removed.
- The paste in the canal is removed by copious and gentle irrigation with 20 mL of 17% EDTA.
- The canal should then be dried with paper points.
- At this point, bleeding into the root canal system should be induced by over-instrumentation. Hemorrhage can be created by passing a precurved K-file 2 mm beyond the apical foramen and using it in a turning motion. The entire canal space should be filled with blood to the level of the cementoenamel junction. Note that bleeding should be stopped at a level that allows placement of 3–4 mm of restorative material.
- A resorbable matrix such as CollaPlug™, Collacote™, or CollaTape™ should be placed over the blood clot followed by white MTA as a capping material.
- A 3–4 mm layer of a glass ionomer cement (e.g., Fuji II LC™, GC America, Alsip, IL, USA) with 3–4 mm thickness should be placed gently over the capping material and light-cured for 40 seconds.
- Since MTA has been associated with staining of coronal tooth structure, using alternatives to MTA should be considered in teeth in the esthetic zone.

  – For anterior teeth and premolars, the use of Collatape/Collaplug and restoration with 3 mm of a bioceramic material (BC) or a resin-modified glass ionomer cement (RMGIC), followed by a resin composite restoration, bonding to beveled enamel margins, should be considered.
  – For molars or for teeth with porcelain fused to metal (PFM) crowns, use of Collatape/Collaplug and restoration with 3 mm of MTA, followed by RMGIC or a silver amalgam alloy, should be considered.

One important step mentioned in most published case reports and guidelines is the coronal placement of a bioactive "coronal plug" using a material with excellent sealing and pro-mineralization properties. MTA [113], Biodentine [114], and BioCeramic Putty [115] have been used and have demonstrated to promote the differentiation of MSCs into an odontoblast-like phenotype while allowing for the greater proliferation of MSCs (Figs. 4, 5 and 6).

8 Can RET Be Performed in All Immature Permanent Teeth with a Necrotic Pulp?

The primary aim of RET is to eliminate clinical signs and symptoms and achieve healing of apical periodontitis. This procedure has been used to manage immature teeth with pulpal necrosis. However, some immature teeth with necrotic pulp may benefit more from RET, while others may be suitable for both MTA apical plug apexification and RET.

Cvek has proposed 5 stages of root development: stage 1, an open apex with root formation less than 1/2; stage 2, an open apex with 1/2 root formation; stage 3, an open apex with 2/3 of root development; stage 4, an open apex with nearly completed root formation; and stage 5, a completed mature root [14].

It is recommended that in immature permanent teeth with a necrotic pulp at stages 1, 2, and 3, RET should be considered as the first option in treatment planning due to the short root, wide-open apex, and thin canal walls. In these stages, apexification is not recommended since the tooth has no potential for root maturation, root canal wall thickening, and/or continued root development.

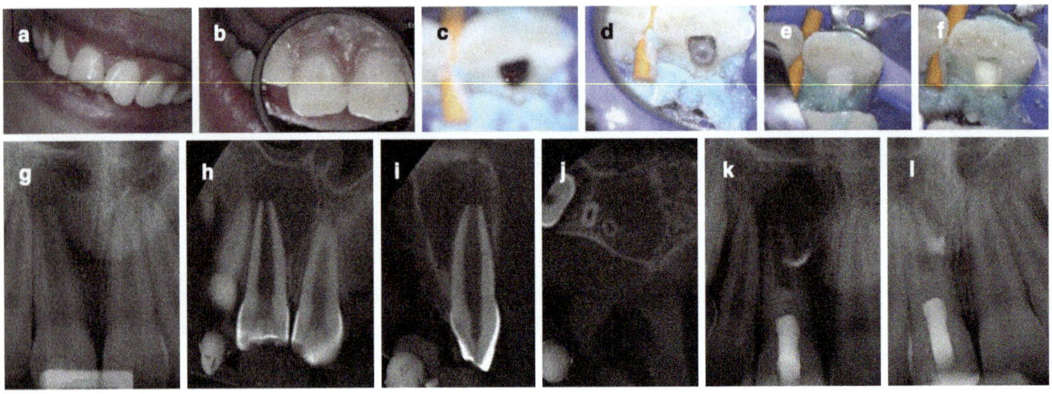

Fig. 4 Case 1 (**a**)–(**b**) A 11-year-old male, ASA I, presented for treatment of tooth #8 with a history of trauma 2 years ago. The tooth was necrotic and had asymptomatic apical periodontitis. (**c**) During the second visit, bleeding was induced by passing a #30 hand file past the apex. (**d**) Emdogain was injected inside the canal as a scaffold. (**e**) Biodentine was placed in the coronal part of the canal. (**f**) Resin composite was placed as the final restoration. (**g**) A preoperative periapical radiograph shows the tooth with an open apex and a periapical lesion. (**h**)–(**j**) Different views of pre-op CBCT. (**k**) The intraoperative periapical radiograph shows that Biodentine has been placed in the coronal third part of the canal. (**l**) The postoperative periapical radiograph shows healing of periapical lesion after 6 months. The photograph and radiograph are courtesy of Dr. Elham Shadmehr

However, in immature permanent teeth at stage 4, with a nearly completed root, both options including an apical MTA plug and root canal filling or RET are viable, because apical sealing is not as challenging. On the other hand, if the tooth requires a post for adequate coronal restoration, it is not a good candidate for RET, and it can be treated more efficiently with an apical MTA plug and a root canal filling.

9 Treatment Outcome

A successful outcome in non-surgical endodontic treatment is defined typically by a lack of symptoms and the absence of radiographic signs of apical periodontitis following the treatment. Unlike conventional approaches, RET can further provide root maturation and development, thickening of dentinal walls, and positive tooth vitality.

Based on the AAE guidelines, a successful outcome after RET can be achieved at three levels:

- *Primary level*: To eliminate symptoms and achieve evidence of bony healing.
- *Secondary level*: To increase root wall thickness and/or root length (desirable, but perhaps not essential).
- *Tertiary level*: To achieve a positive result to pulp sensibility/pulp vitality tests (which if achieved, could indicate a more organized vital pulp tissue has formed).

Recent systematic reviews indicate that the success rates of RET are in the range of 91–94% for resolution of periapical pathology, which is the AAE's primary level outcome [116, 117]. However, the success rates for achieving the secondary level outcomes are more variable, with 80% for increased root development, and 76% for apical closure [116]. Many different studies and systematic reviews have confirmed that RET and MTA apexification is similarly effective in achieving successful primary outcomes for the treatment of immature teeth with pulp necrosis [117, 118].

Diogenes et al. have recommended evaluating the outcomes on three different levels, including patient-centered outcomes, clinician-centered outcomes, and scientist-centered outcomes [3].

Patient-Centered Outcomes These outcomes directly influence the wellbeing of the patient. In

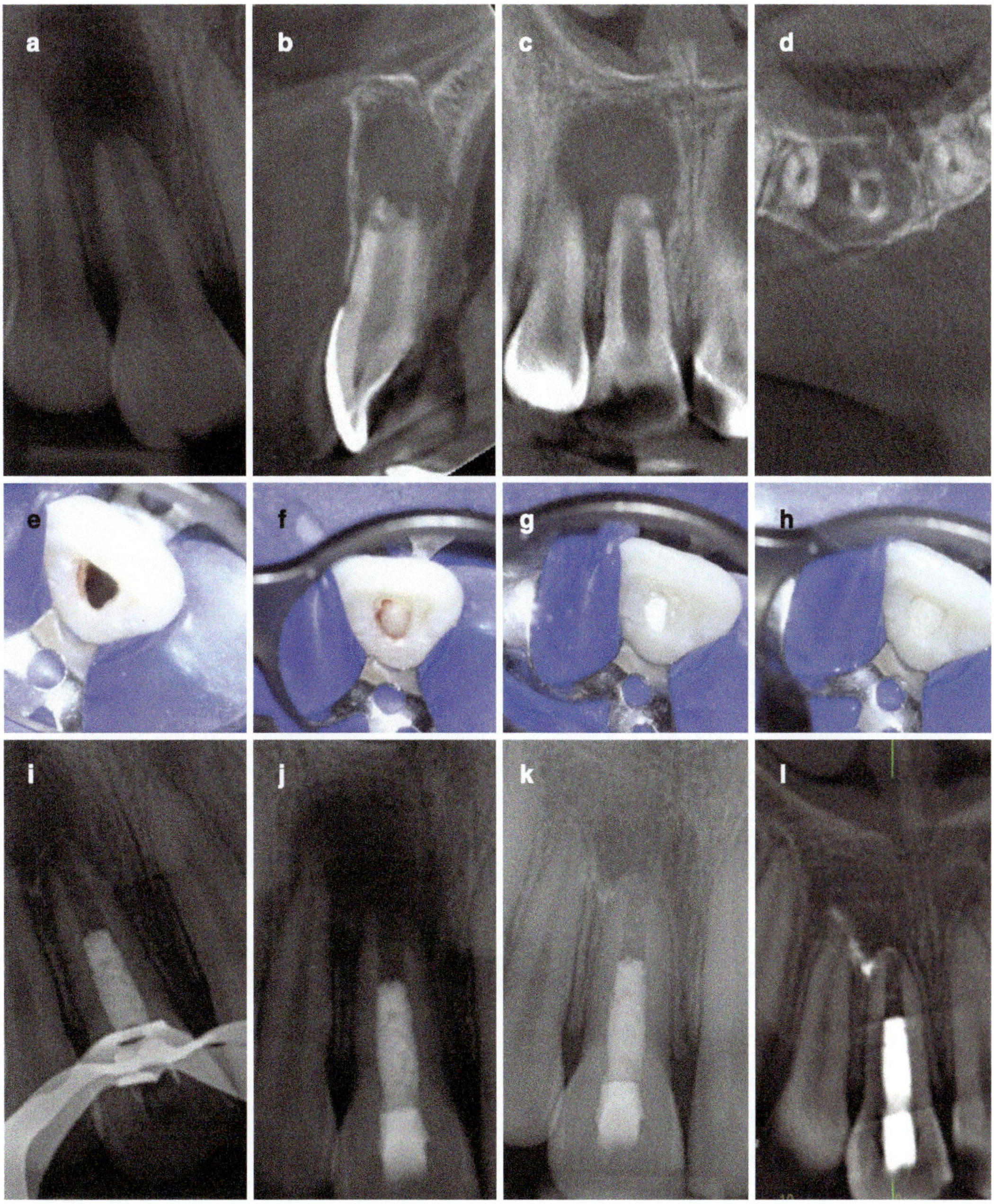

Fig. 5 Case 2 (**a**) A 12-year-old male, ASA I, presented for tooth 11 (#8) with a crown fracture. The tooth was necrotic and had symptomatic apical periodontitis. (**b**)–(**d**) Different views of CBCT show the incisor with an open apex, a periapical lesion, and resorption at the apical third of the root. (**e**) During the second visit, calcium hydroxide was removed and bleeding was induced to the level of CEJ by a hand file. (**f**) Emdogain was placed as a scaffold. (**g**) BC putty was used for a coronal seal. (**h**) Coronal restoration was completed with Fuji II and composite. (**i**) The PA radiograph was taken after Emdogain and BC putty had been placed. (**j**) The post-op radiograph shows Emdogain, BC putty, and coronal restoration with glass ionomer and composite. (**k**) A PA radiograph was taken at a 6-month follow-up visit. An apparent root growth is shown on the PA and incomplete healing of the periapical lesion. Tooth was asymptomatic. (**l**) At the 2-year follow-up visit, the tooth was asymptomatic with absence of a deep pocket or sensitivity to percussion or palpation. CBCT shows that the lesion has been healed. While some level of apex closure is seen, no change in the dentinal wall thickness is detected. Photographs and radiographs are courtesy of Dr. Elham Shadmehr

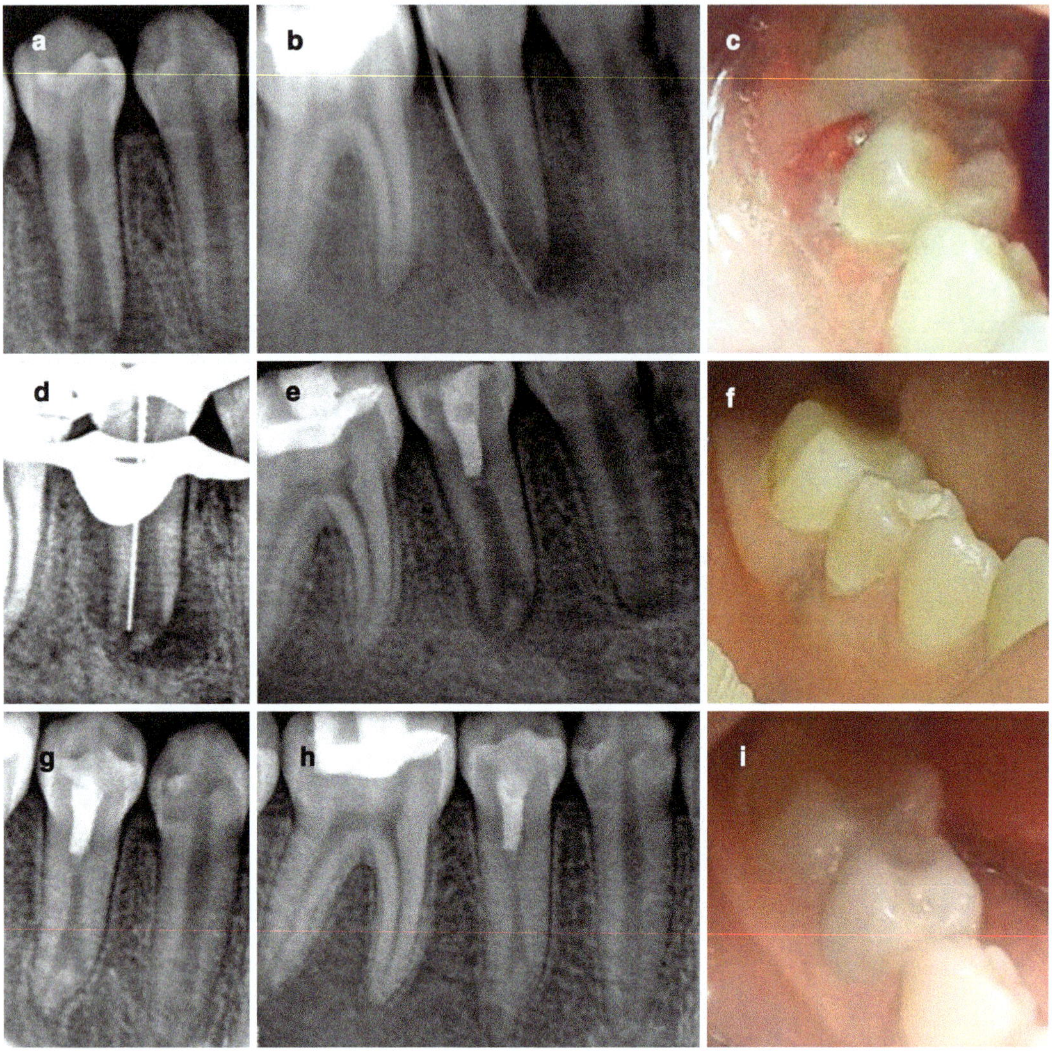

Fig. 6 Case 3 (**a**)–(**c**) A 11-year-old female, ASA I, presented for treatment of tooth 45 (#29) with the presence of a sinus tract (**c**). The tooth was necrotic and had a chronic apical abscess due to the presence of dens evaginatus. (**d**) During the first visit, the working length was determined, and a minimal mechanical instrumentation was done with copious irrigation using 1.5% sodium hypochlorite. A creamy paste of TAP (ciprofloxacin, metronidazole, and minocycline) was placed for 3 weeks. In the second visit, the paste was removed, and the bleeding was induced by passing an ISO #20 hand file past the apex. (**e**), (**f**) 4 mm of MTA was placed in the coronal part of the canal and tooth was temporized with glass ionomer. The temporary filling was replaced with a resin composite final restoration at a separate appointment (**i**). (**g**) This periapical radiograph shows healing of the periapical lesion and the evidence of apical maturation and dentin wall thickening after 12 months. (**h**) At the 18-month follow-up, the apex is fully formed, and root elongation and thickening of the root canal walls are more pronounced. Clinically, the tooth is asymptomatic and sensibility testing shows a mild positive response to cold. The photographs and radiographs are courtesy of Dr. Azar Heydari

other words, they capture the patient's concerns such as pain and tooth discoloration in the esthetic zone. Both patients and guardians appreciate the fact that the tooth is rendered pain-free, and that it remains functional and survives in the long term. From the patient's perspective, an ideal treatment is one that prolongs the life of a functional asymptomatic tooth. In a systematic review, Torabinejad et al. showed a survival rate of 97.8% for RET [117].

Clinician-Centered Outcomes In addition to the primary goal which is to enhance healing of the diseased tissue and also to ensure the patient's wellbeing (patient-centered outcomes), clinicians further assess the outcome by evaluating if the treatment has provided the added benefit of continued root development and increased thickness of the root canal walls, as manifested on radiographs. Reports of continued root development are often subjective, and measurement of increased root width and root length by non-standardized radiographs is prone to error. However, a novel standardized method has been introduced and modified to measure radiographic changes after the treatment of immature teeth [119, 120]. This technique measures changes in the radiographic root area (RRA). Cases treated with RET showed a 31.6% increase in RRA, whereas cases treated with apexification did not show any such increase [120]. Overall, studies using quantitative analyses have presented evidence that RET increases radiographic root length in many cases, but not in all cases. The application of cone-beam computed tomographic (CBCT) imaging (also called cone-beam volumetric tomography (CBVT)) is a useful tool for assessing the outcomes of RET. Its reliability and accuracy have been confirmed [121]. CBCT quantitative measurements are more reliable and accurate than periapical radiographs for outcome assessment [122].

A positive nociceptive response is also one of the secondary goals of RET. Positive pulpal responses have been reported in almost 50% of all published case reports on RET [123]. Positive responses to an electric pulp tester (EPT) are more commonly reported than responses to cold testing. Achieving a positive response to either EPT or to a cold test may suggest the presence of a more structurally organized functioning tissue inside the root canal space. This is a desirable goal, because normal nociception is a protective mechanism. However, a lack of response to the sensibility test should not be interpreted as the failure of the overall treatment.

Scientist-Centered Outcomes This level of the outcome can be achieved when histologic evidence of complete regeneration is seen. Interestingly, animal studies as well as histological evaluation of teeth that were previously treated with RET and extracted later due to fractures or recurring trauma have demonstrated that the newly formed tissue inside the canal space is not the same as normal dental pulp tissue. Different histological studies have demonstrated the presence of loose connective tissues that are vascularized and innervated, and which show some characteristics of a pulp-like tissue. However, the presence of ectopic calcifications such as bone islands within the canal lumen and mineralized tissue resembling cementum along the dentinal walls suggest that the approaches used currently do not achieve complete lineage-specific control of stem cell differentiation. Scientist-centered outcomes can be a strong driver of further advances in the field, as more is known regarding the complex interplay between stem cells, scaffolds, growth factors, and inflammation, which have to be optimized in order to achieve all desired outcomes.

10 Levels of Evidence for Regenerative Endodontics

A recent publication by Shamszadeh et al. reported that some 694 articles had been published in the field of regenerative endodontics up until 2017 [124]. The overall percentage of studies with the highest level of evidence (LOE 1) was low, contributing only 8.34% of all publications. Most published articles were either in vitro or ex vivo (n = 342, 49.27%). Only a few randomized clinical trials (RCT) with short follow-up periods are available [51, 125–127]. The relative distribution of published articles within the pyramid of evidence is presented in Fig. 7. All researchers are encouraged to focus their efforts on performing high-quality research to address the clinical need for better evidence.

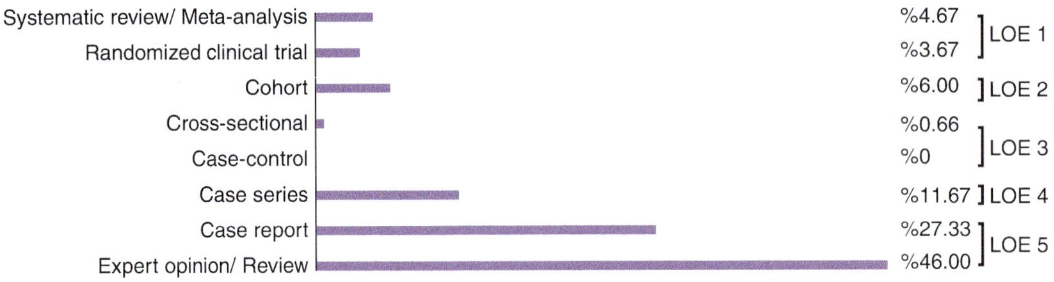

Fig. 7 The relative distribution of articles on regenerative endodontics based on the level of evidence (LOE 1–5). In vitro, in vivo and ex vivo studies are not included [124]

11 Impact of Apical Foramen Size on Treatment Outcome

It was once thought that if the size of the apical foramen was smaller than 1 mm, the chance of revascularization was lower, and the outcome was unpredictable [128]. However, animal studies and clinical trials on RET did not confirm this view. A clinical study showed that regenerative treatment was successful in cases with an apical diameter of only 0.5 mm or larger, however, immature permanent teeth with apical diameter larger than 1 mm show a higher chance of root maturation [129]. Some recent studies have confirmed the feasibility of the regenerative method even when used for cases with mature apices and an apical foramen size of only 0.3 mm [130, 131]. Nevertheless, enlargement of the apical foramen size may facilitate or accelerate the ingrowth of new tissue into the root canal space from the periradicular tissues after RET of mature permanent teeth.

12 Side Effects of RET

Tooth discoloration has been an important and common adverse event associated with RET [60]. This problem is more important for anterior teeth as appearance and esthetics are patient-centered outcomes. Crown discoloration or staining is mainly associated with minocycline which is found in TAP paste [72], and with MTA [132]. To eliminate the risk of discoloration, TAP should be replaced by DAP or calcium hydroxide, and MTA can be replaced with Biodentine™ or another non-staining bioceramic material [133, 134].

Another important side effect of RET is intracanal calcification. This may occur in up to 62% of cases [135]. The main concern is about how to treat root canal calcification of immature permanent teeth when revascularization has failed. Application of the dental operating microscope (DOM), using fine ultrasonic tips and long-shank burs and CBCT can be helpful in localizing and opening cases with calcification [136–138].

Crown-root fracture is another possible complication [139]. In addition, direct failures of the treatment may occur. They may be caused primarily by inadequate removal of bacterial biofilm, possibly due to inadequate irrigation or minimal mechanical instrumentation [96, 140]. Failure may also be due to recontamination of the root canal system [141], from failed restorations that allowed coronal leakage to occur.

13 Follow-Up

To date, most published articles on RET in the endodontic literature have had short-term follow-up periods, since RET is still considered a relatively new treatment approach. The necessity for long-term follow-up seems even more important in RET than with conventional nonsurgical root canal treatments, especially because the regenerated tissue formed in the root canal after RET is more like periodontal tissue rather than dental pulp tissue, according to multiple histologic studies. It is unclear how periodontal tissue in the root canal will behave biologically over the long term. In successful cases, by the second appointment, there should be resolution of problems such as pain, soft tissue swelling, and sinus tracts. Radiographic healing of apical periodontitis is typically seen 12 months after treatment, followed by thickening of root canal walls and the root elongation respectively at 12–24 months after the treatment.

14 Future of Regenerative Endodontics: What Can We Improve?

14.1 Microbial Control

14.1.1 Irrigation Activation

Performing adequate disinfection is paramount for achieving microbial control, as it is a prerequisite for enabling recruitment of stem cells into the root canal space and for maintaining the viability and proliferative capacity of these stem cells [66]. In order to achieve this goal, as discussed earlier, studies have tested various concentrations of different irrigant solutions, and in 2018, the AAE announced its guideline for clinicians. A solution of 1.5% NaOCl is efficient for bacterial reduction and safe for stem cell viability and differentiation. However, previous studies have reported that success rates for the complete elimination of bacteria after the cleaning of the canal are low [142–146].

In order to further increase the efficiency of irrigation, numerous irrigation fluid activation techniques have been introduced, including sonic, ultrasonic, negative pressure irrigation, and laser activation [147, 148]. Four irrigation systems are used commonly in RET [147]: (1) standard needle irrigation; (2) EndoActivator; (3) XP Endo Finisher; (4) laser activation using photon-induced photoacoustic streaming (PIPS).

Standard Needle Irrigation

Standard needle irrigation uses a 30-gauge side-vented needle, placed within 2 mm from the working length [147]. In order to prevent the needle from becoming locked in the canal, it should be moved in a vertical motion. Using this conventional method, irrigation may not be very effective at the apical part of the canal [149]. In addition, the positive pressure used can lead to extrusion of the irrigant, [150] with an associated risk of postoperative pain and tissue damage.

EndoActivator

The EndoActivator (Dentsply Tulsa Dental Specialties, Tulsa, OK, USA) is a sonic handpiece which generates a mechanical oscillation and agitates the irrigant solution vigorously [147]. It uses flexible polymer tips that come in three different sizes. Agitation occurs at 2000–10,000 cycles/min. It is usually used after cleaning and shaping of the canal, to activate the irrigant [151]. Due to its design, it has been reported to allow a safer irrigant activation [151]. By agitating the irrigant solution within the canal, the EndoActivator effectively removes debris and smear layer and reduces the bacterial load [147, 152].

XP Endo Finisher

The XP Endo Finisher (FKG Dentaire SA, La Chaux-de-Fonds, Switzerland) is an endodontic file that is used to disturb the bacterial biofilm in the final disinfection step [147]. Having no taper and being highly flexible, it simply touches the walls and removes debris and smear layer without damaging the dentinal walls of the root canal [108]. By mechanically disturbing the biofilm and agitating the irrigant solution, XP Endo Finisher can improve bacterial elimination [147, 152].

PIPS

Photon-induced photoacoustic streaming (PIPS) is another irrigation technique to enhance the disinfection of the root canal system [153–155]. Using an erbium:yttrium-aluminum-garnet laser, middle infrared laser pulses cause cavitation and transfer energy to the irrigant solution. Rapid and powerful shock waves are generated [156, 157], causing three-dimensional movement of the irrigant solution, and forcing it to penetrate through the entire root canal system [157]. PIPS is considered a more efficient irrigation technique than standard needle irrigation due to its superior ability to reduce the microbial load [147, 153, 157]. SWEEPS™ is a more recent variation on the same PIPS laser agitation concept.

15 Conclusion

In recent years, regenerative endodontics has developed in response to existing limitations of conventional approaches. It provides a new approach for treating immature permanent teeth with necrotic pulps. The most widely used approach is cell homing, and this approach is supported by an extensive literature and by guidelines from the AAE. Future studies should focus on long-term outcome assessment for regenerative endodontic treatments.

References

1. Bhaskar SN, Orban BJ. Orban's Oral histology and embryology. Mosby Year Book: St. Louis; 1991.
2. Torneck CD. Effects and clinical significance of trauma to the developing permanent dentition. Dent Clin N Am. 1982;26(3):481–504.
3. Diogenes A, Ruparel NB, Shiloah Y, Hargreaves KM. Regenerative endodontics: a way forward. J Am Dent Assoc. 2016;147(5):372–80. https://doi.org/10.1016/j.adaj.2016.01.009.
4. Flanagan TA. What can cause the pulps of immature, permanent teeth with open apices to become necrotic and what treatment options are available for these teeth. Aust Endod J. 2014;40(3):95–100. https://doi.org/10.1111/aej.12087.
5. Trope M. Treatment of the immature tooth with a non-vital pulp and apical periodontitis. Dent Clin N Am. 2010;54(2):313–24. https://doi.org/10.1016/j.cden.2009.12.006.
6. Frank AL. Therapy for the divergent pulpless tooth by continued apical formation. J Am Dent Assoc. 1966;72(1):87–93. https://doi.org/10.14219/jada.archive.1966.0017.
7. Heithersay GS. Calcium hydroxide in the treatment of pulpless teeth with associated pathology. J Br Endod Soc. 1975;8(2):74–93. https://doi.org/10.1111/j.1365-2591.1975.tb01000.x.
8. Rafter M. Apexification: a review. Dent Traumatol. 2005;21(1):1–8. https://doi.org/10.1111/j.1600-9657.2004.00284.x.
9. American Association of Endodontists. Glossary of Endodontic Terms. 10th ed. Chicago: American Association of Endodontists. 2020. https://www.aae.org/specialty/clinical-resources/glossary-endodontic-terms. Accessed 6 June 2020.
10. Parirokh M, Torabinejad M, Dummer PMH. Mineral trioxide aggregate and other bioactive endodontic cements: an updated overview – part I: vital pulp therapy. Int Endod J. 2018;51(2):177–205. https://doi.org/10.1111/iej.12841.
11. Maroto M, Barbería E, Planells P, Vera V. Treatment of a non-vital immature incisor with mineral trioxide aggregate (MTA). Dent Traumatol. 2003;19(3):165–9. https://doi.org/10.1034/j.1600-9657.2003.00106.x.
12. Tran D, He J, Glickman GN, Woodmansey KF. Comparative analysis of calcium silicate-based root filling materials using an open apex model. J Endod. 2016;42(4):654–8. https://doi.org/10.1016/j.joen.2016.01.015.
13. Kerekes K, Heide S, Jacobsen I. Follow-up examination of endodontic treatment in traumatized juvenile incisors. J Endod. 1980;6(9):744–8. https://doi.org/10.1016/S0099-2399(80)80186-5.
14. Cvek M. Prognosis of luxated non-vital maxillary incisors treated with calcium hydroxide and filled with gutta-percha. A retrospective clinical study. Endod Dent Traumatol. 1992;8(2):45–55. https://doi.org/10.1111/j.1600-9657.1992.tb00228.x.
15. Finucane D, Kinirons MJ. Non-vital immature permanent incisors: factors that may influence treatment outcome. Endod Dent Traumatol. 1999;15(6):273–7. https://doi.org/10.1111/j.1600-9657.1999.tb00787.x.
16. Kleier DJ, Barr ES. A study of endodontically apexified teeth. Endod Dent Traumatol. 1991;7(3):112–7. https://doi.org/10.1111/j.1600-9657.1991.tb00194.x.
17. Sheehy EC, Roberts GJ. Use of calcium hydroxide for apical barrier formation and healing in non-vital immature permanent teeth: a review. Br Dent J. 1997;183(7):241–6. https://doi.org/10.1038/sj.bdj.4809477.
18. Andreasen JO, Farik B, Munksgaard EC. Long-term calcium hydroxide as a root canal dressing may increase risk of root fracture. Dent Traumatol. 2002;18(3):134–7. https://doi.org/10.1034/j.1600-9657.2002.00097.x.
19. Baldassari-Cruz LA, Walton RE, Johnson WT. Scanning electron microscopy and histologic analysis of an apexification "cap": a case report. Oral Surg Oral Med Oral Pathol Oral Radiol Endod. 1998;86(4):465–8. https://doi.org/10.1016/s1079-2104(98)90375-4.
20. Walia T, Chawla HS, Gauba K. Management of wide open apices in non-vital permanent teeth with Ca(OH)2 paste. J Clin Pediatr Dent. 2000;25(1):51–6. https://doi.org/10.17796/jcpd.25.1.n224g827014n02n2.
21. Giuliani V, Baccetti T, Pace R, Pagavino G. The use of MTA in teeth with necrotic pulps and open apices. Dent Traumatol. 2002;18(4):217–21. https://doi.org/10.1034/j.1600-9657.2002.02107.x.
22. Shabahang S, Torabinejad M. Treatment of teeth with open apices using mineral trioxide aggregate. Pract Periodontics Aesthet Dent. 2000;12(3):315–20.

23. Witherspoon DE, Ham K. One-visit apexification: technique for inducing root-end barrier formation in apical closures. Pract Proced Aesthet Dent. 2001;13(6):455–60.
24. Huang GT, Garcia-Godoy F. Missing concepts in de novo pulp regeneration. J Dent Res. 2014;93(8):717–24. https://doi.org/10.1177/0022034514537829.
25. Kim SG, Malek M, Sigurdsson A, Lin LM, Kahler B. Regenerative endodontics: a comprehensive review. Int Endod J. 2018;51(12):1367–88. https://doi.org/10.1111/iej.12954.
26. Iwaya SI, Ikawa M, Kubota M. Revascularization of an immature permanent tooth with apical periodontitis and sinus tract. Dent Traumatol. 2001;17(4):185–7. https://doi.org/10.1034/j.1600-9657.2001.017004185.x.
27. Banchs F, Trope M. Revascularization of immature permanent teeth with apical periodontitis: new treatment protocol? J Endod. 2004;30(4):196–200. https://doi.org/10.1097/00004770-200404000-00003.
28. Ostby BN. The role of the blood clot in endodontic therapy. An experimental histologic study. Acta Odontol Scand. 1961;19:324–53.
29. American Association of Endodontists. AAE Clinical Considerations for a Regenerative Procedure. American Association of Endodontists. 2018. https://www.aae.org/specialty/wp-content/uploads/sites/2/2018/06/ConsiderationsForRegEndo_AsOfApril2018.pdf. Accessed 6 June 2020.
30. Huang GT, Yamaza T, Shea LD, Djouad F, Kuhn NZ, Tuan RS, et al. Stem/progenitor cell-mediated de novo regeneration of dental pulp with newly deposited continuous layer of dentin in an in vivo model. Tissue Eng Part A. 2010;16(2):605–15. https://doi.org/10.1089/ten.TEA.2009.0518.
31. Galler KM, Krastl G, Simon S, Van Gorp G, Meschi N, Vahedi B, et al. European society of endodontology position statement: revitalization procedures. Int Endod J. 2016;49(8):717–23. https://doi.org/10.1111/iej.12629.
32. Reynolds K, Johnson JD, Cohenca N. Pulp revascularization of necrotic bilateral bicuspids using a modified novel technique to eliminate potential coronal discolouration: a case report. Int Endod J. 2009;42(1):84–92. https://doi.org/10.1111/j.1365-2591.2008.01467.x.
33. Thomson A, Kahler B. Regenerative endodontics—biologically-based treatment for immature permanent teeth: a case report and review of the literature. Aust Dent J. 2010;55(4):446–52. https://doi.org/10.1111/j.1834-7819.2010.01268.x.
34. Hahn CL, Liewehr FR. Innate immune responses of the dental pulp to caries. J Endod. 2007;33(6):643–51. https://doi.org/10.1016/j.joen.2007.01.001.
35. Kaufmann SH, Schaible UE. Antigen presentation and recognition in bacterial infections. Curr Opin Immunol. 2005;17(1):79–87. https://doi.org/10.1016/j.coi.2004.12.004.
36. Steinman RM. The dendritic cell system and its role in immunogenicity. Annu Rev Immunol. 1991;9:271–96. https://doi.org/10.1146/annurev.iy.09.040191.001415.
37. Waite JC, Leiner I, Lauer P, Rae CS, Barbet G, Zheng H, et al. Dynamic imaging of the effector immune response to listeria infection in vivo. PLoS Pathog. 2011;7(3):e1001326. https://doi.org/10.1371/journal.ppat.1001326.
38. Janeway CA Jr, Travers P, Walport M. Immunobiology: the immune system in health and disease. 5th ed. New York: Garland Science; 2001.
39. Ganz T. Defensins: antimicrobial peptides of innate immunity. Nat Rev Immunol. 2003;3(9):710–20. https://doi.org/10.1038/nri1180.
40. Kontakiotis EG, Filippatos CG, Tzanetakis GN, Agrafioti A. Regenerative endodontic therapy: a data analysis of clinical protocols. J Endod. 2015;41(2):146–54. https://doi.org/10.1016/j.joen.2014.08.003.
41. Langer R, Vacanti JP. Tissue engineering. Science. 1993;260(5110):920–6. https://doi.org/10.1126/science.8493529.
42. Nakashima M. Tissue engineering in endodontics. Aust Endod J. 2005;31(3):111–3. https://doi.org/10.1111/j.1747-4477.2005.tb00317.x.
43. Chan CP, Lan WH, Chang MC, Chen YJ, Lan WC, Chang HH, et al. Effects of TGF-beta s on the growth, collagen synthesis and collagen lattice contraction of human dental pulp fibroblasts in vitro. Arch Oral Biol. 2005;50(5):469–79. https://doi.org/10.1016/j.archoralbio.2004.10.005.
44. Ishimatsu H, Kitamura C, Morotomi T, Tabata Y, Nishihara T, Chen KK, et al. Formation of dentinal bridge on surface of regenerated dental pulp in dentin defects by controlled release of fibroblast growth factor-2 from gelatin hydrogels. J Endod. 2009;35(6):858–65. https://doi.org/10.1016/j.joen.2009.03.049.
45. Fawzy El-Sayed KM, Ahmed GM, Abouauf EA, Schwendicke F. Stem/progenitor cell-mediated pulpal tissue regeneration: a systematic review and meta-analysis. Int Endod J. 2019;52(11):1573–85. https://doi.org/10.1111/iej.13177.
46. Xuan K, Li B, Guo H, Sun W, Kou X, He X, et al. Deciduous autologous tooth stem cells regenerate dental pulp after implantation into injured teeth. Sci Transl Med. 2018;10(455):eaaf3227. https://doi.org/10.1126/scitranslmed.aaf3227.
47. Kim JY, Xin X, Moioli EK, Chung J, Lee CH, Chen M, et al. Regeneration of dental-pulp-like tissue by chemotaxis-induced cell homing. Tissue Eng Part A. 2010;16(10):3023–31. https://doi.org/10.1089/ten.TEA.2010.0181.
48. Lovelace TW, Henry MA, Hargreaves KM, Diogenes A. Evaluation of the delivery of mesenchymal stem cells into the root canal space of necrotic immature teeth after clinical regenerative endodontic procedure. J Endod. 2011;37(2):133–8. https://doi.org/10.1016/j.joen.2010.10.009.

49. Jadhav G, Shah N, Logani A. Revascularization with and without platelet-rich plasma in nonvital, immature, anterior teeth: a pilot clinical study. J Endod. 2012;38(12):1581–7. https://doi.org/10.1016/j.joen.2012.09.010.
50. Torabinejad M, Turman M. Revitalization of tooth with necrotic pulp and open apex by using platelet-rich plasma: a case report. J Endod. 2011;37(2):265–8. https://doi.org/10.1016/j.joen.2010.11.004.
51. Bezgin T, Yilmaz AD, Celik BN, Kolsuz ME, Sonmez H. Efficacy of platelet-rich plasma as a scaffold in regenerative endodontic treatment. J Endod. 2015;41(1):36–44. https://doi.org/10.1016/j.joen.2014.10.004.
52. Narang I, Mittal N, Mishra N. A comparative evaluation of the blood clot, platelet-rich plasma, and platelet-rich fibrin in regeneration of necrotic immature permanent teeth: a clinical study. Contemp Clin Dent. 2015;6(1):63–8. https://doi.org/10.4103/0976-237X.1492.
53. Keswani D, Pandey RK. Revascularization of an immature tooth with a necrotic pulp using platelet-rich fibrin: a case report. Int Endod J. 2013;46(11):1096–104. https://doi.org/10.1111/iej.12107.
54. Mishra N, Narang I, Mittal N. Platelet-rich fibrin-mediated revitalization of immature necrotic tooth. Contemp Clin Dent. 2013;4(3):412–5. https://doi.org/10.4103/0976-237X.118379.
55. Shivashankar VY, Johns DA, Vidyanath S, Kumar MR. Platelet rich fibrin in the revitalization of tooth with necrotic pulp and open apex. J Conserv Dent. 2012;15(4):395–8. https://doi.org/10.4103/0972-0707.101926.
56. Bakhtiar H, Esmaeili S, Fakhr Tabatabayi S, Ellini MR, Nekoofar MH, Dummer PM. Second-generation platelet concentrate (platelet-rich fibrin) as a scaffold in regenerative endodontics: a case series. J Endod. 2017;43(3):401–8. https://doi.org/10.1016/j.joen.2016.10.016.
57. Ray HL Jr, Marcelino J, Braga R, Horwat R, Lisien M, Khaliq S. Long-term follow up of revascularization using platelet-rich fibrin. Dent Traumatol. 2016;32(1):80–4. https://doi.org/10.1111/edt.12189.
58. Jung IY, Lee SJ, Hargreaves KM. Biologically based treatment of immature permanent teeth with pulpal necrosis: a case series. J Endod. 2008;34(7):876–87. https://doi.org/10.1016/j.joen.2008.03.023.
59. Nygaard-Ostby B, Hjortdal O. Tissue formation in the root canal following pulp removal. Scand J Dent Res. 1971;79(5):333–49. https://doi.org/10.1111/j.1600-0722.1971.tb02019.x.
60. Nosrat A, Homayounfar N, Oloomi K. Drawbacks and unfavorable outcomes of regenerative endodontic treatments of necrotic immature teeth: a literature review and report of a case. J Endod. 2012;38(10):1428–34. https://doi.org/10.1016/j.joen.2012.06.025.
61. Petrino JA, Boda KK, Shambarger S, Bowles WR, McClanahan SB. Challenges in regenerative endodontics: a case series. J Endod. 2010;36(3):536–41. https://doi.org/10.1016/j.joen.2009.10.006.
62. Martin G, Ricucci D, Gibbs JL, Lin LM. Histological findings of revascularized/revitalized immature permanent molar with apical periodontitis using platelet-rich plasma. J Endod. 2013;39(1):138–44. https://doi.org/10.1016/j.joen.2012.09.015.
63. Khademi AA, Dianat O, Mahjour F, Razavi SM, Younessian F. Outcomes of revascularization treatment in immature dog's teeth. Dent Traumatol. 2014;30(5):374–9. https://doi.org/10.1111/edt.12100.
64. Dianat O, Mashhadi Abas F, Paymanpour P, Eghbal MJ, Haddadpour S, Bahrololumi N. Endodontic repair in immature dogs' teeth with apical periodontitis: blood clot vs plasma rich in growth factors scaffold. Dent Traumatol. 2017;33(2):84–90. https://doi.org/10.1111/edt.12306.
65. Ruparel NB, Teixeira FB, Ferraz CC, Diogenes A. Direct effect of intracanal medicaments on survival of stem cells of the apical papilla. J Endod. 2012;38(10):1372–5. https://doi.org/10.1016/j.joen.2012.06.018.
66. Trevino EG, Patwardhan AN, Henry MA, Perry G, Dybdal-Hargreaves N, Hargreaves KM, et al. Effect of irrigants on the survival of human stem cells of the apical papilla in a platelet-rich plasma scaffold in human root tips. J Endod. 2011;37(8):1109–15. https://doi.org/10.1016/j.joen.2011.05.013.
67. Byström A, Sundqvist G. Bacteriologic evaluation of the efficacy of mechanical root canal instrumentation in endodontic therapy. Scand J Dent Res. 1981;89(4):321–8. https://doi.org/10.1111/j.1600-0722.1981.tb01689.x.
68. Gluskin AH, Peters CI, Peters OA. Minimally invasive endodontics: challenging prevailing paradigms. Br Dent J. 2014;216(6):347–53. https://doi.org/10.1038/sj.bdj.2014.201.
69. Donlan RM, Costerton JW. Biofilms: survival mechanisms of clinically relevant microorganisms. Clin Microbiol Rev. 2002;15(2):167–93. https://doi.org/10.1128/cmr.15.2.167-193.2002.
70. Lee SM. Infection control in regenerative endodontic procedures. American Association of Endodontists. 2019. https://www.aae.org/specialty/2019/11/04/infection-control-in-regenerative-endodontic-procedures. Accessed 6 June 2020.
71. Bao P, Shen Y, Lin J, Haapasalo M. In vitro efficacy of XP-endo finisher with 2 different protocols on biofilm removal from apical root canals. J Endod. 2017;43(2):321–5. https://doi.org/10.1016/j.joen.2016.09.021.
72. Nagata JY, Gomes BP, Rocha Lima TF, Murakami LS, de Faria DE, Campos GR, et al. Traumatized immature teeth treated with 2 protocols of pulp revascularization. J Endod. 2014;40(5):606–12. https://doi.org/10.1016/j.joen.2014.01.032.
73. Chueh LH, Ho YC, Kuo TC, Lai WH, Chen YH, Chiang CP. Regenerative endodontic treatment for necrotic

immature permanent teeth. J Endod. 2009;35(2):160–4. https://doi.org/10.1016/j.joen.2008.10.019.
74. Kahler B, Mistry S, Moule A, Ringsmuth AK, Case P, Thomson A, et al. Revascularization outcomes: a prospective analysis of 16 consecutive cases. J Endod. 2014;40(3):333–8. https://doi.org/10.1016/j.joen.2013.10.032.
75. Becerra P, Ricucci D, Loghin S, Gibbs JL, Lin LM. Histologic study of a human immature permanent premolar with chronic apical abscess after revascularization/revitalization. J Endod. 2014;40(1):133–9. https://doi.org/10.1016/j.joen.2013.07.017.
76. Bezgin T, Yilmaz AD, Celik BN, Sönmez H. Concentrated platelet-rich plasma used in root canal revascularization: 2 case reports. Int Endod J. 2014;47(1):41–9. https://doi.org/10.1111/iej.12144.
77. Miller EK, Lee JY, Tawil PZ, Teixeira FB, Vann WF Jr. Emerging therapies for the management of traumatized immature permanent incisors. Pediatr Dent. 2012;34(1):66–9.
78. Paryani K, Kim SG. Regenerative endodontic treatment of permanent teeth after completion of root development: a report of 2 cases. J Endod. 2013;39(7):929–34. https://doi.org/10.1016/j.joen.2013.04.029.
79. Amit V, Jain A, Nayak UA, Bhat M. Maturogenesis by revascularization in an infected immature permanent tooth. J Indian Soc Pedod Prev Dent. 2014;32(2):172–5. https://doi.org/10.4103/0970-4388.130992.
80. Baumgartner JC, Cuenin PR. Efficacy of several concentrations of sodium hypochlorite for root canal irrigation. J Endod. 1992;18(12):605–12. https://doi.org/10.1016/S0099-2399(06)81331-2.
81. Hand RE, Smith ML, Harrison JW. Analysis of the effect of dilution on the necrotic tissue dissolution property of sodium hypochlorite. J Endod. 1978;4(2):60–4. https://doi.org/10.1016/S0099-2399(78)80255-6.
82. Siqueira JF Jr, Rôças IN, Favieri A, Lima KC. Chemomechanical reduction of the bacterial population in the root canal after instrumentation and irrigation with 1%, 2.5%, and 5.25% sodium hypochlorite. J Endod. 2000;26(6):331–4. https://doi.org/10.1097/00004770-200006000-00006.
83. Martin DE, De Almeida JF, Henry MA, Khaing ZZ, Schmidt CE, Teixeira FB, et al. Concentration-dependent effect of sodium hypochlorite on stem cells of apical papilla survival and differentiation. J Endod. 2014;40(1):51–5. https://doi.org/10.1016/j.joen.2013.07.026.
84. Forough Reyhani M, Rezagholizadeh Y, Narimani MR, Rezagholizadeh L, Mazani M, Barhaghi MHS, et al. Antibacterial effect of different concentrations of sodium hypochlorite on *Enterococcus faecalis* biofilms in root canals. J Dent Res Dent Clin Dent Prospects. 2017;11(4):215–21. https://doi.org/10.15171/joddd.2017.038.
85. de Freitas CS, Diniz HF, Gomes JB, Sinisterra RD, Cortés ME. Evaluation of the substantivity of chlorhexidine in association with sodium fluoride in vitro. Pesqui Odontol Bras. 2003;17(1):78–81. https://doi.org/10.1590/s1517-74912003000100015.
86. Calt S, Serper A. Time-dependent effects of EDTA on dentin structures. J Endod. 2002;28(1):17–9. https://doi.org/10.1097/00004770-200201000-00004.
87. Crumpton BJ, Goodell GG, McClanahan SB. Effects on smear layer and debris removal with varying volumes of 17% REDTA after rotary instrumentation. J Endod. 2005;31(7):536–8. https://doi.org/10.1097/01.don.0000148871.72896.1d.
88. Seidberg BH, Schilder H. An evaluation of EDTA in endodontics. Oral Surg Oral Med Oral Pathol. 1974;37(4):609–20. https://doi.org/10.1016/0030-4220(74)90294-1.
89. Galler KM, Buchalla W, Hiller KA, Federlin M, Eidt A, Schiefersteiner M, et al. Influence of root canal disinfectants on growth factor release from dentin. J Endod. 2015;41(3):363–8. https://doi.org/10.1016/j.joen.2014.11.021.
90. Zhao S, Sloan AJ, Murray PE, Lumley PJ, Smith AJ. Ultrastructural localisation of TGF-beta exposure in dentine by chemical treatment. Histochem J. 2000;32(8):489–94. https://doi.org/10.1023/a:1004100518245.
91. Mathieu S, Jeanneau C, Sheibat-Othman N, Kalaji N, Fessi H, About I. Usefulness of controlled release of growth factors in investigating the early events of dentin-pulp regeneration. J Endod. 2013;39(2):228–35. https://doi.org/10.1016/j.joen.2012.11.007.
92. Kalyva M, Papadimitriou S, Tziafas D. Transdentinal stimulation of tertiary dentine formation and intratubular mineralization by growth factors. Int Endod J. 2010;43(5):382–92. https://doi.org/10.1111/j.1365-2591.2010.01690.x.
93. Melin M, Joffre-Romeas A, Farges JC, Couble ML, Magloire H, Bleicher F. Effects of TGFbeta1 on dental pulp cells in cultured human tooth slices. J Dent Res. 2000;79(9):1689–96. https://doi.org/10.1177/00220345000790090901.
94. He H, Yu J, Liu Y, Lu S, Liu H, Shi J, et al. Effects of FGF2 and TGFbeta1 on the differentiation of human dental pulp stem cells in vitro. Cell Biol Int. 2008;32(7):827–34. https://doi.org/10.1016/j.cellbi.2008.03.013.
95. Mullane EM, Dong Z, Sedgley CM, Hu JC, Botero TM, Holland GR, et al. Effects of VEGF and FGF2 on the revascularization of severed human dental pulps. J Dent Res. 2008;87(12):1144–8. https://doi.org/10.1177/154405910808701204.
96. Lin LM, Shimizu E, Gibbs JL, Loghin S, Ricucci D. Histologic and histobacteriologic observations of failed revascularization/revitalization therapy: a case report. J Endod. 2014;40(2):291–5. https://doi.org/10.1016/j.joen.2013.08.024.
97. Hargreaves KM, Diogenes A, Teixeira FB. Treatment options: biological basis of regenerative endodontic procedures. J Endod. 2013;39(Suppl 3):S30–43. https://doi.org/10.1016/j.joen.2012.11.025.

98. Tagelsir A, Yassen GH, Gomez GF, Gregory RL. Effect of antimicrobials used in regenerative endodontic procedures on 3-week-old *Enterococcus faecalis* biofilm. J Endod. 2016;42(2):258–62. https://doi.org/10.1016/j.joen.2015.09.023.
99. Cetinkaya BO, Keles GC, Pamuk F, Balli U, Keles ZP. Long-term clinical results on the use of platelet concentrate in the treatment of intrabony periodontal defects. Acta Odontol Scand. 2014;72(2):92–8. https://doi.org/10.3109/00016357.2013.775668.
100. Keles GC, Cetinkaya BO, Baris S, Albayrak D, Simsek SB. Comparison of platelet pellet with or without guided tissue regeneration in the treatment of class II furcation defects in dogs. Clin Oral Investig. 2009;13(4):393–400. https://doi.org/10.1007/s00784-008-0245-1.
101. Ulusoy AT, Turedi I, Cimen M, Cehreli ZC. Evaluation of blood clot, platelet-rich plasma, platelet-rich fibrin, and platelet pellet as scaffolds in regenerative endodontic treatment: a prospective randomized trial. J Endod. 2019;45(5):560–6. https://doi.org/10.1016/j.joen.2019.02.002.
102. Brass L. Understanding and evaluating platelet function. Hematology Am Soc Hematol Educ Program. 2010;2010:387–96. https://doi.org/10.1182/asheducation-2010.1.387.
103. Marx RE, Carlson ER, Eichstaedt RM, Schimmele SR, Strauss JE, Georgeff KR. Platelet-rich plasma: growth factor enhancement for bone grafts. Oral Surg Oral Med Oral Pathol Oral Radiol Endod. 1998;85(6):638–46. https://doi.org/10.1016/s1079-2104(98)90029-4.
104. Kotsovilis S, Markou N, Pepelassi E, Nikolidakis D. The adjunctive use of platelet-rich plasma in the therapy of periodontal intraosseous defects: a systematic review. J Periodontal Res. 2010;45(3):428–43. https://doi.org/10.1111/j.1600-0765.2009.01236.x.
105. Marx RE. Platelet-rich plasma (PRP): what is PRP and what is not PRP? Implant Dent. 2001;10(4):225–8. https://doi.org/10.1097/00008505-200110000-00002.
106. Dohan DM, Choukroun J, Diss A, Dohan SL, Dohan AJ, Mouhyi J, et al. Platelet-rich fibrin (PRF): a second-generation platelet concentrate. Part I: technological concepts and evolution. Oral Surg Oral Med Oral Pathol Oral Radiol Endod. 2006;101(3):e37–44. https://doi.org/10.1016/j.tripleo.2005.07.008.
107. Keles GC, Cetinkaya BO, Albayrak D, Koprulu H, Acikgoz G. Comparison of platelet pellet and bioactive glass in periodontal regenerative therapy. Acta Odontol Scand. 2006;64(6):327–33. https://doi.org/10.1080/00016350600758651.
108. Jayakumaar A, Ganesh A, Kalaiselvam R, Rajan M, Deivanayagam K. Evaluation of debris and smear layer removal with XP-endo finisher: a scanning electron microscopic study. Indian J Dent Res. 2019;30(3):420–3. https://doi.org/10.4103/ijdr.IJDR_655_17.
109. Zeichner-David M. Is there more to enamel matrix proteins than biomineralization? Matrix Biol. 2001;20(5–6):307–16. https://doi.org/10.1016/s0945-053x(01)00155-x.
110. Matsumoto N, Minakami M, Hatakeyama J, Haruna C, Morotomi T, Izumi T, et al. Histologic evaluation of the effects of Emdogain gel on injured root apex in rats. J Endod. 2014;40(12):1989–94. https://doi.org/10.1016/j.joen.2014.08.024.
111. Wang HH, Sarmast ND, Shadmehr E, Angelov N, Shabahang S, Torabinejad M. Application of enamel matrix derivative (Emdogain) in endodontic therapy: a comprehensive literature review. J Endod. 2018;44(7):1066–79. https://doi.org/10.1016/j.joen.2018.02.012.
112. Scarparo RK, Dondoni L, Böttcher DE, Grecca FS, Figueiredo JA, Batista EL Jr. Apical periodontium response to enamel matrix derivative as an intracanal medication in rat immature teeth with pulp necrosis: radiographic and histologic findings. J Endod. 2012;38(4):449–53. https://doi.org/10.1016/j.joen.2011.12.041.
113. Zhao X, He W, Song Z, Tong Z, Li S, Ni L. Mineral trioxide aggregate promotes odontoblastic differentiation via mitogen-activated protein kinase pathway in human dental pulp stem cells. Mol Biol Rep. 2012;39(1):215–20. https://doi.org/10.1007/s11033-011-0728-z.
114. Laurent P, Camps J, About I. Biodentine(TM) induces TGF-β1 release from human pulp cells and early dental pulp mineralization. Int Endod J. 2012;45(5):439–48. https://doi.org/10.1111/j.1365-2591.2011.01995.x.
115. Edrees HY, Abu Zeid STH, Atta HM, AlQriqri MA. Induction of osteogenic differentiation of mesenchymal stem cells by bioceramic root repair material. Materials (Basel). 2019;12(14):2311. https://doi.org/10.3390/ma12142311.
116. Tong HJ, Rajan S, Bhujel N, Kang J, Duggal M, Nazzal H. Regenerative endodontic therapy in the management of nonvital immature permanent teeth: a systematic review-outcome evaluation and meta-analysis. J Endod. 2017;43(9):1453–64. https://doi.org/10.1016/j.joen.2017.04.018.
117. Torabinejad M, Nosrat A, Verma P, Udochukwu O. Regenerative endodontic treatment or mineral trioxide aggregate apical plug in teeth with necrotic pulps and open apices: a systematic review and meta-analysis. J Endod. 2017;43(11):1806–20. https://doi.org/10.1016/j.joen.2017.06.029.
118. Kahler B, Rossi-Fedele G, Chugal N, Lin LM. An evidence-based review of the efficacy of treatment approaches for immature permanent teeth with pulp necrosis. J Endod. 2017;43(7):1052–7. https://doi.org/10.1016/j.joen.2017.03.003.
119. Bose R, Nummikoski P, Hargreaves K. A retrospective evaluation of radiographic outcomes in immature teeth with necrotic root canal systems treated with regenerative endodontic procedures. J Endod.

2009;35(10):1343–9. https://doi.org/10.1016/j.joen.2009.06.021.
120. Flake NM, Gibbs JL, Diogenes A, Hargreaves KM, Khan AA. A standardized novel method to measure radiographic root changes after endodontic therapy in immature teeth. J Endod. 2014;40(1):46–50. https://doi.org/10.1016/j.joen.2013.09.025.
121. EzEldeen M, Van Gorp G, Van Dessel J, Vandermeulen D, Jacobs R. 3-dimensional analysis of regenerative endodontic treatment outcome. J Endod. 2015;41(3):317–24. https://doi.org/10.1016/j.joen.2014.10.023.
122. Meschi N, EzEldeen M, Torres Garcia AE, Jacobs R, Lambrechts P. A retrospective case series in regenerative endodontics: trend analysis based on clinical evaluation and 2- and 3-dimensional radiology. J Endod. 2018;44(10):1517–25. https://doi.org/10.1016/j.joen.2018.06.015.
123. Diogenes A, Henry MA, Teixeira FB, Hargreaves KM. An update on clinical regenerative endodontics. Endod Top. 2013;28(1):2–23. https://doi.org/10.1111/etp.12040.
124. Shamszadeh S, Asgary S, Nosrat A. Regenerative endodontics: a scientometric and bibliometric analysis. J Endod. 2019;45(3):272–80. https://doi.org/10.1016/j.joen.2018.11.010.
125. Lin J, Zeng Q, Wei X, Zhao W, Cui M, Gu J, et al. Regenerative endodontics versus apexification in immature permanent teeth with apical periodontitis: a prospective randomized controlled study. J Endod. 2017;43(11):1821–7. https://doi.org/10.1016/j.joen.2017.06.023.
126. Nagata JY, Soares AJ, Souza-Filho FJ, Zaia AA, Ferraz CC, Almeida JF, et al. Microbial evaluation of traumatized teeth treated with triple antibiotic paste or calcium hydroxide with 2% chlorhexidine gel in pulp revascularization. J Endod. 2014;40(6):778–83. https://doi.org/10.1016/j.joen.2014.01.038.
127. Nagy MM, Tawfik HE, Hashem AA, Abu-Seida AM. Regenerative potential of immature permanent teeth with necrotic pulps after different regenerative protocols. J Endod. 2014;40(2):192–8. https://doi.org/10.1016/j.joen.2013.10.027.
128. Andreasen JO, Paulsen HU, Yu Z, Bayer T, Schwartz O. A long-term study of 370 autotransplanted premolars. Part II. Tooth survival and pulp healing subsequent to transplantation. Eur J Orthod. 1990;12(1):14–24. https://doi.org/10.1093/ejo/12.1.14.
129. Estefan BS, El Batouty KM, Nagy MM, Diogenes A. Influence of age and apical diameter on the success of endodontic regeneration procedures. J Endod. 2016;42(11):1620–5. https://doi.org/10.1016/j.joen.2016.06.020.
130. Saoud TM, Sigurdsson A, Rosenberg PA, Lin LM, Ricucci D. Treatment of a large cystlike inflammatory periapical lesion associated with mature necrotic teeth using regenerative endodontic therapy. J Endod. 2014;40(12):2081–6. https://doi.org/10.1016/j.joen.2014.07.027.
131. Shah N, Logani A. SealBio: a novel, non-obturation endodontic treatment based on concept of regeneration. J Conserv Dent. 2012;15(4):328–32. https://doi.org/10.4103/0972-0707.101889.
132. Parirokh M, Torabinejad M. Mineral trioxide aggregate: a comprehensive literature review--part III: clinical applications, drawbacks, and mechanism of action. J Endod. 2010;36(3):400–13. https://doi.org/10.1016/j.joen.2009.09.009.
133. Kohli MR, Yamaguchi M, Setzer FC, Karabucak B. Spectrophotometric analysis of coronal tooth discoloration induced by various bioceramic cements and other endodontic materials. J Endod. 2015;41(11):1862–6. https://doi.org/10.1016/j.joen.2015.07.003.
134. Shokouhinejad N, Nekoofar MH, Pirmoazen S, Shamshiri AR, Dummer PM. Evaluation and comparison of occurrence of tooth discoloration after the application of various calcium silicate-based cements: an ex vivo study. J Endod. 2016;42(1):140–4. https://doi.org/10.1016/j.joen.2015.08.034.
135. Song M, Cao Y, Shin SJ, Shon WJ, Chugal N, Kim RH, et al. Revascularization-associated intracanal calcification: assessment of prevalence and contributing factors. J Endod. 2017;43(12):2025–33. https://doi.org/10.1016/j.joen.2017.06.018.
136. Cohenca N, Shemesh H. Clinical applications of cone beam computed tomography in endodontics: a comprehensive review. Quintessence Int. 2015;46(6):465–80. https://doi.org/10.3290/j.qi.a33990.
137. de Carvalho MC, Zuolo ML. Orifice locating with a microscope. J Endod. 2000;26(9):532–4. https://doi.org/10.1097/00004770-200009000-00012.
138. Kiefner P, Connert T, ElAyouti A, Weiger R. Treatment of calcified root canals in elderly people: a clinical study about the accessibility, the time needed and the outcome with a three-year follow-up. Gerodontology. 2017;34(2):164–70. https://doi.org/10.1111/ger.12238.
139. Shimizu E, Ricucci D, Albert J, Alobaid AS, Gibbs JL, Huang GT, et al. Clinical, radiographic, and histological observation of a human immature permanent tooth with chronic apical abscess after revitalization treatment. J Endod. 2013;39(8):1078–83. https://doi.org/10.1016/j.joen.2013.04.032.
140. Žižka R, Buchta T, Voborná I, Harvan L, Šedý J. Root maturation in teeth treated by unsuccessful revitalization: 2 Case reports. J Endod. 2016;42(5):724–9. https://doi.org/10.1016/j.joen.2016.02.004.
141. Alobaid AS, Cortes LM, Lo J, Nguyen TT, Albert J, Abu-Melha AS, et al. Radiographic and clinical outcomes of the treatment of immature permanent teeth by revascularization or apexification: a pilot retrospective cohort study. J Endod. 2014;40(8):1063–70. https://doi.org/10.1016/j.joen.2014.02.016.

142. McGurkin-Smith R, Trope M, Caplan D, Sigurdsson A. Reduction of intracanal bacteria using GT rotary instrumentation, 5.25% NaOCl, EDTA, and Ca(OH)2. J Endod. 2005;31(5):359–63. https://doi.org/10.1097/01.don.0000145035.85272.7c.
143. Siqueira JF Jr, Guimarães-Pinto T, Rôças IN. Effects of chemomechanical preparation with 2.5% sodium hypochlorite and intracanal medication with calcium hydroxide on cultivable bacteria in infected root canals. J Endod. 2007;33(7):800–5. https://doi.org/10.1016/j.joen.2006.11.023.
144. Siqueira JF Jr, Magalhães KM, Rôças IN. Bacterial reduction in infected root canals treated with 2.5% NaOCl as an irrigant and calcium hydroxide/camphorated paramonochlorophenol paste as an intracanal dressing. J Endod. 2007;33(6):667–72. https://doi.org/10.1016/j.joen.2007.01.004.
145. Sjögren U, Figdor D, Persson S, Sundqvist G. Influence of infection at the time of root filling on the outcome of endodontic treatment of teeth with apical periodontitis. Int Endod J. 1997;30(5):297–306. https://doi.org/10.1046/j.1365-2591.1997.00092.x.
146. Sjögren U, Figdor D, Spångberg L, Sundqvist G. The antimicrobial effect of calcium hydroxide as a short-term intracanal dressing. Int Endod J. 1991;24(3):119–25. https://doi.org/10.1111/j.1365-2591.1991.tb00117.x.
147. Azim AA, Aksel H, Zhuang T, Mashtare T, Babu JP, Huang GT. Efficacy of 4 irrigation protocols in killing bacteria colonized in dentinal tubules examined by a novel confocal laser scanning microscope analysis. J Endod. 2016;42(6):928–34. https://doi.org/10.1016/j.joen.2016.03.009.
148. Gu LS, Kim JR, Ling J, Choi KK, Pashley DH, Tay FR. Review of contemporary irrigant agitation techniques and devices. J Endod. 2009;35(6):791–804. https://doi.org/10.1016/j.joen.2009.03.010.
149. Ballal V, Rao S, Al-Haj Husain N, Özcan M. Evaluation of smear layer removal using different irrigation methods in root canals. Eur J Prosthodont Restor Dent. 2019;27(3):97–102. https://doi.org/10.1922/EJPRD_01817Husain06.
150. Desai P, Himel V. Comparative safety of various intracanal irrigation systems. J Endod. 2009;35(4):545–9. https://doi.org/10.1016/j.joen.2009.01.011.
151. Ruddle CJ. Hydrodynamic disinfection: tsunami endodontics. Dent Today. 2007;26(5):110, 112, 114–7.
152. Elnaghy AM, Mandorah A, Elsaka SE. Effectiveness of XP-endo Finisher, EndoActivator, and File agitation on debris and smear layer removal in curved root canals: a comparative study. Odontology. 2017;105(2):178–83. https://doi.org/10.1007/s10266-016-0251-8.
153. Al Shahrani M, DiVito E, Hughes CV, Nathanson D, Huang GT. Enhanced removal of *Enterococcus faecalis* biofilms in the root canal using sodium hypochlorite plus photon-induced photoacoustic streaming: an in vitro study. Photomed Laser Surg. 2014;32(5):260–6. https://doi.org/10.1089/pho.2014.3714.
154. Koch JD, Jaramillo DE, DiVito E, Peters OA. Irrigant flow during photon-induced photoacoustic streaming (PIPS) using particle image velocimetry (PIV). Clin Oral Investig. 2016;20(2):381–6. https://doi.org/10.1007/s00784-015-1562-9.
155. Mathew J, Emil J, Paulaian B, John B, Raja J, Mathew J. Viability and antibacterial efficacy of four root canal disinfection techniques evaluated using confocal laser scanning microscopy. J Conserv Dent. 2014;17(5):444–8. https://doi.org/10.4103/0972-0707.139833.
156. Arslan H, Akcay M, Capar ID, Ertas H, Ok E, Uysal B. Efficacy of needle irrigation, EndoActivator, and photon-initiated photoacoustic streaming technique on removal of double and triple antibiotic pastes. J Endod. 2014;40(9):1439–42. https://doi.org/10.1016/j.joen.2014.02.013.
157. DiVito E, Peters OA, Olivi G. Effectiveness of the erbium:YAG laser and new design radial and stripped tips in removing the smear layer after root canal instrumentation. Lasers Med Sci. 2012;27(2):273–80. https://doi.org/10.1007/s10103-010-0858-x.

Tooth Bioengineering and Whole Tooth Regeneration

Ning Cheng, Juan Wen, Rita Hitching, Chang Lei, and Chun Xu

1 Introduction

Tooth loss is a common problem, and affects nearly 3.5 billion people worldwide, according to the Global Burden of Disease Study in 2017 [1]. Teeth are complex organs compositing of both hard tissues (enamel, dentine, and cementum) and the associated soft tissues (the dental pulp and the periodontal ligament). Due to their unique anatomy and embryology, teeth have a limited self-repair ability after sustaining damage [2]. Tooth loss can be caused by untreated dental caries, severe periodontal diseases, traumatic dental injury, developmental defects (including congenital cleft and palate defects), and other conditions [1, 3–5].

Tooth loss can be categorized into partial tooth loss (partial edentulism) and the complete loss of all teeth (edentulism). The loss of some teeth can leave an edentulous space that causes esthetic issues, as well as functional problems such as drifting of adjacent teeth, over-eruption of opposing teeth, and impaired oral hygiene [6, 7]. Complete tooth loss is commonly associated with impaired mastication and poor nutrition, and it substantially increases the incidence of obesity, diabetes mellitus, gastrointestinal disease, cardiovascular disease, kidney disease, cancer, dementia, disability, and death [8, 9].

Clinical treatments that are considered the standard approaches for total tooth loss include traditional dentures, and more recently implant stabilized dentures, and implant-supported bridges. As well as these prosthodontic developments, numerous efforts have been made to create novel and advanced methods for tooth bioengineering and tooth regeneration. The aim of these therapies is to restore in part or in whole the structure and function of a tooth. While there is an extensive literature on remineralization of enamel and dentine, and on tooth-colored dental materials for restoring teeth, this chapter focuses on stem cell-based, bioengineered, scaffold biomaterials, used alone or in a combina-

N. Cheng
Department of Otolaryngology-Head and Neck Surgery, University of California San Francisco, San Francisco, CA, USA
e-mail: Ning.Cheng@ucsf.edu

J. Wen
Department of Orthodontics, Nanjing Stomatological Hospital, Medical School of Nanjing University, Nanjing, Jiang Su, China

R. Hitching
Palo Alto Veterans' Institute for Research (PAVIR), VA Palo Alto Health Care System, Palo Alto, CA, USA

C. Lei
Australian Institute for Bioengineering and Nanotechnology, The University of Queensland, Brisbane, QLD, Australia
e-mail: c.lei1@uq.edu.au

C. Xu (✉)
School of Dentistry, The University of Queensland, Herston, QLD, Australia
e-mail: chun.xu@uq.edu.au

© Springer Nature Switzerland AG 2021
S. Hosseinpour et al. (eds.), *Regenerative Approaches in Dentistry*,
https://doi.org/10.1007/978-3-030-59809-9_5

tion to leverage current clinical and translational research in tooth regeneration.

2 Stem Cell-Based Strategies for Tooth Regeneration

Stem cells are characterized by their dual abilities of self-renewal and multi-lineage differentiation. They are considered as "multipotent" or "unipotent," and are derived from almost all specialized adult/postnatal tissues, including teeth. Dental stem cells (DSCs) exhibit mesenchymal stem cell-like properties, and so are of interest for partial and whole tooth regeneration.

DSCs include dental pulp stem cells (DPSCs), stem cells from human exfoliated deciduous teeth (SHED), periodontal ligament stem cells (PDLSCs), stem cells from the apical papilla (SCAP), dental follicle stem cells (DFSCs), and gingival fibroblastic stem cells (GFSCs) (Fig. 1).

Compared to other types of adult stem cells, DSCs can be obtained easily. They exhibit a high proliferation rate, making them an attractive source for autologous tooth bioengineering. In addition to DSCs, Yamanaka's group reported in 2006 the generation of induced pluripotent stem cells (iPSCs) as an alternative source of stem cells for tooth regeneration [10]. In this section, we introduce the properties of various dental stem cells and iPSCs, and discuss their use for tooth regeneration.

2.1 Dental Stem Cells

2.1.1 Dental Pulp Stem Cells

Dental pulp stem cells (DPSCs) are the first type of dental stem cells, and were isolated from permanent third molar teeth by Shi and co-workers [11]. Similar to bone marrow stromal cells (BMSCs), DPSCs express typical MSCs' biomarkers, such as CD90, CD29, CD73, CD44, and CD105 [12, 13]. Moreover, DPSCs can produce sporadic calcified modules [11], and can differentiate into odontoblast-like cells. After implantation into immunocompromised (SCID) mice, DPSCs can form a dentine-pulp like structure, with pulp-like interstitial tissue inside and odontoblast-like tissue

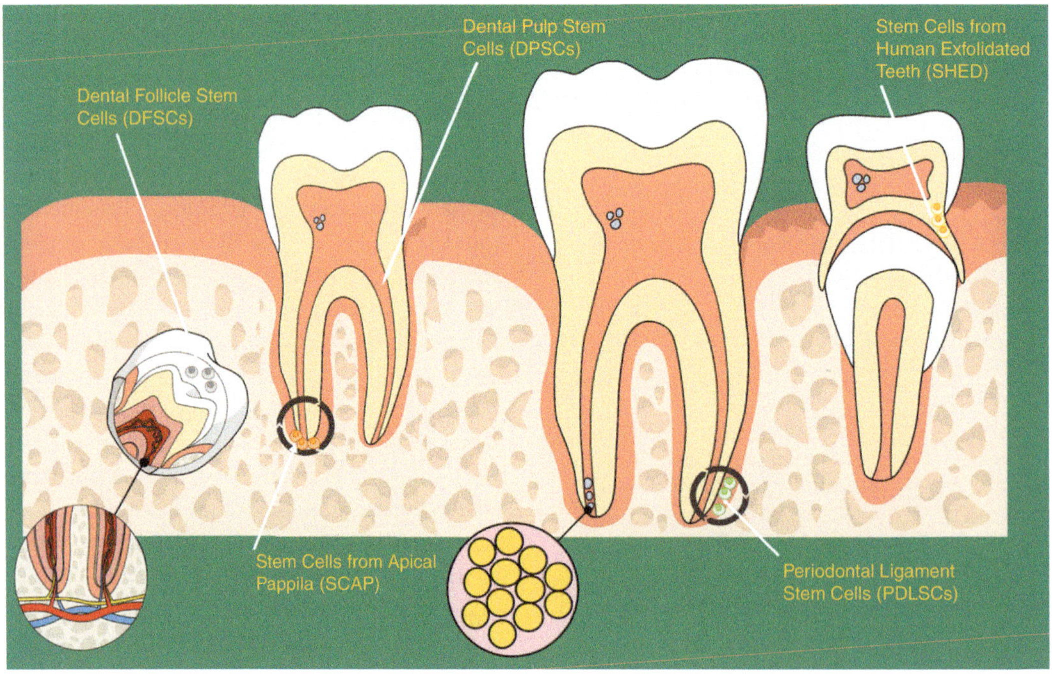

Fig. 1 The sources of adult dental stem cells used for tooth regeneration

outside [11]. In addition to prominent osteogenic differentiation *in vitro* and *in vivo*, DPSCs can differentiate into other mesodermal lineage cells such as chondrocytes, adipocyte, hepatocytes, and neurocytes [14, 15].

2.1.2 Stem Cells from Human Exfoliated Deciduous Tooth

Stem cells from a human exfoliated deciduous tooth (SHED) are easily accessible by standard collection methods, and they can be more rapidly propagated than DPSCs and used for autologous treatments [16, 17]. To explore the mechanisms behind their high proliferation rate, a large-scale gene expression profile analysis has been carried out. 4386 genes associated with cell proliferation and extracellular matrix formation are expressed differently between DPSC and SHED [17]. One such gene that is strongly expressed in SHED is transforming growth factor-beta (TGF-β), which is involved in odontoblastic differentiation [17]. Furthermore, SHED can differentiate into odontoblast-like cells when transplanted into SCID mice, and can generate tubular dentine and differentiate into endothelial-like cells [18].

2.1.3 Periodontal Ligament Stem Cells

Periodontal ligament stem cells (PDLSCs) are isolated from the periodontal ligament, which serves as the interface between the alveolar bone of the tooth socket, and the cementum on the tooth. This ligament structure contains mostly fibrous connective tissue, the fibers of which insert into the bone and the cementum. PDLSCs have been shown to express MSC-associated biomarkers including STRO-1, CD146, and MUC18 [19]. They can differentiate into cementoblast-like cells, adipocytes, fibroblasts, osteoblasts, chondrocytes, neurons, and myocytes [19–21]. After implantation into immunocompromised mice, PDLSCs have been shown to generate cementum and a periodontal ligament-like tissue [19]. Furthermore, when PDLSCs have been transplanted into a preclinical miniature pig model with periodontitis, they were able to form periodontal tissues in surgically created periodontal defects [22, 23].

2.1.4 Root Apical Papilla Stem Cells

Root apical papilla stem cells (SCAP) are isolated from the apical papilla, which located at the toot tip of a growing tooth. SCAP can be obtained from extracted third molar teeth with incomplete roots. SCAP exhibit much higher *in vitro* cell proliferation rates than DPSCs [20], and they can differentiate into odontoblasts and adipocytes [24]. Additionally, co-transplantation of SCAP and PDLSCs into tooth sockets of mini pigs can form a dental root structure containing both dentine and periodontal ligament [2].

2.1.5 Dental Follicle Stem Cells

The dental follicle is the fibrous tissue that surrounds the tooth germ during tooth development. Dental follicle stem cells (DFSCs) are typically isolated from third molar teeth that are removed while still being formed. During culture, DFSCs attach to plastic culture plates rapidly. They express stem cell markers such as nestin, and can produce compact calcified nodules when grown in an osteogenic culture medium [25–27]. After transplantation into SCID mice, DFSCs have been shown to form fibrous tissue and a cementum-like matrix, and to express relevant proteins and genes (e.g., OC, OP, COLI, and BSP) [26].

2.2 Induced Pluripotent Stem Cells

Although somatic stem cells (such as DSCs) have been studied extensively for tooth regeneration, their application could be limited by the source of the specific stem cell line. In contrast, embryonic stem cells (ESCs) are pluripotent. They possess the capacity to differentiate into all three primary germ layers, however, the use of human ESCs for therapeutic application is limited by ethical debates such as the destruction of human embryos to obtain these cells.

The generation of iPSCs overcomes these disadvantages and satisfy the needs for translational research of tooth regeneration. With introduction of specific reprogramming genes

(Oct4, Sox2, cMyc, and Klf4), somatic cells re-gain pluripotent abilities [10].

iPSCs have been investigated extensively in regenerative dentistry. For instance, iPSCs have been induced into becoming enamel-secreting ameloblast-like cells [28, 29], dentine-secreting odontoblast-like cells [30], and osteoblast-like cells [31], for the regeneration of dental enamel, dentine-pulp complex and periodontal tissues, respectively [32, 33]. We discuss the research findings from stem cell-based strategies for tooth regeneration in the following part.

2.3 Stem Cell-Based Strategy for Dental Pulp Regeneration

Dental pulp is a composite of soft fibrous connective tissue with various cells, blood vessels, and nerves. In the circumstance of dental caries or mechanical irritation, the odontoblasts in the dental pulp can be stimulated to form reactive (tertiary) dentine for aiding repair. In the clinical setting, traditional endodontics involves removing the infected dental pulp and then after cleaning and disinfecting the canal, filling it with a material such as gutta percha with a sealer paste. In contrast, dental pulp regeneration holds great therapeutic potential as a way to re-grow dental pulp tissues using DSCs [34]. One example is implanting a biocompatible scaffold seeded with human DPSCs and SCAPs from third molars into the canal space of roots. The proof of concept for this was seen after subcutaneous transplantation of a scaffold with stem cells into SCID mice, where the pulp-like tissue formed within the root canal space and was well vascularized after 4 months [33]. This topic is discussed further in chapter "Dentine-Pulp Complex Regeneration."

2.4 Stem Cell-Based Strategy for Enamel Regeneration

Tooth enamel regeneration remains challenging due to the lack of dental epithelial stem cells or ameloblasts in the region of the crown of adult teeth, since ameloblasts are required for enamel formation. However, the recently developed iPSCs provide new opportunities. Arakaki et al. co-cultured iPSCs with dental epithelium and differentiated them into ameloblast-like cells [35]. Furthermore, they also demonstrated that epithelial-mesenchymal interactions are essential for regulating tooth development, and hence co-culture models are essential. iPSCs are a novel source of mesenchymal cells for tooth formation, and co-culturing of iPSCs with epithelial cells can lead to the formation of cells with an epithelial-like morphology. Further characterization confirmed the expression of ameloblast markers such as ameloblastin and enamelin, and epithelial cell markers including p63 and cytokeratin-14 [36].

2.5 Stem Cell-Based Strategy for Periodontal Tissue Regeneration

PDLSCs, as a typical example of adult stem cells, have been applied directly to repair periodontal defects [37, 38]. Direct local injection of stem cells can be an effective approach for periodontal regeneration, using DPSCs or PDLSCs [39]. The regeneration of periodontal ligament has been observed in a rat periodontitis model [40].

In larger animal models, both allogeneic and autologous PDLSCs have been shown to regenerate periodontal tissues after being transplanted into periodontal defects, for example, in minipigs [22, 41]. This indicates that there is great therapeutic potential using this approach for treating periodontitis. Further studies have shown that PDLSCs have immunomodulatory actions. They can induce T-lymphocyte anergy by secreting prostaglandin E2 [41], and they can suppress lymphocytes activation mediated by the interaction of PD1 and PDL 1 [42].

Recently, a cell sheet technique has been used for periodontal regeneration. This technique has the ability to maintain the extracellular matrix and cell–cell junctions. The sheets can also be degraded using proteolytic enzymes (e.g., trypsin and/or dispase) [43]. Cell sheets can be prepared in unique scaffold-free methods, such as culturing cells with the addition of ascorbic acid, or using special equipment such as temperature-responsive culture vessels [40].

Thus far, a variety of stem cells have been investigated for cell sheet-based periodontal regeneration, such as PDLSCs [41, 44], DPSCs [45], SHEDs [46], and MSCs [40]. For example, transplantation of PDLSCs cell sheets into rats resulted in the regeneration of periodontal ligament-like tissues, which contained an acellular cementum-like layer, with fibers inserting into this layer [47]. Similar studies have been performed in large animals, including in dogs [48, 49].

2.6 Stem Cell-Based Strategy for Whole Tooth Regeneration

Since the tooth is an intact organ, successful regeneration of individual tooth components does not replace an entire missing tooth. As a stem cell-based strategy for whole tooth regeneration, Ikeda et al. generated a functional whole tooth unit, by mixing immature dental DESCs and dental mesenchymal stem cells (DMSCs) to form a tooth germ. This bioengineered tooth germ was then transplanted into the recipient's alveolar bone socket. It developed into a mature tooth that was integrated in the normal way with the surrounding alveolar bone, and it achieved masticatory function. Thus, the reconstituted tooth germ could be induced into an ex vivo whole tooth unit, including surrounding tissues such as periodontal ligaments and alveolar bone [50]. Similarly, a bioengineered tooth was generated from transplanted permanent tooth germs that were reconstituted by autologous DESCs and DMSCs in a postnatal canine model [51].

Overall, dental stem cells represent an attractive source for the repair of each individual damaged component of the dental structure, as well as for the whole tooth unit. Biocompatible scaffolds can provide growing niches for stem cell proliferation and differentiation, and can direct tooth tissue regeneration. This aspect is discussed in details in Sect. 3.

3 Biomaterials-Based Approaches for Tooth Engineering

Another key factor for tooth engineering relies on the development of biomaterials, based on the understanding of tooth development and the biological process of self-repair. Several natural and synthetic materials have been investigated for regenerating enamel, the dentine-pulp complex, periodontal tissue, and the whole tooth, that provide a favorable microenvironment for specific cellular functions.

3.1 Biomaterials for Enamel Regeneration

Enamel is the hardest tissue of the human body and has unique nano- and micro-structures. During natural tooth development, enamel formation (known as amelogenesis) is a highly orchestrated process involving the deposition of inorganic apatite in a manner that is controlled precisely by ameloblasts. For enamel regeneration, inorganic precursors are stabilized by phosphoproteins and then mineralize, to form orientated apatite crystals.

An interesting study conducted by Shao et al. showed the stabilized calcium phosphates cluster with a removable small organic molecule, triethylamine, to establish a crystalline-amorphous mineralization frontier, and induce the growth of HAP. Notably, the newly grown HAP layer can integrate into the native enamel. This provides the potential of a new approach for treating mineral that has been lost by enamel erosion [52].

In addition to small molecules, phosphopeptides guide the crystallization of calcium phosphate. These have been used in dental products for enamel remineralization. In the laboratory, they can regrow enamel crystals in a controlled shape and direction. By adding a leucine-rich amelogenin peptide, a dense, structural orientated enamel-like apatite layer could be formed on the enamel surfaces, in alignment with the enamel rods [53].

P_{11}-4 is another promising peptide for enamel regeneration. This peptide has previously been shown to be biocompatible in a phase I clinical trial. It can modulate mineral behavior in situ during enamel regeneration [54, 55].

3.2 Biomaterials for Dentine-Pulp Complex Regeneration

Organic biomaterials, including both natural and synthesized polymers, have been used to fabricate scaffolds for regenerating the dentine-pulp complex. Fujiwara and co-workers transplanted alginate scaffolds to deliver DPSCs *in vivo* to induce the subcutaneous formation of dentine-like tissues [56, 57].

Chitosan and hyaluronic acid derived scaffolds have also been shown to support DPSC growth and differentiation. Moreover, a carboxymethyl cellulose/chitosan composite scaffold has been shown to improve the adhesion and proliferation of pulp cells [58]. Other naturally derived proteins, including fibrin, silk, and collagen, have been extensively used for dentine-pulp tissue engineering [59, 60]. The addition of growth factors, such as bFGF, could significantly enhance pulp cell proliferation and vascularity [61].

On the other hand, synthetic polymers, such as polylactide-co-glycolide and polylactic acid (PLA), are used for dental pulp tissue engineering. SHED cells have been seeded into tooth slice/PLA biodegradable scaffolds, and the regenerated tissues from this recapture the architecture and cellularity of native dental pulp [62]. Similarly, Huang et al. used PLGA scaffolds containing a mixture of dental stem cells, DPSCs and SCAP, and this formed dental pulp- and dentine-like tissues in the root canal of tooth roots [34].

To facilitate dentine regeneration, inorganic materials have been used to induce dental stem cell differentiation. One such example is the use of porous hydroxyapatite/beta-tricalcium phosphate seeded with DPSCs. This led to the formation of a dentine-like hard tissue and adjacent odontoblast-like cells when it was transplanted into the dorsal space of immunodeficient mice [63].

Bioactive glass, another widely used inorganic scaffold, has been reported to support DPSC growth and osteogenic gene expression. Implantation of such cell-scaffold constructs *in vivo* can form bone-like calcified tissue [64].

Composite scaffolds synthesized from organic and inorganic components can induce the regeneration of dental pulp and dentine. For instance, PLGA/TCP scaffolds promote the proliferation and osteogenic differentiation of DPSCs. When seeded with rat tooth bud cells, these PLGA/TCP scaffolds produce tissues that resemble dentine and pulp [65]. Hybrid scaffolds of mineral trioxide aggregate/PCL (MTA/PCL) have also been used to enhance HDPC adhesion, proliferation, and osteogenesis [66].

3.3 Biomaterials Applied for Periodontal Regeneration

Scaffold biomaterials are also suitable for the regeneration of periodontal tissues (Fig. 2). Polymeric materials, such as collagen [67], gelatin [68, 69], and chitosan [70], have relatively low mechanical strength, and seem appropriate for PDL regeneration. On the other hand, inorganic materials, such as hydroxyapatite (HA) [71, 72], tricalcium phosphate (TCP) [73, 74], biphasic calcium phosphate (BCP) [75, 76], and bioactive glass (BG) [77, 78] are used commonly used to repair cementum and alveolar bone defects due to their relatively high mechanical strength.

Importantly, composite scaffolds made up of both polymers and inorganic components are promising candidates for the regeneration of the cementum–PDL–alveolar bone complex [79–81].

Surface modifications of functional scaffolds may be beneficial for periodontal regeneration. Recently, a chitosan-based scaffold modified with a tri-layer porous structure has been shown to achieve simultaneous healing of bone, gingiva, and PDL regeneration in three porous compartments [70]. Also, HA bioceramics with a micro-nano-hybrid surface (mnHA) can promote osteogenic/cementogenic differentiation of

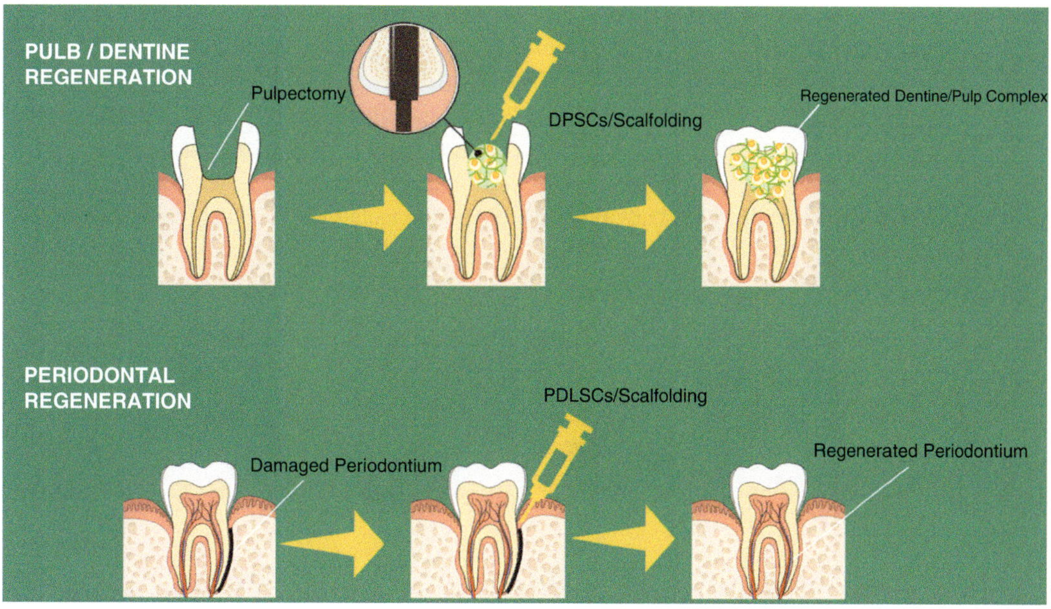

Fig. 2 Schematic representation of biomaterials-based regenerative approaches in dentistry

human PLSCs, in a manner superior to unmodified HA bioceramics [82].

Scaffolds can also be used for drug delivery. As one example, ibuprofen (IBU) is clinically approved for the treatment of postoperative pain after periodontal surgery. A recent study reported success in developing a functionalized poly-ε-caprolactone scaffold with IBU, to enhance its anti-inflammatory effects and promote periodontal healing [83].

Growth factors also enhance outcomes. These can be delivered by scaffold biomaterials, for example, when treating periodontal diseases. A biodegradable gelatin/β-tricalcium phosphate sponge has been used to deliver recombinant human fibroblast growth factor-2 in a dog model of buccal gingival recession defects. Complete root coverage of newly formed cementum-PDL-alveolar bone tissues was achieved [81].

Another study demonstrated spatially delivered growth factors, such as amelogenin, BMP-2, and connective tissue growth factor, with a 3D printed multiphase scaffold. Interestingly, all three growth factors were time-released at each phase from each compartment, which yielded cementum-PDL-alveolar bone when seeded with DPSCs [84].

3.4 Biomaterials for Whole Tooth Regeneration

To advance the concept of whole tooth regeneration, either a decellularized tooth or a tooth-shaped scaffold can be seeded with dental cells alone or with growth factors. Several studies have been conducted with the goal being to form bioengineered teeth of comparable size and histological structure to natural teeth, with well-organized dentine and enamel-like tissues, for use in *in vivo* transplantation [85–87].

In other studies, tooth bud constructs were grown from GelMA hydrogel containing postnatal dental cells. These showed key features involved in natural tooth development, such as the formation of a dental epithelial stem cell niche, transient amplifying cells, and mineralization of dental tissue [88]. Another similar example is the use of biodegradable PGA/PLGA scaffolds seeded with 4-day-old postnatal rat tooth bud cells, to repeatedly generate bioengineered whole tooth tissues [89]. Mao's group has incorporated two growth factors, SDF-1 and BMP-7, into tooth-shaped scaffolds to induce host multi-lineage cell recruitment, and to achieve regeneration of an entire anatomically shaped tooth. They referred to this translatable process as cell homing [90, 91].

4 Bioengineered Tooth Germ-Based Methods for Functional Tooth Replacement

Autologous transplantation of a tooth germ, such as that from an impacted third molar, into an edentulous area, has been investigated in clinical practice. This approach is limited by feasibility, and by problems such as pulp necrosis and tooth ankylosis.

A bioengineered tooth germ provides a more promising method (Fig. 3). Takashi's group first introduced this bioengineered tooth germ transplantation technique in 2007. They incubated dissociated cells from epithelia and mesenchymal

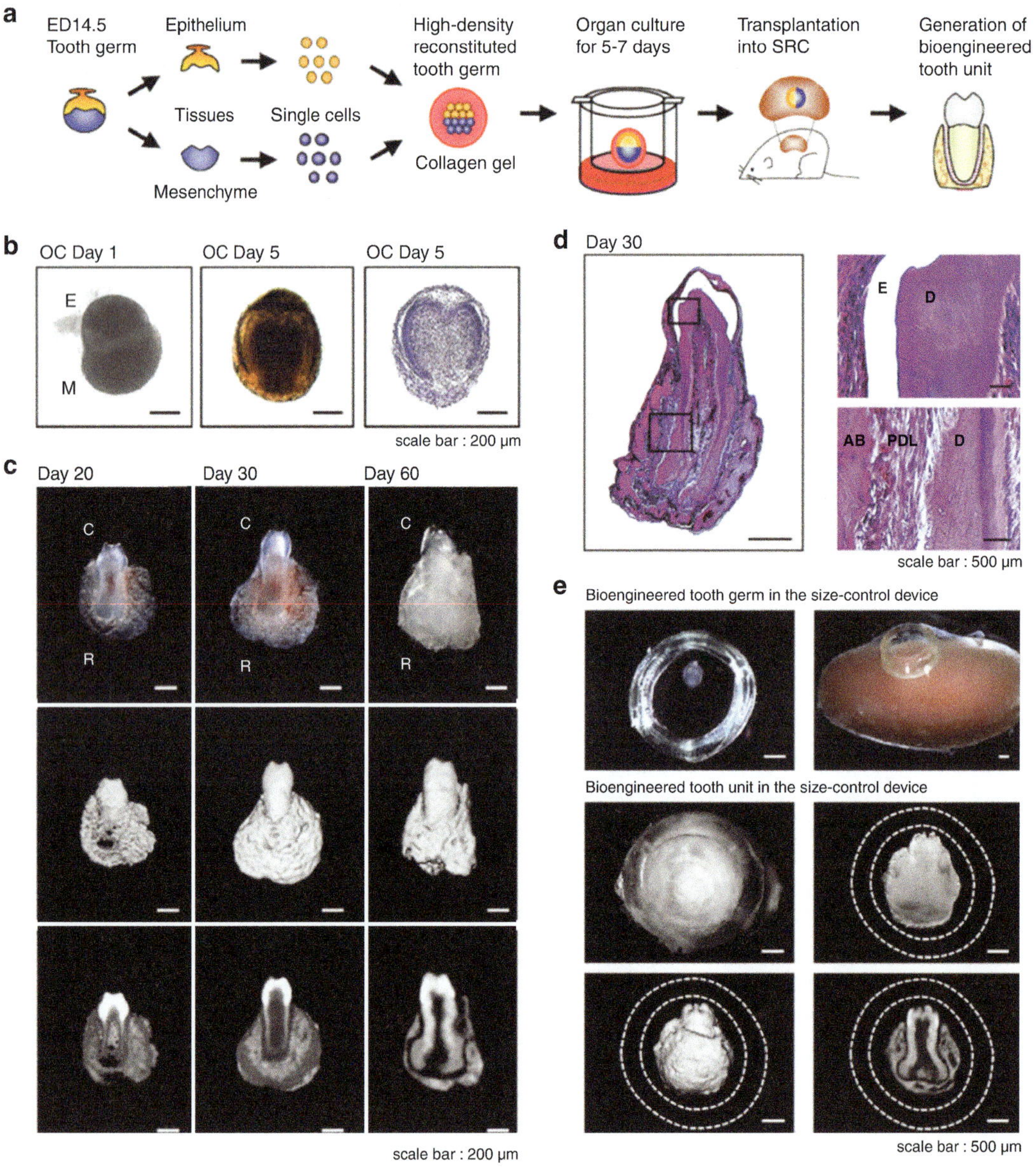

Fig. 3 Bioengineered tooth germ-based methods for functional tooth replacement. (**a**) Schematic drawing of the process of bioengineered tooth germ. Phase-contrast images (**b**), photos and micro-CT images (**c**), and histological analysis (**d**) of bioengineered tooth germ on different days. (**e**) Photos and micro-CT images of bioengineered tooth germs cultured in a size-control device. Reprinted with permission from [50]

tissues and used a 3D organ-germ culture method [92]. After transplantation into the subcutaneous space of a rodent, the bioengineered tooth germ developed into a functional tooth unit. It established an adequate supply of nerves and blood vessels, and exhibited mechanical properties that were similar to a natural tooth [92]. An erupted tooth from a bioengineered tooth germ can also respond in the normal way to orthodontic treatment, which indicates that they have proper periodontal ligament function [50, 93].

The bioengineered tooth germ technique has also been evaluated in a large animal model. In dogs, these transplanted tooth germs created teeth that eventually erupted in the oral cavity and came into function [51]. Numerous efforts have been made to optimize methods of reconstructing a tooth germ for whole tooth regeneration. Such attempts include the combination of epithelial tissues with mesenchymal tissues, or epithelial cells with mesenchymal cells, to promote tooth germ formation [50].

5 Conclusion and Future Perspectives

In summary, with the development and recent advances in stem cell and biomaterial research, considerable progress has been achieved toward the goal of tooth engineering and whole tooth regeneration. Rationally designed scaffolds together with advanced culture methods provide a suitable environment for the development of stem cells, including DSCs and iPSCs, to form different parts of the tooth, or even a whole tooth.

Despite the impressive progress in the field, several major obstacles remain before the concept of an engineered tooth can be applied in clinical practice. The first challenge is how to achieve the correct type, size, shape, and color of the original teeth, in order to achieve both an esthetic outcome and a functional occlusion. Previous studies have reported the use of a three-dimensional size- and shape-controlled scaffold device to guide the generation of bioengineered tooth unit as it forms, before its transplantation into the oral cavity [50]. Future efforts should be placed on ways to control the various aspects of tooth parameters, to meet the objective of restoring the missing tooth faithfully and accurately. Current attempts include inducible scaffolds [94], 3D printing biotechnologies [95], and computer-aided design/computer-aided manufacturing [96]. Other works are focusing on regulating morphogenesis-related genes, and cytokines that are involved in tooth eruption and movement [97], to apply these approaches to regulate the morphology of the regenerated tooth. Further work is needed around how well the bioengineered enamel responds to the loads and acid attacks that it will experience in the oral cavity, and whether it behaves in the same manner as natural dental enamel.

A second obstacle is how to achieve the full function of the regenerated tooth with vascularization, innervation, and supporting tissues. Similar to a naturally growing tooth, a successful transplanted bioengineered tooth germ should develop into an intact viable tooth, and be supported by proper vascularization and innervation, to provide the necessary nutrients and metabolites, and the sensory response to various stimuli. It must also show similar features in how it responds to infection and inflammation. The vascularization and innervation of a regenerated tooth make it unique when compared to a dental implant or a denture restoration. The latter are inert. They also lack the connective tissues that provide mechanical resistance against fracture.

Moreover, periodontal supporting tissues, including the periodontal ligament and/or bone tissue, are essential for the proper functional integration of a transplanted tooth within the host jaw. Recent progress has been made to generate highly vascularized and innervated dental pulp, such as mixing cells with a microstructural hydrogel [98], or adding bone marrow-derived cells [99]. Interestingly, a light sheet microscopy-based 3D imaging method has been reported to visualize the vasculature and innervation of the dental pulp. This provides remarkable insights that can help progress endodontic regeneration and whole tooth regeneration therapies [100]. Further work on whole tooth regeneration should always explore whether there has been simultaneous growth of vessels, nerves, and supporting periodontal tissues.

A third challenge is how to maintain oral health and prevent the failure of the regenerated tooth. Common and well-known causes of tooth loss are microbial diseases such as dental caries and periodontitis. Dental care of patients remains challenging, especially for those who suffer from systemic disorders including hematologic conditions, cardiovascular diseases, cancer, endocrine and neurologic disorders, and radiation-induced damage to the oral tissues [101]. Therefore, oral hygiene should be improved, and further maintenance should be conducted on a regular basis to prevent the failure of a bioengineered tooth. One promising strategy has involved the study of natural salivary defenses and the oral microbiota as an ecosystem, which needs to be maintained for a symbiotic relationship with the oral microflora [102, 103]. The long-term success of a regenerated whole tooth relies on multifaceted factors. The future of whole tooth regeneration will certainly benefit from improved knowledge of stem cell therapy, and advances in bio-scaffold engineering.

References

1. Peres MA, Macpherson LMD, Weyant RJ, Daly B, Venturelli R, Mathur MR, et al. Oral diseases: a global public health challenge. Lancet. 2019;394(10194):249–60. https://doi.org/10.1016/S0140-6736(19)31146-8.
2. Volponi AA, Pang Y, Sharpe PT. Stem cell-based biological tooth repair and regeneration. Trends Cell Biol. 2010;20(12):715–22. https://doi.org/10.1016/j.tcb.2010.09.012.
3. Mehanna H, Beech T, Nicholson T, El-Hariry I, McConkey C, Paleri V, et al. Prevalence of human papillomavirus in oropharyngeal and nonoropharyngeal head and neck cancer—systematic review and meta-analysis of trends by time and region. Head Neck. 2013;35(5):747–55. https://doi.org/10.1002/hed.22015.
4. Petti S, Glendor U, Andersson L. World traumatic dental injury prevalence and incidence, a meta-analysis-one billion living people have had traumatic dental injuries. Dent Traumatol. 2018;34(2):71–86. https://doi.org/10.1111/edt.12389.
5. Mossey PA, Little J, Munger RG, Dixon MJ, Shaw WC. Cleft lip and palate. Lancet. 2009;374(9703):1773–85. https://doi.org/10.1016/S0140-6736(09)60695-4.
6. Jeyapalan V, Krishnan CS. Partial edentulism and its correlation to age, gender, socio-economic status and incidence of various Kennedy's classes – a literature review. J Clin Diagn. 2015;9(6):ZE14–7. https://doi.org/10.7860/JCDR/2015/13776.6124.
7. Pozhitkov AE, Leroux BG, Randolph TW, Beikler T, Flemmig TF, Noble PA. Towards microbiome transplant as a therapy for periodontitis: an exploratory study of periodontitis microbial signature contrasted by oral health, caries and edentulism. BMC Oral Health. 2015;15:125. https://doi.org/10.1186/s12903-015-0109-4.
8. Payne AG, Alsabeeha NH, Atieh MA, Esposito M, Ma S, Anas El-Wegoud M. Interventions for replacing missing teeth: attachment systems for implant overdentures in edentulous jaws. Cochrane Database Syst Rev. 2018;10:CD008001. https://doi.org/10.1002/14651858.CD008001.pub2.
9. Maiorana C, Andreoni D, Polacco P, Poli PP. Multidisciplinary oral rehabilitation of a severely compromised dentition. Case Rep Dent. 2020;2020:2429505. https://doi.org/10.1155/2020/2429505.
10. Takahashi K, Yamanaka S. Induction of pluripotent stem cells from mouse embryonic and adult fibroblast cultures by defined factors. Cell. 2006;126(4):663–76. https://doi.org/10.1016/j.cell.2006.07.024.
11. Gronthos S, Mankani M, Brahim J, Robey PG, Shi S. Postnatal human dental pulp stem cells (DPSCs) in vitro and in vivo. Proc Natl Acad Sci USA. 2000;97(25):13625–30. https://doi.org/10.1073/pnas.240309797.
12. Volpe G, Bernstock JD, Peruzzotti-Jametti L, Pluchino S. Modulation of host immune responses following non-hematopoietic stem cell transplantation: translational implications in progressive multiple sclerosis. J Neuroimmunol. 2019;331:11–27.
13. Luo L, Albashari AA, Wang X, Jin L, Zhang Y, Zheng L, et al. Effects of transplanted heparin-poloxamer hydrogel combining dental pulp stem cells and bFGF on spinal cord injury repair. Stem Cells Int. 2018;2398521:1. https://doi.org/10.1155/2018/2398521.
14. Wang H, Zhong Q, Yang T, Qi Y, Fu M, Yang X, et al. Comparative characterization of SHED and DPSCs during extended cultivation in vitro. Mol Med Rep. 2018;17(5):6551–9.
15. Ballini A, Cantore S, Scacco S, Perillo L, Scarano A, Aityan S, et al. A comparative study on different stemness gene expression between dental pulp stem cells vs. dental bud stem cells. Eur Rev Med Pharmacol Sci. 2019;23(4):1626–33.
16. Mautner K, Carr D, Whitley J, Bowers R. Allogeneic versus autologous injectable mesenchymal stem cells for knee osteoarthritis: review and current status. Tech Orthopaedics. 2019;34(4):244–56.
17. Nakamura S, Yamada Y, Katagiri W, Sugito T, Ito K, Ueda M. Stem cell proliferation pathways comparison between human exfoliated deciduous teeth and dental

pulp stem cells by gene expression profile from promising dental pulp. J Endod. 2009;35(11):1536–42. https://doi.org/10.1016/j.joen.2009.07.024.
18. Sakai VT, Zhang Z, Dong Z, Neiva KG, Machado MAAM, Shi S, et al. SHED differentiate into functional odontoblasts and endothelium. J Dent Res. 2010;89(8):791–6. https://doi.org/10.1177/0022034510368647.
19. Seo BM, Miura M, Gronthos S, Bartold PM, Batouli S, Brahim J, et al. Investigation of multipotent postnatal stem cells from human periodontal ligament. Lancet. 2004;364(9429):149–55. https://doi.org/10.1016/S0140-6736(04)16627-0.
20. Sonoyama W, Liu Y, Fang DAJ, Yamaza T, Seo BM, Zhang CM, et al. Mesenchymal stem cell-mediated functional tooth regeneration in swine. PLoS One. 2006;1(1):e79. https://doi.org/10.1371/journal.pone.0000079.
21. Pizzicannella J, Gugliandolo A, Orsini T, Fontana A, Ventrella A, Mazzon E, et al. Engineered extracellular vesicles from human periodontal-ligament stem cells increase VEGF/VEGFR2 expression during bone regeneration. Front Phys. 2019;10:512. https://doi.org/10.3389/fphys.2019.00512.
22. Liu Y, Zheng Y, Ding G, Fang DJ, Zhang CM, Bartold PM, et al. Periodontal ligament stem cell-mediated treatment for periodontitis in miniature swine. Stem Cells. 2008;26(4):1065–73. https://doi.org/10.1634/stemcells.2007-0734.
23. Blufstein A, Behm C, Gahn J, Uitz O, Naumovska I, Moritz A, et al. Synergistic effects triggered by simultaneous toll-like receptor-2 and -3 activation in human periodontal ligament stem cells. J Periodontol. 2019;90(10):1190–201. https://doi.org/10.1002/JPER.19-0005.
24. Huang GTJ, Sonoyama W, Liu Y, Liu H, Wang SL, Shi ST. The hidden treasure in apical papilla: the potential role in pulp/dentin regeneration and bioroot engineering. J Endod. 2008;34(6):645–51. https://doi.org/10.1016/j.joen.2008.03.001.
25. Handa K, Saito M, Tsunoda A, Yamauchi M, Hattori S, Sato S, et al. Progenitor cells from dental follicle are able to form cementum matrix in vivo. Connect Tissue Res. 2002;43(2–3):406–8. https://doi.org/10.1080/03008200290001023.
26. Handa K, Saito M, Yamauchi M, Kiyono T, Sato S, Teranaka T. Cementum matrix formation by cultured dental follicle cells in vivo. J Dent Res. 2003;82:403.
27. Zhang J, Ding H, Liu X, Sheng Y, Liu X, Jiang C. Dental follicle stem cells: tissue engineering and immunomodulation. Stem Cells Dev. 2019;28(15):986–94. https://doi.org/10.1089/scd.2019.0012.
28. Cai J, Zhang Y, Liu P, Chen S, Wu X, Sun Y, et al. Generation of tooth-like structures from integration-free human urine induced pluripotent stem cells. Cell Regen. 2013;2(1):6. https://doi.org/10.1186/2045-9769-2-6.
29. Abdullah AN, Miyauchi S, Onishi A, Tanimoto K, Kato K. Differentiation of mouse-induced pluripotent stem cells into dental epithelial-like cells in the absence of added serum. In Vitro Cell Dev Biol Anim. 2019;55(2):130–7. https://doi.org/10.1007/s11626-019-00320-z.
30. Xie H, Dubey N, Shim W, Ramachandra CJA, Min KS, Cao T, et al. Functional odontoblastic-like cells derived from human iPSCs. J Dent Res. 2018;97(1):77–83. https://doi.org/10.1177/0022034517730026.
31. Li F, Niyibizi C. Cells derived from murine induced pluripotent stem cells (iPSC) by treatment with members of TGF-beta family give rise to osteoblasts differentiation and form bone in vivo. BMC Cell Biol. 2012;13. https://doi.org/10.1186/1471-2121-13-35.
32. Duan XJ, Tu QS, Zhang J, Ye JH, Sommer C, Mostoslavsky G, et al. Application of induced pluripotent stem (iPS) cells in periodontal tissue regeneration. J Cell Physiol. 2011;226(1):150–7. https://doi.org/10.1002/jcp.22316.
33. Hynes K, Menicanin D, Han J, Marino V, Mrozik K, Gronthos S, et al. Mesenchymal stem cells from iPS cells facilitate periodontal regeneration. J Dent Res. 2013;92(9):833–9. https://doi.org/10.1177/0022034513498258.
34. Huang GTJ, Yamaza T, Shea LD, Djouad F, Kuhn NZ, Tuan RS, et al. Stem/progenitor cell-mediated de novo regeneration of dental pulp with newly deposited continuous layer of dentin in an in vivo model. Tissue Eng Part A. 2010;16(2):605–15. https://doi.org/10.1089/ten.tea.2009.0518.
35. Arakaki M, Ishikawa M, Nakamura T, Iwamoto T, Yamada A, Fukumoto E, et al. Role of epithelial-stem cell interactions during dental cell differentiation. J Biol Chem. 2012;287(13):10590–601. https://doi.org/10.1074/jbc.M111.285874.
36. Otsu K, Kishigami R, Oikawa-Sasaki A, Fukumoto S, Yamada A, Fujiwara N, et al. Differentiation of induced pluripotent stem cells into dental mesenchymal cells. Stem Cells Dev. 2012;21(7):1156–64. https://doi.org/10.1089/scd.2011.0210.
37. Yamada Y, Ueda M, Hibi H, Baba S. A novel approach to periodontal tissue regeneration with mesenchymal stem cells and platelet-rich plasma using tissue engineering technology: a clinical case report. Int J Periodont Restorative Dent. 2006;26(4):363–9.
38. Bartold PM, Shi ST, Gronthos S. Stem cells and periodontal regeneration. Periodontol 2000. 2006;40:164–72. https://doi.org/10.1111/j.1600-0757.2005.00139.x.
39. Baik HS, Park J, Lee KJ, Chung C. Local application of periodontal ligament stromal cells promotes soft tissue regeneration. Oral Dis. 2014;20(6):574–81. https://doi.org/10.1111/odi.12175.
40. Du J, Shan Z, Ma P, Wang S, Fan Z. Allogeneic bone marrow mesenchymal stem cell transplantation for periodontal regeneration. J Dent Res. 2014;93(2):183–8. https://doi.org/10.1177/0022034513513026.
41. Ding G, Liu Y, Wang W, Wei FL, Liu DY, Fan ZP, et al. Allogeneic periodontal ligament stem cell therapy for periodontitis in swine. Stem Cells. 2010;28(10):1829–38. https://doi.org/10.1002/stem.512.

42. Liu OS, Xu JJ, Ding G, Liu DY, Fan ZP, Zhang CM, et al. Periodontal ligament stem cells regulate B lymphocyte function via programmed cell death protein 1. Stem Cells. 2013;31(7):1371–82. https://doi.org/10.1002/stem.1387.
43. Owaki T, Shimizu T, Yamato M, Okano T. Cell sheet engineering for regenerative medicine: current challenges and strategies. Biotechnol J. 2014;9(7):904–14. https://doi.org/10.1002/biot.201300432.
44. Ding G, Wang W, Liu Y, An YQ, Zhang CM, Shi ST, et al. Effect of cryopreservation on biological and immunological properties of stem cells from apical papilla. J Cell Physiol. 2010;223(2):415–22. https://doi.org/10.1002/jcp.22050.
45. Hu JC, Cao Y, Xie YL, Wang H, Fan ZP, Wang JS, et al. Periodontal regeneration in swine after cell injection and cell sheet transplantation of human dental pulp stem cells following good manufacturing practice. Stem Cell Res Ther. 2016;7:130. https://doi.org/10.1186/s13287-016-0362-8.
46. Fu XR, Jin LY, Ma P, Fan ZP, Wang SL. Allogeneic stem cells from deciduous teeth in treatment for periodontitis in miniature swine. J Periodontol. 2014;85(6):845–51. https://doi.org/10.1902/jop.2013.130254.
47. Hasegawa M, Yamato M, Kikuchi A, Okano T, Ishikawa I. Human periodontal ligament cell sheets can regenerate periodontal ligament tissue in an athymic rat model. Tissue Eng. 2005;11(3–4):469–78. https://doi.org/10.1089/ten.2005.11.469.
48. Akizuki T, Oda S, Komaki M, Tsuchioka H, Kawakatsu N, Kikuchi A, et al. Application of periodontal ligament cell sheet for periodontal regeneration: a pilot study in beagle dogs. J Periodontal Res. 2005;40(3):245–51. https://doi.org/10.1111/j.1600-0765.2005.00799.x.
49. Iwata T, Yamato M, Tsuchioka H, Takagi R, Mukobata S, Washio K, et al. Periodontal regeneration with multilayered periodontal ligament-derived cell sheets in a canine model. Biomaterials. 2009;30(14):2716–23. https://doi.org/10.1016/j.biomaterials.2009.01.032.
50. Oshima M, Mizuno M, Imamura A, Ogawa M, Yasukawa M, Yamazaki H, et al. Functional tooth regeneration using a bioengineered tooth unit as a mature organ replacement regenerative therapy. PLoS One. 2011;6(7):e21531. https://doi.org/10.1371/journal.pone.0021531.
51. Ono M, Oshima M, Ogawa M, Sonoyama W, Hara ES, Oida Y, et al. Practical whole-tooth restoration utilizing autologous bioengineered tooth germ transplantation in a postnatal canine model. Sci Rep. 2017;(7):44522. https://doi.org/10.1038/srep44522.
52. Shao C, Jin B, Mu Z, Lu H, Zhao Y, Wu Z, et al. Repair of tooth enamel by a biomimetic mineralization frontier ensuring epitaxial growth. Sci Adv. 2019;5(8):eaaw9569. https://doi.org/10.1126/sciadv.aaw9569.
53. Kwak SY, Litman A, Margolis HC, Yamakoshi Y, Simmer JP. Biomimetic enamel regeneration mediated by leucine-rich amelogenin peptide. J Dent Res. 2017;96(5):524–30. https://doi.org/10.1177/0022034516688659.
54. Kirkham J, Firth A, Vernals D, Boden N, Robinson C, Shore RC, et al. Self-assembling peptide scaffolds promote enamel remineralization. J Dent Res. 2007;86(5):426–30. https://doi.org/10.1177/154405910708600507.
55. Kind L, Stevanovic S, Wuttig S, Wimberger S, Hofer J, Muller B, et al. Biomimetic remineralization of carious lesions by self-assembling peptide. J Dent Res. 2017;96(7):790–7. https://doi.org/10.1177/0022034517698419.
56. Fujiwara S, Kumabe S, Iwai Y. Isolated rat dental pulp cell culture and transplantation with an alginate scaffold. Okajimas Folia Anat Jpn. 2006;83(1):15–24. https://doi.org/10.2535/ofaj.83.15.
57. Kumabe S, Nakatsuka M, Kim GS, Jue SS, Aikawa F, Shin JW, et al. Human dental pulp cell culture and cell transplantation with an alginate scaffold. Okajimas Folia Anat Jpn. 2006;82(4):147–55. https://doi.org/10.2535/ofaj.82.147.
58. Chen H, Fan M. Chitosan/carboxymethyl cellulose polyelectrolyte complex scaffolds for pulp cells regeneration. J Bioact Compat Polym. 2013;22(5):475–91.
59. Galler KM, Hartgerink JD, Cavender AC, Schmalz G, D'Souza RN. A customized self-assembling peptide hydrogel for dental pulp tissue engineering. Tissue Eng Part A. 2012;18(1–2):176–84. https://doi.org/10.1089/ten.TEA.2011.0222.
60. Suzuki T, Lee CH, Chen M, Zhao W, Fu SY, Qi JJ, et al. Induced migration of dental pulp stem cells for in vivo pulp regeneration. J Dent Res. 2011;90(8):1013–8. https://doi.org/10.1177/0022034511408426.
61. Yang JW, Zhang YF, Sun ZY, Song GT, Chen Z. Dental pulp tissue engineering with bFGF-incorporated silk fibroin scaffolds. J Biomater Appl. 2015;30(2):221–9. https://doi.org/10.1177/0885328215577296.
62. Cordeiro MM, Dong Z, Kaneko T, Zhang Z, Miyazawa M, Shi S, et al. Dental pulp tissue engineering with stem cells from exfoliated deciduous teeth. J Endod. 2008;34(8):962–9. https://doi.org/10.1016/j.joen.2008.04.009.
63. Tonomura A, Mizuno D, Hisada A, Kuno N, Ando Y, Sumita Y, et al. Differential effect of scaffold shape on dentin regeneration. Ann Biomed Eng. 2010;38(4):1664–71. https://doi.org/10.1007/s10439-010-9910-z.
64. El-Gendy R, Yang XB, Newby PJ, Boccaccini AR, Kirkham J. Osteogenic differentiation of human dental pulp stromal cells on 45S5 bioglass(R) based scaffolds in vitro and in vivo. Tissue Eng Part A. 2013;19(5–6):707–15. https://doi.org/10.1089/ten.TEA.2012.0112.
65. Zheng L, Yang F, Shen H, Hu X, Mochizuki C, Sato M, et al. The effect of composition of calcium phosphate composite scaffolds on the formation of tooth tissue from human dental pulp stem cells. Biomaterials. 2011;32(29):7053–9. https://doi.org/10.1016/j.biomaterials.2011.06.004.

66. Chiu YC, Fang HY, Hsu TT, Lin CY, Shie MY. The characteristics of mineral trioxide aggregate/polycaprolactone 3-dimensional scaffold with osteogenesis properties for tissue regeneration. J Endod. 2017;43(6):923–9. https://doi.org/10.1016/j.joen.2017.01.009.
67. Momose T, Miyaji H, Kato A, Ogawa K, Yoshida T, Nishida E, et al. Collagen hydrogel scaffold and fibroblast growth factor-2 accelerate periodontal healing of class II furcation defects in dog. Open Dent J. 2016;10:347–59. https://doi.org/10.2174/1874210601610010347.
68. Chen X, Bai S, Li B, Liu H, Wu G, Liu S, et al. Fabrication of gelatin methacrylate/nanohydroxyapatite microgel arrays for periodontal tissue regeneration. Int J Nanomedicine. 2016;11:4707–18. https://doi.org/10.2147/IJN.S111701.
69. Li Z, Qu T, Ding C, Ma C, Sun H, Li S, et al. Injectable gelatin derivative hydrogels with sustained vascular endothelial growth factor release for induced angiogenesis. Acta Biomater. 2015;13:88–100. https://doi.org/10.1016/j.actbio.2014.11.002.
70. Varoni EM, Vijayakumar S, Canciani E, Cochis A, De Nardo L, Lodi G, et al. Chitosan-based trilayer scaffold for multitissue periodontal regeneration. J Dent Res. 2018;97(3):303–11. https://doi.org/10.1177/0022034517736255.
71. Kawase T, Okuda K, Kogami H, Nakayama H, Nagata M, Sato T, et al. Human periosteum-derived cells combined with superporous hydroxyapatite blocks used as an osteogenic bone substitute for periodontal regenerative therapy: an animal implantation study using nude mice. J Periodontol. 2010;81(3):420–7. https://doi.org/10.1902/jop.2009.090523.
72. Lee JS, Park WY, Cha JK, Jung UW, Kim CS, Lee YK, et al. Periodontal tissue reaction to customized nano-hydroxyapatite block scaffold in one-wall intrabony defect: a histologic study in dogs. J Periodontal Implant Sci. 2012;42(2):50–8. https://doi.org/10.5051/jpis.2012.42.2.50.
73. Matsuura T, Akizuki T, Hoshi S, Ikawa T, Kinoshita A, Sunaga M, et al. Effect of a tunnel-structured beta-tricalcium phosphate graft material on periodontal regeneration: a pilot study in a canine one-wall intrabony defect model. J Periodontal Res. 2015;50(3):347–55. https://doi.org/10.1111/jre.12213.
74. Maroo S, Murthy KR. Treatment of periodontal intrabony defects using beta-TCP alone or in combination with rhPDGF-BB: a randomized controlled clinical and radiographic study. Int J Periodontics Restorative Dent. 2014;34(6):841–7. https://doi.org/10.11607/prd.2030.
75. Santos PS, Cestari TM, Paulin JB, Martins R, Rocha CA, Arantes RVN, et al. Osteoinductive porous biphasic calcium phosphate ceramic as an alternative to autogenous bone grafting in the treatment of mandibular bone critical-size defects. J Biomed Mater Res B Appl Biomater. 2018;106(4):1546–57. https://doi.org/10.1002/jbm.b.33963.
76. Bansal R, Patil S, Chaubey KK, Thakur RK, Goyel P. Clinical evaluation of hydroxyapatite and beta-tricalcium phosphate composite graft in the treatment of intrabony periodontal defect: a clinico-radiographic study. J Indian Soc Periodontol. 2014;18(5):610–7. https://doi.org/10.4103/0972-124X.142455.
77. Carvalho SM, Oliveira AA, Jardim CA, Melo CB, Gomes DA, de Fatima Leite M, et al. Characterization and induction of cementoblast cell proliferation by bioactive glass nanoparticles. J Tissue Eng Regen Med. 2012;6(10):813–21. https://doi.org/10.1002/term.488.
78. Wu C, Zhou Y, Lin C, Chang J, Xiao Y. Strontium-containing mesoporous bioactive glass scaffolds with improved osteogenic/cementogenic differentiation of periodontal ligament cells for periodontal tissue engineering. Acta Biomater. 2012;8(10):3805–15. https://doi.org/10.1016/j.actbio.2012.06.023.
79. Liu Z, Yin X, Ye Q, He W, Ge M, Zhou X, et al. Periodontal regeneration with stem cells-seeded collagen-hydroxyapatite scaffold. J Biomater Appl. 2016;31(1):121–31. https://doi.org/10.1177/0885328216637978.
80. Brown A, Zaky S, Ray H Jr, Sfeir C. Porous magnesium/PLGA composite scaffolds for enhanced bone regeneration following tooth extraction. Acta Biomater. 2015;11:543–53. https://doi.org/10.1016/j.actbio.2014.09.008.
81. Shujaa Addin A, Akizuki T, Hoshi S, Matsuura T, Ikawa T, Fukuba S, et al. Biodegradable gelatin/beta-tricalcium phosphate sponges incorporating recombinant human fibroblast growth factor-2 for treatment of recession-type defects: a split-mouth study in dogs. J Periodontal Res. 2017;52(5):863–71. https://doi.org/10.1111/jre.12456.
82. Mao L, Liu J, Zhao J, Chang J, Xia L, Jiang L, et al. Effect of micro-nano-hybrid structured hydroxyapatite bioceramics on osteogenic and cementogenic differentiation of human periodontal ligament stem cell via Wnt signaling pathway. Int J Nanomedicine. 2015;10:7031–44. https://doi.org/10.2147/IJN.S90343.
83. Batool F, Morand DN, Thomas L, Bugueno IM, Aragon J, Irusta S, et al. Synthesis of a novel electrospun polycaprolactone scaffold functionalized with ibuprofen for periodontal regeneration: an in vitro and in vivo study. Materials (Basel). 2018;11(4):580. https://doi.org/10.3390/ma11040580.
84. Lee CH, Hajibandeh J, Suzuki T, Fan A, Shang P, Mao JJ. Three-dimensional printed multiphase scaffolds for regeneration of periodontium complex. Tissue Eng Part A. 2014;20(7–8):1342–51. https://doi.org/10.1089/ten.TEA.2013.0386.
85. Traphagen S, Yelick PC. Reclaiming a natural beauty: whole-organ engineering with natural extracellular materials. Regen Med. 2009;4(5):747–58. https://doi.org/10.2217/rme.09.38.
86. Traphagen SB, Fourligas N, Xylas JF, Sengupta S, Kaplan DL, Georgakoudi I, et al. Characterization of natural, decellularized and reseeded porcine tooth bud

matrices. Biomaterials. 2012;33(21):5287–96. https://doi.org/10.1016/j.biomaterials.2012.04.010.
87. Zhang W, Vazquez B, Oreadi D, Yelick PC. Decellularized tooth bud scaffolds for tooth regeneration. J Dent Res. 2017;96(5):516–23. https://doi.org/10.1177/0022034516689082.
88. Smith EE, Angstadt S, Monteiro N, Zhang W, Khademhosseini A, Yelick PC. Bioengineered tooth buds exhibit features of natural tooth buds. J Dent Res. 2018;97(10):1144–51. https://doi.org/10.1177/0022034518779075.
89. Duailibi MT, Duailibi SE, Young CS, Bartlett JD, Vacanti JP, Yelick PC. Bioengineered teeth from cultured rat tooth bud cells. J Dent Res. 2004;83(7):523–8. https://doi.org/10.1177/154405910408300703.
90. Lee CH, Cook JL, Mendelson A, Moioli EK, Yao H, Mao JJ. Regeneration of the articular surface of the rabbit synovial joint by cell homing: a proof of concept study. Lancet. 2010;376(9739):440–8. https://doi.org/10.1016/S0140-6736(10)60668-X.
91. Kim K, Lee CH, Kim BK, Mao JJ. Anatomically shaped tooth and periodontal regeneration by cell homing. J Dent Res. 2010;89(8):842–7. https://doi.org/10.1177/0022034510370803.
92. Nakao K, Morita R, Saji Y, Ishida K, Tomita Y, Ogawa M, et al. The development of a bioengineered organ germ method. Nat Methods. 2007;4(3):227–30. https://doi.org/10.1038/nmeth1012.
93. Ikeda E, Morita R, Nakao K, Ishida K, Nakamura T, Takano-Yamamoto T, et al. Fully functional bioengineered tooth replacement as an organ replacement therapy. Proc Natl Acad Sci USA. 2009;106(32):13475–80. https://doi.org/10.1073/pnas.0902944106.
94. Moussa DG, Aparicio C. Present and future of tissue engineering scaffolds for dentin-pulp complex regeneration. J Tissue Eng Regen Med. 2019;13(1):58–75. https://doi.org/10.1002/term.2769.
95. Han J, Kim DS, Jang H, Kim HR, Kang HW. Bioprinting of three-dimensional dentin-pulp complex with local differentiation of human dental pulp stem cells. J Tissue Eng. 2019;10:2041731419845849. https://doi.org/10.1177/2041731419845849.
96. Joda T, Bragger U, Zitzmann NU. CAD/CAM implant crowns in a digital workflow: five-year follow-up of a prospective clinical trial. Clin Implant Dent Res. 2019;21(1):169–74. https://doi.org/10.1111/cid.12681.
97. Nakayama M, Kondo O, Pesonen P, Alvesalo L, Lahdesmaki R. Influence of long and short arms of X chromosome on maxillary molar crown morphology. PLoS One. 2018;13(11):e0207070. https://doi.org/10.1371/journal.pone.0207070.
98. Athirasala A, Lins F, Tahayeri A, Hinds M, Smith AJ, Sedgley C, et al. A novel strategy to engineer pre-vascularized full-length dental pulp-like tissue constructs. Sci Rep. 2017;7(1):3323. https://doi.org/10.1038/s41598-017-02532-3.
99. Strub M, Keller L, Idoux-Gillet Y, Lesot H, Clauss F, Benkirane-Jessel N, et al. Bone marrow stromal cells promote innervation of bioengineered teeth. J Dent Res. 2018;97(10):1152–9. https://doi.org/10.1177/0022034518779077.
100. Franca CM, Riggers R, Muschler JL, Widbiller M, Lococo PM, Diogenes A, et al. 3D-Imaging of whole neuronal and vascular networks of the human dental pulp via clarity and light sheet microscopy. Sci Rep. 2019;9(1):10860. https://doi.org/10.1038/s41598-019-47221-5.
101. Kim J, Amar S. Periodontal disease and systemic conditions: a bidirectional relationship. Odontology. 2006;94(1):10–21. https://doi.org/10.1007/s10266-006-0060-6.
102. Lynge Pedersen AM, Belstrom D. The role of natural salivary defences in maintaining a healthy oral microbiota. J Dent. 2019;80(Suppl 1):S3–S12. https://doi.org/10.1016/j.jdent.2018.08.010.
103. Kilian M, Chapple IL, Hannig M, Marsh PD, Meuric V, Pedersen AM, et al. The oral microbiome – an update for oral healthcare professionals. Br Dent J. 2016;221(10):657–66. https://doi.org/10.1038/sj.bdj.2016.865.

Regenerative Approaches in Periodontics

Necla Asli Kocak Oztug,
Srinivas Sulugodu Ramachandra,
Cagdas Caglar Lacin, Aya Alali, and Amelia Carr

1 Introduction

Periodontal diseases are some of the most common diseases of humanity. According to the Global Burden of Disease Study in 2010, severe periodontal disease was the sixth most common health condition in the world. This disease had a worldwide prevalence of 11.2%, and affected some 743 million people, with that number increasing every year [1, 2].

Damage to the periodontium can occur due to trauma, gingivitis, periodontitis, or age-related loss of tissue. Destruction of the periodontium eventually may result in the loss of teeth and surrounding tissues. The ultimate goal of regenerative periodontal treatment is to prevent tooth loss by achieving complete regeneration of the periodontium [3].

Conservative periodontal therapy, which continues to be used effectively today, encompasses measures to control biofilm. It focusses on the removal of both supragingival and subgingival dental plaque biofilm, the removal of infected cementum, and the correction of problematic aspects of restorations such as overhangs [4–6]. Conservative periodontal treatment and supportive periodontal therapy (SPT) is essential as a foundation for any surgical interventions, and good control of biofilm must ideally be achieved prior to all attempts at periodontal reconstructive and regenerative surgery.

Periodontal regeneration can only be achieved by creating the appropriate microenvironment for the differentiation of specific cell types that constitute the periodontium. Epithelial cells (keratinocytes) and fibroblasts can proliferate faster than other cells in the periodontium, and controlling the proliferation of fibroblasts and epithelial cells poses a major challenge after surgical periodontal treatment [7]. All regenerative treatment approaches are based on maintaining a separation that controls the tendency of epithelial cells and fibroblasts to overgrow during the healing period and to give an opportunity for others cells from the periodontium to proliferate and differentiate [4].

In recent decades, deficiencies in the methods and outcomes of traditional surgical treat-

N. A. Kocak Oztug (✉)
Department of Periodontology, Faculty of Dentistry, Istanbul University, Istanbul, Turkey

School of Dentistry, The University of Queensland, Herston, QLD, Australia
e-mail: asli.kocak@istanbul.edu.tr, n.kocakoztug@uq.edu.au

S. S. Ramachandra · A. Alali · A. Carr
School of Dentistry, The University of Queensland, Herston, QLD, Australia
e-mail: s.sulugoduramachandra@uq.edu.au; a.alali@uq.net.au; amelia.carr@uq.net.au

C. C. Lacin
Department of Periodontology, Faculty of Dentistry, Istanbul Kent University, Istanbul, Turkey

© Springer Nature Switzerland AG 2021
S. Hosseinpour et al. (eds.), *Regenerative Approaches in Dentistry*,
https://doi.org/10.1007/978-3-030-59809-9_6

ment approaches for periodontal disease have led researchers to investigate new regenerative treatment methods involving tissue engineering technologies [8]. Tissue engineering is defined by Groll et al. as "an interdisciplinary field that provides the restoration and function of biological units using the principles of engineering and life sciences" [9].

A combination of conservative periodontal therapy and new techniques in tissue engineering has created a new regenerative direction that is the focus of much current periodontal research. Amongst the disciplines within dentistry, periodontology has arguably gained the most benefited from advances in tissue engineering. Ongoing developments in this field have been very promising, and they underpin the success of periodontal regeneration methods used in clinical treatments [4, 8].

2 Periodontium

The periodontium is a specialized multi-tissue structure in the oral cavity. It contains both mineralized and soft tissues, and it comprises the cementum, gingiva, periodontal ligament and alveolar bone. The periodontal apparatus is responsible for providing the maxillary and mandibular teeth with the necessary support and protection for their normal functions [10]. Each of the four components of the periodontium has its own architecture, composition, and differentiation, yet they all work in function together as one entity in the oral cavity [11].

Gingiva

This is the most superficial part of the periodontium, and it includes the free gingiva, the attached gingiva and the interdental gingiva. This soft tissue overlies the alveolar bone. Its epithelial layer lines the sulcus of teeth and is the first barrier that prevents microbial invasion of the underlying structures.

Periodontal Ligament

This is a vascular fibrous structure positioned as the interface between two hard tissue structures, the cementum and alveolar bone. It supports the tooth in its socket. The stem cells that are present in the periodontal ligament have an astonishing ability to differentiate into other cell types, especially in cases of injury.

Cementum

This is an avascular calcified tissue that covers the dentine of the tooth root. It consists of a primary acellular type and a secondary cellular type. The cementum is attached to the alveolar bone through periodontal ligament fibers (Sharpey's fibers). Cementoblasts produce the intrinsic collagenous matrix of cementum [10].

Alveolar Bone

Alveolar bone or the alveolar process is the mineralized part of the periodontium, that is attached to the cementum of the root through the Sharpey's fibers of the periodontal ligament [12].

Many different factors that can cause destruction and loss of the periodontium. These include both systemic, developmental, and acquired diseases (genetic, endocrine, connective tissue diseases, acquired immunodeficiency, neoplasms) as well as oral diseases (periodontal diseases and oral malignancies). Periodontal destruction can also be caused by non-disease factors such as traumatic occlusal forces. Since these various conditions and diseases can cause the overall oral health of patients to deteriorate, a corrective approach or therapeutic intervention is necessary to stop the process of damage, limit its extent, and maximize the health status of the affected area [13].

3 Nature of Periodontal Healing and Regeneration

"*Periodontal regeneration*" is the concept where materials and procedures are used to induce reconstruction of a part or the whole of the periodontal tissues [14]. It involves complex biological cooperation, where different cells and bioactive proteins are responsible for interacting, with the goal being to reproduce the previous normal function and architecture of the tissues. In an ideal scenario, coordination of the four compartments would give rise to a connection between the new cementum and new alveolar bone through Sharpey's fibers, giving a functional periodontal ligament [15].

A true visualization of the repaired tissue can only be achieved at the histological level, where all the periodontal structures can be viewed. At the clinical level, comparing probing and attachment levels before and after treatment can provide an indication of treatment outcomes. Conventional intraoral radiography for evaluating bone fill in the defect sites is not considered reliable, since a certain amount of mineralization needs to occur before it can be visible on a radiograph. Methods such as digital subtraction radiology, cone beam volumetric tomography, and computer-assisted densitometric image analysis can provide more information on changes to bone. In some cases, re-entry surgery to the treated site is used to visualize the healing, however even this method cannot show fully what type of outcome has occurred [12].

Periodontal regeneration is the ultimate outcome of any periodontal therapy. Conventional surgical and non-surgical periodontal therapy debrides the root surface to prepare it for the healing process. There are usually two main paths for this healing—either by regeneration, where complete renewal of the tissue function and structure takes place or by repair, which results in compromised clinical outcomes [10]. At the microscopic level, the repair is the most common and default outcome with conventional surgical and non-surgical periodontal therapy.

Successful reconstruction of the lost tissue requires collaborative efforts from progenitor cells in order to deposit new tissue and allow this to mature. Some of these events require careful coordination, for example, growth of alveolar bone should occur in a coronal direction towards the soft tissue, with no adjacent bony part at the other end. This is a unique situation for bone that is not found elsewhere in the body [11].

The healing of periodontal wounds is a complicated process, for many reasons, including the proximity to an avascular tooth surface, and the potential ingress of pathogenic microorganisms into the surgical site [11]. It is also challenging because the goal is to form new attachment, but there is a loss in the regulatory messages needed to direct this process [16].

During the normal healing process, as can be shown in surgically induced bone defect models in rats, bone formation starts at the bony part of the wound, and after that gradual thickening and mineralization of cementum can be observed around the apex of the tooth. Periodontal ligament fibers are the last to integrate into both cementum and bone [17].

Like any other wound in the body, in a periodontal wound a fibrin clot forms between the wound edges. In a periodontal wound, this clot forms between the wound margin and the root surface. It is replaced later by granulation tissue, and hopefully by a new connective tissue attachment. However, as the clot sits between a hard tissue (the root) and a soft tissue, microscopic movements of these create tensile forces, which can readily displace the clot. Rapid migration of epithelium causes space to becomes epithelialized. The net result is that the root surface has a long junctional epithelium (LJE) on its surface, rather than periodontal ligament fibers. The outcome of an LJE is the most common form of repair in periodontal defects after open or closed periodontal debridement. The epithelium has the highest migration rate, so will always dominate in the wound healing sequence if it is not excluded. This is why in periodontal regeneration, the epithelium must be excluded. If the clot can be maintained in a stable position against the root, it is more likely that a new connective tissue attachment will emerge on the root surface [18].

Ideally, the outcome of periodontal healing is to reconstruct the periodontium, and eliminate periodontal pockets (Fig. 1). Reconstruction involves restoring the proper anatomical and functional relationships between the junctional epithelium, connective tissue, the periodontal ligament, the cementum, and the bone. Any lost tissue should be replaced. This is especially challenging in terms of the alveolar bone [12].

In 1976, Melcher et al. proposed the tissue compartment hypothesis [19]. According to this, there is a "first mover" advantage, so the type of cell which is the pioneer at the defect area will determine the outcome. Therefore, there is usu-

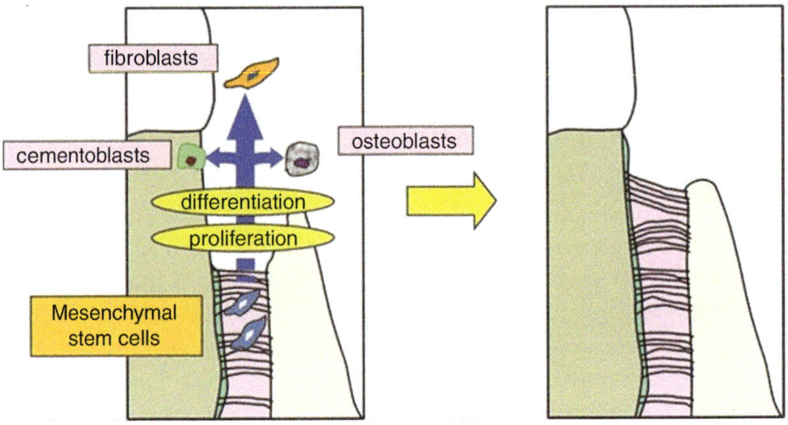

Fig. 1 Periodontal ligament cells differentiation and regenerative capacity [23]

ally one or a combination of four possible healing outcomes that can occur after conventional periodontal treatments:

- *Long Junctional Epithelium*, where epithelial cells are the first dominators.
- *Recession*, where new connective tissue attachment is formed apically to the cemento-enamel junction (CEJ) but without regeneration of periodontal ligament.
- *Ankylosis*, where the ligament is lost and there is the union of bone and tooth, with resultant tooth resorption.
- *Recurrence of the pocket* is also a possibility in cases of repair [20].

4 Regenerative Capacity of Periodontal Ligament Cells

Various experiments have revealed that neither osteoblasts nor gingival connective tissue cells have the ability to produce new connective tissue attachment and ankylosis will result in the areas where the periodontal ligament does not exist. As the periodontal ligament contains mesenchymal stem cells and periodontal stem cells, it should have the ability to produce other cell lineages. This regenerative capacity is regulated by several growth factors, including fibroblast growth factor, transforming growth factor-β, insulin growth factors 1 and 2, platelet-derived growth factor, and bone morphogenic proteins. These can activate cells to divide. Signals from growth factors and cytokines drive the differentiation of stem cells [21].

Mesenchymal cells of the periodontal ligament are of particular interest as they express high levels of markers for bone, cementum, and fibroblasts. When grown in culture in various conditioned media, periodontal ligament mesenchymal cells show elevated levels of markers for an osteoblastic lineage (RUNX2, ALP, OPN and COL1), a cementoblastic lineage (CEMP1, CAP), and increased content of both fibronectin and collagen. These extracellular matrix components increase the ability of the cells to attach. They also regulate how the cells interact with other cells and tissues in the periodontium [22].

5 Types of Periodontal Defects

Achieving periodontal regeneration for bone defects of various shapes is one of the most important challenges of surgical periodontal therapy. The approach used for reconstructive periodontal management varies according to the type of tissue that has been lost. Periodontal defects can be classified as soft tissue defects, hard tissue defects, or a combination of both [24].

Osseous (hard tissue) defects are classified according to the level of the periodontal pocket base in reference to the alveolar crest, into either suprabony defects (supracrestal) or infrabony defects (subcrestal).

The latter includes;

(A) Intrabony defects are classified according to the number of remaining osseous walls (one, two, or three-wall defects). The lesser the number of residual walls, the more difficult it is to get bone fill to restore the defect. Some defects have less walls remaining coronally than at the basal part. These are termed "combination intrabony defects."
(B) Craters. These occur frequently in the mandibular posterior segment. The buccolingual concave shape of bone resorption involves two neighboring teeth, or the septum between the buccal and lingual walls is resorbed. When the osseous defect is between the roots of multirooted teeth, these are known as inter-radicular defects.

Glickman et al. classified osseous defects according to their vertical and horizontal dimensions. Later, a classification of the vertical dimension of furcation defects was introduced [25]. Hamp et al. classified horizontal bone loss numerically, where Class I has less than 3 mm loss of periodontal tissue; Class II has more than 3 mm loss of periodontal tissue, but the probe cannot go through the defect, and Class III is a through-and-through horizontal loss of tissue [26]. This classification is still one of the most commonly used in clinical practice [25].

Soft tissue defects can present as gingival recession, where the gingival margins have become apical to the CEJ. It can affect the total width of the attached gingiva, on one tooth or many teeth. Classically, marginal soft tissue recession has been classified descriptively as shallow or deep, or wide or narrow. In 1985, Miller et al. proposed four categories for gingival recession, in relation to the mucogingival relationship and the alveolar bone status. Class I refers to recessions in which the gingival margin is coronal to the mucogingival margin. In Class II, the gingival margin reaches the mucogingival margin or beyond, with no bony defect. Class III is when the gingival margin is at the mucogingival margin, and is accompanied by interproximal alveolar bone loss and/or tooth malposition, while Class IV describes severe alveolar bone loss and/or considerable tooth malposition [27]. This classification, although very popular, is currently seen as inadequate, as some clinical cases do not belong to either Class I nor II (e.g., when interproximal bone loss occurs while the gingival margin is still coronal to the mucogingival margin). Another drawback of this classification is that it ignores the issue of the recession on the palate, which can be of significance to clinical management [27].

A newer classification of the gingival recession that was adopted by the World Workshop of Periodontology, classifies the gingival recession with respect to clinical attachment loss at the interproximal area. In Type 1, there is no attachment loss interproximal, and the interproximal CEJ cannot be seen. In both Type 2 and Type 3, there is the loss of interproximal attachment (from the CEJ to the most apical point of the pocket at both the mesial and distal sides) and is compared to the buccal attachment loss (from the CEJ to the most apical point of the pocket at the buccal side). If it is less than or equal to the buccal attachment loss, then it is classified as Type 2, and if it is higher it comes under Type 3 gingival recession. This classification is seen as treatment directed and can be combined with assessment of other gingival parameters, such as the width of keratinized gingiva and the gingival phenotype or biotype [28].

6 Current Surgical Treatment Protocols in Regenerative Periodontology

Periodontitis results from the host response to the accumulation of dental plaque on the roots of the teeth. In a susceptible patient, long-standing inflammation of the supporting structures of the teeth results in loss of alveolar bone. Subsequently, apical migration of the soft tissue can manifest as gingival recession, with loss of interdental papilla.

Comprehensive treatment of advanced periodontitis involves the regeneration of both hard

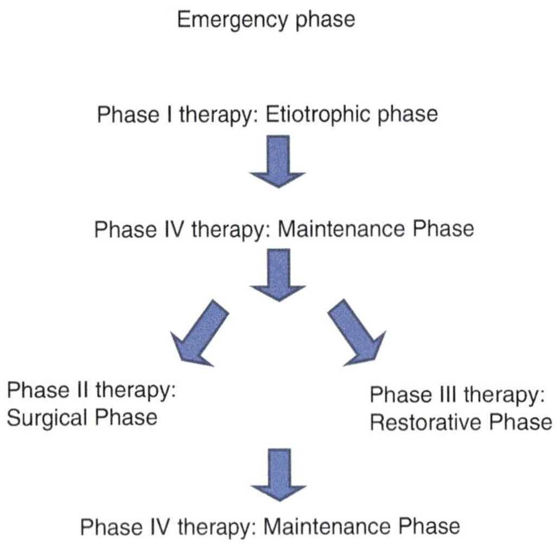

Fig. 2 Flow chart of comprehensive periodontal treatment divided into four phases. Phase I therapy involves the removal of all aetiologic factors. Following re-evaluation, patients with a healthy periodontium are placed in the maintenance phase. Patients not responding adequately to phase I therapy are treated with surgical periodontal therapy

and soft tissue components of the periodontium. As discussed earlier, non-surgical periodontal therapies (NSPT) remain the cornerstone of periodontal treatment, and the majority of cases are treated with NSPT. Periodontitis is initially treated with NSPT (phase I) either as a preparatory phase for further treatment or as a definitive treatment. The decision to move the patient towards either surgical periodontal therapy (phase II) or maintenance therapy (also known as supportive periodontal treatment (SPT)) (phase IV) is made following re-evaluation of the clinical situation (Fig. 2). However, a portion of cases in advanced stages do not respond to NSPT and will require surgical periodontal therapy [29].

7 Team of the Periodontist and the General Dental Practitioner

Around the world, in many countries, there is an aging population. Many older patients have retained a significant portion of their natural teeth, and this has resulted in an increased demand for comprehensive periodontal treatment. Advanced periodontitis is best treated by a shared approach, wherein both the general dentist and the periodontist work together towards improving and maintaining the periodontal health of the patient [30]. The general dentist screens and identifies cases of advanced periodontitis, and refers them to a periodontist. They also coordinate the patient's care and provide their routine restorative and preventive oral health care. By the general dentist explaining the various available treatment options, pointing out the advantages and disadvantages, the patient is able to make their own choices based on evidence [29, 30]. The periodontist carries out active and comprehensive periodontal treatment, whereas the general dentist maintains the treated periodontium by providing supporting periodontal treatment.

7.1 Indications for Surgical Periodontal Therapies

1. The presence of advanced stages of periodontitis. In the current classification of periodontitis, this would translate into stage III and stage IV, grade C periodontitis [31].
2. Patients not responding to NSPT, as evidenced by the presence of multiple residual periodontal pockets ≥6 mm, would be ideal candidates for surgical periodontal therapy [32]. The decision to use surgical techniques for a patient is dependent on the response of the patient towards NSPT at the re-evaluation visit.
3. The presence of bone defects, including intrabony defects (three-wall defects) and craters that would benefit from the use of regenerative bone grafts or membranes (regenerative periodontal surgery) [33].
4. The presence of bony contours or abnormalities that need re-contouring to establish an ideal periodontal architecture (resective periodontal surgery) [34].

7.2 Contraindications for Surgical Periodontal Therapies

1. Poor oral hygiene or patients in whom maintenance of optimal oral hygiene is difficult.
2. The presence of risk factors including uncontrolled diabetes mellitus (≥7% HbA1C), and continued use of tobacco (smoking ≥20 cigarettes/day), since these are known to impair healing and to result in poor surgical outcomes.

The main goals of surgical therapy are to: enhance accessibility to root deposits for root debridement and scaling, reduce or eliminate pockets by resection of the pocket wall, expose the diseased zone to perform regenerative techniques, and gain access for resection of periodontal defects [29–32].

8 Periodontal Flaps

As discussed above, periodontal surgical treatment is applied where preventive periodontal treatment is not sufficient. It aims to completely eliminate the cause of the disease and to ensure as much regeneration of the destroyed tissue as possible. Periodontal flaps can be designed in different ways to provide access to the periodontal defect region.

"A periodontal flap is a section of mucosa and/or gingiva surgically separated from the underlying tissues to provide access and visibility to the bone and root surface" [35]. Periodontal flaps are classified according to the exposure of bone, after flap reflection, into either partial thickness flaps or full-thickness flaps. Flaps can also be classified based on their placement after the operation, whether displaced (coronally, apically or laterally) or undisplaced. Also, according to the management of the interdental papilla, flaps can be classified as conventional flaps or papilla preservation flaps [32]. Table 1 lists the major features of the main types of periodontal flaps.

Various techniques and methods have been utilized in surgical periodontal procedures, including the modified Widman flap, the papilla preservation flap, and various modifications of these [32].

9 Classification of Periodontal Surgical Procedures

Periodontal surgerical procedures are broadly classified into resective and regenerative periodontal surgical procedures.

9.1 Resective Periodontal Surgical Procedures

Resective periodontal surgery involves eliminating bony irregularities, to create a stable peri-

Table 1 Various classifications of periodontal flaps and their main features [32]

	Based on	Flap type	
1	Bone exposure	**Full-thickness flaps/Mucoperiosteal flaps** all the soft tissue, including the periosteum, is reflected to expose the underlying bone	**Partial thickness flaps/Mucosal flaps/Split thickness flap** includes only the epithelium and a layer of the underlying connective tissue. The bone remains covered by a layer of connective tissue, including the periosteum
2	Placement of the flap after surgery	**Displaced flaps**: after the completion of the surgery, flaps are moved either coronally or apically	**Undisplaced flaps**: after completion of the surgery, flaps are placed back in the same position and sutured
3	Management of the papilla	**Conventional flaps** are used when the interdental spaces are too narrow. In this procedure, the papilla is split into two (labial and lingual) portions	**Papilla preservation flaps** are indicated when there are wide interdental spaces. One papilla is incorporated into either the buccal and lingual flaps. This flap is commonly used whenever regenerative periodontal products are used

odontal architecture that will assist the patient in maintaining their oral hygiene. Essentially, this approach sacrifices some bone for the creation of a stable periodontal architecture. In recent decades, the scientific community has realized the importance of regenerating and conserving bone. Therefore, resective periodontal surgical approaches have become less popular, and today increased emphasis is directed towards regenerating bone [33–35].

9.2 Regenerative Periodontal Surgical Procedures

Regenerative periodontal therapy aims to reconstruct and reconstitute the tooth-supporting structures that have been lost through periodontitis or because of other causes. Regenerative periodontal procedures involve the use of various regenerative materials and techniques for the regeneration of the lost portions of the periodontium. The most commonly used products in periodontal regeneration are bone grafts and barrier membranes. Each of these materials has certain distinct advantages and disadvantages [33–35].

10 Regeneration of Bone and Bone Grafts

After the placement of bone grafts into osseous defects, several steps of healing follow. The process finally culminates in the complete integration of the bone graft into the surrounding native bone, with the goal being that the new bone displays similar structural and biochemical properties to the native bone. Bone regeneration requires three main components including cells (osteoblasts), a scaffold (which provides the structural framework for the development of clotting, maturation, remodelling, and for the recreation of the Haversian canal systems), and signals (such as from bone morphogenic proteins and other signals that can induce the formation of bone) [36].

The regenerative potential of a bone graft is based on the properties of osteogenesis, osteoinduction, and osteoconduction.

- *Osteogenesis* is the formation of bone by osteoblasts.
- *Osteoinduction* is the process by which proteins and cellular signalling molecules present in the graft will induce neighbouring mesenchymal stem cells to differentiate into osteoblasts and thereby produce bone.
- *Osteoconduction* is the presence of a matrix/scaffold that helps with the organization of the population of cells in the scaffold and the creation of a system similar to native bone.

Apart from these three main criteria for ideal bone grafts, other relevant criteria include: the ease with which the graft can be procured, its cost, its biocompatibility (e.g., being free from antigenic properties that would evoke an inflammatory response), and the provision for an adequate amount of the graft. Table 2 lists the various criteria for an ideal bone graft. However, an ideal bone graft will likely never exist, and any bone graft will be a compromise in terms of its various properties. The selection of the bone graft to be used by the clinician will be based on the circumstances and needs of the individual case [37].

11 Classification of Bone Grafts

Bone grafts are classified based on their source, as follows: autografts, allografts, xenografts and alloplastic materials [38].

Table 2 Features of an ideal bone graft [36, 37]

An ideal bone graft should	
1	Have osteogenic potential, meaning it should contain some osteoblasts that can directly produce bone.
2	Have a scaffold framework (structural framework for the development, maturation, and remodelling of tissue).
3	Have signals for the induction of mesenchymal stem cells to differentiate into osteoblasts.
4	Be biocompatible, and unable to elicit an inflammatory immunological response.
5	Be free from any virus or prion-related infection.
6	Be economical.
7	Be available in sufficient quantity to fill or replace the lost portion of the bone.

11.1 Autografts

Autogenous bone grafting involves harvesting bone grafts from a donor site, and then using this bone fill the bone defect within the same individual from whom the graft was taken, thereby reducing the concerns of allergenicity. Autogenous bone grafting is predictable and is considered to be the gold standard because it has the three essential criteria: osteogenic cells, osteoinductive growth factors, and a scaffold for the growth of cells. Autogenous grafting materials include cortical bone, cancellous bone, and bone marrow transpirates [37]. Among these, cancellous and bone marrow transpirates are considered to be more osteogenic due to their higher osteogenic potential and vascularity. Iliac crest, tibia, fibula, and ribs are the most favoured extraoral sites, whereas the symphysis of the mandible, tori/exostoses, the retromolar pad, healing wounds, extraction sites, and the region of the maxillary tuberosity are the most common intraoral sites [38].

With autogenous bone grafts, the main disadvantages are the necessity of a second surgical site and the limited amount of bone that can be harvested. In patients with immunocompromised conditions or elderly patients, the creation of a second surgical site may result in considerable donor site morbidity and further complications, including postoperative infections. Autografts also need to be done by an experienced clinician who is capable of harvesting from the donor tissue while causing minimal damage to that site. Although autogenous bone is inexpensive as it is retrieved from the same patient, there are additional expenses because of the additional effort required by the surgeon. Such factors make autografts less appealing than other options [38, 39].

11.2 Allografts

Allografts are grafts taken from the same species, i.e., from humans, however, the donor is genetically different from the recipient [39]. All allografts carry a risk of blood-borne virus transmission, which is eliminated by rigorous screening and testing of potential cadavers for such transmissible diseases. There are three main types of allografts: (1) frozen, (2) freeze-dried bone allograft (FDBA) and (3) demineralized freeze-dried bone allograft (DFDBA). Fresh unprocessed bone and frozen bone from cadavers are not used due to the chance of spreading an infection.

FDBAs contain the mineralized skeleton for the population of native cells. FDBAs require a long period for neovascularization of the mineralized skeleton and conversion into new bone. This mineralized framework, without any growth factors, makes FDBAs osteoconductive, but not osteoinductive. The majority of periodontal defects treated with FDBA have shown complete bone fill or more than 50% bone fill [40]. A mixture of autogenous grafts and recombinant human bone morphogenetic protein 2 (rhBMP2) with FDBA provides improved results with complete regeneration of bone defects [40, 41].

DFDBAs or demineralized bone matrix (DBM) has the inorganic content removed, retaining the organic portion and its osteoinductive properties. The DBM or a DFDBA contains bone morphogenic proteins (BMPs). These are acidic polypeptides from the transforming growth factor-β family, and they initiate the differentiation of mesenchymal stem cells into osteoblasts. Higher levels of BMPs in the graft favours increased oxygen tension, thereby helping in the production of bone [41]. To date, DFDBAs are the only bone graft material that has been shown to provide complete regeneration using histological methods [42]. The American Association of Tissue Banks (AATB) is a scientific organization that provides guidelines for standard setting and the use of donated human tissues. The majority of nations have developed their own tissue bank associations following the guidelines of the AATB. The British Association for Tissue Banking for the United Kingdom and Australian Tissue Banking Association for Australia are examples of some of the tissue banks that are now involved in regulating harvesting and processing of allografts [42–44].

11.3 Xenografts

Bone grafts transferred from other species, including animals, to humans are known as xenografts. Commonly used products come from cows, pigs, or from natural marine coral [44]. Tissue banks procure bone grafts from animals, and process them using a battery of intense physical and chemical purification methods to remove components that may be antigenic, and also components that may spread infection, such as prions [43]. Testing of donor animals for bovine spongiform encephalopathy reduces the risk of disease transmission from bone grafts. Additionally, removal of the entire organic portion virtually eliminates the chances of prion-related infections, since prions are proteinaceous in nature. The remaining inorganic portion of the bone serves as a scaffold, within which neovascularization takes place [39].

Bio-Oss®, manufactured by Geistlich, is one of the most commonly used xenograft materials of bovine origin. Bovine bone has an inorganic structure that is similar to that of human bone, thus favouring its osteoconductive activity. Additionally, inorganic bovine bone has a high degree of porosity, thus increasing the chances of osteoconductivity and angiogenesis. Other similar products are available from various manufacturers.

The combination of P-15, a synthetic cell-binding peptide, with a bovine-derived bone graft, has been shown to provide improved clinical results when compared to DFDBA alone in the treatment of human periodontal intrabony defects [45]. Biocoral, on the other hand, contains calcium carbonate derived from natural coral and is this material is resorbable. Biocoral has a large amount of porosity, and it does not cause fibrous encapsulation, thus giving it high osteoconductivity [46].

11.4 Alloplasts

These are synthetic or semi-synthetic inorganic bone graft materials. Hydroxyapatite, calcium phosphate, β-tricalcium phosphate, and bioactive glasses are the most commonly used alloplastic materials [43, 44]. Alloplasts are often combined with growth factors or antimicrobial agents to improve their efficacy. The addition of autografts in small quantities to alloplasts provides a mixture with osteoinductive potential. However, this approach involves the harvesting of autografts, with the possibility of complications as discussed earlier.

Alloplasts with active additive agents have two major functions: the alloplast itself acts as a replacement material, while the added active agent helps induce bone formation, or prevents secondary infection [47, 48].

Bone grafts are also classified based on their regenerative potential, as outlined in Table 3.

Table 3 Classification of bone grafts based on their regenerative potential [43, 44]

1. Osteogenic/osteoproliferative bone grafts	Possess osteoblasts, thus having the highest amount of regenerative potential. These would be considered as the gold standard among bone grafts
2. Osteoinductive bone grafts	Possess certain molecules, which could induce neighbouring mesenchymal stem cells to convert into osteoblasts and thereby lay down bone. For example, bone morphogenic proteins have the ability to stimulate mesenchymal stem cells to convert them to osteoblasts and produce bone
3. Osteoconductive bone grafts	Provide a meshwork for the formation of bone. These grafts are dense and non-resorbable. They act as a scaffold wherein bone from the adjacent area can form new bone. However, the majority of osteoconductive grafts are non-resorbable, so the use of these grafts for implant site development is not recommended

12 Guided Tissue Regeneration

Periodontitis causes the loss of both the hard and the soft tissue of the periodontium (Fig. 3a). Any periodontal surgical procedure involves the removal of the diseased pocket epithelium, resulting in a surgical wound. The

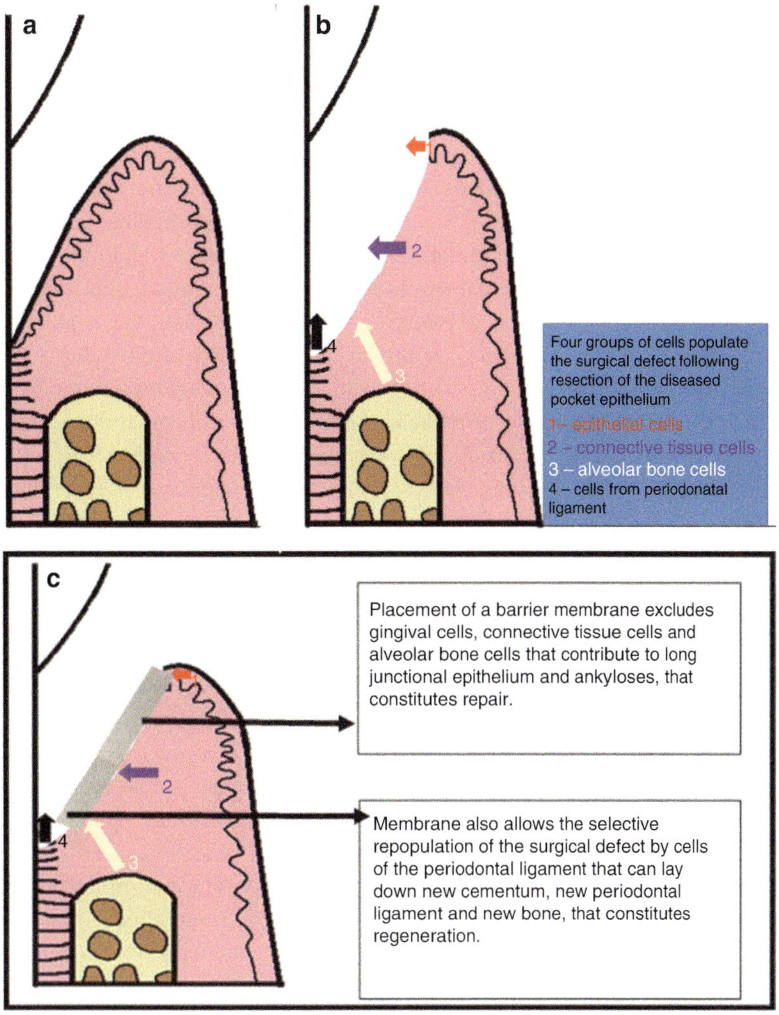

Fig. 3 (**a**) Periodontitis results in loss of supporting tissues of the periodontium. (**b**) Surgical excision or removal of the diseased pocket epithelium results in the repopulation of the surgical area by four groups of cells: (1) epithelial cells, (2) connective tissue cells, (3) alveolar bone cells, and (4) cells from periodontal ligament. (**c**) Placement of the barrier membrane, leading to cell exclusion and selective repopulation of the surgical defect

surgical wound can contribute four types of cells for possible repair/regeneration: epithelial cells, connective tissue cells, alveolar bone cells, and cells of the periodontal ligament (Fig. 3b) [49].

As discussed earlier, the repopulation of the defect by epithelial cells and fibroblasts results in the formation of an LJE, which constitutes repair rather than regeneration. The LJE situation remains susceptible to the recurrence of periodontitis. The population of the surgical site by alveolar bone cells results in ankylosis, which subsequently leads to replacement resorption, tooth mobility, and finally tooth loss. On the other hand, repopulation of the defect by cells of the periodontal ligament can result in the formation of new periodontal ligament, new bone, and new cementum.

Since the migratory rate of epithelial and connective tissue cells is faster than that of cells of the periodontal ligament, periodontal regeneration requires specific methods to prevent the migration of the epithelial and connective tissue cells into the periodontal defect, otherwise, it will heal by repair and not by regeneration as desired [49, 50].

The concept of isolating/preventing epithelial, connective tissue cells and osteoblasts (all of which are non-desirable cells for regeneration) by means of a barrier membrane, and at the same time guiding the periodontal ligament cells into the defect, results in the formation of new periodontal ligament, new cementum and new bone (Fig. 3c). This concept is known as "*guided tissue regeneration*" or *GTR* [50, 51].

12.1 Barrier Membranes

Various regenerative barrier membranes are used for the purposes of periodontal regeneration. The ideal barrier membrane should possess various properties including being nontoxic, non-carcinogenic, non-antigenic, and sterile. The membrane should be able to maintain space, so that it can withstand the forces of suturing, and the weight of the overlying flap, so that under normal function including masticatory forces it does not collapse into the periodontal defect. At the same time, the membrane should be sufficiently flexible and easy to use, so that the clinician can mould and shape the membrane to adapted to the particular architecture of the periodontal defect that is being treated. Systematic reviews have shown that the use of barrier membranes and bone grafts results in improved results compared to open flap debridement alone [51].

Membranes should be bioresorbable, as this eliminates the need for a second stage surgical procedure to remove the membrane. In cases where they are non-resorbable, they should be easily retrievable. Additionally, membranes should have a long shelf life, and they should be inexpensive. Currently, the ideal barrier membrane does not exist, and the existing membranes that are available on the market represent a compromise of these various properties.

Barrier membranes are classified into either non-resorbable and resorbable types [52]. Expanded polytetrafluoroethylene (ePTFE) was used for the earliest work in guided tissue regeneration. Reinforcement of these membranes with a mesh or exoskeleton of very thin titanium, greatly improved their mechanical strength, and their ability to preserve the space at the surgical site. However, such non-resorbable membranes need to be removed after the initial healing period, necessitating additional surgical procedure, trauma to the tissues, and additional expenses to the patient [52].

Among the resorbable membranes, collagen membranes have been studied extensively. These are biocompatible, and also somewhat hemostatic in nature. They possess chemostatic properties, allowing them to induce the migration of host cells and to facilitate primary wound closure. This reduces the chances of membrane exposure and the associated secondary infection of the surgical site. Their hemostatic properties promote the establishment of a clot and improve the overall stability of the wound site. However, collagen membranes are delicate, and they lack mechanical strength and therefore can collapse into the defect. Hence, placement of bone grafts beneath a collagen membrane provides improved clinical outcomes [53].

There are many disadvantages of using membranes for periodontal regeneration. Exposure of the membranes leads to localized infection which causes failure of the procedure. Exposed membranes have been shown to harbour significantly higher levels of *Aggregatibacter actinomycetemcomitans,* a bacterial species that produces potent leukotoxins, resulting in worse clinical outcomes. Membranes must be placed in areas with sufficient soft tissue, to ensure complete coverage of the membrane. In larger defects, placement of membranes without underlying bone grafts may result in the collapse of the membrane [54].

As well as these available materials for periodontal regeneration, enamel matrix proteins have become an important addition to the armamentarium for guided tissue regeneration in recent years. Before the formation of cementum, a thin layer of enamel matrix proteins, known as amelogenins, is deposited onto the root surface. This layer of enamel matrix proteins plays a vital role in the formation of cementum, as they initiate differentiation of periodontal ligament cells into cementoblasts. Following the isolation of enamel matrix proteins from various animal and human sources, they have been investigated extensively. The use of enamel matrix proteins or amelogenins enhances the formation of periodontal tissues in osseous defects. Enamel matrix proteins have been widely used alone and combination with other materials [55].

13 Periodontal Plastic Surgery

"Periodontal (and peri-implant) plastic surgery encompasses the surgical procedures performed to prevent or correct anatomical, developmental, traumatic or disease-induced defects of the masticatory mucosa (gingiva), lining mucosa (alveolar mucosa) or bone" [56]. Clinicians need to consider the thickness of the gingival tissue during the planning of periodontal plastic surgeries, taking into account a phenomenon termed as "periodontal phenotype," previously known as the periodontal biotype [57].

Patients with thin friable gingival tissue, a minimal amount of attached gingiva, thin underlying alveolar bone and long, narrow, conical crowns are known as having "thin periodontal phenotype." This tissue is more prone to gingival recession and interdental papilla loss, which results in poorer clinical outcomes after periodontal operations.

Patients with thick fibrotic gingival tissue, wide zones of attached gingiva, thick underlying alveolar bone and short-wide crowns are considered to have a "thick periodontal phenotype." This gingival tissue type is resistant to gingival recession and provides better clinical outcomes following periodontal plastic surgical procedures [58]. Placement of soft tissue grafts can convert a thin periodontal phenotype to a thick periodontal phenotype [59]. Figure 4 lists the major differences between thin and thick periodontal phenotype.

Soft tissue defects include gingival recession, interdental papilla loss, and mucogingival deformities. These defects can be treated, with several options available including autografts, allografts, and xenografts.

Autografts include a lateral pedicle flap, free gingival grafts, and sub-epithelial connective tissue grafts. Free gingival grafts are easier to harvest, however, they result in a colour mismatch between the grafted site and the adjacent tissues. Sub-epithelial connective tissue grafts are considered as the gold standard among soft tissue grafts [60].

For the treatment of periodontal soft tissue defects, gingival tissue can be harvested from the hard palate [61]. However, these procedures are technique sensitive and can result in a number of postoperative complications including pain, bleeding from the palate, exposure of the underlying bone and its necrosis, paresthesia, and permanent anesthesia due to damage to the greater palatine nerve [62]. Periodontal dressings are placed in the palate to protect the donor site and held in place by an acrylic stent that must be worn by the patient during the immediate postoperative period to retain the dressing. Additionally, in some individuals, autografts harvested from

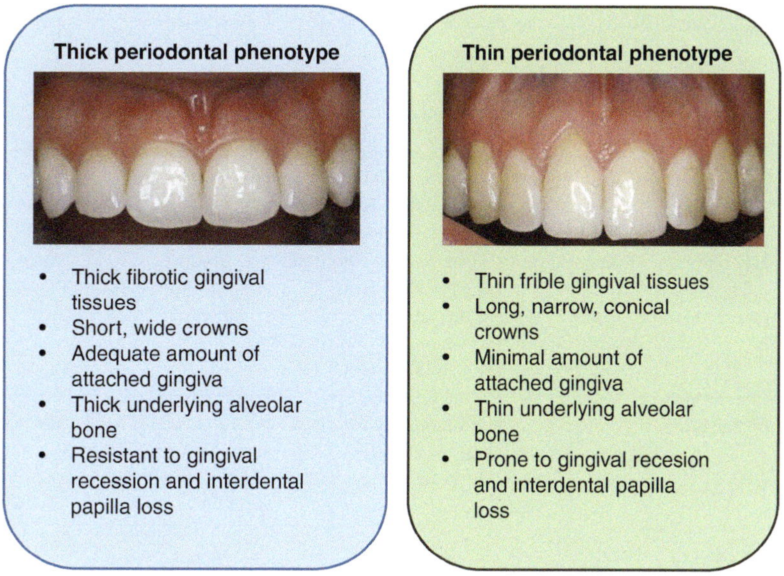

Fig. 4 Differences between thick and thin periodontal phenotypes [61] (Courtesy of Dr. Aya Alali and Dr. Salah Mortaja from the University of Queensland School of Dentistry, Australia)

the palate do not have an adequate thickness of tissue.

Because the majority of these complications are related to the creation of a second surgical site while harvesting the graft, various techniques have been developed to avoid the need for a second surgical site. Such options include allografts and xenografts. Table 4 lists the various soft tis-

Table 4 Various soft tissue grafts, their sources, advantages, and disadvantages [63]

Name of the graft	Type of graft	Source	Advantages	Disadvantages
Lateral pedicle flap	Autograft	Gingiva from adjacent teeth	Technique sensitive	Adjacent area may have an inadequate amount of tissue, so the technique can be used in very limited instances
Sub-epithelial connective tissue grafts	Autograft	Palatal mucosa, rarely from retromolar pad and adjacent edentulous site	Considered as the gold standard among soft tissue grafts. Can be harvested from the palate, so is economical	Creates a second surgical site with related possible complications including: impaired wound healing, severe pain, necrosis of the bone, profuse bleeding from the palate, paresthesia, and permanent anesthesia of the palate
Free gingival/ epithelial grafts	Autograft	Palatal mucosa, rarely from retromolar pad and adjacent edentulous site	Can be harvested from the palate, so is economical	Similar disadvantages associated with harvesting donor graft from the second surgical site, as mentioned above. Mismatch in the color between the grafted area and normal gingival tissue. Donor site heals by secondary intention, so donor site is painful for long period
Alloderm® regenerative tissue matrix	Allograft	Tissue banks obtain tissue from the skin of donated human cadavers. Following thorough processing, antigenicity is removed to prevent transmission of virus and infections	Will be useful in multiple recession defects wherein more quantity is essential. Correction of cancer surgeries. Correction and repair of burn injuries. Used instead of autografts	Expensive
MucomatrixX®	Xenograft	Consists of collagen and elastin derived from animal tissue, to remove all antigenic properties	Provides a stable three-dimensional matrix consisting of collagen and elastin	Expensive
Mucograft® collagen matrix	Xenograft Porcine origin	Consists of pure porcine collagen, following sterilization and processing, to remove prion-related infections	The product contains pure collagen	Expensive
Platelet-rich fibrin (PRF)	Autologous	PRF is obtained by autologous means and is compressed as a membrane	Preparation is easy with a simple centrifuge machine and is inexpensive. Contains growth factors that assist in the regeneration	PRF membranes are thin, so suturing may lead to tear of the membrane

sue graft options, with their relevant advantages and disadvantages [63].

14 Concept of Periodontal Tissue Engineering

As discussed in the introduction, the periodontium and the tooth form a highly developed functional structure [64, 65]. Periodontal diseases (such as gingivitis and periodontitis) are one of the most common inflammatory diseases seen in adults, and they cause high financial costs for their direct treatment [2].

Advance periodontitis can cause numerous problems such as increased mobility and loss of teeth, aesthetic problems, halitosis, and loss of masticatory efficiency, leading to changes in the diet and in food selection. When one takes into consideration the cumulative effect of untreated periodontitis on systemic health conditions, such as diabetes, periodontal disease is the oral disease that has potentially the most negative impact on the quality of life of an individual [66].

In line with developments in technology and science, periodontal treatment has also changed over time. In the early stages of periodontology, non-surgical and surgical treatment was directed towards removing the biofilm and the tissue that was thought to be infected. Subsequent histological examinations of these interventions led to a better understanding of periodontal regeneration. It soon became apparent that traditional closed debridement allowed epithelial cells to migrate to the root surface, reaching this before other cell types, and thus providing by creating a long junctional epithelium. Using this traditional approach meant that full regeneration of periodontal tissues could not be achieved [67].

Once the concept of guided tissue regeneration (GTR) had been developed, and begin to be used more widely as a surgical procedure for regeneration of the periodontium, it was realized that preventing epithelial cell migration using barrier membranes was only one of the first parts of the overall solution. Later work showed the importance of bone substitutes, root surface demineralization procedures, and the need to combine tissue regeneration protocols with periodontal treatment [68]. The sequential development of the concepts is shown in Fig. 5.

The term periodontology and tissue engineering were used together for the first time about 20 years ago [68, 69]. Modern periodontal tissue engineering targets the treatment of damaged periodontium using cells, bioactive molecules, and scaffolds together, in various combinations to achieve periodontal regeneration [65].

14.1 Stem Cells and Cell Sheet Technology

The concept of tissue engineering for periodontal regeneration can be classified into scaffold-free or scaffold-based approaches, depending on whether biomaterials are used or not. In the scaffold-free concept, cells can be placed directly into a periodontal defect without a cell carrier. This regenerative approach, which includes a single stem cell type or a combination of stem cell types, is called "cell sheet" technology [8, 70–72].

Iwaka et al. stated that according to the cell sheet engineering principle, at 37 °C cells adhere to and proliferate on the surface of a temperature-sensitive polymer comfortably, because the polymer is anhydrous and compact at this temperature. At temperatures below 32 °C, layers of cells with extracellular matrix proteins are spontaneously separated from the temperature-sensitive polymer [73]. This approach to release the cell sheets from the culture flask without the need to use any enzymes or other chemicals opened up a broader research field.

Periodontal ligament stem cells (PDLSCs), bone marrow-derived mesenchymal stem cells (MSCs), dental pulp stem cells (DPSCs) and fat-derived stem cells (ADSCs) have been investigated to determine their ability to induce periodontal regeneration [70–72]. There are now several studies showing promising results for the application of cell sheets in various in vivo models. Raju et al. recently reported successful in vivo periodontal regeneration, in a large peri-

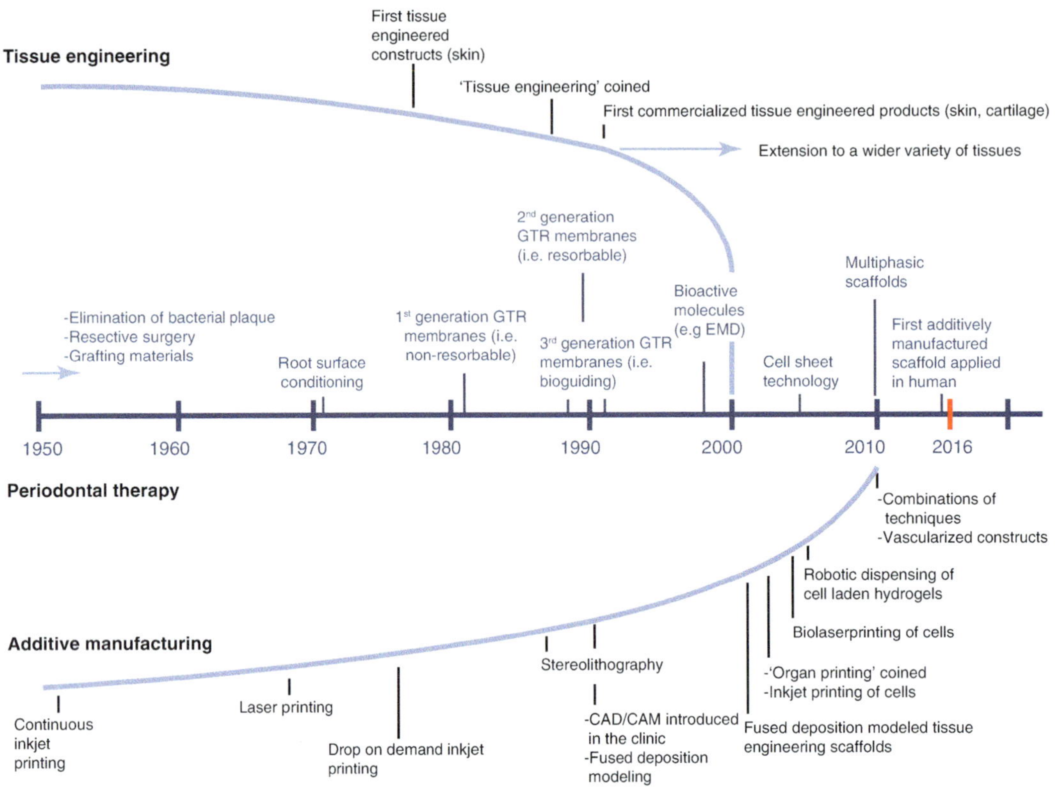

Fig. 5 Reflection on historical developments in tissue engineering and additive manufacturing to periodontal regenerative therapy. In the past 20 years, applications in the periodontology field of tissue engineering have focused on enabling regeneration [69]

odontal defect, fabricating a three-dimensional cell sheet, including a bone-ligament component, by layering together periodontal ligament cells and osteoblast-like cells [74].

Despite the enormous promise of the cell sheet approach, the use of robotic systems for sterile cell culture in the transplantation of allogeneic frozen cells significantly increases the cost of cell therapies, limiting the widespread use of this technology [73, 75].

14.2 Biomaterials for Periodontal Tissue Engineering

In periodontal tissue engineering, various biomaterials with different targets can be used alone or in combinations. As discussed earlier, while those materials with relatively high mechanical strength, such as hydroxyapatite (HA), are used for filling defects in cases involving the regeneration of alveolar bone and cementum, softer polymeric materials, such as collagen, which have relatively low mechanical strength, are used for periodontal ligament and gingiva regeneration [8].

Ceramic biomaterials are used particularly in hard tissue engineering, due to their similar structures and chemical composition. HA, calcium phosphate (CaP), calcium sulfate [9], and bioactive glass (BG) are the most extensively researched bioceramics in periodontal regeneration [69].

HA was one of the first biomaterials used in periodontal tissue engineering. It can be derived from natural bovine bone or used as a pure synthetic material [76]. Besides their advantages such as slow degradation, and their ability to stimulate and enhance the proliferation of osteoblasts, bioceramics also have disadvantages, such as fragility and low bioactivity [69, 76].

Polymers used in periodontal tissue engineering are divided into two groups: synthetic polymers and natural polymers. Synthetic polymers such as polylactic acid (PLA), polycaprolactone (PCL), copolymer poly lactic-co-glycolic acid (PLGA) and polyglycolic acid (PGA) have properties that be adjusted readily, and they can be manufactured with impeccable repeatability. Nevertheless, during the printing process of synthetic polymers, they pass through stages that are inherently harmful to the viability of cells, such as high temperature. These procedures complicate attempts to combine cells and polymers [69]. Natural polymers include biomaterials such as albumin, hyaluronic acid, cellulose, chitosan, and collagen [76].

These various biomaterials can be altered or unified into composite materials to create a suitable microenvironment and used in scaffolding systems to induce periodontal regeneration. However, complete imitation of the unique architecture of the periodontium remains challenging, even with currently available biomaterials, and this area still requires much further research [8].

14.3 Scaffolds

The ideal scaffold used in tissue engineering should prevent the collapse of the site during wound healing; it should promote the ingrowth of cells and blood vessels, and be easy to manage. The main aspects of scaffolds include their morphology (including porosity), mechanical strength, and chemical composition. The latter affects their rate of degradation [4, 77–79]. Manufacturing technologies, such as three-dimensional (3D) printing, provide precise control over the architectural and dimensional aspects of scaffolds, to ensure that they are conducive for reproducing the unique structure of periodontium [80].

Scaffolds for use in periodontal regeneration are inspired by guided tissue regeneration. Specially designed biomaterial scaffolds protect the surgical site and promote the formation of periodontal tissues during healing [3].

Scaffolds for periodontal tissue engineering can be monophasic or multiphasic. In monophasic scaffolds, the cells are encapsulated in various hydrogels, or they are planted on carrier scaffolds and then placed in periodontal defects. During wound healing, monophasic scaffolds play a protective role in maintaining the shape of the periodontal defect, and transporting the necessary cells as a carrier. Another role of monophasic scaffolds in periodontal regeneration is to protect the wound area against the challenge of bacterial infection from the oral environment, for example, by releasing antibacterial agents. Furthermore, monophasic scaffolds can be used to deliver growth factors to the defect region [3, 77]. Although monophasic scaffold technology is theoretically a simple concept, such scaffolds can be utilized for many purposes, by using different combinations of biomaterials and cell types [14].

The periodontium has a unique layout and contains many different structures, cells, and tissues. In order to perform periodontal regeneration successfully, each sub-tissue group that constitutes the periodontium must create coordinated sub-regenerations within a certain sequence (Fig. 6). One of the most important developments in periodontal tissue engineering is to emulate this complex regeneration system, by designing multiphasic scaffolds, each representing a different periodontium tissue [14].

In periodontal regeneration, multiphasic scaffolds mainly target the regeneration of soft-hard tissue interfaces, that is, between the alveolar bone and periodontal ligament, and between the periodontal ligament and the cementum. Thus far, biphasic and triphasic scaffolds have been introduced to periodontal tissue engineering [77].

Park et al. reported the usage of a biphasic scaffold, produced with polyglycolic acid for the bone phase and polycaprolactone for the periodontal ligament phase [81]. Following this study, the same technique was tested in vivo in a periodontal defect model. Perpendicularly oriented microchannels guided the periodontal fibers [82].

Lee et al. fabricated a scaffold from a mixture of PCL and hydroxyapatite (90%/10%), and used this to create a triphasic scaffold. These three

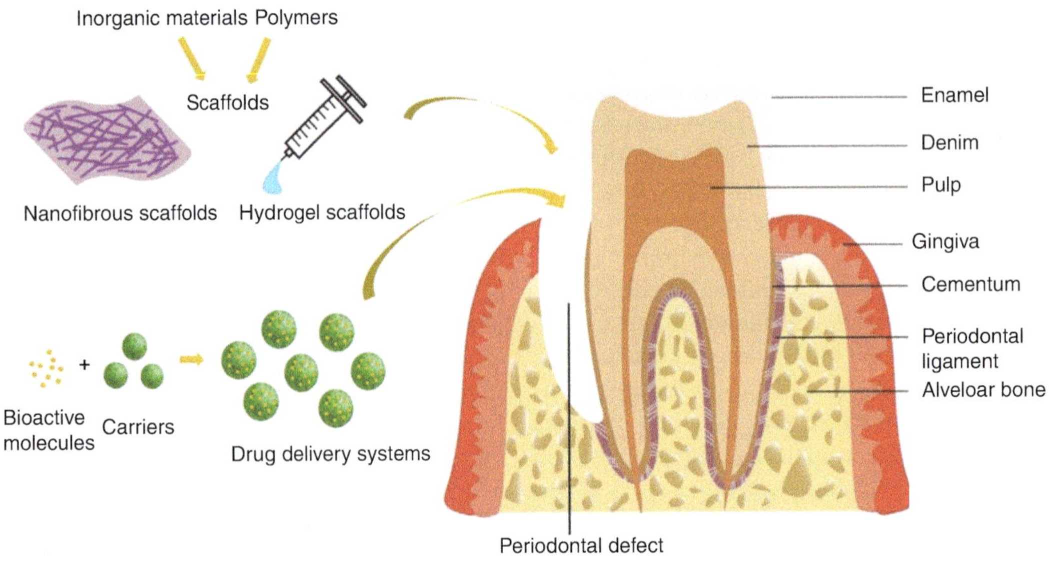

Fig. 6 Different approaches used in modern periodontal tissue engineering. Various combinations of cells, biomaterials and bioactive molecules can be the key to achieve complex periodontal regeneration in the future [8]

compartments consisted of 100 μm, 600 μm and 300 μm microchannel architecture, representing the cementum-dentine interface, the periodontal ligament, and the alveolar bone respectively. In this study, distinctive tissue phenotypes were formed after 4 weeks of separate in vitro incubation with periodontal ligament stem cells (PDLSCs), alveolar bone stem cells (ABSCs) or dental pulp stem cells (DPSCs). PDLSCs made up tissues that were rich in collagen I, while mineralized tissues were produced by DPSCs, PDLSCs, and ABSCs. This approach is a promising strategy for complete periodontal regeneration involving different tissue types [83].

14.4 Growth Factors

Growth factors are proteins that direct cellular behaviour in the relevant tissue, during physiological remodelling. In order to fulfill their duties, they must be released to the region at specific times and in sufficient doses. They can be delivered to the site directly, from cells, or transported there by a carrier. To date, growth factors that have been used for periodontal regeneration include bone morphogenic proteins, enamel matrix derivatives, vascular endothelial growth factor, platelet-derived growth factor, transforming growth factor and fibroblast growth factor [65].

Enamal Matrix Derivatives

The main component of Enamal Matrix Derivative (EMD) is amelogenin. Amelogenin is a unique extracellular matrix protein that directs mineralization of the enamel. It stimulates new bone formation and wound healing, when used under appropriate physiological conditions. There are studies showing that EMD can mimic odontogenesis and promote the stimulation of cementoblasts on the root surface [4, 55, 84].

The largest controversy around EMD is regarding the gel structure of the material, and whether or not it will remain in the wound environment after surgery. However, immunohistochemical evaluations have shown that EMD does continue to stay on the root surface for some 4 weeks after application. EMD is commercially available for clinical usage (Emdogain®) in more than 100 countries [84, 85].

Bone Morphogenic Proteins

Bone Morphogenic Proteins (BMPs) cover more than 20 growth and differentiation factors. These proteins are also members of the TGF-β superfamily [11]. BMPs, along with their receptors, have been proven to exist in periodontal tissues. Due to the potential use of BMPs to improve periodontal wound healing and regeneration, most research has been done on BMP-2, BMP-7, and BMP-3. The United States, Food and Drug Administration (FDA) approved the use of BMP-2 for maxillary sinus augmentation and localized ridge augmentation. Animals studies using BMP-2 in alveolar ridge augmentation have provided promising results [65].

Fibroblast Growth Factor-2

Fibroblast Growth Factor-2 (FGF-2) has been investigated for periodontal treatment due to its angiogenic and mitogenic effects in wound healing. FGF-2 increases the proliferation of fibroblasts. Additionally, it increases the release of osteoblast differentiation factors, thus accelerating bone formation. This protein has been approved for use in Asia and in the USA for periodontal clinical research [86].

Platelet-Derived Growth Factor

Platelet-Derived Growth Factor (PDGF) is arguably the most well-researched growth factor in dentistry and it plays an important role in wound healing. PDGF is chemotactic for periodontal ligament cells, causing them to migrate into the defect area. Additionally, PDGF enhances fibroblast proliferation and collagen synthesis. PDGF has several isotypes (AA, AB, BB). Recombinant human PDGF isotype BB (rh PDGF-BB) is available commercially in the United States for use in periodontal therapy [4, 87].

15 Soft Tissue Engineering

The simultaneous exposure of the root surface in the oral cavity and the apical relocation of the gingival margin from the cemento-enamel junction constitutes gingival recession. In most cases, there is resorption of buccal alveolar bone. While GTR techniques were used initially to treat gingival recession, it became clear that this approach had numerous limitations in thin phenotype cases and in Miller type 1 situations [88, 89].

As discussed earlier, the total renewal of the periodontium involves the regeneration of both soft and hard tissues. Surgical applications that aim to achieve only the regeneration of soft tissues can be performed for cases of gingival recession and mucogingival defects where there are keratinized tissue deficiencies [88, 90], to address problems with aesthetics and cervical dentinal hypersensitivity due to the exposure of the root surface in the oral cavity [89]. For mucogingival surgical procedures, autogenous grafts (free gingival grafts and connective tissue grafts) remain the "gold standard," even though both have drawbacks including the necessity for a secondary surgical procedure at the donor site a risk of postoperative complications such as paraesthesia and hemorrhage, and the limited amount of material that can be harvested.

There is a wide range of biomaterials on the market, designed for use with different technologies for mucogingival applications. These biomaterials may be allogenic (e.g., AlloDerm®), xenogeneic (e.g., Mucograft®), or tissue engineering materials (e.g., living cell constructs). Only Alloderm and Mucograft have been examined extensively. Both have been relatively successful as alternatives to autogenous grafts for mucogingival surgery [90].

In past years, various biomaterials have been used in periodontal regenerative medicine as a connective tissue scaffold, involving various tissue culturing techniques and tissue engineering approaches. Recent tissue engineering research has investigated the possibility of using synthetic biomaterials with live-cell constructs as an alternative to autografts for mucogingival surgery [91]. However, the performance of live cell constructs has not yet been fully explored in human trials. Live cell constructs have been assumed to improve the wound environment through wound coverage, growth factor interactions, and enhanced matrix accumulation [92], however such mechanism remains largely speculative at present.

16 Challenges in Periodontal Regenerative Therapies

One of the most thought-provoking points regarding clinical applications of regenerative periodontal medicine is the various limitations. To achieve optimal outcomes, every clinician should appreciate the challenges, contraindications, and limitations related to the technique being used, and the patient and site that the technique is being used on [11].

16.1 Variables Influencing Periodontal Regeneration

Predictable regeneration of periodontal tissues remains as a major challenge. Anatomically, the periodontium is a combination of hard and soft tissues. The hard tissue compartment consists of alveolar bone and root cementum, whereas the soft tissue compartment consists of the periodontal ligament and the gingiva [14]. A range of regenerative approaches have been developed to regenerate the periodontium, including GTR, bioactive molecules, bone grafts, soft tissue grafts, and tissue engineering. The results of these diverse applications are still not predictable [93–97].

On the other hand, the progress of periodontal regenerative medicine has been increasing year by year, with many recently invented materials and techniques. All bioactive materials also need progenitor cells to fulfill their purpose. The various progenitor cells (pre-fibroblasts, pre-cementoblasts, and pre-osteoblasts) must first migrate to the defect site, proliferate, and then mature into functional periodontium tissues, in the correct sequence. The success of this depends on the presence of suitable growth factors and the control of the progenitor cell gene expression [98, 99].

From this standpoint, it can be concluded that periodontal regeneration depends on the following variables:

(a) A sufficient supply of suitable progenitor cells, with the ability for polarization into the required mature tissue-forming phenotypes (fibroblasts, cementoblasts, and osteoblasts).
(b) An adequate cavity or space for regeneration of the tissues to take place, which is maintained by a physiological or therapeutically placed biomaterial.
(c) The proper signals to coordinate cellular polarization and tissue maturation [11].

In addition to these variables, patient factors (systemic factors, personal habits), selecting the appropriate surgical technique, local factors (tooth type and defect type) and the surgeon's experience also influence the outcomes. All these variables must be considered together during the planning of periodontal regeneration therapy [100].

16.2 Patient-Related Factors

Patient-related variables that affect the outcome of periodontal regenerative approaches include oral hygiene habits, systemic conditions, and smoking. As well, variables such as age, genetics, and stress have been suggested to have negative effects on periodontal regenerative therapy, but there is not sufficient scientific data at present to support this.

During the first few weeks after initial periodontal therapy, observing a patient's compliance and motivation in regards to changing personal habits is highly informative [100]. If the patient manages to perform satisfactory oral hygiene, this provides some optimism, since poor plaque control will hinder all regenerative periodontal processes. Many studies have demonstrated the value of professional maintenance and high standards of plaque control, both for non-surgical conservative periodontal approaches, as well as periodontal surgery, to maintain gains in clinical attachment and to reduce probing depths over the long term [100, 101].

Smoking is a substantial modifying factor for periodontal disease [102]. It impairs wound healing after surgical therapy and after non-surgical treatment [103, 104]. During wound healing, smoking reduces fibroblast cell attachment and blood flow through the tissues [105]. Smokers have worse periodontal inflammatory conditions compared to non-smokers [106–108]. Furthermore, smoking causes worse clinical outcomes from regenerative periodontal treatment [102, 109], with less gain of alveolar bone in periodontal defects [110–114].

Other patient factors include systemic health. It is now recognized that there is a complex interplay between oral health, periodontal health, and systemic health. The concept of "periodontal medicine" is well researched, with various associations (mostly due to common risk factors) between periodontal health and 57 different systemic conditions (including in descending order of importance diabetes mellitus, pregnancy, rheumatoid arthritis, respiratory diseases, cardiovascular diseases, psoriasis, and chronic kidney diseases). The periodontium can be an important aspect of the total body burden of infection [115], however, effects of periodontal treatment on the progression of these systemic conditions has yet to be clearly documented. One notable exception is diabetes mellitus, which has a bidirectional relationship with uncontrolled periodontitis. There is a higher prevalence and greater severity of periodontal disease in poorly controlled diabetic patients than in individuals without diabetes [116–118].

16.3 Surgical Technique

Periodontal intrabony defects vary in topography from wide to narrow and very deep to shallow. Most surgical periodontal protocols are effective in the treatment of intrabony 1, 2 and 3 wall defects or a combination of them [33]. Appropriate surgical techniques and materials selection are important. Surgical approaches that use a papilla preservation flap design, and suture techniques that ensure tissue integrity, stability, and primary closure of tissues, are associated with better regenerative results [119–121]. In 2019, a meta-analysis reported that papilla preservation flaps improve the clinical outcome of regeneration procedures and should be considered a surgical pre-requisite [122].

The selection of the correct material in periodontal regenerative therapy is equally as important as choosing the correct surgical technique. However, even with the proper material, complications can still occur, such as membrane exposure after surgical intervention. Exposure of the membrane negatively affects the outcome of regenerative surgical procedures, especially when it is related to a non-resorbable membrane.

This has led to the design of modified surgical approaches to maintain interdental periodontal tissues and reduce the chances of membrane exposure [123, 124]. Cortellini et al. demonstrated that the prevalence of membrane exposure could be decreased with the utilization of specifically designed flaps that protect the interdental tissues by using the modified papilla preservation technique [119, 121, 125]. Numerous studies have demonstrated that membranes that become exposed to the oral area are readily infected with bacteria, and this bacterial contamination (whether of non-resorbable or resorbable membranes) lowers attachment gains in intrabony defects [126–130].

Better periodontal regenerative results can be achieved if the patient is worked up with a meticulous pre-surgical examination, to explore the anatomy of the defect, the interdental space, and determine which type of material is going to be used [131]. A multicenter randomized study has shown that shallow and deep bone defects are likely to give similar regenerative outcomes [132]. However, each patient has their unique characteristics, and each defect is unique. No one periodontal regenerative protocol can address all possible defects by itself. Consequently, there must be a clinical decision pathway that leads the surgeon to select the best surgical approach for periodontal regeneration of each individual defect [33, 34].

16.4 Local Factors

The defect and the tooth condition have an important impact on the success of periodontal regeneration locally. These aspects should be evaluated clinically and radiographically.

Defect Factors

The predictability of periodontal regenerative therapy is influenced by the local anatomy and the nature of the periodontal defect. Tonetti et al. showed that intrabony periodontal defects deeper than 3 mm show superior probing attachment and alveolar bone gain, than shallow defects after GTR [133]. Tonetti et al. also reported that the morphology of the defect can influence the results, with greater regeneration of deep, narrow defects with more remaining bone walls [133, 134]. In a different study, Tonetti et al. showed that better regeneration was achieved in non-smokers and for intrabony defects with three walls [135].

Tooth Factors

Endodontic status and tooth mobility are potential factors of interest. Ehnevid et al. demonstrated that inadequately treated and endodontically compromised teeth respond less favorably to periodontal regenerative approaches [136, 137]. On the other hand, when performed precisely, endodontic therapy does not impact the healing outcome and the long-term results when treating deep intrabony periodontal defects with regenerative protocols [138].

According to a multicenter clinical trial, greater tooth mobility is related to worse clinical outcomes for periodontal regenerative protocols [120]. In particularly, Miller grade III mobility adversely affects the outcomes. However, teeth with mobility lower than 1 mm horizontally can successfully be treated with periodontal regenerative protocols [139].

Aside from these various factors, there is one further very important factor that has a strong impact on the outcomes of regenerative therapy: *operator experience*. Predictable clinical outcomes require meticulous diagnosis and treatment, hence the clinical skills of the operator influence the success of periodontal regenerative treatments [131, 140].

17 Future of Regenerative Periodontics: What Is Next?

Current developments in the field of regenerative periodontology are exciting. While many techniques in tissue engineering have produced promising results from in vitro and in vivo studies, few methods have been applied to humans. In 2015, Rasperini et al. reported the first human case of surgical treatment of a periodontal defect with a 3D printed scaffold. A customized 3D laser sintered approach was used in the production of this scaffold, which incorporated PCL and a combination of HA powders. The scaffold was used to treat a very large periodontal defect. It did not cause any inflammatory reaction in the first 6 months. However, due to exposure of the scaffold after 12 months, it was found to be clinically inadequate [141].

One of the most important developments in biological fabrication is the production of a whole tissue or organ by utilizing bioprinting techniques. The goal of biological printing is to build a particular tissue or a whole organ by using the living cells of the individual, placed in an extracellular matrix environment. Improvements in biological printing have led to progress in some branches of medicine, whereas the biological production of oral tissues using the same method is still in its early stages.

Raveendran et al. reported an optimization study involving the 3D bioprinting of periodontal ligament cells, where a gelatin methacryloyl hydrogel (GelMA) was used as the carrier. Various parameters such as the concentration of the hydrogel, the printing pressure, and the aperture of the needle used for printing influenced the resolution and dimensional stability of the bioprinting process [80].

Gene therapy, on the other hand, had led to new approaches to treat hereditary dental diseases [4]. Gene therapy usually involves placing relevant genes in an individual's cells to achieve an increase in the release of a specific growth factor. Gene therapy can be performed by two methods. In the first method, the gene vector is placed directly into the target region (in vivo). Alternatively, in a second approach, the selected cells are harvested, expanded, and then genetically transduced, for example, using an adenovirus vector or another vector (ex vivo) [142].

Although gene therapy is a promising method, it is debatable whether the use of adenovirus for this purpose is safe [143]. Variations in immune responses to gene therapy may also expose problems. However, the approach of stimulating the genes of the cells towards targeted tissue regeneration may emerge as a very important treatment alternative in the future, not only in periodontology, but in other disciplines [4, 143].

As discussed earlier, the combined use of growth factors and 3D scaffolds with biological materials for periodontal regeneration is one of the most important developments in tissue engineering. However, the biomolecules used in the production of growth factors are expensive, and the production of 3D scaffolds is a highly demanding process. Furthermore, even the use of a low dose of biomolecules can cause side effects in some individuals. These reasons make it difficult for innovative regenerative treatment protocols to be accessible for everyday clinical usage [75]. However, various research groups from all around the world are working to make these protocols more user-friendly for clinicians.

18 Conclusion

The ultimate goal of periodontal regenerative medicine is to prevent tooth loss by reproducing the supporting tissue of the tooth. Regenerative treatment with the combination of available surgical techniques and materials is generally limited to defined types of periodontal defects. Periodontium regeneration must rest on a solid biological rationale, histological evidence, as well as evidence of achieving the desired clinical outcomes [144].

Recent developments have combined cells, appropriate biomaterials, and growth factors for periodontal regeneration. There have been many other positive developments in the regeneration of complex alveolar bone defects. Techniques for the production and design of scaffolds are becoming more predictable [3].

It is an extremely difficult task to completely renew periodontal tissues, both functionally and morphologically. Current scientific studies state that any single regenerative approach is unlikely to be successful for every purpose. A key point is that migration of soft tissue into the periodontal defect should be prevented, and appropriate pioneering signals should be released. Therefore, the use of multiple layer manufacturing systems for tissue engineering seems to hold the greatest promise [145].

Any regenerative approaches must ensure effective control of local infections caused by microbial pathogens during healing. Therefore, a suitable strategy must consider how best to limit bacterial growth, [4] and ensure that patient's maintain good levels of plaque control. Clinical improvements after regenerative therapy can be maintained for a long period of time, in most treated areas, if patients do not smoke and continue to maintain high standards of oral hygiene [89]. A further factor that influences outcomes is the surgeon's ability to choose and then apply the best surgical technique, based on patient-specific and region-specific criteria.

New periodontal regenerative approaches seem promising for the transition to clinical practice in the not too distant future. Each method that has been successful in in vitro approaches, however, needs to be tested extensively in vivo in animal models and then in human clinical trials. Any biological products that are to be used in daily treatments must be free of pathogens and of high quality and meet the required regulatory approvals.

References

1. Murray CJL, Vos T, Lozano R, Naghavi M, Flaxman AD, Michaud C, et al. Disability-adjusted life years (DALYs) for 291 diseases and injuries in 21 regions, 1990–2010: a systematic analysis for the global burden of disease study 2010. Lancet. 2012;380(9859):2197–223. https://doi.org/10.1016/S0140-6736(12)61689-4.
2. Tonetti MS, Jepsen S, Jin L, Otomo-Corgel J. Impact of the global burden of periodontal diseases on health, nutrition and wellbeing of mankind: a call for global action. J Clin Periodontol. 2017;44(5):456–62. https://doi.org/10.1111/jcpe.12732.
3. Vaquette C, Pilipchuk SP, Bartold PM, Hutmacher DW, Giannobile WV, Ivanovski S. Tissue engineered constructs for periodontal regeneration: current status and future perspectives. Adv Healthc Mater. 2018;7(21):e1800457. https://doi.org/10.1002/adhm.201800457.

4. Iviglia G, Kargozar S, Baino F. Biomaterials, current strategies, and novel nano-technological approaches for periodontal regeneration. J Funct Biomater. 2019;10(1):3. https://doi.org/10.3390/jfb10010003.
5. Alpiste Illueca FM, Buitrago Vera P, de Grado Cabanilles P, Fuenmayor Fernandez V, Gil Loscos FJ. Periodontal regeneration in clinical practice. Med Oral Patol Oral Cir Bucal. 2006;11(4): E382–92.
6. Rosling B, Nyman S, Lindhe J, Jern B. The healing potential of the periodontal tissues following different techniques of periodontal surgery in plaque-free dentitions. A 2-year clinical study. J Clin Periodontol. 1976;3(4):233–50. https://doi.org/10.1111/j.1600-051x.1976.tb00042.x.
7. Engler WO, Ramfjord SP, Hiniker JJ. Healing following simple gingivectomy. A tritiated thymidine radioautographic study. I. Epithelialization. J Periodontol. 1966;37(4):298–308. https://doi.org/10.1902/jop.1966.37.4.298.
8. Liang Y, Luan X, Liu X. Recent advances in periodontal regeneration: a biomaterial perspective. Bioact Mater. 2020;5(2):297–308. https://doi.org/10.1016/j.bioactmat.2020.02.012.
9. Groll J, Boland T, Blunk T, Burdick JA, Cho DW, Dalton PD, et al. Biofabrication: reappraising the definition of an evolving field. Biofabrication. 2016;8(1):013001. https://doi.org/10.1088/1758-5090/8/1/013001.
10. Lindhe J, Lang NP. Clinical periodontology and implant dentistry. Chichester: Wiley Blackwell; 2015.
11. Hughes FJ, Ghuman M, Talal A. Periodontal regeneration: a challenge for the tissue engineer? Proc Inst Mech Eng H J Eng Med. 2010;224(12):1345–58. https://doi.org/10.1243/09544119jeim820.
12. Newman MG, Takei HH, Klokkevold PR, Carranza FA. Newman and Carranza's clinical periodontology. London: Elsevier; 2019.
13. Jepsen S, Caton JG, Albandar JM, Bissada NF, Bouchard P, Cortellini P, et al. Periodontal manifestations of systemic diseases and developmental and acquired conditions: consensus report of workgroup 3 of the 2017 world workshop on the classification of periodontal and peri-implant diseases and conditions. J Clin Periodontol. 2018;45(Suppl 20):S219–s29. https://doi.org/10.1111/jcpe.12951.
14. Ivanovski S. Periodontal regeneration. Aust Dent J. 2009;54(Suppl 1):S118–28. https://doi.org/10.1111/j.1834-7819.2009.01150.x.
15. Aaron JE. Periosteal Sharpey's fibers: a novel bone matrix regulatory system? Front Endocrinol. 2012;3:98. https://doi.org/10.3389/fendo.2012.00098.
16. McCulloch CA. Basic considerations in periodontal wound healing to achieve regeneration. Periodontology 2000. 1993;1(1):16–25.
17. McCulloch CA. Historical perspective of periodontal progenitor cells: early studies that clarified identity and function. Curr Oral Health Rep. 2015;2(4):227–35. https://doi.org/10.1007/s40496-015-0061-z.
18. Wikesjö UM, Nilvéus RE, Selvig KA. Significance of early healing events on periodontal repair: a review. J Periodontol. 1992;63(3):158–65. https://doi.org/10.1902/jop.1992.63.3.158.
19. Sam G, Pillai BRM. Evolution of barrier membranes in periodontal regeneration-"are the third generation membranes really here?". J Clin Diagn Res. 2014;8(12):ZE14–ZE7. https://doi.org/10.7860/JCDR/2014/9957.5272.
20. Caton JG. Overview of clinical trials on periodontal regeneration. Ann Periodontol. 1997;2(1): 215–22. https://doi.org/10.1902/annals.1997.2.1.215.
21. Cochran DL, Wozney JM. Biological mediators for periodontal regeneration. Periodontology 2000. 1999;19:40–58. https://doi.org/10.1111/j.1600-0757.1999.tb00146.x.
22. Liu J, Zhao Z, Ruan J, Weir MD, Ma T, Ren K, et al. Stem cells in the periodontal ligament differentiated into osteogenic, fibrogenic and cementogenic lineages for the regeneration of the periodontal complex. J Dent. 2020;92:103259. https://doi.org/10.1016/j.jdent.2019.103259.
23. Murakami S, editor. Emerging regenerative approaches for periodontal regeneration: the future perspective of cytokine therapy and stem cell therapy. Interface Oral Health Science 2016. Singapore: Springer; 2017.
24. Schüpbach P, Gaberthüel T, Lutz F, Guggenheim B. Periodontal repair or regeneration: structures of different types of new attachment. J Periodontal Res. 1993;28(4):281–93. https://doi.org/10.1111/j.1600-0765.1993.tb02095.x.
25. Pilloni A, Rojas MA. Furcation involvement classification: a comprehensive review and a new system proposal. Dent J. 2018;6(3):34. https://doi.org/10.3390/dj6030034.
26. Hamp SE, Nyman S, Lindhe J. Periodontal treatment of multirooted teeth. Results after 5 years. J Clin Periodontol. 1975;2(3):126–35. https://doi.org/10.1111/j.1600-051x.1975.tb01734.x.
27. Pini-Prato G. The Miller classification of gingival recession: limits and drawbacks. J Clin Periodontol. 2011;38(3):243–5. https://doi.org/10.1111/j.1600-051X.2010.01655.x.
28. Cairo F, Nieri M, Cincinelli S, Mervelt J, Pagliaro U. The interproximal clinical attachment level to classify gingival recessions and predict root coverage outcomes: an explorative and reliability study. J Clin Periodontol. 2011;38(7):661–6. https://doi.org/10.1111/j.1600-051X.2011.01732.x.
29. Lang NP, Salvi GE, Sculean A. Nonsurgical therapy for teeth and implants-when and why? Periodontology 2000. 2019;79(1):15–21. https://doi.org/10.1111/prd.12240.
30. Kraatz J, Hoang H, Ivanovski S, Ware RS, Crocombe LA. Periodontal diagnosis, treatment, and referral patterns of general dental practitioners. J Investig Clin Dent. 2019;10(3):e12411. https://doi.org/10.1111/jicd.12411.

31. Tonetti MS, Greenwell H, Kornman KS. Staging and grading of periodontitis: framework and proposal of a new classification and case definition. J Periodontol. 2018;89(Suppl 1):S159–s72. https://doi.org/10.1002/jper.18-0006.
32. Graziani F, Karapetsa D, Mardas N, Leow N, Donos N. Surgical treatment of the residual periodontal pocket. Periodontology 2000. 2018;76(1):150–63. https://doi.org/10.1111/prd.12156.
33. Cortellini P, Tonetti MS. Clinical concepts for regenerative therapy in intrabony defects. Periodontology 2000. 2015;68(1):282–307. https://doi.org/10.1111/prd.12048.
34. Carnevale G, Kaldahl WB. Osseous resective surgery. Periodontology 2000. 2000;22:59–87. https://doi.org/10.1034/j.1600-0757.2000.2220106.x.
35. Heitz-Mayfield LJ, Lang NP. Surgical and nonsurgical periodontal therapy. Learned and unlearned concepts. Periodontology 2000. 2013;62(1):218–31. https://doi.org/10.1111/prd.12008.
36. Boyne PJ. Methods of osseous reconstruction of the mandible following surgical resection. J Biomed Mater Res. 1973;7(3):195–204. https://doi.org/10.1002/jbm.820070313.
37. Fernandez de Grado G, Keller L, Idoux-Gillet Y, Wagner Q, Musset AM, Benkirane-Jessel N, et al. Bone substitutes: a review of their characteristics, clinical use, and perspectives for large bone defects management. J Tissue Eng. 2018;9:2041731418776819. https://doi.org/10.1177/2041731418776819.
38. AlGhamdi AS, Shibly O, Ciancio SG. Osseous grafting part I: autografts and allografts for periodontal regeneration—a literature review. J Int Acad Periodontol. 2010;12(2):34–8.
39. Nasr HF, Aichelmann-Reidy ME, Yukna RA. Bone and bone substitutes. Periodontology 2000. 1999;19: 74–86. https://doi.org/10.1111/j.1600-0757.1999.tb00148.x.
40. Mellonig JT. Freeze-dried bone allografts in periodontal reconstructive surgery. Dent Clin N Am. 1991;35(3):505–20.
41. Jain AP, Pundir S, Sharma A. Bone morphogenetic proteins: the anomalous molecules. J Indian Soc Periodontol. 2013;17(5):583–6. https://doi.org/10.4103/0972-124x.119275.
42. Bowers GM, Chadroff B, Carnevale R, Mellonig J, Corio R, Emerson J, et al. Histologic evaluation of new attachment apparatus formation in humans. Part II. J Periodontol. 1989;60(12):675–82. https://doi.org/10.1902/jop.1989.60.12.675.
43. Kim Y, Rodriguez AE, Nowzari H. The risk of prion infection through bovine grafting materials. Clin Implant Dent Relat Res. 2016;18(6):1095–102. https://doi.org/10.1111/cid.12391.
44. AlGhamdi AS, Shibly O, Ciancio SG. Osseous grafting part II: xenografts and alloplasts for periodontal regeneration—a literature review. J Int Acad Periodontol. 2010;12(2):39–44.
45. Yukna RA, Callan DP, Krauser JT, Evans GH, Aichelmann-Reidy ME, Moore K, et al. Multi-center clinical evaluation of combination anorganic bovine-derived hydroxyapatite matrix (ABM)/cell binding peptide (P-15) as a bone replacement graft material in human periodontal osseous defects. 6-month results. J Periodontol. 1998;69(6):655–63. https://doi.org/10.1902/jop.1998.69.6.655.
46. Molly L, Vandromme H, Quirynen M, Schepers E, Adams JL, van Steenberghe D. Bone formation following implantation of bone biomaterials into extraction sites. J Periodontol. 2008;79(6):1108–15. https://doi.org/10.1902/jop.2008.070476.
47. Boyne PJ. Application of bone morphogenetic proteins in the treatment of clinical oral and maxillofacial osseous defects. J Bone Joint Surg Am. 2001;83-A(Suppl 1(Pt 2)):S146–50.
48. Boyne PJ, Lilly LC, Marx RE, Moy PK, Nevins M, Spagnoli DB, et al. De novo bone induction by recombinant human bone morphogenetic protein-2 (rhBMP-2) in maxillary sinus floor augmentation. J Oral Maxillofac Surg. 2005;63(12):1693–707. https://doi.org/10.1016/j.joms.2005.08.018.
49. Nyman S, Gottlow J, Lindhe J, Karring T, Wennstrom J. New attachment formation by guided tissue regeneration. J Periodontal Res. 1987;22(3):252–4. https://doi.org/10.1111/j.1600-0765.1987.tb01581.x.
50. Nyman S. Bone regeneration using the principle of guided tissue regeneration. J Clin Periodontol. 1991;18(6):494–8. https://doi.org/10.1111/j.1600-051x.1991.tb02322.x.
51. Murphy KG, Gunsolley JC. Guided tissue regeneration for the treatment of periodontal intrabony and furcation defects. A systematic review. Ann Periodontol. 2003;8(1):266–302. https://doi.org/10.1902/annals.2003.8.1.266.
52. Wang HL, Modarressi M, Fu JH. Utilizing collagen membranes for guided tissue regeneration-based root coverage. Periodontology 2000. 2012;59(1):140–57. https://doi.org/10.1111/j.1600-0757.2011.00438.x.
53. Stoecklin-Wasmer C, Rutjes AW, da Costa BR, Salvi GE, Jüni P, Sculean A. Absorbable collagen membranes for periodontal regeneration: a systematic review. J Dent Res. 2013;92(9):773–81. https://doi.org/10.1177/0022034513496428.
54. Ling LJ, Hung SL, Lee CF, Chen YT, Wu KM. The influence of membrane exposure on the outcomes of guided tissue regeneration: clinical and microbiological aspects. J Periodontal Res. 2003;38(1): 57–63. https://doi.org/10.1034/j.1600-0765.2003.01641.x.
55. Hammarström L. Enamel matrix, cementum development and regeneration. J Clin Periodontol. 1997;24(9 Pt 2):658–68. https://doi.org/10.1111/j.1600-051x.1997.tb00247.x.
56. Wennström JL. Mucogingival therapy. Ann Periodontol. 1996;1(1):671–701. https://doi.org/10.1902/annals.1996.1.1.671.
57. Lindhe J. Textbook of clinical periodontology. Copenhagen: Munksgaard; 1989.
58. Kao RT, Fagan MC, Conte GJ. Thick vs. thin gingival biotypes: a key determinant in treatment

planning for dental implants. J Calif Dent Assoc. 2008;36(3):193–8.
59. Ramachandra SS, Patil M, Mehta DS. Implants placed into extraction sockets: a literature review. Dent Implantol Updat. 2009;20(2):1–8.
60. Fischer KR, Richter T, Kebschull M, Petersen N, Fickl S. On the relationship between gingival biotypes and gingival thickness in young Caucasians. Clin Oral Implants Res. 2015;26(8):865–9. https://doi.org/10.1111/clr.12356.
61. Thoma DS, Benić GI, Zwahlen M, Hämmerle CH, Jung RE. A systematic review assessing soft tissue augmentation techniques. Clin Oral Implants Res. 2009;20(Suppl 4):146–65. https://doi.org/10.1111/j.1600-0501.2009.01784.x.
62. Griffin TJ, Cheung WS, Zavras AI, Damoulis PD. Postoperative complications following gingival augmentation procedures. J Periodontol. 2006;77(12):2070–9. https://doi.org/10.1902/jop.2006.050296.
63. Ramachandra SS, Rana R, Reetika S, Jithendra KD. Options to avoid the second surgical site: a review of literature. Cell Tissue Bank. 2014;15(3):297–305. https://doi.org/10.1007/s10561-013-9395-8.
64. Tran SD, Bakkar MO, Sumita Y, Kishimoto N. Regenerative dentistry in periodontics. Saudi Dent J. 2019;31(3):301–2. https://doi.org/10.1016/j.sdentj.2019.05.002.
65. Carmagnola D, Pellegrini G, Dellavia C, Rimondini L, Varoni E. Tissue engineering in periodontology: biological mediators for periodontal regeneration. Int J Artif Organs. 2019;42(5):241–57. https://doi.org/10.1177/0391398819828558.
66. Bui FQ, Almeida-da-Silva CLC, Huynh B, Trinh A, Liu J, Woodward J, et al. Association between periodontal pathogens and systemic disease. Biom J. 2019;42(1):27–35. https://doi.org/10.1016/j.bj.2018.12.001.
67. Sculean A, Chapple IL, Giannobile WV. Wound models for periodontal and bone regeneration: the role of biologic research. Periodontology 2000. 2015;68(1):7–20. https://doi.org/10.1111/prd.12091.
68. Bartold PM, McCulloch CA, Narayanan AS, Pitaru S. Tissue engineering: a new paradigm for periodontal regeneration based on molecular and cell biology. Periodontology 2000. 2000;24:253–69. https://doi.org/10.1034/j.1600-0757.2000.2240113.x.
69. Carter SD, Costa PF, Vaquette C, Ivanovski S, Hutmacher DW, Malda J. Additive biomanufacturing: an advanced approach for periodontal tissue regeneration. Ann Biomed Eng. 2017;45(1):12–22. https://doi.org/10.1007/s10439-016-1687-2.
70. Requicha JF, Viegas CA, Muñoz F, Azevedo JM, Leonor IB, Reis RL, et al. A tissue engineering approach for periodontal regeneration based on a biodegradable double-layer scaffold and adipose-derived stem cells. Tissue Eng Part A. 2014;20(17–18):2483–92. https://doi.org/10.1089/ten.TEA.2013.0360.
71. Seo BM, Miura M, Gronthos S, Bartold PM, Batouli S, Brahim J, et al. Investigation of multipotent postnatal stem cells from human periodontal ligament. Lancet (London, England). 2004;364(9429):149–55. https://doi.org/10.1016/s0140-6736(04)16627-0.
72. Liu Y, Zheng Y, Ding G, Fang D, Zhang C, Bartold PM, et al. Periodontal ligament stem cell-mediated treatment for periodontitis in miniature swine. Stem Cells (Dayton, Ohio). 2008;26(4):1065–73. https://doi.org/10.1634/stemcells.2007-0734.
73. Iwata T, Washio K, Yoshida T, Ishikawa I, Ando T, Yamato M, et al. Cell sheet engineering and its application for periodontal regeneration. J Tissue Eng Regen Med. 2015;9(4):343–56. https://doi.org/10.1002/term.1785.
74. Raju R, Oshima M, Inoue M, Morita T, Huijiao Y, Waskitho A, et al. Three-dimensional periodontal tissue regeneration using a bone-ligament complex cell sheet. Sci Rep. 2020;10(1):1656. https://doi.org/10.1038/s41598-020-58222-0.
75. Xu XY, Li X, Wang J, He XT, Sun HH, Chen FM. Concise review: periodontal tissue regeneration using stem cells: strategies and translational considerations. Stem Cells Transl Med. 2019;8(4):392–403. https://doi.org/10.1002/sctm.18-0181.
76. Dabra S, Chhina K, Soni N, Bhatnagar R. Tissue engineering in periodontal regeneration: a brief review. Dent Res J. 2012;9(6):671–80.
77. Ivanovski S, Vaquette C, Gronthos S, Hutmacher DW, Bartold PM. Multiphasic scaffolds for periodontal tissue engineering. J Dent Res. 2014;93(12):1212–21. https://doi.org/10.1177/0022034514544301.
78. Motamedian SR, Hosseinpour S, Ahsaie MG, Khojasteh A. Smart scaffolds in bone tissue engineering: a systematic review of literature. World J Stem Cells. 2015;7(3):657.
79. Hosseinpour S, Ahsaie MG, Rad MR, taghi Baghani M, Motamedian SR, Khojasteh A. Application of selected scaffolds for bone tissue engineering: a systematic review. Oral Maxillofac Surg. 2017;21(2):109–29.
80. Thattaruparambil Raveendran N, Vaquette C, Meinert C, Samuel Ipe D, Ivanovski S. Optimization of 3D bioprinting of periodontal ligament cells. Dent Mater. 2019;35(12):1683–94. https://doi.org/10.1016/j.dental.2019.08.114.
81. Park CH, Rios HF, Jin Q, Bland ME, Flanagan CL, Hollister SJ, et al. Biomimetic hybrid scaffolds for engineering human tooth-ligament interfaces. Biomaterials. 2010;31(23):5945–52. https://doi.org/10.1016/j.biomaterials.2010.04.027.
82. Park CH, Rios HF, Jin Q, Sugai JV, Padial-Molina M, Taut AD, et al. Tissue engineering bone-ligament complexes using fiber-guiding scaffolds. Biomaterials. 2012;33(1):137–45. https://doi.org/10.1016/j.biomaterials.2011.09.057.
83. Lee CH, Hajibandeh J, Suzuki T, Fan A, Shang P, Mao JJ. Three-dimensional printed multiphase scaffolds for regeneration of periodontium complex. Tissue Eng Part A. 2014;20(7–8):1342–51. https://doi.org/10.1089/ten.TEA.2013.0386.

84. Stavropoulos A, Sculean A. Current status of regenerative periodontal treatment. Curr Oral Health Rep. 2017;4(1):34–43. https://doi.org/10.1007/s40496-017-0122-6.
85. Sculean A, Windisch P, Keglevich T, Fabi B, Lundgren E, Lyngstadaas PS. Presence of an enamel matrix protein derivative on human teeth following periodontal surgery. Clin Oral Investig. 2002;6(3):183–7. https://doi.org/10.1007/s00784-002-0171-6.
86. Nagayasu-Tanaka T, Anzai J, Takaki S, Shiraishi N, Terashima A, Asano T, et al. Action mechanism of fibroblast growth factor-2 (FGF-2) in the promotion of periodontal regeneration in beagle dogs. PLoS One. 2015;10(6):e0131870. https://doi.org/10.1371/journal.pone.0131870.
87. Nevins M, Garber D, Hanratty JJ, McAllister BS, Nevins ML, Salama M, et al. Human histologic evaluation of anorganic bovine bone mineral combined with recombinant human platelet-derived growth factor BB in maxillary sinus augmentation: case series study. Int J Periodontics Restorative Dent. 2009;29(6):583–91.
88. Tavelli L, McGuire MK, Zucchelli G, Rasperini G, Feinberg SE, Wang HL, et al. Biologics-based regenerative technologies for periodontal soft tissue engineering. J Periodontol. 2020;91(2):147–54. https://doi.org/10.1002/jper.19-0352.
89. Trombelli L, Minenna L, Farina R, Scabbia A. Guided tissue regeneration in human gingival recessions. A 10-year follow-up study. J Clin Periodontol. 2005;32(1):16–20. https://doi.org/10.1111/j.0303-6979.2004.00625.x.
90. Lam L, Lee RSB, Ivanovski S. Periodontal soft tissue reconstruction. In: Tayebi L, Moharamzadeh K, editors. Biomaterials for oral and dental tissue engineering. Cambridge: Woodhead Publishing; 2017. p. 257–78.
91. Nevins ML. Tissue-engineered bilayered cell therapy for the treatment of oral mucosal defects: a case series. Int J Periodontics Restorative Dent. 2010;30(1):31–9.
92. McGuire MK, Scheyer ET, Nevins ML, Neiva R, Cochran DL, Mellonig JT, et al. Living cellular construct for increasing the width of keratinized gingiva: results from a randomized, within-patient, controlled trial. J Periodontol. 2011;82(10):1414–23. https://doi.org/10.1902/jop.2011.100671.
93. Villar CC, Cochran DL. Regeneration of periodontal tissues: guided tissue regeneration. Dent Clin N Am. 2010;54(1):73–92. https://doi.org/10.1016/j.cden.2009.08.011.
94. Kao RT, Nares S, Reynolds MA. Periodontal regeneration – intrabony defects: a systematic review from the AAP regeneration workshop. J Periodontol. 2015;86(Suppl 2):S77–104. https://doi.org/10.1902/jop.2015.130685.
95. Lin Z, Rios HF, Cochran DL. Emerging regenerative approaches for periodontal reconstruction: a systematic review from the AAP regeneration workshop. J Periodontol. 2015;86(Suppl 2):S134–52. https://doi.org/10.1902/jop.2015.130689.
96. Li F, Yu F, Xu X, Li C, Huang D, Zhou X, et al. Evaluation of recombinant human FGF-2 and PDGF-BB in periodontal regeneration: a systematic review and meta-analysis. Sci Rep. 2017;7(1):65. https://doi.org/10.1038/s41598-017-00113-y.
97. Esposito M, Grusovin MG, Papanikolaou N, Coulthard P, Worthington HV. Enamel matrix derivative (Emdogain(R)) for periodontal tissue regeneration in intrabony defects. Cochrane Database Syst Rev. 2009;2009(4):Cd003875. https://doi.org/10.1002/14651858.CD003875.pub3.
98. Ivanovski S, Gronthos S, Shi S, Bartold PM. Stem cells in the periodontal ligament. Oral Dis. 2006;12(4):358–63. https://doi.org/10.1111/j.1601-0825.2006.01253.x.
99. Aukhil I. Biology of wound healing. Periodontology 2000. 2000;22:44–50. https://doi.org/10.1034/j.1600-0757.2000.2220104.x.
100. Lindhe J, Nyman S. Long-term maintenance of patients treated for advanced periodontal disease. J Clin Periodontol. 1984;11(8):504–14. https://doi.org/10.1111/j.1600-051x.1984.tb00902.x.
101. Casarin RC, Del Peloso Ribeiro E, Nociti FH Jr, Sallum AW, Sallum EA, Ambrosano GM, et al. A double-blind randomized clinical evaluation of enamel matrix derivative proteins for the treatment of proximal class-II furcation involvements. J Clin Periodontol. 2008;35(5):429–37. https://doi.org/10.1111/j.1600-051X.2008.01202.x.
102. Patel RA, Wilson RF, Palmer RM. The effect of smoking on periodontal bone regeneration: a systematic review and meta-analysis. J Periodontol. 2012;83(2):143–55. https://doi.org/10.1902/jop.2011.110130.
103. Kinane DF, Chestnutt IG. Smoking and periodontal disease. Crit Rev Oral Biol Med. 2000;11(3):356–65. https://doi.org/10.1177/10454411000110030501.
104. Preshaw PM, Heasman L, Stacey F, Steen N, McCracken GI, Heasman PA. The effect of quitting smoking on chronic periodontitis. J Clin Periodontol. 2005;32(8):869–79. https://doi.org/10.1111/j.1600-051X.2005.00779.x.
105. Palmer RM, Wilson RF, Hasan AS, Scott DA. Mechanisms of action of environmental factors—tobacco smoking. J Clin Periodontol. 2005;32(Suppl 6):180–95. https://doi.org/10.1111/j.1600-051X.2005.00786.x.
106. Javed F, Al-Askar M, Samaranayake LP, Al-Hezaimi K. Periodontal disease in habitual cigarette smokers and nonsmokers with and without prediabetes. Am J Med Sci. 2013;345(2):94–8. https://doi.org/10.1097/MAJ.0b013e31824d5337.
107. Rheu GB, Ji S, Ryu JJ, Lee JB, Shin C, Lee JY, et al. Risk assessment for clinical attachment loss of periodontal tissue in Korean adults. J Adv Prosthodont. 2011;3(1):25–32. https://doi.org/10.4047/jap.2011.3.1.25.
108. Abdul Rahman N, Nickles K, Gallenbach K, Dannewitz B, Ramich T, Scharf S, et al. Five-year

stability of clinical attachment after regenerative treatment of infrabony defects compared to controls. J Clin Periodontol. 2019;46(6):650–8. https://doi.org/10.1111/jcpe.13105.
109. Tonetti MS, Pini-Prato G, Cortellini P. Effect of cigarette smoking on periodontal healing following GTR in infrabony defects. A preliminary retrospective study. J Clin Periodontol. 1995;22(3):229–34. https://doi.org/10.1111/j.1600-051x.1995.tb00139.x.
110. Ehmke B, Rüdiger SG, Hommens A, Karch H, Flemmig TF. Guided tissue regeneration using a polylactic acid barrier. J Clin Periodontol. 2003;30(4):368–74. https://doi.org/10.1034/j.1600-051x.2003.00312.x.
111. Heden G. A case report study of 72 consecutive Emdogain-treated intrabony periodontal defects: clinical and radiographic findings after 1 year. Int J Periodontics Restorative Dent. 2000;20(2):127–39.
112. Loos BG, Louwerse PH, Van Winkelhoff AJ, Burger W, Gilijamse M, Hart AA, et al. Use of barrier membranes and systemic antibiotics in the treatment of intraosseous defects. J Clin Periodontol. 2002;29(10):910–21. https://doi.org/10.1034/j.1600-051x.2002.291006.x.
113. Trombelli L, Kim CK, Zimmerman GJ, Wikesjö UM. Retrospective analysis of factors related to clinical outcome of guided tissue regeneration procedures in intrabony defects. J Clin Periodontol. 1997;24(6):366–71. https://doi.org/10.1111/j.1600-051x.1997.tb00199.x.
114. Yilmaz S, Cakar G, Ipci SD, Kuru B, Yildirim B. Regenerative treatment with platelet-rich plasma combined with a bovine-derived xenograft in smokers and non-smokers: 12-month clinical and radiographic results. J Clin Periodontol. 2010;37(1):80–7. https://doi.org/10.1111/j.1600-051X.2009.01509.x.
115. Mattila KJ, Nieminen MS, Valtonen VV, Rasi VP, Kesäniemi YA, Syrjälä SL, et al. Association between dental health and acute myocardial infarction. BMJ (Clin Res ed). 1989;298(6676):779–81. https://doi.org/10.1136/bmj.298.6676.779.
116. Llambés F, Arias-Herrera S, Caffesse R. Relationship between diabetes and periodontal infection. World J Diabetes. 2015;6(7):927–35. https://doi.org/10.4239/wjd.v6.i7.927.
117. Falcao A, Bullón P. A review of the influence of periodontal treatment in systemic diseases. Periodontology 2000. 2019;79(1):117–28. https://doi.org/10.1111/prd.12249.
118. Bourgeois D, Inquimbert C, Ottolenghi L, Carrouel F. Periodontal pathogens as risk factors of cardiovascular diseases, diabetes, rheumatoid arthritis, cancer, and chronic obstructive pulmonary disease-is there cause for consideration? Microorganisms. 2019;7(10):424. https://doi.org/10.3390/microorganisms7100424.
119. Cortellini P, Prato GP, Tonetti MS. The modified papilla preservation technique. A new surgical approach for interproximal regenerative procedures. J Periodontol. 1995;66(4):261–6. https://doi.org/10.1902/jop.1995.66.4.261.
120. Cortellini P, Tonetti MS, Lang NP, Suvan JE, Zucchelli G, Vangsted T, et al. The simplified papilla preservation flap in the regenerative treatment of deep intrabony defects: clinical outcomes and postoperative morbidity. J Periodontol. 2001;72(12):1702–12. https://doi.org/10.1902/jop.2001.72.12.1702.
121. Cortellini P, Buti J, Pini Prato G, Tonetti MS. Periodontal regeneration compared with access flap surgery in human intra-bony defects 20-year follow-up of a randomized clinical trial: tooth retention, periodontitis recurrence and costs. J Clin Periodontol. 2017;44(1):58–66. https://doi.org/10.1111/jcpe.12638.
122. Nibali L, Koidou VP, Nieri M, Barbato L, Pagliaro U, Cairo F. Regenerative surgery versus access flap for the treatment of intrabony periodontal defects. A systematic review and meta-analysis. J Clin Periodontol. 2019;47:320–51. https://doi.org/10.1111/jcpe.13237.
123. Murphy KG. Postoperative healing complications associated with gore-tex periodontal material. Part I. Incidence and characterization. Int J Periodont Restorative Dent. 1995;15(4):363–75.
124. Selvig KA, Kersten BG, Chamberlain AD, Wikesjö UM, Nilvéus RE. Regenerative surgery of intrabony periodontal defects using ePTFE barrier membranes: scanning electron microscopic evaluation of retrieved membranes versus clinical healing. J Periodontol. 1992;63(12):974–8. https://doi.org/10.1902/jop.1992.63.12.974.
125. Cortellini P, Pini Prato G, Tonetti MS. Periodontal regeneration of human intrabony defects with titanium reinforced membranes. A controlled clinical trial. J Periodontol. 1995;66(9):797–803. https://doi.org/10.1902/jop.1995.66.9.797.
126. De Sanctis M, Zucchelli G, Clauser C. Bacterial colonization of barrier material and periodontal regeneration. J Clin Periodontol. 1996;23(11):1039–46. https://doi.org/10.1111/j.1600-051x.1996.tb00534.x.
127. De Sanctis M, Zucchelli G, Clauser C. Bacterial colonization of bioabsorbable barrier material and periodontal regeneration. J Periodontol. 1996;67(11):1193–200. https://doi.org/10.1902/jop.1996.67.11.1193.
128. Machtei EE, Cho MI, Dunford R, Norderyd J, Zambon JJ, Genco RJ. Clinical, microbiological, and histological factors which influence the success of regenerative periodontal therapy. J Periodontol. 1994;65(2):154–61. https://doi.org/10.1902/jop.1994.65.2.154.
129. Nowzari H, Slots J. Microorganisms in polytetrafluoroethylene barrier membranes for guided tissue regeneration. J Clin Periodontol. 1994;21(3):203–10. https://doi.org/10.1111/j.1600-051x.1994.tb00305.x.
130. Nowzari H, Matian F, Slots J. Periodontal pathogens on polytetrafluoroethylene membrane for guided tis-

sue regeneration inhibit healing. J Clin Periodontol. 1995;22(6):469–74. https://doi.org/10.1111/j.1600-051x.1995.tb00179.x.
131. Cortellini P, Tonetti MS. Clinical performance of a regenerative strategy for intrabony defects: scientific evidence and clinical experience. J Periodontol. 2005;76(3):341–50. https://doi.org/10.1902/jop.2005.76.3.341.
132. Cortellini P, Carnevale G, Sanz M, Tonetti MS. Treatment of deep and shallow intrabony defects. A multicenter randomized controlled clinical trial. J Clin Periodontol. 1998;25(12):981–7. https://doi.org/10.1111/j.1600-051x.1998.tb02402.x.
133. Tonetti MS, Prato GP, Cortellini P. Factors affecting the healing response of intrabony defects following guided tissue regeneration and access flap surgery. J Clin Periodontol. 1996;23(6):548–56. https://doi.org/10.1111/j.1600-051x.1996.tb01823.x.
134. Tonetti MS, Pini-Prato G, Cortellini P. Periodontal regeneration of human intrabony defects. IV. Determinants of healing response. J Periodontol. 1993;64(10):934–40. https://doi.org/10.1902/jop.1993.64.10.934.
135. Tonetti MS, Lang NP, Cortellini P, Suvan JE, Adriaens P, Dubravec D, et al. Enamel matrix proteins in the regenerative therapy of deep intrabony defects. J Clin Periodontol. 2002;29(4):317–25. https://doi.org/10.1034/j.1600-051x.2002.290407.x.
136. Ehnevid H, Jansson L, Lindskog S, Blomlöf L. Periodontal healing in teeth with periapical lesions. A clinical retrospective study. J Clin Periodontol. 1993;20(4):254–8. https://doi.org/10.1111/j.1600-051x.1993.tb00354.x.
137. Ehnevid H, Jansson LE, Lindskog SF, Blomlöf LB. Periodontal healing in relation to radiographic attachment and endodontic infection. J Periodontol. 1993;64(12):1199–204. https://doi.org/10.1902/jop.1993.64.12.1199.
138. Cortellini P, Tonetti MS. Evaluation of the effect of tooth vitality on regenerative outcomes in infrabony defects. J Clin Periodontol. 2001;28(7):672–9. https://doi.org/10.1034/j.1600-051x.2001.028007672.x.
139. Trejo PM, Weltman RL. Favorable periodontal regenerative outcomes from teeth with presurgical mobility: a retrospective study. J Periodontol. 2004;75(11):1532–8. https://doi.org/10.1902/jop.2004.75.11.1532.
140. Cortellini P, Tonetti MS. Focus on intrabony defects: guided tissue regeneration. Periodontology 2000. 2000;22:104–32. https://doi.org/10.1034/j.1600-0757.2000.2220108.x.
141. Rasperini G, Pilipchuk SP, Flanagan CL, Park CH, Pagni G, Hollister SJ, et al. 3D-printed bioresorbable scaffold for periodontal repair. J Dent Res. 2015;94(Suppl 9):153s–7s. https://doi.org/10.1177/0022034515588303.
142. Jin Q, Anusaksathien O, Webb SA, Printz MA, Giannobile WV. Engineering of tooth-supporting structures by delivery of PDGF gene therapy vectors. Mol Therapy. 2004;9(4):519–26. https://doi.org/10.1016/j.ymthe.2004.01.016.
143. Jin QM, Anusaksathien O, Webb SA, Rutherford RB, Giannobile WV. Gene therapy of bone morphogenetic protein for periodontal tissue engineering. J Periodontol. 2003;74(2):202–13. https://doi.org/10.1902/jop.2003.74.2.202.
144. Palmer RM, Cortellini P. Periodontal tissue engineering and regeneration: consensus report of the sixth European workshop on periodontology. J Clin Periodontol. 2008;35(Suppl 8):83–6. https://doi.org/10.1111/j.1600-051X.2008.01262.x.
145. Babo PS, Reis RL, Gomes ME. Periodontal tissue engineering: current strategies and the role of platelet rich hemoderivatives. J Mater Chem B. 2017;5(20):3617–28. https://doi.org/10.1039/c7tb00010c.

Regeneration for Implant Dentistry

Tulio Fernandez-Medina and Ashwin Nanda

1 Guided Bone Regeneration

The concept of incorporating titanium implants into a prepared socket, that can then integrate with the surrounding bone, has revolutionised the concept of oral rehabilitation, and this has led to numerous treatment modalities for patients affected by complete or partial edentulism. Prior to implant placement, bone width and height, and the location of surrounding nerves and blood vessels must be examined, to ensure a healthy environment for osteogenesis around the implant. For rigid fixation of the implant, the dimensions of the surrounding bone are critical. In a number of conditions, these dimensions are inadequate, for example because of periodontal diseases or anatomical discrepancies. To rectify the problem, many studies have been carried out over the past three decades, although each approach has its own pros and cons.

The strategies that have been proposed include inlay/onlay bone graft techniques, distraction osteogenesis, and guided bone regeneration, with the common aim of establishing sufficient bone volume around the implant, thereby re-establishing bone integrity to sustain an adequate functional load and to regain aesthetics. Amongst all the methods that have been used to increase bone dimensions, guided bone regeneration (GBR) is the most commonly used method for the reconstruction of alveolar bone. The basic principle of this method is to selectively allow osteoprogenitor cells (stem cells) to differentiate into osteoblasts, controlling the local environment using a membrane, to promote osteogenesis under a controlled environment [1].

The ideal properties required in a barrier membrane are high biocompatibility, positive cell occlusivity, controlled space maintenance, and adequate mechanical and physical sustainability. The long term stability and success of an implant increases when GBR is used [2]. Short-term studies reveal implant placement with GBR or without GBR show similar success in bone growth around the implant (both horizontally and vertically).

GBR provides a positive support for implants to integrate at a desired site, although the major challenge here is to overcome the challenges of native bone, such as the configuration of the residual bone and the severity of bone loss in the specific individual [3]. Regaining bone dimension is possible by using advanced techniques and materials to initiate bone growth right from the molecular level. Attaining a good bone height and then maintaining the same is the more difficult part of the whole process. Any bone defect $\geq$4 mm normally requires the use of an autologous bone block graft to regain the lost bone height.

T. Fernandez-Medina (✉) · A. Nanda
School of Dentistry, The University of Queensland, Herston, QLD, Australia
e-mail: t.fernandez@uq.edu.au; a.nanda@uq.net.au

© Springer Nature Switzerland AG 2021

S. Hosseinpour et al. (eds.), *Regenerative Approaches in Dentistry*,
https://doi.org/10.1007/978-3-030-59809-9_7

Several techniques lack long-term clinical documentation, making it challenging to compare the effectiveness of individual techniques, due to the frequent combination with particulate bone and the use of non-resorbable/absorbable membranes. Clinically, it is convenient to place a resorbable membrane for GBR and secure its positive outcomes, hence many studies have focussed on the benefits of the resorbable membrane technique. Technically, non-resorbable membranes tend to give a superior outcome in regaining bone. Such non-resorbable membranes have to be removed once its purpose is served, thereby complicating the overall process with a second surgical procedure [1].

1.1 Membranes

The membrane is the most important part of the GBR technique as it holds the material in place, allows the bone to grow, prevents contamination of the site, and supports osteogenesis at the site. The membranes used for GBR can be classified as either non-resorbable or resorbable, based on their properties. Expanded-polytetrafluoroethylene (e-PTFE) membranes were the first generation technology for membranes used clinically in the GBR technique [4]. Key characteristics are low immunogenicity, resistance to enzymatic degradation by the host tissue or microbes, and a synthetic porous polymer structure.

Membranes can be reinforced with a titanium mesh to provide the necessary physical and mechanical properties to support the space, and enough flexibility to adapted over an irregular bony defect. The major disadvantage of the GBR technique is progressive soft tissue complications due to premature membrane exposure, that increases tension on the flaps and reduces vascular perfusion, eventually causing the whole system to fail [4]. Membrane exposure often leads to infection of the adjacent tissues and the need for the membrane's removal, which hampers the outcome of bone regeneration.

Alternative approaches have been proposed to manage membrane exposure, with an intention being to attain the best outcomes, although the results vary [5–7]. The most important drawback of this technique using non-resorbable membranes is the need for an additional surgical stage to remove the membrane. This additional surgery increases the morbidity and subjects the patient to further possible complications such as pain and infection. Absorbable membranes have been proposed to overcome such drawbacks.

Membranes manufactured from native collagen exhibit functional tissue integration, reduce the foreign body reaction, and increase vascularisation and biodegradation. The elimination of second-stage surgery, better cost-effectiveness, and decreased patient morbidity are the most well-known advantages of the use of collagen-based membranes [8]. In the case of a mucosal dehiscence and early exposure, the membrane induces epithelialisation and secondary wound closure. These reduce morbidity and avoid the need for any further surgical intervention.

On the other hand, the major drawback of membranes manufactured from native collagen are their poor mechanical properties, and their rapid degradation, which drives a natural loss of barrier function. In recent years, the development of multilayer collagen-based membranes (or techniques) [9, 10] has aimed to increase the life-span of these membranes, slowing their reabsorption rate, to retain the membrane in the body for a longer period to enhance bone regeneration.

Synthetic resorbable membranes can be manufactured from aliphatic polyesters such as polylactic acid, polyglycolic acid and their copolymers. These have been proposed to overcome the disadvantages of e-PTFE and collagen-based membranes [11–13]. Such new materials also offer the possibility of increasing the life-span of the membrane, improving its mechanical properties, and incorporating a drug delivery system [14]. However, the use of these membranes is frequently associated with inflammatory foreign body reactions due to their degradation products [15].

1.2 Bone Graft Substitutes

Autologous bone has been considered the gold standard for bone regeneration, due to its osteogenic, osteoinductive, and osteoconductive properties [16]. However, the amount of bone tissue that can be harvested intraorally at any oral site (i.e. symphysis and mandibular ramus) or extraoral site (i.e. iliac crest, calvarium) is often insufficient for treating large bone defects, especially when these are bilateral, as is usually the case. Moreover, morbidity, pain, and discomfort from the donor site, and unpredictable graft resorption are the most important limitations of the use of autogenous grafts [17, 18]. Bone grafts substitutes have been developed to augment or replace bone autografts in bone augmentation procedures. These different alternative materials include allografts, xenografts and alloplastic biomaterials [19–21].

Allografts are tissue obtained from the same species. Samples are treated to reduce their antigenicity and infectivity using freeze-drying and irradiation methods. Such products have been commercialised by licensed tissue banks, and their availability depends on a particular country's regulations, donor intents, and ethical regulations.

Xenografts are tissue obtained from a different species (e.g. bovine, equine, and swine). A different protocol is applied to generate a collagen-rich residual scaffold, by using complete or partial thermo-chemical removal of the organic material. Synthetic bone substitutes (alloplastic materials) are biomaterials synthesised from different components that are mostly inorganic in origin [22]. They are classified according to their porosity, structure, and performance. Biomaterials such as calcium phosphate, bioactive glass, tricalcium phosphate and calcium sulphate aim to replace the inorganic component of bones, to mimic its mechanical and biological properties [23].

The clinical application or combination of different materials used in bone regeneration must consider the type of bone defect, the vascular supply, and the amount of tissue to be regenerated [24]. The graft properties must include biomaterials with excellent mechanical properties, to maintain the space, and induce angiogenic growth factors to facilitate the proliferation of new blood vessels from the periphery to reach the inner core of the graft. The biomaterials ideally should have osteogenic properties, to invoke de novo bone formation [25]. To meet these requirements, it is commonplace to combine different biomaterials with various proteins and growth factors, such as platelet-rich plasma, to increase osteoinductivity of graft materials.

1.3 Bioactive Molecules

1.3.1 Bone Morphogenetic Proteins

The use of biological activate molecules that are partially responsible for regenerating bone formation was initially described by Urist [26], and subsequently named as bone morphogenetic proteins (BMPs) [27]. These constitute a large family of regulatory factors that are related to the transforming growth factor-β (TGF-β) superfamily, with the ability to initiate de novo endochondral bone formation by stimulating undifferentiated pluripotent cells to differentiate into cartilage-forming and bone-forming cells [28].

Only a small number of BMPs (BMP-2, BMP-4, BMP-6, BMP-7, and BMP-14) seem to have osteoinductive functions [29, 30]. Some current commercially available products combine a collagen matrix with a BMP (BMP-2 and BMP-7) synthesised using recombinant DNA technology. Such factors have been shown to enhance the formation of new cementum, new alveolar bone and new periodontal ligament [31]. It has also been hypothesised that lower concentrations of rhBMP-2 can boost initial chondrification, while higher concentrations are supposed to enhance osteogenesis [32].

The role of the vehicle in tailoring the delivery of BMPs has historically been underestimated. More recently, attention has focussed on improving controlled delivery systems to match the biological timing for bone tissue regeneration. The main drawback of using a collagen carrier is its rapid degradation, leading into an early boost of BMP release. This molecule has also been related to the formation of seromas, which are a sterile accumulation of serum within the tissue [33–35]. Due to the ability to predictably promote osteogenesis using BMP, there is an ongoing need for studies to optimising dose, delivery technologies, and conditions for stimulation of bone growth [36].

1.3.2 Growth Differentiation Factor 5 (GDF-5)

This molecule is US FDA approved for periodontal regeneration, alveolar bone regeneration, and sinus augmentation. GDF-5 is a member of the TGF-β/BMP superfamily [37] that plays a critical role in mesenchymal cell differentiation and in morphogenesis of skeletal tissue. For periodontal regeneration, GDF-5 stimulates periodontal ligament cell proliferation, osteoblast differentiation (in the early stages), and extracellular matrix synthesis, by both cell types [38]. For implant site development, GDF-5 has been demonstrated in vivo to induce bone in ectopic muscle pouches, by improving mineralised tissue formation [39]. GDF-5 has been proven effective when used in high concentrations (800 μg) for sinus lift procedures. By the 12th week of the follow-up, it has shown good growth of bone with adequate density [40]. In combination with *β*-TCP, it could enhance bone formation, comparable to what happens with an autologous bone graft [41]. However, the ideal carrier and quantity to be delivered to achieve optimal bioactivity are unclear, hence the need for further research.

1.3.3 Teriparatide (Human Recombinant Parathyroid Hormone)

The US FDA has approved parathyroid hormone (PTH) to treat osteoporosis. It causes osteoblast-like behaviour, with increased osteoprotegerin expression [42]. This polypeptide (34 amino acids) has been tested in animal models, where it shows bone formation in extraction sockets, maintaining the three-dimensional aspects of alveolar bone [43]. Although PTH has shown a positive short-term effect on bone formation, the delivery route needed to maintain sufficient tissue levels needs more work to make this acceptable for patients.

1.3.4 Hemoderivates: Platelet-Fibrin (PRF), Platelet-Rich Plasma (PRP), and Leukocyte-Platelet-Rich Fibrin

Hemoderivates are preparations obtained via blood plasma fractionation through centrifugation after peripheral phlebotomy. Their empirical utilisation for improving tissue healing in a wide range of clinical applications has been reported in the literature since the early 1950s [44]. Advantages such as an autologous origin and ready availability via relatively simple chairside extraction methods have generated considerable interest in both the clinical setting and more recently in regenerative medicine (Fig. 1) [45, 46].

The action of these preparations relies on the secretion of a cocktail of proteins from the platelet-α-granules. Growth factors such as insulin-like growth factor-1 (IGF-I), platelet-derived growth factor (PDGF), vascular endothelial growth factor (VEGF), fibroblast growth factor (FGF), epidermal growth factor (EGF), platelet-derived epidermal growth factor (PDGF) and fibrin matrix proteins are found in these preparations at concentrations greater than those in blood, which is why they can contribute directly to accelerating tissue regeneration [47, 48]. Hemoderivates have been clinically used as a substitute for connective tissue [49], as a graft material in sinus lifts [50] and in guided bone regeneration [51, 52]. Although the application of blood concentrates for boosting regeneration is an attractive approach, the mechanisms responsible for evoking specific biological outputs remain to be elucidated.

2 Posterior Mandible Bone Regeneration

Regeneration of bone in the posterior region of the mandible is one of the most challenging procedures, since this region has minimal blood flow, is distracted by muscle movements, and is burdened by occlusal loading. The structural fea-

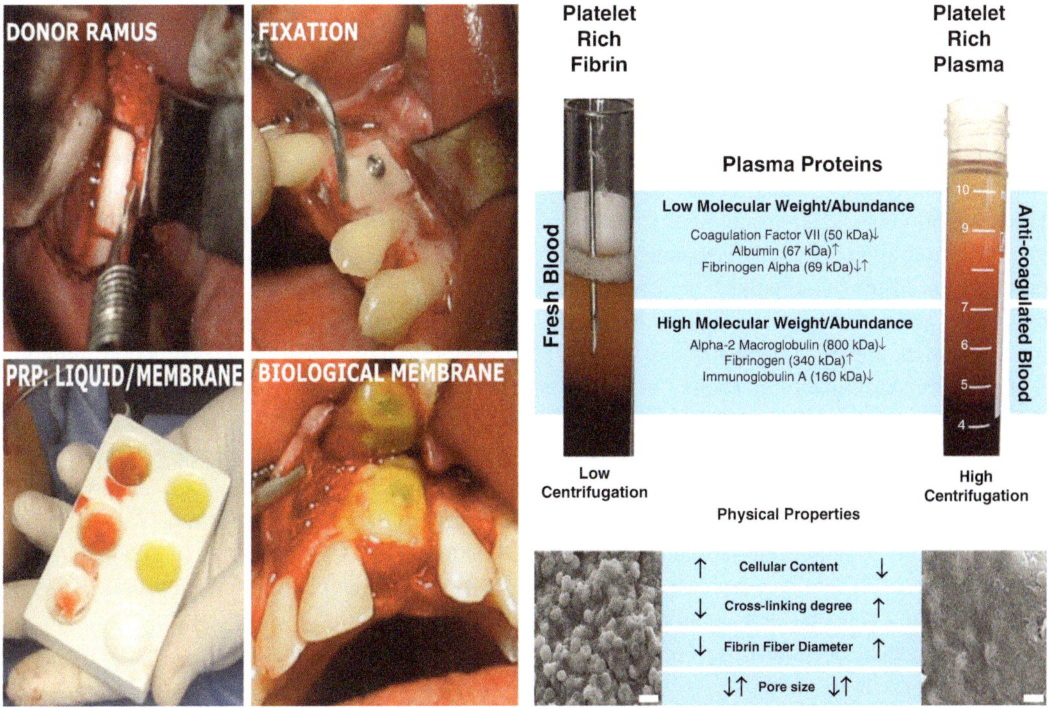

Fig. 1 (Left) An inlay bone block graft from the ascending mandibular ramus that has been fixed in the premaxilla. Platelet-rich plasma (PRP) was prepared from peripheral blood (liquid and membrane) to improve soft tissue adaptation. (Right) A general overview of fresh and anticoagulant blood protocols, protein content, and the physical properties of the fibrin network produced by centrifugation. Based on the work of author TF

tures of this region such as thicker cortical bone and the presence of the inferior alveolar foramen which serves as the entry point for nerves and blood vessels are major reasons causing failure in achieving horizontal and vertical bone regeneration. These factors influence implant placement as clinicians tend to place shorter implants and restore those with a broader crown, giving an abnormal crown to root ratio. The same region is difficult to access for oral hygiene, and implant failures compromise the whole treatment plan.

Different techniques have been proposed for vertical ridge augmentation, including block bone grafting, distraction osteogenesis and guided bone regeneration. The use of an autogenous block bone graft is still considered the 'gold standard' method, but such grafts may not be stable over long periods of follow-up, and this could compromise dental implant success. Additionally, the amount of bone tissue required frequently requires that for bilateral reconstructions an extra-oral donor site (such as the iliac crest) is used, which increases the overall morbidity of the procedure and reduces its acceptability for patients. The crucial period for an autologous bone block graft is within the first year, and after this point, the situation stabilises. As a result, clinicians are tempted to over-graft the surgical site, to compensate for the loss of hard tissue within the first year after grafting. To address these issues, the use of alternative materials has been proposed, however, a lack of evidence, especially for extended period of follow-up, limit their use.

2.1 Bone Block Grafting

In cases of severe ridge atrophy large defects (>7 mm), guided bone regeneration (GBR) or onlay/inlay bone grafts are used to re-establish volumetric posterior mandibular dimensions. However, bone blocks give outcomes that are not very predictable. This may be associated with the composition of the mandibular bone itself, when compared with the maxilla (dense trabecular bone with a thick cortical layer).

Bone microarchitecture (bone quality) is determined by the combination of factors associated with trabecular morphology and porosity. The usual primary sources for autologous bone blocks are the iliac crest, tibia, and calvarium. Intraorally, the body and the ramus of the mandible are used frequently to harvest uni-cortical blocks of living bone tissue that can be fixed into crestal and buccal bone defects. The mean gain when using bone block techniques is 3.47 ± 0.41 mm (95% CI 2.67 to 4.27 mm), regardless of the donor site [25]. Complications of the block grafting technique are sensory disorders at the donor site, especially in the mandibular symphysis region, suture dehiscence, graft exposure and graft contamination. If the bone block is not properly stabilised by titanium screws, this leads to the fibrotic encapsulation and tissue sequestration.

2.2 Distraction Osteogenesis (DO)

Distraction osteogenesis (DO) was first described in the early 1900s, and used by Ilizarov in more than 15,000 patients [53]. DO gains its regenerative capabilities by the separation of two bone segments during the bone healing process, allowing bone to grow longitudinally. Bone segment separation is achieved using a titanium distractor that is fixed in place using screws. It was not until McCarthy reported the successful application of DO in the mandible [54] that this technique was used as an alternative treatment for vertical bony defects. DO is associated with intra-operative and post-operative complications that are related primarily to the vector of distraction [55].

The DO procedure starts with a bone osteotomy and installation of the distractor. This is followed by a latency stage during which the distractor device remains static without activation, to allow osteogenic cells in the osteotomized location to proliferate. Later, the activation phase begins. The bone segment moves through a predetermined linear path (the distraction vector) towards the desired position, to fil the defect. The device is activated once or twice a day at a rate of 0.5–1 mm/day. Finally, there is a consolidation stage without any activation, to allow the bone to mature and mineralise fully. In a second surgical procedure, the distractor device is removed, and dental implants are placed. DO can provide an average gain of 6.84 ± 0.61 mm (95% CI 5.64–8.05 mm) [25]. Complications of DO include lingual vector inclination and loosening of the distractor.

2.3 Guided Bone Regeneration

Guided bone regeneration (GBR) is one of the most common methods used to reconstruct alveolar bone deficiencies [56, 57]. Since a membrane is an essential component of this technique, different materials have been used as membranes, with the goal of providing suitable mechanical and physical properties to maintain space. The membrane must be sufficiently rigid, to withstand the compression of the overlying soft tissue of the posterior mandible. Membranes can be resorbable or non-resorbable. They should also possess a degree of plasticity so they can be adapted to irregular bone defects. Membranes also need biocompatibility, and the ability to occlude the migration of cells.

Titanium mesh structures have excellent mechanical properties and are used to prevent membrane collapse, so that they provide the required level of space maintenance. These meshes can be bent, adapted, and contoured to match the surgical site. Rough cut edges of the mesh may cause mucosal irritation, leading to early exposure of the membrane with subsequent infection. Adding different biomaterials such as PTFE and collagen into the titanium mesh can reduce such difficulties [23]. The average gain for GBR with of PTFE occlusive membranes, when used in the posterior mandible, is 3.83 ± 0.49 mm (95% CI 2.85–4.80 mm) [25]. Outcomes vary according to the type of membrane used, and the type of bone defects being treated.

Experimental GBR approaches with promising results have been reported with the inclusion of bioactive molecules such as fibrinogen [58], PRP [59] and PDGF-BB [60], or the incorporation of cell seeding strategies including MSCs [61–65] or autologous osteoblasts [66]. These results are summarised in Table 1.

3 Sinus Floor Elevation for Bone Regeneration

Dental implants offer an effective way to replace missing teeth. Adequate bone depth is essential to integrate an implant and functionalise it in the oral cavity. The available bone dimension is often compromised in the posterior region of the maxilla,

Table 1 Regenerative biomaterials for Implant Dentistry

Bio material	Group		Origin	Advantages	Disadvantages
Membranes	Natural polymer	• Collagen and extracellular matrix • Chitosan • Alginate	• Human • Bovine • Porcine	• Reduce immunogenicity • Resorbable • Tailor degradability • Carrier for drug delivery or bioactive molecules • Low complication membrane exposure	• Reduced mechanical properties • Poor space maintenance properties due to increased resorbable property • Possible cross-contamination between species
	Synthetic polymer	• Aliphatic (PCL, PLA, PGA)		• Biocompatibility • Manufacturing standardisation • Tailor mechanical properties • Bioactive molecule carries • Blend with complementary substitutes and cells • Tailor specific degradation time • 3D Printability	• Lack of self-osteoinductivity capacities • Some acidic by-products upon degradation are reported • Regulatory approvals need in some countries • Needs for long-term studies • Manufacturing cost
		• Polytetrafluoroethylene (PTFE)		• Bio inert polymer with low degradability • Selective permeable barrier • Regenerative compartmentalisation	• Non-resorbable • Complications related to early membrane exposure • Reduce mechanical properties
	Metal	• Titanium and titanium alloys • Cobalt-Chromium alloys		• Excellent mechanical properties • Reduce immunogenicity • Space maintenance • 3D Printability	• Non-resorbable biomaterials • Complication related to early exposure difficult to treat • Lack of self-osteoinductivity capacities • Need for screw fixation

(continued)

Table 1 (continued)

Bio material	Group		Origin	Advantages	Disadvantages
Alloplastic	Inorganic	• Calcium phosphate • Hydroxyapatite • Calcium sulphate • Bio-glass		• Biocompatibility • Osteoinductivity • Favourable to tailor particle size, porosity and surface modification	• Reduced mechanical properties • High solubility may hamper ions release at the longer term
Xenograft	Natural tissue	• Fresh-frozen, freeze-dried, demineralised freeze-dried bone block or particles	• Bovine • Porcine • Equine	• Biocompatibility • Osteoconductivity • Resorbable • Low immunogenicity • Availability • Favourable tailor of particle size and dimensions • CAD-CAM custom	• Possible cross-contamination between species • Regulatory issues in some countries • Variable predictability • Needs for screw fixation (blocks) • Needs for long-term studies
Allograft	Natural tissue	• Fresh-frozen, freeze-dried, demineralised freeze-dried bone block or particles	• Human	• Biocompatibility • Osteoconductivity, low osteoinductivity • Resorbable • Low immunogenicity • Favourable tailor of particle size and dimensions • CAD-CAM custom	• Possible cross-contamination • Reduced availability for insufficient donor schemes and ethical issue in some countries • Need for screw fixation (blocks)
Autologous	Live natural tissue	• Particulate. Blocks	• Patient's own tissue	• Osteoconductivity • Osteoinductivity • Osteogenic • High predictability when combined with other biomaterials • Considered the gold standard	• High morbidity and discomfort from donor site • Reduced availability: intraoral from Chin/ Mandible body (≈5 cc) • Extra-oral: Tibia(≈25 cc), Calvarias (≈30 cc), Iliac Crest (≈70 cc)

because of pneumatisation of the maxilla by the maxillary sinus. This problem increases when the duration of edentulism is longer, due to alveolar bone resorption over time once teeth are lost. Bone loss in the posterior maxilla can be treated using the patient's own bone (autogenous bone grafts), biomaterials, a combination of both, or a technique that uses a blood clot as a foundation to hold the upcoming graft, to eventually increase the bone height.

3.1 Techniques of Sinus Augmentation (Sinus Lift)

In 1980, an innovative technique to re-establish the dimensions of the posterior maxilla was described in which a buccal window was made in the posterior maxillary bone, that allowed careful elevation of the sinus epithelium, to create a space for placing particulate bone from the iliac crest [67]. After a healing period of 6 months, dental implants may be placed. The technique was subsequently described in detail by Tatum [68], including different variables within the technique such as tissue incisions, bone access, types of biomaterials used, and the combination of sinus augmentation for implant placement as a one-stage or two-stage technique. The lateral window sinus lift is now a widely used technique. It is considered reliable, especially when used with the autologous bone (≥50%).

To provide a minimally invasive approach, a one-stage technique is advised for sinus elevation. Summers [69] described the use of concave tipped osteotomes via a crestal approach, to advance a mass of bone beyond the level of the original sinus floor. This hydrostatic pressure elevates the sinus epithelium, resulting in the creation of a space that is filled with a bone graft material, with the subsequent insertion of a dental implant. This requires a minimal height of 6 mm between the floor of the sinus and the crest of the residual alveolar bone, to ensure dental implant stability. For cases with less than 6 mm of residual bone, a two-stage technique has been proposed, with the first surgery to increase bone quantity, and then dental implant placement 6 months later [70].

Although the trans-alveolar technique has many advantages, the amount of bone height that is gained is usually less when compared to the lateral window technique. Moreover, in the event of complications such sinus membrane perforation, the resolution requires the lateral window approach. Additionally, the anatomical macro and microstructural characteristics of the posterior maxillary bone make it difficult to achieve dental implant stability whenever the bone is less than 3 mm in thickness [71].

3.2 Sinus Lift Outcomes

Although the sinus lift technique is a predictable procedure for increasing the amount of bone in the posterior maxilla (>90%) [72], there is insufficient evidence whether sinus lift procedures in bone with a residual height between 4 and 9 mm are more or less successful than alternative procedures such as the placement of short implants (5–8.5 mm in length) for reducing prosthesis or implant failure [24]. Most studies comparing these techniques have focussed on complications and the amount of regenerated bone, but without taking into consideration the implications for long-term implant survival (at >10 years). Complications of sinus lifts will in turn affect the success of dental implants. The survival rate of implants placed into the sinus cavity is 95% at 52.7 months of follow-up. The more common complications are epistaxis (3.4%) and thickening of the Schneiderian membrane (14.8%) [73].

4 Bone Tissue Engineering (BTE)

In the last decade, tissue-engineered strategies have emerged as promising solutions for the reconstruction of different types of oral, maxillofacial, and periodontal tissues. The ability to offer a safe and standard technique with predictable results to partially or completely edentulous patients requiring prosthetic implant solutions is the tissue engineering chimera of the future.

Bone Tissue Engineering (BTE) involves the application of biomaterials (scaffolds), cells, and bioactive molecules (such as grown factors, hormones, and peptides), to promote regeneration of lost bone tissue. In the conventional tissue engineering paradigm, combinations of cells and bioactive molecules are seeded onto three-dimensional biomaterial scaffolds to promote an implantable 'osteogenic' scaffold.

The BTE concept aims to mimic natural tissue characteristics, by using a scaffold that closely substitutes for extracellular matrix, and provides structural stability, allowing for vascularisation of the site from the surrounding tissue. The new blood vessels bring in oxygen and nutrients necessary for cell proliferation. The concept promotes morphogenetic signalling, to direct cells to the most phenotypically desirable type. The scaffold is typically enriched with progenitor cells (such as MSCs), or bioactive molecules such as growth factors and peptides. Furthermore, the construct is degradable, leaving the new tissue behind.

4.1 Biomaterials

4.1.1 Osteoinductive Materials

These materials induce bone formation [74]. Natural and synthetic materials such as hydroxyapatite (HA) and calcium phosphate (CaP) have been used in a variety of forms including ceramics [75], cements [76], and coatings [77]. Calcium phosphates and apatites can be derived from natural sources such as marine coral [78]. Adding such materials onto scaffolds can increase their bioactivity, to induce ectopic bone formation [79, 80].

It has been hypothesised that the biomaterial surface can absorb and present osteoinductive factors to the surrounding tissue, and can also release the calcium and phosphate ions which act on undifferentiated cells to evoke bone-cell phenotype differentiation [74]. A major drawback is poor mechanical properties, which impair their clinical application for the regeneration of sites under considerable mechanical loads (such as femur, tibia, and mandible) [81].

Bioactive glasses (BGs) are a group of silica-based osteoconductive and osteoinductive glass biomaterials containing SiO_2-CaO-P_2O_5 networks. BGs have good biocompatibility when used in bone and in soft tissue. They stimulate osteogenesis by triggering cellular proliferation and osteogenic differentiation [82]. However, their poor mechanical properties make them difficult to use clinically for large bone defects where there is significant mechanical loading.

4.1.2 Polymers

Polymers of two or more monomeric structures, such as polylactide-co-glycolide (PLGA), can show a desirable degradation rate. PLGA biomaterials form acidic by-products upon degradation, and this may result in tissue necrosis over the long term [83]. On the other hand, polycaprolactone (PCL) is a biodegradable polyester that is non-toxic and tissue compatible. It has a longer degradation time (2–3 years) and has been used widely in resorbable sutures, as well as in scaffolds for regenerative therapy and for drug delivery. It degrades by hydrolysis of its aliphatic ester linkage under physiological conditions [84]. It has recently been used for 3D printing, to produce highly porous resorbable custom 3D printed scaffolds for regeneration of large volume alveolar bone defects, with the aid of CT-scan data from the patient [85].

4.1.3 Collagen-Based Composite Scaffolds

Pure collagen scaffolds have insufficient mechanical properties to be applied as core materials for bone regeneration. Moreover, biomaterials made from pure collagen lack sufficient bioactivity to stimulate cells to infiltrate during bone formation [86]. The incorporation of a bioactive component improves the mechanical strength, bioactivity, and osteogenesis by increasing dimensional stability and the surface area for cell attachment, and has shown promising results [87].

Two methods are used widely to fabricate collagen/bioceramic composite scaffolds: an immersion method (co-precipitation) [88] and a suspension method (direct mixing) [89]. Collagen/HA composite biomaterial scaffolds have been investigated intensely, followed by β-Tricalcium Phosphate (β-TCP). These inorganic materials not only improves cellular adhesion but also accelerate cell differentiation and proliferation [90]. The ratio of collagen to the inorganic material can be altered to tailor the degradation rate of the scaffold to the clinical situation [91].

4.2 Cells

An essential requirement for the cells used in tissue engineering is that they have sufficient plasticity to be modified by the local microenvironment provided by the scaffold and the surrounding tissue. In the past, the usual approach has been to incorporate stem cells, that can then differentiate into multiple cell lineages. Mesenchymal stem cells (MSCs) are non-hematopoietic stem cells that were initially isolated from bone marrow. They have a morphology similar to fibroblasts, and can readily be found in adulty bone marrow. They can be grown in plastic culture flasks, are self-renewable, and can differentiate into osteoblasts, adipocytes, or chondrocytes in vitro [92].

The use of MSCs to repair bone defects could involve implanted cells alone injected into the site, or used in combination with a scaffold. For maxillofacial applications, MSCs are typically derived from bone marrow concentrates and then expanded. They are used in combination with scaffolds containing β-TCP or synthetic hydroxyapatite, or with decellularized bone powders or granules, or they are embedded in hemoderivates (i.e. platelet-rich plasma/fibrin) [93–95].

A significant regulatory issue exists in many countries (especially in Europe) for the use of MSCs. From the translational point of view, a chairside strategy would be ideal for reducing the possible risks associated with cross-contamination or with immunogenicity of allogeneic cells. Even when autologous MSCs are used, the amount recovered, even from the iliac crests, is too small in most cases to accomplish the reconstruction of significant bone defects. Thus, human studies on the reconstruction of complex bone defects have proposed the different strategies of tissue engineering.

Earlier attempts have followed the guided bone regeneration approach [96–98]. From the tissue engineering perspective, these studies have demonstrated the need for space maintenance to evoke guide bone regeneration, and this necessitates the use of non-resorbable membranes to isolate the bone chamber from the surrounding soft tissue. The addition of bioactive molecules to stimulate bone regeneration can provide additional advantages over the normal processes of bone healing [99, 100]. Custom-made scaffolds and personalised bone graft substitutes can reduce operative time and increase predictability [101, 102]. Poly D, L-lactide meshes (made using a box design) have demonstrated promising results for 3D bone reconstruction in totally edentulous patients with severe resorption [103].

4.3 Protein Corona on Biomaterial Surfaces

Previous literature has focussed on the potential of nanoparticle biomaterials that that would replace the bone material to form new bone, while in recent years, the biomaterials that interact with blood and the adjacent tissue to form native bone have attracted attention [104]. After surgical implantation, biomaterials are exposed to various physiological fluids, such as blood. Many of the proteins found in blood (e.g. albumin, fibrinogen, fibronectin, vitronectin, gamma-globulins) may be bound onto the surface of the biomaterial [105].

Depending on factors such as size, surface charge, fluid composition, and physicochemical properties, the surface of the biomaterial may create a complex interface that has loosely bound proteins, and this is termed the protein corona [106]. Proteins present at high concentrations bind first and are then replaced gradually by proteins that bind with higher affinity. This is known as the Vroman effect [107].

The protein corona is responsible for further recruitment and adhesion of pro-inflammatory cells. Blood clot formation defines the provisional matrix around the biomaterial, and the type of tissue that ultimately will form on the surface [108, 109]. The variable rates of success or failure in cases reported in the literature can be explained in part by variations in the macro, micro, and physicochemical composition of materials used for guided bone regeneration and for dental implantology, as each will have a different protein corona (Fig. 2).

The incorporation of hemoderivates (i.e. PRP, PRF, L-PRF) as co-adjuvant for bone regeneration [45, 46, 110] uses blood components that have been separated by centrifugation/time, to alter the amount and type of plasma proteins, giving a specific protein corona around bone graft materials. Optimising the protein corona on the biomaterial surface provides a new way to control osseointegration of dental implants at the molecular level, and the same concept could be used in bone tissue engineering. The field is in its infancy, and a substantial amount of research must be conducted to understand how to optimise the corona for bone regeneration.

5 Peri-Implantitis

Ever since Branemark [111] introduced the concept of osseointegration, oral rehabilitation treatment planning for tooth replacement has shifted

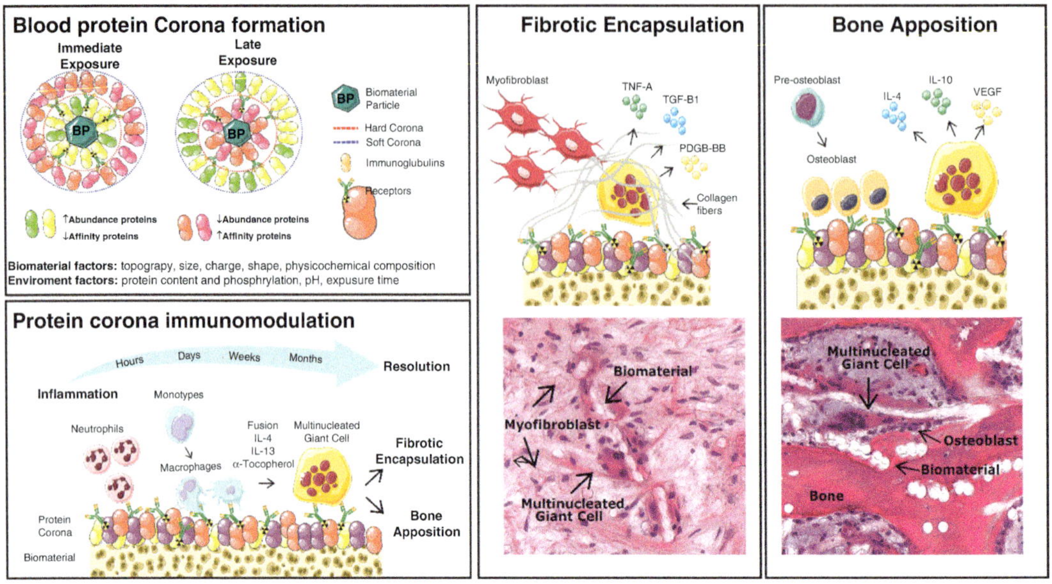

Fig. 2 Schematic representation development of blood protein corona on the surface of biomaterials, giving this surface an immune-modulatory role

from dentures that rely on undercuts and salivary cohesion for retention, to crowns, bridges or dentures that are stabilised or supported by dental implants, being held in place by special connectors. The concept of an implant-retained prosthesis has increased the outcomes that can be achieved in terms of the restoration of masticatory and phonetic functions and aesthetics in edentulous patients. The long-term success for an implant-supported or implant-stabilised prosthesis is in many patients similar to the longevity of the natural dentition, on the proviso that healthy soft tissues are maintained around the dental implants. This is a challenge because of the continuous presence of high levels of bacteria in the oral cavity, creating problems with the control of bacterial plaque. Prolonged accumulation of dental plaque biofilm leads to inflammatory conditions that hampers the long-term survival of the implants and the associated dental prostheses [112].

The first inflammatory stage is known as peri-implant mucositis. This resembles gingivitis, and inflammation is restricted to the tissues around the dental implant, but there is no loss of the adjacent bone. If not treated, in a susceptible patient this condition may progress to a more severe condition known as peri-implantitis, where bone loss occurs, and may threaten the longevity of the dental implant [113]. Although bone loss around dental implants may also be caused by overload [114], in most cases, bone loss is due to the host response to the accumulation of dental plaque, and the accompanying peri-implant inflammation [115].

The treatment of peri-implant bone defects is complex because of the topography of the implant surface as well as the three-dimensional shape of the defect. Relevant variables include the type of bone defect, its location and extent, the patient's medical background, and the quality of supportive periodontal care as well as the patient's own habits of oral hygiene. At the present time, there is no 'gold standard' treatment for peri-implantitis, and the published evidence does not suggest whether surgical or non-surgical intervention is the most effective [116].

A foundation of current treatment approaches involves decontamination of the implant surface, using hand or powered instruments. This debridement may be accompanied by, in some cases, the use of topical antimicrobial agents, or the local or systemic administration of antibiotics. Because of the complexity of the defects that are encountered, in many cases open surgical debridement is necessary. Despite this, the extent of improvement in probing attachment level (PAL) and probing pocket depth (PPD) in sites that have lost more than 50% of the supporting bone is rather limited [117]. The use of particle beams, and pulsed middle infrared lasers such as the Er: YAG laser, have attracted interest because these methods can potentially reach better into the threads of dental implants, then can traditional methods such as using plastic curettes. Reported outcomes of treatment for mechanical debridement alone vary considerably [118–120].

The treatment of peri-implant bony defects using guided bone regeneration has been reported. Such techniques include a membrane combined with a bovine-derived xenograft or with resorbable nano-crystalline hydroxyapatite. The GBR approach seems to provide greater improvement in PAL and PPD after 4 years of follow-up [121]. There are mixed results reported in terms of whether an occlusive membrane is included or not. In one study, bone regeneration associated with the inclusion or exclusion of a membrane was evaluated in 38 patients, where 29 implants were treated with a bone substitute and a membrane, while 36 implants were treated with only a bone substitute. After 1 year of follow-up, there were no statistically significant differences between the two interventions, and both therapies achieved a bone gain of around 1.5 mm [122]. Known complications include barrier membrane exposure and subsequent infection at the site [123].

Although effective at-home plaque control, regular post-operative maintenance for supportive periodontal therapy, and reduction of risk factors such as smoking are always advised to patients, these foundation for the success of surgical treatment may not always in practice be achieved. This may explain why surgical treatments based on the concept of bone regeneration may not always be effective in the long run. In response to this, resective modalities such implantoplasty have been developed, to eliminate the areas that are difficult to debride, and these seem able to cause a significant reduction in BOP and PD [124], LB it at the cost of weakening the implant fixture itself. Such approaches may only be widely acceptable in cases without strong aesthetic requirements, such as over-dentures or where hybrid prostheses are being worn.

Further studies are needed to explore the reasons for discrepancies in the outcomes of surgical treatments for bone defects around dental implants, to better explain the considerable variability that is seen clinically. A better understanding of those factors that determine the success or failure of regenerative surgical methods used in the treatment of peri-implant defects would better inform the selection of methods for specific clinical scenarios, following the principle of personalised medicine.

6 Conclusions and Future Direction

Bone and soft tissue reconstruction have been developed in parallel to dental implants in the rehabilitation of edentulous patients. It is well reported in the literature that, in many cases, there is a need to perform some kind of regenerative procedure associate with fixture placement. This is related to the preceding destruction resorption of the alveolar bone. Autologous bone is still considered the gold standard in bone regeneration due to its capacity for osteoconduction, osteoinduction and osteogenicity. A limited amount of bone volume from donor sites (either intra or extra-oral) and the morbidity at the donor site limits the applicability of autogenous bone grafting in the clinical setting. On the other hand, the use of different biomaterials and membranes for bone regeneration can give useful outcomes, and the approach can be customised to the situation of the individual site. Variability in the nature of the bone defect and in the systemic background of the patient, as well as in the particular surgical skills from operator, makes it difficult to compare success rates between different techniques.

In recent years, a better understanding of what is happening at the molecular level during bone regeneration has renewed interest in the treatment of bone defects of the jaws using regenerative approaches. At the present time, the ideal biomaterial and technique remain to be elucidated, however many promising avenues of research are being explored, including the use of various composite scaffolds, and the use of particular blood extracts to alter the protein corona. More clinical trials are needed to compare the available techniques, to better inform the selection of different types of autogenous bone graft substitutes, and the clinical decision around whether or not to use a membrane barrier. Future research must clarify at the molecular level what are the mechanisms to induce three-dimensional reconstruction of alveolar bone.

References

1. Retzepi M, Donos N. Guided bone regeneration: biological principle and therapeutic applications. Clin Oral Implants Res. 2010;21(6):567–76.
2. Bornstein MM, et al. A retrospective analysis of patients referred for implant placement to a specialty clinic: indications, surgical procedures, and early failures. Int J Oral Maxillofac Implants. 2008;23(6):1109–16.
3. Aghaloo TL, Moy PK. Which hard tissue augmentation techniques are the most successful in furnishing bony support for implant placement? Int J Oral Maxillofac Implants. 2007;22(Suppl):49–70.
4. Zitzmann NU, Naef R, Scharer P. Resorbable versus nonresorbable membranes in combination with bio-Oss for guided bone regeneration. Int J Oral Maxillofac Implants. 1997;12(6):844–52.
5. Garcia J, et al. Effect of membrane exposure on guided bone regeneration: a systematic review and meta-analysis. Clin Oral Implants Res. 2018;29(3):328–38.
6. Wessing B, Lettner S, Zechner W. Guided bone regeneration with collagen membranes and particulate graft materials: a systematic review and meta-analysis. Int J Oral Maxillofac Implants. 2018;33(1):87–100.
7. Lim G, et al. Wound healing complications following guided bone regeneration for ridge augmentation: a systematic review and meta-analysis. Int J Oral Maxillofac Implants. 2018;33(1):41–50.
8. Sbricoli L, et al. Selection of collagen membranes for bone regeneration: a literature review. Materials (Basel). 2020;13(3):786.
9. Abou Fadel R, Samarani R, Chakar C. Guided bone regeneration in calvarial critical size bony defect using a double-layer resorbable collagen membrane covering a xenograft: a histological and histomorphometric study in rats. Oral Maxillofac Surg. 2018;22(2):203–13.
10. Kozlovsky A, et al. Bio-degradation of a resorbable collagen membrane (Bio-Gide) applied in a double-layer technique in rats. Clin Oral Implants Res. 2009;20(10):1116–23.
11. Zwahlen RA, et al. Comparison of two resorbable membrane systems in bone regeneration after removal of wisdom teeth: a randomized-controlled clinical pilot study. Clin Oral Implants Res. 2009;20(10):1084–91.
12. Jung RE, et al. Evaluation of an in situ formed synthetic hydrogel as a biodegradable membrane for guided bone regeneration. Clin Oral Implants Res. 2006;17(4):426–33.
13. Lutolf MP, et al. Repair of bone defects using synthetic mimetics of collagenous extracellular matrices. Nat Biotechnol. 2003;21(5):513–8.
14. Gentile P, et al. Multilayer nanoscale encapsulation of biofunctional peptides to enhance bone tissue regeneration in vivo. Adv Healthc Mater. 2017;6(8).
15. von Arx T, et al. Evaluation of a prototype trilayer membrane (PTLM) for lateral ridge augmentation: an experimental study in the canine mandible. Int J Oral Maxillofac Surg. 2002;31(2):190–9.
16. Hjorting-Hansen E. Bone grafting to the jaws with special reference to reconstructive preprosthetic surgery. A historical review. Mund Kiefer Gesichtschir. 2002;6(1):6–14.
17. Johansson B, et al. A clinical study of changes in the volume of bone grafts in the atrophic maxilla. Dentomaxillofac Radiol. 2001;30(3):157–61.
18. Nkenke E, et al. Morbidity of harvesting of chin grafts: a prospective study. Clin Oral Implants Res. 2001;12(5):495–502.
19. Palmer P, Palmer R. Dental implants. 8. Implant surgery to overcome anatomical difficulties. Br Dent J. 1999;187(10):532–40.
20. Moussa NT, Dym H. Maxillofacial bone grafting materials. Dent Clin N Am. 2020;64(2):473–90.
21. Shamsoddin E, Houshmand B, Golabgiran M. Biomaterial selection for bone augmentation in implant dentistry: a systematic review. J Adv Pharm Technol Res. 2019;10(2):46–50.
22. Benic GI, Hammerle CH. Horizontal bone augmentation by means of guided bone regeneration. Periodontol 2000. 2014;66(1):13–40.
23. Elgali I, et al. Guided bone regeneration: materials and biological mechanisms revisited. Eur J Oral Sci. 2017;125(5):315–37.
24. Esposito M, Felice P, Worthington HV. Interventions for replacing missing teeth: augmentation procedures of the maxillary sinus. Cochrane Database Syst Rev. 2014;5:CD008397.
25. Elnayef B, et al. Vertical ridge augmentation in the atrophic mandible: a systematic review and meta-analysis. Int J Oral Maxillofac Implants. 2017;32(2):291–312.
26. Urist MR. Bone: formation by autoinduction. Science. 1965;150(3698):893–9.
27. Urist MR, Strates BS. Bone morphogenetic protein. J Dent Res. 1971;50(6):1392–406.
28. Wozney JM. The bone morphogenetic protein family and osteogenesis. Mol Reprod Dev. 1992;32(2):160–7.

29. Wise GE, et al. Requirement of alveolar bone formation for eruption of rat molars. Eur J Oral Sci. 2011;119(5):333–8.
30. Schwarz F, et al. Guided bone regeneration using rhGDF-5- and rhBMP-2-coated natural bone mineral in rat calvarial defects. Clin Oral Implants Res. 2009;20(11):1219–30.
31. Schwartz Z, et al. Addition of human recombinant bone morphogenetic protein-2 to inactive commercial human demineralized freeze-dried bone allograft makes an effective composite bone inductive implant material. J Periodontol. 1998;69(12):1337–45.
32. zur Nieden NI, et al. Induction of chondro-, osteo- and adipogenesis in embryonic stem cells by bone morphogenetic protein-2: effect of cofactors on differentiating lineages. BMC Dev Biol. 2005;5:1.
33. Hunt DR, et al. Hyaluronan supports recombinant human bone morphogenetic protein-2 induced bone reconstruction of advanced alveolar ridge defects in dogs. A pilot study. J Periodontol. 2001;72(5):651–8.
34. Wikesjo UM, et al. rhBMP-2 significantly enhances guided bone regeneration. Clin Oral Implants Res. 2004;15(2):194–204.
35. Jovanovic SA, et al. Bone reconstruction following implantation of rhBMP-2 and guided bone regeneration in canine alveolar ridge defects. Clin Oral Implants Res. 2007;18(2):224–30.
36. Wikesjo UM, et al. Bone morphogenetic proteins for periodontal and alveolar indications; biological observations – clinical implications. Orthod Craniofac Res. 2009;12(3):263–70.
37. Hotten G, et al. Cloning and expression of recombinant human growth/differentiation factor 5. Biochem Biophys Res Commun. 1994;204(2):646–52.
38. Lee J, Wikesjo UM. Growth/differentiation factor-5: pre-clinical and clinical evaluations of periodontal regeneration and alveolar augmentation—review. J Clin Periodontol. 2014;41(8):797–805.
39. Kakudo N, et al. Analysis of osteochondro-induction using growth and differentiation factor-5 in rat muscle. Life Sci. 2007;81(2):137–43.
40. Brockmeyer P, et al. Increase of homogenous new bone formation using osteoinductive factor rhGDF-5 during sinus floor augmentation in Goettingen Minipigs. Clin Oral Implants Res. 2015;26(11):1321–7.
41. Koch FP, et al. A prospective, randomized pilot study on the safety and efficacy of recombinant human growth and differentiation factor-5 coated onto beta-tricalcium phosphate for sinus lift augmentation. Clin Oral Implants Res. 2010;21(11):1301–8.
42. Lossdorfer S, Gotz W, Jager A. PTH(1-34) affects osteoprotegerin production in human PDL cells in vitro. J Dent Res. 2005;84(7):634–8.
43. Kawane T, et al. Anabolic effects of recombinant human parathyroid hormone (1 - 84) and synthetic human parathyroid hormone (1 - 34) on the mandibles of osteopenic ovariectomized rats with maxillary molar extraction. Horm Metab Res. 2002;34(6):293–302.
44. Kingsley CS. Blood coagulation; evidence of an antagonist to factor VI in platelet-rich human plasma. Nature. 1954;173(4407):723–4.
45. Del Fabbro M, et al. Healing of postextraction sockets preserved with autologous platelet concentrates. A systematic review and meta-analysis. J Oral Maxillofac Surg. 2017;75(8):1601–15.
46. Anitua E, et al. Autologous fibrin scaffolds: when platelet- and plasma-derived biomolecules meet fibrin. Biomaterials. 2019;192:440–60.
47. Weibrich G, et al. Comparison of platelet, leukocyte, and growth factor levels in point-of-care platelet-enriched plasma, prepared using a modified Curasan kit, with preparations received from a local blood bank. Clin Oral Implants Res. 2003;14(3):357–62.
48. Montanari M, et al. A new biological approach to guided bone and tissue regeneration. BMJ Case Rep. 2013;2013:bcr2012008240.
49. Shepherd N, et al. Root coverage using acellular dermal matrix and comparing a coronally positioned tunnel with and without platelet-rich plasma: a pilot study in humans. J Periodontol. 2009;80(3):397–404.
50. Khairy NM, et al. Effect of platelet rich plasma on bone regeneration in maxillary sinus augmentation (randomized clinical trial). Int J Oral Maxillofac Surg. 2013;42(2):249–55.
51. Kawase T, et al. The heat-compression technique for the conversion of platelet-rich fibrin preparation to a barrier membrane with a reduced rate of biodegradation. J Biomed Mater Res B Appl Biomater. 2015;103(4):825–31.
52. Dori F, et al. Effect of platelet-rich plasma on the healing of intrabony defects treated with an anorganic bovine bone mineral: a pilot study. J Periodontol. 2009;80(10):1599–605.
53. Ilizarov GA. The principles of the Ilizarov method. Bull Hosp Jt Dis Orthop Inst. 1988;48(1):1–11.
54. McCarthy JG, et al. Lengthening the human mandible by gradual distraction. Plast Reconstr Surg. 1992;89(1):1–8. discussion 9-10
55. Hatefi S, et al. Review of automatic continuous distraction osteogenesis devices for mandibular reconstruction applications. Biomed Eng Online. 2020;19(1):17.
56. Hammerle CH, Jung RE. Bone augmentation by means of barrier membranes. Periodontol 2000. 2003;33:36–53.
57. Rakhmatia YD, et al. Current barrier membranes: titanium mesh and other membranes for guided bone regeneration in dental applications. J Prosthodont Res. 2013;57(1):3–14.
58. Groger A, et al. Tissue engineering of bone for mandibular augmentation in immunocompetent minipigs: preliminary study. Scand J Plast Reconstr Surg Hand Surg. 2003;37(3):129–33.
59. Ito K, et al. Osteogenic potential of injectable tissue-engineered bone: a comparison among autogenous bone, bone substitute (Bio-oss), platelet-rich plasma, and tissue-engineered bone with respect to their mechanical properties and histological findings. J Biomed Mater Res A. 2005;73(1):63–72.

60. Khojasteh A, et al. The osteoregenerative effects of platelet-derived growth factor BB cotransplanted with mesenchymal stem cells, loaded on freeze-dried mineral bone block: a pilot study in dog mandible. J Biomed Mater Res B Appl Biomater. 2014;102(8):1771–8.
61. Liao HT, et al. Combination of guided osteogenesis with autologous platelet-rich fibrin glue and mesenchymal stem cell for mandibular reconstruction. J Trauma. 2011;70(1):228–37.
62. Park JH, et al. Periimplant bone regeneration in hydroxyapatite block grafts with mesenchymal stem cells and bone morphogenetic protein-2. Tissue Eng Regen Med. 2016;13(4):437–45.
63. Khojasteh A, et al. The effect of PCL-TCP scaffold loaded with mesenchymal stem cells on vertical bone augmentation in dog mandible: a preliminary report. J Biomed Mater Res B Appl Biomater. 2013;101(5):848–54.
64. Kuznetsov SA, et al. Long-term stable canine mandibular augmentation using autologous bone marrow stromal cells and hydroxyapatite/tricalcium phosphate. Biomaterials. 2008;29(31):4211–6.
65. Zhao J, et al. Apatite-coated silk fibroin scaffolds to healing mandibular border defects in canines. Bone. 2009;45(3):517–27.
66. Wang S, et al. Vertical alveolar ridge augmentation with beta-tricalcium phosphate and autologous osteoblasts in canine mandible. Biomaterials. 2009;30(13):2489–98.
67. Boyne PJ, James RA. Grafting of the maxillary sinus floor with autogenous marrow and bone. J Oral Surg. 1980;38(8):613–6.
68. Tatum H Jr. Maxillary and sinus implant reconstructions. Dent Clin N Am. 1986;30(2):207–29.
69. Summers RB. A new concept in maxillary implant surgery: the osteotome technique. Compendium. 1994;15(2):152, 154–6, 158 passim; quiz 162.
70. Summers RB. The osteotome technique: Part 4—Future site development. Compend Contin Educ Dent. 1995;16(11):1090, 1092 passim; 1094–1096, 1098, quiz 1099.
71. Cosci F, Luccioli M. A new sinus lift technique in conjunction with placement of 265 implants: a 6-year retrospective study. Implant Dent. 2000;9(4):363–8.
72. Al-Nawas B, Schiegnitz E. Augmentation procedures using bone substitute materials or autogenous bone – a systematic review and meta-analysis. Eur J Oral Implantol. 2014;7(Suppl 2):S219–34.
73. Ragucci GM, et al. Influence of exposing dental implants into the sinus cavity on survival and complications rate: a systematic review. Int J Implant Dent. 2019;5(1):6.
74. Barradas AM, et al. Osteoinductive biomaterials: current knowledge of properties, experimental models and biological mechanisms. Eur Cell Mater. 2011;21:407–29. discussion 429
75. Klein C, et al. Osseous substance formation induced in porous calcium phosphate ceramics in soft tissues. Biomaterials. 1994;15(1):31–4.
76. Gosain AK, et al. A 1-year study of osteoinduction in hydroxyapatite-derived biomaterials in an adult sheep model: part II. Bioengineering implants to optimize bone replacement in reconstruction of cranial defects. Plast Reconstr Surg. 2004;114(5):1155–63. discussion 1164-5
77. Habibovic P, et al. Influence of octacalcium phosphate coating on osteoinductive properties of biomaterials. J Mater Sci Mater Med. 2004;15(4):373–80.
78. Pollick S, et al. Bone formation and implant degradation of coralline porous ceramics placed in bone and ectopic sites. J Oral Maxillofac Surg. 1995;53(8):915–22. discussion 922-3
79. Devin JE, Attawia MA, Laurencin CT. Three-dimensional degradable porous polymer-ceramic matrices for use in bone repair. J Biomater Sci Polym Ed. 1996;7(8):661–9.
80. Hasegawa S, et al. In vivo evaluation of a porous hydroxyapatite/poly-DL-lactide composite for bone tissue engineering. J Biomed Mater Res A. 2007;81(4):930–8.
81. Jeong J, et al. Bioactive calcium phosphate materials and applications in bone regeneration. Biomater Res. 2019;23:4.
82. Rawlings RD. Bioactive glasses and glass-ceramics. Clin Mater. 1993;14(2):155–79.
83. Bhattacharyya S, et al. Biodegradable polyphosphazene-nanohydroxyapatite composite nanofibers: scaffolds for bone tissue engineering. J Biomed Nanotechnol. 2009;5(1):69–75.
84. Bartnikowski M, et al. Degradation mechanisms of polycaprolactone in the context of chemistry, geometry and environment. Prog Polym Sci. 2019;96:1–20.
85. Bartnikowski M, Vaquette C, Ivanovski S. Workflow for highly porous resorbable custom 3D printed scaffolds using medical grade polymer for large volume alveolar bone regeneration. Clin Oral Impl Res. 2020;31:431–41.
86. Otsuka M, et al. Effect of geometrical structure on the in vivo quality change of a three-dimensionally perforated porous bone cell scaffold made of apatite/collagen composite. J Biomed Mater Res B Appl Biomater. 2013;101B(2):338–45.
87. Yunus Basha R, Sampath Kumar TS, Doble M. Design of biocomposite materials for bone tissue regeneration. Mater Sci Eng C. 2015;57:452–63.
88. Yunoki S, et al. Three-dimensional porous hydroxyapatite/collagen composite with rubber-like elasticity. J Biomater Sci Polym Ed. 2007;18(4):393–409.
89. Xia L, et al. Akermanite bioceramics promote osteogenesis, angiogenesis and suppress osteoclastogenesis for osteoporotic bone regeneration. Sci Rep. 2016;6:22005.
90. Yunoki S, et al. Control of pore structure and mechanical property in hydroxyapatite/collagen composite using unidirectional ice growth. Mater Lett. 2006;60(8):999–1002.
91. Murakami S, et al. Dose effects of beta-tricalcium phosphate nanoparticles on biocompatibility and bone

conductive ability of three-dimensional collagen scaffolds. Dent Mater J. 2017;36(5):573–83.
92. Dominici M, et al. Minimal criteria for defining multipotent mesenchymal stromal cells. The International Society for Cellular Therapy position statement. Cytotherapy. 2006;8(4):315–7.
93. Fuerst G, et al. Are culture-expanded autogenous bone cells a clinically reliable option for sinus grafting? Clin Oral Implants Res. 2009;20(2):135–9.
94. Rickert D, et al. Maxillary sinus floor elevation with bovine bone mineral combined with either autogenous bone or autogenous stem cells: a prospective randomized clinical trial. Clin Oral Implants Res. 2011;22(3):251–8.
95. Shayesteh YS, et al. Sinus augmentation using human mesenchymal stem cells loaded into a beta-tricalcium phosphate/hydroxyapatite scaffold. Oral Surg Oral Med Oral Pathol Oral Radiol Endod. 2008;106(2):203–9.
96. Simion M, Trisi P, Piattelli A. Vertical ridge augmentation using a membrane technique associated with osseointegrated implants. Int J Periodontics Restorative Dent. 1994;14(6):496–511.
97. Simion M, Rocchietta I, Dellavia C. Three-dimensional ridge augmentation with xenograft and recombinant human platelet-derived growth factor-BB in humans: report of two cases. Int J Periodontics Restorative Dent. 2007;27(2):109–15.
98. Simion M, et al. Vertical ridge augmentation by expanded-polytetrafluoroethylene membrane and a combination of intraoral autogenous bone graft and deproteinized anorganic bovine bone (Bio Oss). Clin Oral Implants Res. 2007;18(5):620–9.
99. Kohal RJ, et al. Evaluation of guided bone regeneration around oral implants over different healing times using two different bovine bone materials: a randomized, controlled clinical and histological investigation. Clin Implant Dent Relat Res. 2015;17(5):957–71.
100. Marx R, et al. rhBMP-2/ACS grafts versus autogenous cancellous marrow grafts in large vertical defects of the maxilla: an unsponsored randomized open-label clinical trial. Int J Oral Maxillofac Implants. 2013;28:e243–51.
101. Mangano F, et al. Maxillary ridge augmentation with custom-made CAD/CAM scaffolds. A 1-year prospective study on 10 patients. J Oral Implantol. 2014;40(5):561–9.
102. Luongo F, et al. Custom-made synthetic scaffolds for bone reconstruction: a retrospective, multicenter clinical study on 15 patients. Biomed Res Int. 2016;(Dic 14:2016):5862586.
103. Menoni A, et al. Full-arch vertical reconstruction of an extremely atrophic mandible with "box technique". A novel surgical procedure: a clinical and histologic case report. Implant Dent. 2013;22(1):2–7.
104. Mariani E, et al. Biomaterials: foreign bodies or tuners for the immune response? Int J Mol Sci. 2019;20(3):636.
105. Corbo C, et al. The impact of nanoparticle protein corona on cytotoxicity, immunotoxicity and target drug delivery. Nanomedicine (Lond). 2016;11(1):81–100.
106. Nguyen VH, Lee BJ. Protein corona: a new approach for nanomedicine design. Int J Nanomed. 2017;12:3137–51.
107. Vroman L, et al. Interaction of high molecular weight kininogen, factor XII, and fibrinogen in plasma at interfaces. Blood. 1980;55(1):156–9.
108. Wilson CJ, et al. Mediation of biomaterial-cell interactions by adsorbed proteins: a review. Tissue Eng. 2005;11(1–2):1–18.
109. Tang L, Eaton JW. Fibrin(ogen) mediates acute inflammatory responses to biomaterials. J Exp Med. 1993;178(6):2147–56.
110. Dohan Ehrenfest DM, Rasmusson L, Albrektsson T. Classification of platelet concentrates: from pure platelet-rich plasma (P-PRP) to leucocyte- and platelet-rich fibrin (L-PRF). Trends Biotechnol. 2009;27(3):158–67.
111. Branemark PI, et al. Osseointegrated implants in the treatment of the edentulous jaw. Experience from a 10-year period. Scand J Plast Reconstr Surg Suppl. 1977;16:1–132.
112. Pontoriero R, et al. Experimentally induced peri-implant mucositis. A clinical study in humans. Clin Oral Implants Res. 1994;5(4):254–9.
113. Esposito M, et al. Differential diagnosis and treatment strategies for biologic complications and failing oral implants: a review of the literature. Int J Oral Maxillofac Implants. 1999;14(4):473–90.
114. Isidor F. Loss of osseointegration caused by occlusal load of oral implants. A clinical and radiographic study in monkeys. Clin Oral Implants Res. 1996;7(2):143–52.
115. Lindhe J, Meyle J. Peri-implant diseases: consensus report of the sixth European workshop on periodontology. J Clin Periodontol. 2008;35(8 Suppl):282–5.
116. Esposito M, Grusovin MG, Worthington HV. Interventions for replacing missing teeth: treatment of peri-implantitis. Cochrane Database Syst Rev. 2012;1:Cd004970.
117. Buchter A, et al. Sustained release of doxycycline for the treatment of peri-implantitis: randomised controlled trial. Br J Oral Maxillofac Surg. 2004;42(5):439–44.
118. Schwarz F, et al. Clinical evaluation of an Er:YAG laser for nonsurgical treatment of peri-implantitis: a pilot study. Clin Oral Implants Res. 2005;16(1):44–52.
119. Schwarz F, et al. Nonsurgical treatment of moderate and advanced periimplantitis lesions: a controlled clinical study. Clin Oral Investig. 2006;10(4):279–88.
120. Renvert S, et al. Treatment of peri-implantitis using an Er:YAG laser or an air-abrasive device: a randomized clinical trial. J Clin Periodontol. 2011;38(1):65–73.

121. Schwarz F, et al. Healing of intrabony peri-implantitis defects following application of a nanocrystalline hydroxyapatite (Ostim) or a bovine-derived xenograft (Bio-Oss) in combination with a collagen membrane (Bio-Gide). A case series. J Clin Periodontol. 2006;33(7):491–9.
122. Roos-Jansaker AM, et al. Surgical treatment of peri-implantitis using a bone substitute with or without a resorbable membrane: a prospective cohort study. J Clin Periodontol. 2007;34(7):625–32.
123. Schwarz F, et al. Impact of the method of surface debridement and decontamination on the clinical outcome following combined surgical therapy of peri-implantitis: a randomized controlled clinical study. J Clin Periodontol. 2011;38(3):276–84.
124. Khoury F, et al. Surgical treatment of peri-implantitis – consensus report of working group 4. Int Dent J. 2019;69(Suppl 2):18–22.

Regenerative Approaches in Orthodontic and Orthopedic Treatment

Yan He, Fernando Guastaldi, Chun Xu, and Qingsong Ye

1 Bone Remodeling and Regeneration Through Tissue Engineering

Bone is a dynamic tissue that undergoes remodeling all the time. Bone tissue has an intrinsic ability to repair small defects and some fractures. However, if bone defects exceed the critical size (which depends on the location and anatomy, usually >2 cm in humans) [1], the body usually cannot repair them unaided.

In defect situations, the bone needs to be reconstructed, to providing it with the necessary mechanical integrity and o aid rehabilitation. Current clinical treatments of large bone defects using bone grafts, including autografts and allografts, have considerable limitations [2, 3].

Y. He · Q. Ye (✉)
Skeletal Biology Research Center, Massachusetts General Hospital and Harvard School of Dental Medicine, Boston, MA, USA

Institute of Stem Cells and Tissue Engineering, Wenzhou Medical University, Wenzhou, China
e-mail: yhe9@mgh.harvard.edu; qye4@mgh.harvard.edu

F. Guastaldi
Skeletal Biology Research Center, Massachusetts General Hospital and Harvard School of Dental Medicine, Boston, MA, USA

C. Xu
School of Dentistry, The University of Queensland, Herston, QLD, Australia
e-mail: chun.xu@uq.edu.au

Autografts require a second operation to harvest bone from other sites in the body. Associated complications may include donor site morbidity, intraoperative morbidity, and prolonged hospitalization. Limitations also include an inadequate quantity of donor bone and difficulties in shaping the graft to the correct shape to restore complex 3-dimensional defects. The use of allografts is associated with the potential transmission of infection and with host immune responses.

Recently, engineered bone scaffolds have been receiving increasing attention as alternatives to conventional bone grafts [4]. For bone tissue engineering, three factors are crucial for the long-term success: osteogenic cells to produce the bone matrix, biomaterials, or scaffolds with suitable mechanical properties, to provide the desired microenvironment, and biomolecular cues such as growth factors to attract osteogenic cells and/or modify their functions. In the following part, those key factors for bone tissue engineering are discussed in detail.

2 Cell Sources for Bone Regeneration

Mesenchymal Stem Cells (MSCs)

MSCs have long been recognized for their potential use because they can differentiate and form bone during the natural bone development process. MSCs have been defined through the

© Springer Nature Switzerland AG 2021
S. Hosseinpour et al. (eds.), *Regenerative Approaches in Dentistry*, https://doi.org/10.1007/978-3-030-59809-9_8

expression of various markers (i.e., negative for CD34, CD45, CD14, CD11a, CD19, and HLA-DR, and positive for STRO-1, CD29, CD73, CD90, CD105, CD106, CD166, CD146, and CD44) [5, 6]. The high proliferative potential of MSCs, combined with their ability to withstand freezing conditions, allows for their expansion in vitro, to obtain clinically relevant cell numbers.

Dental Stem Cells (DSCs)

In addition to MSCs, various DSCs such as dental pulp stem cells (DPSCs), stem cells from the human exfoliated deciduous tooth (SHED), dental follicle stem cells (DFSCs) and periodontal ligament stem cells (PDLSCs) have been used for bone tissue engineering due to their ease of harvesting and high proliferation rate [7, 8].

DPSCs express typical MSC biomarkers, such as CD90, CD29, CD73, CD44, and CD105 [9, 10], and they differentiate into odontoblast-like cells [11]. In vivo transplantation of such DPSCs into nude rats generates living fibrous lamellar bone tissues containing osteocytes [12].

Induced Pluripotent Stem Cells (iPSCs)

The generation of iPSCs was first reported by the Yamanaka group and others in 2006. This approach involves directly introducing specific reprogramming genes (Oct4, Sox2, cMyc, and Klf4) into somatic cells to give them pluripotent abilities [13]. iPSCs have been induced into osteoblast-like cells [14, 15] for bone regeneration.

Other Cells

Other stem cells such as adipose-derived stem cells [16], and peripheral blood-derived stem cells [17] are also used for bone and cartilage repair.

3 Biomaterials for Bone Regeneration

Bioceramics

Bone contains an inorganic component of carbonated apatite minerals. Bioceramics are a class of inorganic biomaterials which have similar composition to the mineral parts of natural bone. Calcium phosphates such as β-tricalcium phosphate and hydroxyapatite are the most common types of bioactive ceramics used for bone tissue engineering. They are widely used in clinical practice as bone cements or as coatings on implants [18].

Bioactive glass is another important type of bioceramic. It is composed of silicates, calcium, and phosphate [19]. Compared to hydroxyapatite, bioactive glass is reported to have a faster bone regeneration rate in vivo as a bone graft [20]. Recently, the capacity of releasing bioactive ions such as Ca^{2+}, Si^{4+}, Mg^{2+}, Cu^{2+}, Sr^{2+}, Li^{+}, and Ag^{+} from bioceramics has been receiving interest [21].

Polymers

Various naturally derived and synthetic polymers are used for bone tissue engineering. Natural polymers, such as collagen, gelatine, and alginate exhibit several desirable characteristics such as good biocompatibility, degradability, and cell attachment. Collagen is the main protein component of natural bone. These biopolymers contain aminoacid sequences (specifically, the adhesion ligand arginine–glycine–aspartic acid (RGD)) to which cells readily attach. For natural polymers, concerns exist over their immunogenicity and their relatively weak mechanical properties.

Synthetic polymers such as polycaprolactone, polylactic acid (PLA), polyglycolic acid (PGA), copolymers of PLA and PGA (PLGA) offer a versatile alternative. PCL is a popular polymer for use in bone tissue engineering systems. It has high mechanical strength and is included in the list of US Food and Drug Administration (FDA) approved products. Synthetic polymers usually lack features that promote cell adhesion, and they undergo very slow hydrolytic *degradation* in vivo. A combination with natural polymers such as a surface coating can address this particular concern for synthetic polymers.

Composites and Hybrids

Composites are an increasingly important class of biomaterials used for bone tissue engineering, owing to their ability to combine the strength of both bioceramics and polymers. Inorganic-organic composites aiming to "mimic" the composite nature of native bone combine the toughness of a polymer phase with the compressive strength of an inorganic phase, to generate bioactive materials with improved mechanical properties and degradation profiles. These composites and hybrid usually include a biodegradable polymer phase, in which bioceramic particles are incorporated as fillers. Tissue-engineered porous PEO layered polymer-magnesium system, for example, is emerging and are showing promising [22].

3.1 Growth Factors for Bone Regeneration

Bone is a dynamic tissue that constantly undergoes remodeling, with a coupled process of bone formation by osteoblasts and resorption by osteoclasts. When bone defects occur, a bone healing process is triggered. Osteoblasts differentiate and are activated to form new bone. During this process, growth factors particularly bone morphogenetic proteins (BMP) [23] are expressed to promote the differentiation of osteoblasts and to enhance the bone-forming activity [24]. Preclinical and clinical studies have shown that BMP-2 has a strong osteoinductive ability, and can be utilized in therapeutic interventions for bone defects, non-union fractures, spinal fusion, osteoporosis, and root canal surgery [23, 25]. Recombinant human BMP-2 (rhBMP-2) has been approved by the US FDA for clinical use [26].

In addition to BMPs, fibroblast growth factors (FGFs) and vascular endothelial growth factor (VEGF) are also important in bone tissue engineering. VEGF is an angiogenic protein that regulates endothelial cell proliferation. FGFs are group of proteins that induce angiogenesis through endothelial and osteoblast cell proliferation. The design of scaffolds with the localized release of growth factors has attracted significant attention, due to the potential for dose reduction, a controlled release pattern, and lower side effects compared to systemic delivery. Various advanced fabrication techniques such as 3D printing, electro-spinning, and electro-spraying are used for growth factor and drug delivery to enhance bone growth when scaffolds are used [1].

4 Regenerative Approaches for Jaw Discrepancies

A jaw size discrepancy is commonly seen in orthodontic patients and can cause a mild to a severe malocclusion. The latter usually requires a multidisciplinary treatment plan including orthodontic treatment and orthognathic surgery. Conventionally, orthodontic treatment is used to compensate for the size difference between jaws in moderate cases, and to de-compensate the inclined teeth to facilitate orthognathic surgery in severe cases.

As shown in Fig. 1, in the past, when there is no skeletal discrepancy, no treatment or simple orthodontic treatment is needed for an ideal occlusion. When there exists q moderate level of jaw discrepancy, orthodontic treatment can camouflage the skeletal difference via the repositioning of teeth. When severe skeletal problems are identified in patients, usually a combination of orthodontic and orthognathic treatment is planned. At this point, there is not much tissue regeneration involved to correct the jaw discrepancy.

As is now well accepted, orthodontic tooth movement relies on alveolar bone remodeling, which is initiated by force. This bone remodeling process involves both osteoblastic and osteoclastic activities. With the understanding of this process and the development of bone tissue engineering, the treatment efficacy of orthodontic approaches for dealing with jaw discrepancies has been greatly improved.

Unlike moderate jaw discrepancy, when a severe discrepancy exists, this indicates that a large volume of bone is needed. Besides grafting, bone tissue engineering techniques have been used widely in this field. In this section, first, tissue engineering techniques that further developed the capacity of conventional orthodontic treatment will be introduced. This part addresses

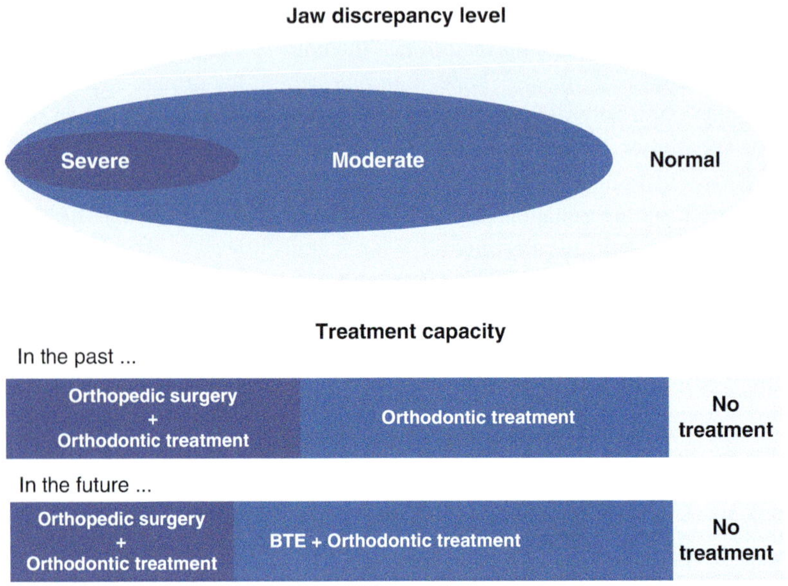

Fig. 1 Comparison of treatment capacity in the past and in the future where bone tissue engineering (BTE) enables more orthodontic treatment approaches to correct jaw discrepancy at moderate and severe levels

moderate jaw discrepancies. Secondly, bone tissue engineering technique advancements for treating severe skeletal jaw discrepancies will be discussed. Currently, with the aid of rapidly developing tissue engineering techniques, less invasive procedures should be possible for treating jaw discrepancies in the future (Fig. 1).

4.1 Orthodontic Treatment Tissue Engineering Approaches for Jaw Discrepancies

Modern orthodontics has adopted so many tissue engineering techniques that the capacity of orthodontists to correct jaw discrepancy has been expanded considerably. The border between camouflage orthodontic treatment and surgical treatment has been pushed outwardly (Fig. 1, lower panel). Implant anchorage, maxillary expansion, micro-osteoperforation, and corticotomy are de facto common surgical techniques that expand the capacity of traditional orthodontic appliances to enable faster tooth movement. These also allow clinicians to move teeth across a longer distance. While most clinicians may not instantly relate these techniques to tissue engineering, they all rely heavily on the bone remodeling capabilities of alveolar bone [27] and/or jaw bone. This section focuses on tissue engineering techniques adopted by orthodontists, and experimental approaches for orthodontic treatment.

4.1.1 Osteoclastic Activity Accelerates Orthodontic Tooth Movement Rate

Surgically facilitated orthodontic treatment is performed widely in modern orthodontic practice. Liou and coworkers showed that orthognathic surgery could create a window period of up to 3 months for active osteoclast activity and alveolar bone metabolism, which resulted in accelerated tooth movement [28]. A regional acceleration phenomenon relates to an active osteosis response around a corticotomy site. Cortical bone is regarded as the main resistance during tooth movement, and the cortical layer is removed in front of the tooth along the track of its movement. After corticotomy, the accelerating effect persists for the first 3 months. This effect is not caused by the removal of resistance, but rather by the activation of resorption and formation processes in the alveolar bone itself. To initiate a burst of acceleration, a corticotomy requires a full-thickness gingival and mucosa flap to provide direct access to the surgical site. As such, a conventional corticotomy is relatively complex, with postoperative complications such as pain and discomfort. A series of minimally invasive modifications are now available, including corticision, piezocision, micro-osteoperforation, and discision [29]. A tissue flap is avoided in all these modified versions, and this reduces soft tissue reaction.

In general, corticotomy is suitable for many types of tooth movement. Its application is limited to patients who take anti-inflammatory medicines (e.g., corticosteroids, non-steroidal anti-inflammatory agents) and bisphosphonates, as these interfere with bone remodeling [30, 31]. There is a delicate balance between the choice of a corticotomy to maintain a satisfactory acceleration of tooth movement, and the risk of harm to the individual. More research is needed in this field, as inconsistent results have been found in animals versus patients. A recent animal study indicated that even remote corticotomy can effectively accelerate tooth movement [32]. A review of clinical studies has pointed out that strong clinical evidence is lacking, mostly due to poorly designed studies [33].

Micro-osteoperforation has been used in both animal and clinical studies, with little or no complications. As micro-osteoperforation avoids raising a flap, irritation and tissue swelling are minimized. Most patients tolerate this procedure very well [34, 35]. In one study, three micro-osteoperforations were performed on the buccal side of the target canine. There was no significant acceleration of tooth movement observed, which may reflect the low frequency of the procedure. A higher frequency may be needed to achieve an obvious acceleration [36]. A 2020 systematic review and meta-analysis concluded that micro-osteoperforation was not effective in enhancing orthodontic tooth movement [37].

4.1.2 Osseointegration Enables Definite Anchorage in Orthodontic Treatment

A mini-screw implant used in orthodontic treatment provides stable anchorage, reduces the need for patient compliance, and avoids unfavorable tooth movement. Osseointegration, as first introduced by Branemark in the 1960s, serves as the biological basis for the min-screw implants used as orthodontic anchorage. The application of mini-screw implants has changed the management of many cases from extraction to non-extraction, and from surgical cases to extraction cases. OTM facilitated by these implants is comparable to what occurs with conventional anchorage [38, 39].

4.1.3 Intramembranous Osteogenesis Corrects Jaw Discrepancy

Maxillary expansion is to address a discrepancy between the upper and lower jaws in the transverse plane. Usually in such cases, the width of the maxilla is below normal, while the width of the mandible is within the normal range. Correction of mandibular width is sometimes carried out at the same time as the expansion of maxilla. The optimal age to undergo expansion is in the teenage years, from 13 to 15 years old. The mechanism behind the maxillary expansion is the same as distraction osteogenesis (DO) (as described in the next section). Patients with skeletal jaw discrepancy also can benefit from the surgically facilitated maxillary expansion, e.g., for mature adult patients.

4.1.4 Factors Regulating Orthodontic Tooth Movement

It is well accepted that a continuous light force can provide an optimal rate of tooth movement with minimum tissue damage, e.g., root resorption. Current research on tooth movement indicates that it involves three separate but interacting osteosis activities: resorption, formation, and remodeling. Bone resorption is regarded as rate-limiting aspect for tooth movement rate [40], and animal studies have proved a direct relationship with bone resorption via osteoclast activation [41, 42]. The sympathetic nervous system regulates bone remodeling (including osteoblast mediated bone formation and osteoclast-mediated bone resorption) through β-2 adrenergic receptors (Adrb2) [43].

A number of chemicals and drugs have been studied for possible use in accelerating the rate of orthodontic tooth movement [44, 45], e.g., hydrogen sulfide [46, 47], triptolide [48], and asperosaponin [49]. On the other hand, resveratrol has been found to reduce the rate of orthodontic tooth movement [50].

Low level laser therapy has been proposed to accelerate orthodontic tooth movement, with most studies using near-infrared gallium–aluminum–arsenic (GaAlAs) diode lasers for photobiomodulation [51–53]. Low level laser therapy increases IL-1β secretion and faster tooth movement is observed [53]. A 2014 systematic review and meta-analysis concluded that low level laser therapy accelerated tooth movement, with a moderate level of evidence. Further research is needed to optimize this technique so that it becomes a part of the normal routine [54]. In 2020, a triple-blind, split-mouth, randomized controlled trial found no accelerating effect on tooth movement [55]. This may be due to sub-optimal laser parameters being used.

4.2 Orthodontic/Orthopedic Treatment Tissue Engineering Approaches for Jaw Discrepancies

Depending on the type of discrepancy, jaw correction sometime requires orthopedic intervention with additional orthodontic treatment. For instance, in an adult patient with a palatal cleft, dentition compensation (including changes to the alveolar bone) always exists. In order to achieve the mastication function, orthodontic treatment is needed to correct the malocclusion (a de-compensation), before treatment of the jaw discrepancy.

Most cases of jaw discrepancy are well compensated with teeth, alveolar bone and soft tissues, which results in a normal facial appearance, progressing to moderate asymmetry and/or malocclusion. Only a small proportion of patients with a moderate to severe jaw discrepancy present with an obviously asymmetric facial appearance and severe malocclusion [56]. This latter situation is seen commonly in cases of cleft palate, or of benign tumors that progressively disrupt bone (e.g., aggressive giant cell lesions, or ameloblastoma). On the other hand, trauma in the maxillofacial area, especially cases of TMJ trauma in young patients, often result in a severe jaw discrepancy.

Conventionally, there are two types of treatment approaches. Distraction osteogenesis (DO) was proposed in 1989 [57], and was first applied in maxillofacial cases in 1998 [58]. DO has now become a standard procedure for elongating the jawbone and the expanding maxilla. DO uses the patient's own bone formation capacity to increase the bone volume. Its application in the maxillofacial area is limited by the complex and irregular shapes of the facial bones in individual cases. Each person has a unique facial contour. Another issue is damage to accompanying nerves, since this can impact significantly on the quality of the newly formed bone [59].

4.2.1 DO in the Maxilla

Distraction osteogenesis as a treatment for skeletal deformities relies on achieving an increase in bone volume. It avoids or reduces the need for bone grafting, and the surgery is less invasive. Because this technique can expand bone in any direction, it has often been used for correcting jaw discrepancies in cases of congenital or acquired deformities, where the bone volume deficit is large.

In DO, an osteotomy separates the bone, the two parts are then fixed a distraction device, after which a gradual distraction period allows intramembranous bone formation to occur. In this process, mesenchymal stem cells from the bone marrow differentiate into osteoblasts to form neo-callus. Bone formation follows the "tension-stress principle," as proposed by Ilizarov [60]. Neovascularization is critically required for successful bone formation in this process and for this reason, systemic factors that impair neovascularization also affect bone regeneration at the DO site [61].

Midface deficiency is common in cleft patients due to unavoidable scar tissue formation during surgical closure of clefts at an early age. When the growth of the maxillofacial complex ends, a severe crossbite and maxillary hypoplasia are often the end result, with a low bone volume. A simple maxillary advancement (Le Fort I) is not enough. DO can provide reliable advancement of the maxilla with substantial bone deposition. Results are stable 12 months after treatment. Using a rigid external distraction device, it is possible to perform a stable advancement of the maxilla in both the horizontal and vertical planes.

4.2.2 DO in the Mandible

In bone tissue, the areas with the highest metabolic activity receive the richest sympathetic innervation [62]. Bone cells express neuronal signal receptors, which mediate neuro-osteogenic interactions [63]. The sympathetic nervous system regulates the bone remodeling process [43, 64, 65], stimulating osteoclastic activity via β-2 adrenergic receptors (Adrb2). In Adrb2 knockout animals, tooth movement is significantly reduced.

Nerve integrity has an impact on bone remodeling. In sagittal DO of the mandible, there is a risk that DO surgery procedures may injure or transect the inferior dental nerve. If this occurs, there would be a reduction in new bone formation following DO [59]. In the mandible, the approach of using DO is limited to cases where correction in the transverse dimension is required, such as in hypoplasia of the mandible [66].

4.3 Bone Tissue Engineering for the Maxillofacial Region

Besides auto-transplantation, heterogeneric osteogenesis has been explored extensively for bone tissue engineering. Bone tissue grafting [67, 68] and the implantation of synthetic bone substitutes have been the two main approaches used for the treatment of jaw discrepancies. A recently review suggested that using allogenic bone chips could be a safe technique, [69] however this approach has limited source material and is not likely to achieve the maximum extent of bone regeneration capacity. This next section will focus on synthetic bone substitutes used in bone tissue engineering for the maxillofacial region [70, 71].

Bone substitutes include both organic and inorganic-metal substitutes. This has been a greatly expanding topic within tissue engineering over recent decades.

4.3.1 Vascularization in Bone Regeneration

Scaffold materials have been applied to repair bone defects in the maxillofacial area. When the bone defect size exceeds a certain limit, or the bone volume needed to correct a jaw discrepancy is large, achieving sufficient tissue in -growth into the scaffold becomes a major challenge. The importance of achieving blood vessel in-growth into scaffolds used in bone regeneration is acknowledged widely. To promote angiogenesis in bone regeneration, many approaches have been reported, including the addition of growth factors (e.g., VEGF and bFGF) [72, 73], incorporating blood vessels [74], pre-vascularizing the scaffold material with a cell sheet [75], and optimizing the scaffold microstructure [76]. In yet another approach, an arteriovenous loop was microsurgically created and introduced inside a scaffold used to repair a critical size bone defect in the mandible. There was a significant blood vessel formation inside that scaffold, and more bone formation throughout the scaffold [74].

4.3.2 Research Models

To study the bone regeneration capacity of bone substitutes used in the maxillofacial region, many in vivo models have been proposed [77, 78]. Bone in the maxillofacial area follows intramembranous formation, like the bones of the cranium. The calvarial bone defect model has become the most common critical size defect model used to evaluate the bone formation capacity of new regeneration techniques and materials used in the maxillofacial region. The calvarial model is typically employed with small animals, e.g., mice [73], rats [79], and rabbits [80, 81]. A mandible bone defect model is often used with large animals, e.g., dogs [82, 83], pigs [84], sheep [85] and non-human primates [86, 87]. Recently, a mandible defect model in small animals has also been reported [88]. In the maxilla, the research model is usually alveolar bone regeneration of clefts [89, 90]. Recently, a rat mandible model has been introduced to study clefts [91].

5 Regenerative Aspects in Orthodontic Treatments for Periodontal Diseases

Orthodontic tooth movement is not limited to biological events within the periodontal ligament. It involves two interacting biological activities: one involves remodeling of alveolar bone, and the other involves remodeling of periodontal tissues. Orthodontic intervention, essentially applied as a light force, initiates mechanotransduction, and triggers a series of osteosis and angiogenesis processes that are yet to be understood fully.

The concept of guided orthodontic regeneration has been proposed by Paolone and coworkers [92, 93]. This recognizes the regenerative potency of orthodontics in periodontal tissues, including both soft and hard tissues. From a tissue engineering standpoint, using orthodontic treatment (with light force) appears to be a feasible approach to tissue regeneration with less complications than surgical methods (Fig. 2) [94, 95]. In this section, the regeneration of periodontal tissue achieved via orthodontic treatment will be discussed.

Tissue responses to applied forces vary depending on the force type that is used. Take alveolar bone, for example. Bone formation can be observed under an expansion force, while bone resorption occurs when a compression force is applied. The periodontium plays a conductor role by transferring the applied mechanical forces into biological signals, to initiate tissue responses. As shown in vivo, periodontal tissue can be reshaped as needed by applying different forces. An orthodontic force can regenerate gingival tissue, alveolar bone and periodontium. It can also aid periodontal treatment to preserve a tooth.

An insufficient alveolar bone volume is a common problem in edentulous ridge areas where teeth have been lost. To prepare the

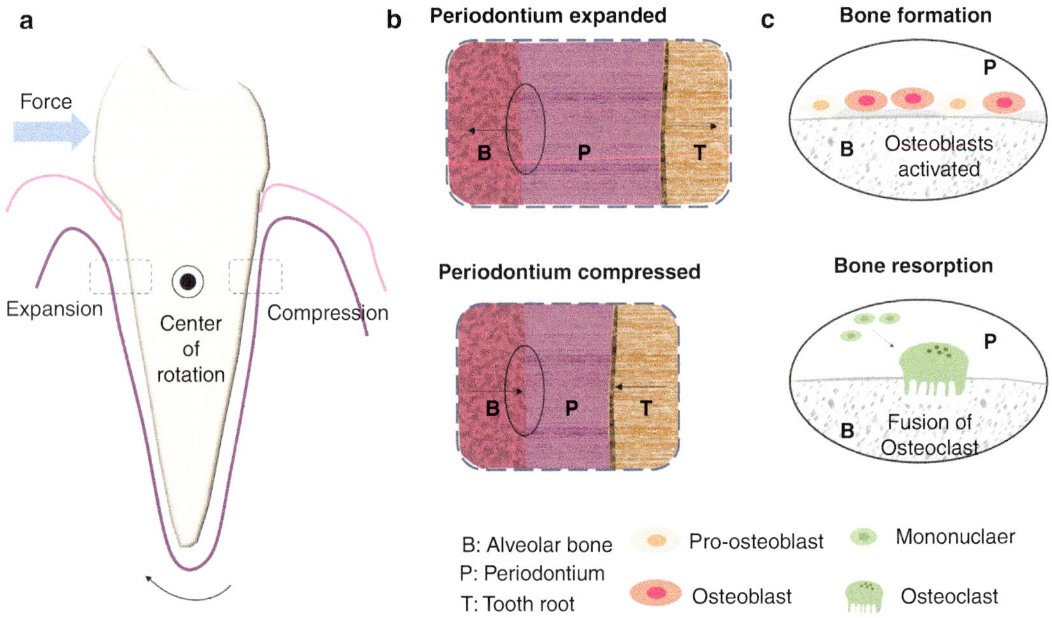

Fig. 2 Orthodontic force-initiated tooth movement and tissue responses. Orthodontic force is applied to the labial side of the tooth crown. The tooth rotates around the center of resistance: the crown moves to the lingual side, while the root moves to the labial side (**a**). At the cervical level of the root, periodontal tissue on the labial side is expanded, while periodontal tissue on the lingual side is compressed (**b**). Expansion stress activates osteoblast precursors, and bone formation is observed; compression stress promotes osteoclast formation via fusion of mononuclear cells, and bone resorption occurs. Usually these mononuclear cells migrate in from nearby blood vessels (**c**)

periodontal tissue to support an implant, orthodontic approaches can be used to increase the alveolar crest height prior to implant insertion.

Besides implants, as an alternative treatment for a missing tooth, tooth auto-transplantation has been undertaken, and this has proved effective in the long term. An active periodontium has been regarded as the most important factor for the success of tooth auto-transplantation.

Orthodontic force can create an activated periodontium. In the following part, orthodontic application in periodontal tissue regeneration is discussed for gingival, alveolar bone and periodontal regeneration.

5.1 Orthodontics Improves Periodontal Esthetics: Gingival Recession/Black Triangle

The etiology of the gingival recession is multifactorial. It is generally regarded that growing patient can gain spontaneous re-growth of gingival while it is almost impossible to regain recessed gingiva in adults. Few studies have reported the regain of recessed labial gingiva and gingival papilla. The mechanism behind the success of gingiva re-growth lies in the dynamically remodeling alveolar bone. Orthodontic force or surgical procedures (for instance, corticotomy introduced in other parts) could activate a series of osteosis process including resorption, formation, and remodeling of the alveolar bone. Research has gradually better understood the relation between orthodontics and regeneration of periodontal tissue reaction. Gingival attachment is determined by supracrestal tissue attachment [96], which is related to the shape of alveolar bone, mostly the height. In practice, treating recessed gingival tissue can be (partially) achieved with orthodontic intervention.

Gingival attachment is determined by the alveolar bone. When the front tooth is labially positioned, it is not surprising to observe the gingival recession. Hence obvious regain of recessed gingiva is seen in repositioning the teeth in the center of the alveolar ridge [97–102]. Depending on the original position of the recessed tooth and the occlusion, common orthodontic approaches include torque (incline the crown while maintain the position of the root), retracting, and intrusion. Orthodontic correction can greatly improve the recession grading, which is correlated to the treatment prognosis of periodontal plastic surgery [103–107]. Hence in the treatment of the gingival recession, the position of the tooth inside the alveolar ridge should be prioritized.

As the cause of gingival recession is multifactorial, there has not been a systematic treatment protocol or guideline. There exists a controversial debate in the relation of orthodontic treatment and gingival recession. Conclusive research on gingival recession is lacking in the literature. Yet some evidence showed it is possible to regrow gingival when dental plaque is well controlled. A multi-disciplinary team management is required for the succeed in the treatment of gingival recession. For instance, under a carefully monitored plaque control, orthodontic treatment could significantly minimize the recessed area followed by gingival reconstructive surgery. This combined strategy shall provide a good prognosis of treatment of gingival recession.

5.2 Orthodontics in Prosthesis

The esthetics of labial gingival contour (height, width, and symmetry) plays an essential role in dentofacial esthetics in the front view. When performing implantation at the anterior maxillary region, gingiva usually recesses to a noticeable level in the first 3–6 months post-surgery [108–111]. To prepare alveolar bone and gingival for esthetic implant placement [112], orthodontic extrusion, a coronal tooth movement, can be performed to activates the osteosis process [93]. As a result, the alveolar process will be remodeled at the coronal direction (increasing crest height) as well as a reduction in periodontal pocket depth could be observed [113]. Orthodontic appliance can be used to develop implant site, which is not restricted by the residual attachment level. The efficacy of gingival and alveolar bone regeneration was reported about 70% and 60% [114]. Due to this plastic nature of periodontal tissue with force, orthodontic techniques have been adopted to grow the periodontal structure prior to an implant placement [115–121].

In some cases, to gain biological distance restoration, the clinical crown (the part of a tooth that is exposed to oral cavity) needs to be elongated. Usually orthodontic extrusion is preferred. Conventional orthodontic extrusion creates bone apposition at the alveolar crest broadening the width of the attached gingival. This tissue regeneration is appreciated in implant placement where there is lacking enough bone to support the implant. On the contrary, in periodontal treatment of crown lengthening, this tissue regeneration becomes undesired. Then gingival fiber resection (fiberotomy) and root surface scaling are needed to avoid the coronal migration of periodontal tissue [122]. The reason is because that alveolar bone grows coronally as the tooth being extruded. And periodontal soft tissue attachment is determined by *biological width*. The term has been adopted as a clinical term that describes the variable dimension of the supracrestal attached periodontal tissue in apicocoronal (vertical) direction. It has been recently replaced by *supracrestal tissue attachment*. Supracrestal tissue attachment is a concise, descriptive definition of the histological structure: junctional epithelium and supracrestal connective tissue attachment [96]. The relation between alveolar bone crest to the periodontal attachment remained acknowledged. This concept of attachment centers on the important role of alveolar height in periodontal treatment.

5.3 Orthodontics Helps Reconstruct Periodontium

Missing tooth brings functional and esthetic issues. Prosthetic implant is commonly used to resolve the problem. Yet it lacks periodontal ligament which provides individuals with a physiologic proprioception and sensory reflection. In growing patients, implants fail to provide physiologic eruption which results with an unleveled gingival margin. Tooth auto-transplantation can be a good alternative as the tooth comes with viable periodontal ligament which enables a high successful rate [123]. Ankylosis and root resorption are common complications. Orthodontic treatment (force) activates physiological tissue response. Upon distortion of collagen fibers within periodontal ligament, mechanical strain transduces to cells inside ligament and neighboring tissues. Macrophages response early and appear in periodontal ligament when orthodontic force is applied. They then response to mechanotransductively released cytokines. Proinflammatory and angiogenic cytokines are produced by macrophages under compression strain, which alter the microenvironment of periodontal ligament [124]. These cytokines are influential to the survival of periodontal ligament. Bone lining cells in periodontal ligament play an important role in tooth movement and osteoclastogenesis in response to mechanical force [125]. Short term application of orthodontic force on donor's tooth can activate the periodontal ligament with upregulated expression of inflammation, osteoclastogenesis. After transplantation, donor tooth activated by orthodontic force showed higher tissue regenerative potency than normal tooth. Genes relating to periodontal ligament regeneration, cell proliferation, osteoblastogenesis, and osteoclastogenesis are highly expressed in orthodontically activated donor tooth in the first-week post-transplantation. Osteoclastogenic gene expression in orthodontically activated donor tooth is reduced to significant lower level than that in normal donor tooth in the fourth-week post-transplantation. This consisted of histologic observation where root resorption was less and inflammatory activity subsided in 4 weeks [126].

The vitality of periodontal ligament is a key factor in the prognosis of transplanted teeth [127]. Hypofunctional, un-occluded, teeth have narrow periodontal ligament. Donor teeth with functional periodontal ligament survived much better than unerupted or partially erupted teeth [128]. They survive longer, have less complications, e.g., root resorption. This updates previous belief that unerupted or partially erupted teeth are preferable in tooth transplantation as they have wide periapical foramen and higher chance to maintain the vitality of pulp. The application of orthodontic force also increases the survival chance of transplanted tooth [127]. Early orthodontic force engagement (within 4 weeks post-transplantation) could reduce the incidence of ankylosis [128].

5.4 Orthodontic Application in Periodontal Treatment

Orthodontic force creates stress of periodontal tissue and receives a complicate set of signaling responses including osteoclastogenic and angiogenic activation. The former results in alveolar bone change and the latter may impact on the alteration in the gingival tissue and periodontal fiber. Orthodontic tooth movement can create periodontal tissue, which benefits perio-restorative patients. As discussed in the previous section "*Orthodontics in prosthesis*," increased amount of soft and hard tissue is generated during extrusion. In perio-restorative patients, these excessive tissues could be used in periodontal regenerative surgery [94].

Aggressive periodontitis featured by disruption in periodontal tissue, extrusion of front teeth, and loss of teeth leaves esthetics and mastication disabilities [129]. In chronic periodontitis, with loss of periodontal tissue, crown/root proportion becomes large, which creates a greater unfavorable force to remained periodontium under normal mastication compared a healthy tooth. Orthodontic tooth intrusion is performed to correct the crown/root proportion [130]. In both aggressive and chronic periodontitis, there is tooth loss and space remained, which compromises mastication efficiency. To restore these spaces, orthodontic treatment is often applied to re-arrange remained teeth and space for a balanced and periodontal tissue-friendly restoration.

Zasciurinskiene and colleagues concluded that no evidence indicate either beneficial or deteriorating role of orthodontic treatment to periodontally compromised dentition [131]. A recent clinical study demonstrated that under strict plaque control periodontal assessment of aggressive periodontitis was similar to that of the orthodontic patient who has healthy periodontal tissue [129]. There have been a few case reports of orthodontic treatment on periodontitis [132].

6 Regenerative Approaches for Temporomandibular Joint Disorders (TMD)

During basic daily functions (i.e., speaking, swallowing, and eating), the temporomandibular joint (TMJ) plays an extremely important role in coordinating the jaw movements. The TMJ is described as a bilateral synovial joint formed by fibrocartilaginous articular surfaces of the mandibular condyle and glenoid fossa, muscles, ligaments, and the articular disc [133]. It is estimated that 10–40% of the population are affected by temporomandibular joint disorders (TMD), with a predominance among young adults under 45 years of age [134].

The main signs and symptoms of TMDs include limited mouth opening, mandibular deviation during the opening, displacement, clicking, locking, and muscle pain during mandibular movements. TMJ pathologies that require clinical treatment are internal derangement, degenerative joint disease, and ankylosis [133, 134].

Three categories are described for the clinical treatment of TMJ pathologies: non-invasive, minimally invasive, and invasive. When in an advanced stage of degenerative disease, the total alloplastic reconstruction of the TMJ is considered the treatment of choice. To possibly eliminate the need for total TMJ replacement, tissue engineering may provide a functional and permanent biological replacement of the TMJ structures [135–139]. Figure 3 shows a general overview of main TMJ pathologies, current therapies, and the tissue engineering approach of the different TMJ structures.

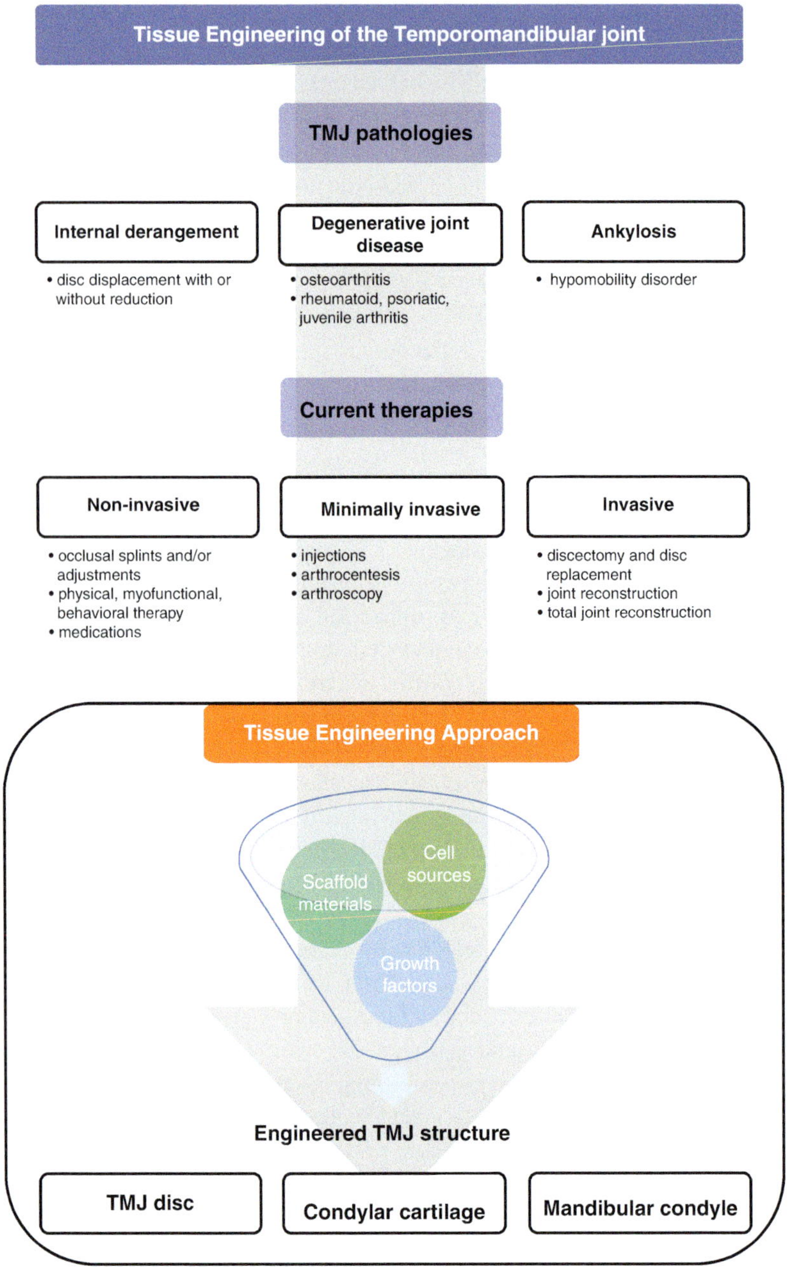

Fig. 3 Overview of main TMJ pathologies, current therapies, and the tissue engineering approach of the different TMJ structures

6.1 Tissue Engineering of the TMJ

Many tissues of the body, after an injury, exhibit an ability to repair themselves, however, some tissues have little or no ability to self-repair. Among these, the TMJ tissues can be included. Given an advanced pathological process of the TMJ (i.e., osteoarthritis), coupled with limited repair capacity of the TMJ tissues (i.e., fibrocartilage, cartilage, bone), the current treatment options for clinicians and surgeons, in order to maintain normal function and eliminate the patient's pain, are limited and considered semi-permanent [133, 134, 137–139].

To overcome the current obstacles, tissue engineering may provide permanent, biomimetic

Table 1 Most used biomaterials, stem cells, and growth factors for bone tissue engineering

Stem cells	Biomaterials	Growth factors
Mesenchymal stem cells	Bioceramics	BMPs
Dental stem cells	Polymers	FGFs
iPSC	Hybrid	VEGF
Adipose-derived stem cells, peripheral blood-derived stem cells	Composites	Others

replacement tissue for the TMJ [136–138]. Thus, scientists have used the tissue engineering paradigm [139]. The first step is to characterize the native and healthy tissues, providing parameters for the appropriate design using the concepts of tissue engineering. Considering the tissues to be regenerated and its characteristics and properties, the parameters obtained will guide the optimal selection of cells, of scaffold materials (extracellular matrix), of proper growth factors and, in specific tissues, the necessity of biomechanical stimulation (mainly for the articular disc), making it possible to obtain an implantable biomimetic tissue [135–139]. Table 1 describes the main cell sources, scaffold materials, and growth factors used to tissue engineer the TMJ disc, the condylar cartilage, and the mandibular condyle.

Over the past decade, advances in the field of biomaterials science, tissue engineering, and stem cell therapies have led to the development of less invasive and alternative treatments for diseased joint tissues [133, 136, 137, 139]. Cell-based therapies involving expansion and transplantation of stem cells combined with different scaffold materials and growth factors have been shown regenerative capabilities to repair diseased TMJ tissues [140, 141]. Also, cell-free scaffolds have been used for TMJ cartilage regeneration, for fibrocartilage (disc) regeneration, and for osteochondral regeneration with regenerative capabilities in animal models [142–144].

Two methods have been extensively described in the literature for bone tissue engineering and for TMJ tissue engineering: (a) in situ tissue engineering, which consists of using cell-free scaffolds to attract local cells (cell homing) that will guide the regeneration process; (b) and scaffolds seeded with competent cells to guide the regeneration process [145].

Table 2 Main cell sources, scaffold materials, and growth factors used to tissue engineer the TMJ disc, condylar cartilage, and the mandibular condyle

	Temporomandibular joint structure		
	TMJ disc	Condylar cartilage	Mandibular condyle
Cell sources	Costal chondrocytes; primary disc cells; multipotent MSCs; umbilical cord matrix stem cells; pluripotent ESCs	Primary cartilage cells; multipotent hUCMSCs	Mature osteoblasts and chondrocytes; bone marrow-derived MSCs
Scaffold materials	Porous collagen scaffold; PGA; PLA; ePTFE; PLLA; alginate hydrogels	PGA	PEG; PCL; PLA; PGA; PLGA; calcium phosphate ceramics (HA, TCP)
Growth factors	PDGF; bFGF; TGF-b1; TGF-b3; IGF-I	bFGF; IGF-I; TGF-b1; EGF	IGF-I; TGF-b1

Abbreviations: mesenchymal stem cells (MSCs); embryonic stem cells (ESCs); polyglycolic acid (PGA); polylactic acid (PLA); expanded polytetrafluoroethylene (ePTFE); poly-l-lactic acid (PLLA); platelet-derived growth factor (PDGF); basic fibroblast growth factor (bFGF); transforming growth factor-b1 (TGF-b1); transforming growth factor-b3 (TGF-b3); insulin-like growth factor-I (IGF-I); human umbilical cord mesenchymal stromal cells (hUCMSCs); epidermal growth factor (EGF); poly(ethylene glycol) (PEG); polycaprolactone (PCL); poly-lactic-co-glycolic acid (PLGA); calcium phosphate ceramics: hydroxyapatite (HA) and tricalcium phosphate (TCP)

Preclinical studies using small and large animals using different cell sources, combined with scaffolds made of a wide range of materials and enriched with a variety of growth factors have been described to regenerate the TMJ disc, the condylar cartilage and the mandibular condyle with promising outcomes [136, 146]. Some clinical trials [147] have demonstrated the efficacy of autologous or allogeneic MSCs in cartilage repair [148] (Table 2).

Besides the promising results from preclinical data, there are still challenges to overcome to bring tissue-engineered TMJ structures to the clinic. Among them, the total restoration and incorporation of fibrocartilage in the articular surfaces, the possible displacement of the implanted material, and the lack of long-term results of the regenerative approaches of the TMJ structures.

6.2 Future Treatment for the TMD Treatment

In view of the challenges that the unique TMJ environment represents (i.e., mechanically demanding and biochemically active), the field of tissue engineering has made significant progress over the past decade with promising results to replace diseased, displaced, or degenerated tissues. Currently, research has focused on biological substitutions of the mandibular disc, adjacent structures of the TMJ, and engineering of craniofacial tissues (i.e., bone, soft tissue, nerve, muscle). In addition, tissue engineering strategies to provide treatment options for the glenoid fossa and the articular eminence should be considered.

The scientific and technological advances available provide a solid basis for scientists and surgeons to overcome the challenges that still exist in the field of TMJ tissue engineering, such as the proper selection of cell sources, scaffold materials, and growth factors. A detailed understanding of the native tissues of the TMJ and their respective complex pathologies are essential for scientists who wish to develop and increase the success of permanent biological TMJ replacements.

7 Conclusions and Future Perspectives

Current Status

The regenerative approaches and tissue engineering combine stem cells with scaffolding biomaterials as well as growth factors. This has potential applications in surgical correction of jaw discrepancies, bone loss due to periodontal disease, congenital bone defect such as cleft lips and palates, TMJ disorders related to bone/cartilage defect, and alveolar bone lesions. Recent advances in stem cells indicate effective treatment and improved clinical outcomes in bone tissue regeneration in orthodontic/orthopedic treatment. Studies also showed that stem cells based regenerative approaches can reduce morbidity and speed up the recovery process comparing to the conventional surgical approaches. Taken together, we can conclude that regenerative approaches through tissue engineering are promising for orthodontic/orthopedic treatments.

Future Perspectives

The contemporary evidence indicates the feasibility of regenerative approaches in orthodontic/orthopedic treatments, yet the majority of the evidence is from preclinical studies with animal models. It is worthwhile to notice that there is still a long distance from bench to bed. The future study in this field need to focus on the followings:

- Advances technologies: Nanoscale biomaterials will be applied as control delivery systems for tissue engineering. The nanomaterials can also regulate the immuno-response as such to enhance the regeneration efficiency. Among all the stem cells, dental pulp stem cells (DPSCs) are an emerging source that has drawn more and more intention. The advantages of DPSCs include but not limited to: low immunogenicity, high differentiative capacity, and easy to access through bio-banking of the deciduous teeth or young adult teeth.
- Clinical translation: More clinical trials are required to assess the safety and efficacy of the stem cell-based regenerative approaches for orthodontic and orthopedic patients. Ideally such clinical trials should be conducted in double-blinded, randomized, and controlled manner in order to produce the high quality of clinical evidence.

References

1. Koons GL, Diba M, Mikos AG. Materials design for bone-tissue engineering. Nat Rev Mater. 2020;5:584. https://doi.org/10.1038/s41578-020-0204-2.
2. Amini AR, Laurencin CT, Nukavarapu SP. Bone tissue engineering: recent advances and challenges. Crit Rev Biomed Eng. 2012;40(5):363–408. https://doi.org/10.1615/critrevbiomedeng.v40.i5.10.
3. Stevens MM. Biomaterials for bone tissue engineering. Mater Today. 2008;11(5):18–25. https://doi.org/10.1016/S1369-7021(08)70086-5.
4. Wei GB, Ma PX. Structure and properties of nano-hydroxyapatite/polymer composite scaffolds for bone tissue engineering. Biomaterials. 2004;25(19):4749–57. https://doi.org/10.1016/j.biomaterials.2003.12.005.
5. Gomez-Barrena E, Rosset P, Muller I, Giordano R, Bunu C, Layrolle P, et al. Bone regeneration: stem cell therapies and clinical studies in orthopaedics and traumatology. J Cell Mol Med. 2011;15(6):1266–86. https://doi.org/10.1111/j.1582-4934.2011.01265.x.
6. Arvidson K, Abdallah BM, Applegate LA, Baldini N, Cenni E, Gomez-Barrena E, et al. Bone regeneration and stem cells. J Cell Mol Med. 2011;15(4):718–46. https://doi.org/10.1111/j.1582-4934.2010.01224.x.
7. Nakamura S, Yamada Y, Katagiri W, Sugito T, Ito K, Ueda M. Stem cell proliferation pathways comparison between human exfoliated deciduous teeth and dental pulp stem cells by gene expression profile from promising dental pulp. J Endod. 2009;35(11):1536–42. https://doi.org/10.1016/j.joen.2009.07.024.
8. Chen X, Zhang T, Shi J, Xu P, Gu Z, Sandham A, et al. Notch1 signaling regulates the proliferation and self-renewal of human dental follicle cells by modulating the G1/S phase transition and telomerase activity. PLoS One. 2013;8(7):e69967.
9. Volpe G, Bernstock JD, Peruzzotti-Jametti L, Pluchino S. Modulation of host immune responses following non-hematopoietic stem cell transplantation: translational implications in progressive multiple sclerosis. J Neuroimmunol. 2019;331:11–27.
10. Luo L, Albashari AA, Wang X, Jin L, Zhang Y, Zheng L, et al. Effects of transplanted heparin-poloxamer hydrogel combining dental pulp stem cells and bFGF on spinal cord injury repair. Stem Cells Int. 2018;2018:1.
11. Gronthos S, Mankani M, Brahim J, Robey PG, Shi S. Postnatal human dental pulp stem cells (DPSCs) in vitro and in vivo. Proc Natl Acad Sci USA. 2000;97(25):13625–30. https://doi.org/10.1073/pnas.240309797.
12. Laino G, d'Aquino R, Graziano A, Lanza V, Carinci F, Naro F, et al. A new population of human adult dental pulp stem cells: a useful source of living autologous fibrous bone tissue (LAB). J Bone Miner Res. 2005;20(8):1394–402. https://doi.org/10.1359/Jbmr.050325.
13. Takahashi K, Yamanaka S. Induction of pluripotent stem cells from mouse embryonic and adult fibroblast cultures by defined factors. Cell. 2006;126(4):663–76. https://doi.org/10.1016/j.cell.2006.07.024.
14. Cai J, Zhang Y, Liu P, Chen S, Wu X, Sun Y, et al. Generation of tooth-like structures from integration-free human urine induced pluripotent stem cells. Cell Regener. 2013;2(1):6. https://doi.org/10.1186/2045-9769-2-6.
15. Abdullah AN, Miyauchi S, Onishi A, Tanimoto K, Kato K. Differentiation of mouse-induced pluripotent stem cells into dental epithelial-like cells in the absence of added serum. In Vitro Cell Dev-An. 2019;55(2):130–7. https://doi.org/10.1007/s11626-019-00320-z.
16. Storti G, Scioli MG, Kim BS, Orlandi A, Cervelli V. Adipose-derived stem cells in bone tissue engineering: useful tools with new applications. Stem Cells Int. 2019;2019:1. https://doi.org/10.1155/2019/3673857.
17. Chen YR, Yan X, Yuan FZ, Ye J, Xu BB, Zhou ZX, et al. The use of peripheral blood-derived stem cells for cartilage repair and regeneration in vivo: a review. Front Pharmacol. 2020;11:404. https://doi.org/10.3389/fphar.2020.00404.
18. Jeong J, Kim JH, Shim JH, Hwang NS, Heo CY. Bioactive calcium phosphate materials and applications in bone regeneration. Biomater Res. 2019;23:4. https://doi.org/10.1186/s40824-018-0149-3.
19. Xie H, Gu Z, Li C, Franco C, Wang J, Li L, Meredith N, Ye Q, Wan C. A novel bioceramic scaffold integrating silk fibroin in calcium polyphosphate for bone tissue-engineering. Ceram Int. 2016;42(2):2386–92.
20. Fujishiro Y, Hench LL, Oonishi H. Quantitative rates of in vivo bone generation for Bioglass(R) and hydroxyapatite particles as bone graft substitute. J Mater Sci-Mater M. 1997;8(11):649–52. https://doi.org/10.1023/A:1018527621356.
21. Zhou YL, Wu CT, Chang J. Bioceramics to regulate stem cells and their microenvironment for tissue regeneration. Mater Today. 2019;24:41–56. https://doi.org/10.1016/j.mattod.2018.07.016.
22. Alabbasi A, Mehjabeen A, Kannan MB, Ye Q, Blawert C. Biodegradable polymer for sealing porous PEO layer on pure magnesium: an in vitro degradation study. Appl Surf Sci. 2014;301:463–7.
23. Chen D, Zhao M, Mundy GR. Bone morphogenetic proteins. Growth Factors. 2004;22(4):233–41. https://doi.org/10.1080/08977190412331279890.
24. Deal C. Future therapeutic targets in osteoporosis. Curr Opin Rheumatol. 2009;21(4):380–5. https://doi.org/10.1097/Bor.0b013e32832cbc2a.
25. Park SB, Park SH, Kim NH, Chung CK. BMP-2 induced early bone formation in spine fusion using rat ovariectomy osteoporosis model. Spine

J. 2013;13(10):1273–80. https://doi.org/10.1016/j.spinee.2013.06.010.
26. Zhang SF, Kucharski C, Doschak MR, Sebald W, Uludag H. Polyethylenimine-PEG coated albumin nanoparticles for BMP-2 delivery. Biomaterials. 2010;31(5):952–63. https://doi.org/10.1016/j.biomaterials.2009.10.011.
27. Kapoor P, Kharbanda OP, Monga N, Miglani R, Kapila S. Effect of orthodontic forces on cytokine and receptor levels in gingival crevicular fluid: a systematic review. Prog Orthod. 2014;15(1):65.
28. Liou EJ, Chen PH, Wang YC, Yu CC, Huang CS, Chen YR. Surgery-first accelerated orthognathic surgery: postoperative rapid orthodontic tooth movement. J Oral Maxillofac Surg. 2011;69(3):781–5.
29. Alfawal AM, Hajeer MY, Ajaj MA, Hamadah O, Brad B. Evaluation of piezocision and laser-assisted flapless corticotomy in the acceleration of canine retraction: a randomized controlled trial. Head Face Med. 2018;14(1):4.
30. Al-Naoum F, Hajeer MY, Al-Jundi A. Does alveolar corticotomy accelerate orthodontic tooth movement when retracting upper canines? A split-mouth design randomized controlled trial. J Oral Maxillofac Surg. 2014;72(10):1880–9.
31. Kundi I, Alam MK, Shaheed S. Micro-osteo perforation effects as an intervention on canine retraction. Saudi Dent J. 2020;32(1):15–20.
32. Dutra EH, Ahmida A, Lima A, Schneider S, Nanda R, Yadav S. The effects of alveolar decortications on orthodontic tooth movement and bone remodelling in rats. Eur J Orthodont. 2018;40(4):423–9.
33. Patterson BM, Dalci O, Darendeliler MA, Papadopoulou AK. Corticotomies and orthodontic tooth movement: a systematic review. J Oral Maxillofac Surg. 2016;74(3):453–73.
34. Tsai CY, Yang TK, Hsieh HY, Yang LY. Comparison of the effects of micro-osteoperforation and corticision on the rate of orthodontic tooth movement in rats. Angle Orthod. 2016;86(4):558–64.
35. Alikhani M, Raptis M, Zoldan B, Sangsuwon C, Lee YB, Alyami B, Corpodian C, Barrera LM, Alansari S, Khoo E, Teixeira C. Effect of micro-osteoperforations on the rate of tooth movement. Am J Orthod Dentofac Orthop. 2013;144(5):639–48.
36. Alkebsi A, Al-Maaitah E, Al-Shorman H, Alhaija EA. Three-dimensional assessment of the effect of micro-osteoperforations on the rate of tooth movement during canine retraction in adults with class II malocclusion: a randomized controlled clinical trial. Am J Orthod Dentofac Orthop. 2018;153(6):771–85.
37. Sivarajan S, Ringgingon LP, Fayed MM, Wey MC. The effect of micro-osteoperforations on the rate of orthodontic tooth movement: a systematic review and meta-analysis. Am J Orthod Dentofac Orthop. 2020;157(3):290–304.
38. Ali D, Mohammed H, Koo SH, Kang KH, Kim SC. Three-dimensional evaluation of tooth movement in class II malocclusions treated without extraction by orthodontic mini-implant anchorage. Korean J Orthod. 2016;46(5):280–9.
39. Antoszewska-Smith J, Sarul M, Łyczek J, Konopka T, Kawala B. Effectiveness of orthodontic miniscrew implants in anchorage reinforcement during en-masse retraction: a systematic review and meta-analysis. Am J Orthod Dentofac Orthop. 2017;151(3):440–55.
40. Roberts WE, Huja SS. Bone physiology, metabolism, and biomechanics in orthodontic practice. Orthodontics: current principles and techniques. 6th ed. Oxford: Elsevier Health Sciences; 2016. p. 99–152.
41. Chen N, Sui BD, Hu CH, Cao J, Zheng CX, Hou R, Yang ZK, Zhao P, Chen Q, Yang QJ, Jin Y. microRNA-21 contributes to orthodontic tooth movement. J Dent Res. 2016;95(12):1425–33.
42. Wu L, Su Y, Lin F, Zhu S, Wang J, Hou Y, Du J, Liu Y, Guo L. MicroRNA-21 promotes orthodontic tooth movement by modulating the RANKL/OPG balance in T cells. Oral Dis. 2020;26(2):370–80.
43. Cao H, Kou X, Yang R, Liu D, Wang X, Song Y, Feng L, He D, Gan Y, Zhou Y. Force-induced Adrb2 in periodontal ligament cells promotes tooth movement. J Dent Res. 2014;93(11):1163–9.
44. Kouskoura T, Katsaros C, von Gunten S. The potential use of pharmacological agents to modulate orthodontic tooth movement (OTM). Front Physiol. 2017;8:67.
45. Makrygiannakis MA, Kaklamanos EG, Athanasiou AE. Does common prescription medication affect the rate of orthodontic tooth movement? A systematic review. Eur J Orthod. 2018;40(6):649–59.
46. Qin J, Hua Y. Effects of hydrogen sulfide on the expression of alkaline phosphatase, osteocalcin and collagen type I in human periodontal ligament cells induced by tension force stimulation. Mol Med Rep. 2016;14(4):3871–7.
47. Pu H, Hua Y. Hydrogen sulfide regulates bone remodeling and promotes orthodontic tooth movement. Mol Med Rep. 2017;16(6):9415–22.
48. Yang F, Wang XX, Ma D, Cui Q, Zheng DH, Liu XC, Zhang J. Effects of Triptolide on tooth movement and root resorption in rats. Drug Des Devel Ther. 2019;13:3963.
49. Ma D, Wang X, Ren X, Bu J, Zheng D, Zhang J. Asperosaponin VI injection enhances orthodontic tooth movement in rats. Med Sci Mon Int Med J Exp Clin Res. 2020;26:e922372–1.
50. Liu XC, Wang XX, Zhang LN, Yang F, Nie FJ, Zhang J. Inhibitory effects of resveratrol on orthodontic tooth movement and associated root resorption in rats. Arch Oral Biol. 2020;111:104642.
51. Youssef M, Ashkar S, Hamade E, Gutknecht N, Lampert F, Mir M. The effect of low-level laser therapy during orthodontic movement: a preliminary study. Lasers Med Sci. 2008;23(1):27–33.
52. Qamruddin I, Alam MK, Mahroof V, Fida M, Khamis MF, Husein A. Effects of low-level laser irradiation

on the rate of orthodontic tooth movement and associated pain with self-ligating brackets. Am J Orthod Dentofac Orthop. 2017;152(5):622–30.
53. Varella AM, Revankar AV, Patil AK. Low-level laser therapy increases interleukin-1β in gingival crevicular fluid and enhances the rate of orthodontic tooth movement. Am J Orthod Dentofac Orthop. 2018;154(4):535–44.
54. Gkantidis N, Mistakidis I, Kouskoura T, Pandis N. Effectiveness of non-conventional methods for accelerated orthodontic tooth movement: a systematic review and meta-analysis. J Dent. 2014;42(10):1300–19.
55. Mistry D, Dalci O, Papageorgiou SN, Darendeliler MA, Papadopoulou AK. The effects of a clinically feasible application of low-level laser therapy on the rate of orthodontic tooth movement: a triple-blind, split-mouth, randomized controlled trial. Am J Orthod Dentofac Orthop. 2020;157(4):444–53.
56. Karamesinis K, Basdra EK. The biological basis of treating jaw discrepancies: an interplay of mechanical forces and skeletal configuration. BBA-Mol Basis Dis. 2018;1864(5):1675–83.
57. Ilizarov GA. The tension stress effect on the genesis and growth of tissues. Part I. The influence of stability of fixation and soft-tissue preservation. Clin Orthop Relat Res. 1989;238:249–81.
58. Rachmiel A, Laufer D, Jackson IT, Lewinson D. Midface membranous bone lengthening: a one-year histological and morphological follow-up of distraction osteogenesis. Calcif Tissue Int. 1998;62:370–6.
59. Cao J, Zhang S, Gupta A, Du Z, Lei D, Wang L, Wang X. Sensory nerves affect bone regeneration in rabbit mandibular distraction osteogenesis. Int J Med Sci. 2019;16(6):831.
60. Ilizarov GA. Clinical application of the tension–stress effect for limb lengthening. Clin Orthop Relat Res. 1990;250:8–26.
61. Weiss S, Zimmermann G, Baumgart R, Kasten P, Bidlingmaier M, Henle P. Systemic regulation of angiogenesis and matrix degradation in bone regeneration—distraction osteogenesis compared to rigid fracture healing. Bone. 2005;37(6):781–90.
62. Chartier SR, Mitchell SA, Majuta LA, Mantyh PW. The changing sensory and sympathetic innervation of the young, adult and aging mouse femur. Neuroscience. 2018;387:178–90.
63. Tomlinson RE, Li Z, Zhang Q, Goh BC, Li Z, Thorek DL, Rajbhandari L, Brushart TM, Minichiello L, Zhou F, Venkatesan A. NGF-TrkA signaling by sensory nerves coordinates the vascularization and ossification of developing endochondral bone. Cell Rep. 2016;16(10):2723–35.
64. Takeda S, Elefteriou F, Levasseur R, Liu X, Zhao L, Parker KL, Armstrong D, Ducy P, Karsenty G. Leptin regulates bone formation via the sympathetic nervous system. Cell. 2002;111(3):305–17.
65. Elefteriou F, Ahn JD, Takeda S, Starbuck M, Yang X, Liu X, Kondo H, Richards WG, Bannon TW, Noda M, Clement K. Leptin regulation of bone resorption by the sympathetic nervous system and CART. Nature. 2005;434(7032):514–20.
66. Kishore SN, Yuvraj I, Ankur T. Mandibular distraction osteogenesis. J Craniofac Surg. 2019;30(8):e743–6.
67. Yoda N, Zheng K, Chen J, Liao Z, Koyama S, Peck C, Swain M, Sasaki K, Li Q. Biomechanical analysis of bone remodeling following mandibular reconstruction using fibula free flap. Med Eng Phys. 2018;56:1–8.
68. Li X, Jiang C, Gao H, Wang C, Wang C, Ji P. Biomechanical analysis of various reconstructive methods for the mandibular body and ramus defect using a free vascularized fibula flap. Biomed Res Int. 2020;13:2020.
69. Stopa Z, Siewert-Gutowska M, Abed K, Szubińska-Lelonkiewicz D, Kamiński A, Fiedor P. Evaluation of the safety and clinical efficacy of allogeneic bone grafts in the reconstruction of the maxilla and mandible. Transplant Proc. 2018;50(7):2199–201.
70. Henkel J, Woodruff MA, Epari DR, Steck R, Glatt V, Dickinson IC, Choong PF, Schuetz MA, Hutmacher DW. Bone regeneration based on tissue engineering conceptions—a 21st century perspective. Bone Res. 2013;1(1):216–48.
71. Jazayeri HE, Tahriri M, Razavi M, Khoshroo K, Fahimipour F, Dashtimoghadam E, Almeida L, Tayebi L. A current overview of materials and strategies for potential use in maxillofacial tissue regeneration. Mater Sci Eng C. 2017;70:913–29.
72. Rumney RM, Lanham SA, Kanczler JM, Kao AP, Thiagarajan L, Dixon JE, Tozzi G, Oreffo RO. In vivo delivery of VEGF RNA and protein to increase osteogenesis and intraosseous angiogenesis. Sci Rep. 2019;9(1):17745.
73. Qu D, Li J, Li Y, Gao Y, Zuo Y, Hsu Y, Hu J. Angiogenesis and osteogenesis enhanced by bFGF ex vivo gene therapy for bone tissue engineering in reconstruction of calvarial defects. J Biomed Mater Res A. 2011;96(3):543–51.
74. Eweida AM, Nabawi AS, Abouarab M, Kayed M, Elhammady H, Etaby A, Khalil MR, Shawky MS, Kneser U, Horch RE, Nagy N. Enhancing mandibular bone regeneration and perfusion via axial vascularization of scaffolds. Clin Oral Investig. 2014;18(6):1671–8.
75. Dong QN, Kanno T, Bai Y, Sha J, Hideshima K. Bone regeneration potential of uncalcined and unsintered hydroxyapatite/poly L-lactide bioactive/osteoconductive sheet used for maxillofacial reconstructive surgery: an in vivo study. Materials. 2019;12(18):2931.
76. Cao Y, Xiao L, Cao Y, Nanda A, Xu C, Ye Q. 3D printed β-TCP scaffold with sphingosine 1-phosphate coating promotes osteogenesis and inhibits inflammation. Biochem Biophys Res Commun. 2019;512(4):889–95.
77. Mardas N, Dereka X, Donos N, Dard M. Experimental model for bone regeneration in oral

and cranio-maxillo-facial surgery. J Investig Surg. 2014;27(1):32–49.
78. Shanbhag S, Pandis N, Mustafa K, Nyengaard JR, Stavropoulos A. Alveolar bone tissue engineering in critical-size defects of experimental animal models: a systematic review and meta-analysis. J Tissue Eng Regen Med. 2017;11(10):2935–49.
79. Liu G, Guo Y, Zhang L, Wang X, Liu R, Huang P, Xiao Y, Chen Z, Chen Z. A standardized rat burr hole defect model to study maxillofacial bone regeneration. Acta Biomater. 2019;86:450–64.
80. Xie H, Wang J, He Y, Gu Z, Xu J, Li L, Ye Q. Biocompatibility and safety evaluation of a silk fibroin-doped calcium polyphosphate scaffold copolymer in vitro and in vivo. RSC Adv. 2017;7(73):46036–44.
81. Deng N, Sun J, Li Y, Chen L, Chen C, Wu Y, Wang Z, Li L. Experimental study of rhBMP-2 chitosan nano-sustained release carrier-loaded PLGA/nHA scaffolds to construct mandibular tissue-engineered bone. Arch Oral Biol. 2019;102:16–25.
82. Wang S, Zhao J, Zhang W, Ye D, Zhang X, Zou D, Zhang X, Sun X, Sun S, Zhang W, Yang C. Comprehensive evaluation of cryopreserved bone-derived osteoblasts for the repair of segmental mandibular defects in canines. Clin Implant Dent Relat Res. 2015;17(4):798–810.
83. Yang C, Liu X, Zhao K, Zhu Y, Hu B, Zhou Y, Wang M, Wu Y, Zhang C, Xu J, Ning Y. miRNA-21 promotes osteogenesis via the PTEN/PI3K/Akt/HIF-1α pathway and enhances bone regeneration in critical size defects. Stem Cell Res Ther. 2019;10(1):65.
84. Konopnicki S, Sharaf B, Resnick C, Patenaude A, Pogal-Sussman T, Hwang KG, Abukawa H, Troulis MJ. Tissue-engineered bone with 3-dimensionally printed β-tricalcium phosphate and polycaprolactone scaffolds and early implantation: an in vivo pilot study in a porcine mandible model. J Oral Maxillofac Surg. 2015;73(5):1016–e1.
85. Schliephake H, Knebel JW, Aufderheide M, Tauscher M. Use of cultivated osteoprogenitor cells to increase bone formation in segmental mandibular defects: an experimental pilot study in sheep. J Oral Maxillofac Surg. 2001;30(6):531–7.
86. Chanchareonsook N, Tideman H, Feinberg SE, Jongpaiboonkit L, Lee S, Flanagan C, Krishnaswamy G, Jansen J. Segmental mandibular bone reconstruction with a carbonate-substituted hydroxyapatite-coated modular endoprosthetic poly (ε-caprolactone) scaffold in *Macaca fascicularis*. J Biomed Mater Res B Appl Biomater. 2014;102(5):962–76.
87. Xie Y, Su Y, Min S, Tang J, Goh BT, Saigo L, Ansari S, Moshaverinia A, Zhang C, Wang J, Liu Y. Collagen sponge functionalized with chimeric anti-BMP-2 monoclonal antibody mediates repair of critical-size mandibular continuity defects in a nonhuman primate model. Biomed Res Int. 2017;16:2017.
88. Trejo-Iriarte CG, Serrano-Bello J, Gutiérrez-Escalona R, Mercado-Marques C, García-Honduvilla N, Buján-Varela J, Medina LA. Evaluation of bone regeneration in a critical size cortical bone defect in rat mandible using microCT and histological analysis. Arch Oral Biol. 2019;101:165–71.
89. Zhang D, Chu F, Yang Y, Xia L, Zeng D, Uludağ H, Zhang X, Qian Y, Jiang X. Orthodontic tooth movement in alveolar cleft repaired with a tissue engineering bone: an experimental study in dogs. Tissue Eng Part A. 2011;17(9–10):1313–25.
90. Ahn G, Lee JS, Yun WS, Shim JH, Lee UL. Cleft alveolus reconstruction using a three-dimensional printed bioresorbable scaffold with human bone marrow cells. J Craniofac Surg. 2018;29(7):1880–3.
91. Sasayama S, Hara T, Tanaka T, Honda Y, Baba S. Osteogenesis of multipotent progenitor cells using the epigallocatechin gallate-modified gelatin sponge scaffold in the rat congenital cleft-jaw model. Int J Mol Sci. 2018;19(12):3803.
92. Kaitsas R, Paolone MG, Paolone G. Guided orthodontic regeneration: a tool to enhance conventional regenerative techniques in implant surgery. Int Orthod. 2015;13(4):539–54.
93. Paolone MG, Kaitsas R. Orthodontic-periodontal interactions: orthodontic extrusion in interdisciplinary regenerative treatments. Int Orthod. 2018;16(2):217–45.
94. Meikle MC. The tissue, cellular, and molecular regulation of orthodontic tooth movement: 100 years after Carl Sandstedt. Eur J Orthod. 2006;28(3):221–40.
95. Liu Z, Yin X, Ye Q, He W, Ge M, Zhou X, Hu J, Zou S. Periodontal regeneration with stem cells-seeded collagen-hydroxyapatite scaffold. J Biomater Appl. 2016;31(1):121–31.
96. Jepsen S, Caton JG, Albandar JM, Bissada NF, Bouchard P, Cortellini P, Demirel K, de Sanctis M, Ercoli C, Fan J, Geurs NC. Periodontal manifestations of systemic diseases and developmental and acquired conditions: consensus report of workgroup 3 of the 2017 world workshop on the classification of periodontal and Peri-implant diseases and conditions. J Clin Periodontol. 2018;45:S219–29.
97. Engelking G, Zachrisson BU. Effects of incisor repositioning on monkey periodontium after expansion through the cortical plate. Am J Orthod. 1982;82(1):23–32.
98. Pazera P, Fudalej P, Katsaros C. Severe complication of a bonded mandibular lingual retainer. Am J Orthod Dentofac Orthop. 2012;142(3):406–9.
99. Machado AW, MacGinnis M, Damis L, Moon W. Spontaneous improvement of gingival recession after correction of tooth positioning. Am J Orthod Dentofac Orthop. 2014;145(6):828–35.
100. Farret MM, Farret MM, da Luz VG, Assaf JH, de Lima EM. Orthodontic treatment of a mandibular incisor fenestration resulting from a broken retainer. Am J Orthod Dentofac Orthop. 2015;148(2):332–7.
101. Laursen MG, Rylev M, Melsen B. Treatment of complications after unintentional tooth displacement by active bonded retainers. J Clin Orthod. 2016;50(5):290–7.

102. Laursen MG, Rylev M, Melsen B. The role of orthodontics in the repair of gingival recessions. Am J Orthod Dentofac Orthop. 2020;157(1):29–34.
103. Holbrook T, Ochsenbein C. Complete coverage of the denuded root surface with a one-stage gingival graft. Int J Periodontics Restorative Dent. 1983;3(3):8.
104. Miller PD Jr. A classification of marginal tissue recession. Int J Periodont Rest Dent. 1985;5:9.
105. Prato GP, Tinti C, Vincenzi G, Magnani C, Cortellini P, Clauser C. Guided tissue regeneration versus mucogingival surgery in the treatment of human buccal gingival recession. J Periodontol. 1992;63(11):919–28.
106. Trombelli L, Schincaglia GP, Scapoli C, Calura G. Healing response of human buccal gingival recessions treated with expanded polytetrafluoroethylene membranes. A retrospective report. J Periodontol. 1995;66(1):14–22.
107. Cortellini P, Bissada NF. Mucogingival conditions in the natural dentition: narrative review, case definitions, and diagnostic considerations. J Periodontol. 2018;89:S204–13.
108. Saadoun AP, LeGall M, Touati B. Selection and ideal tridimensional implant position for soft tissue aesthetics. Pract Periodontics Aesthet Dent. 1999;11(9):1063.
109. Small PN, Tarnow DP. Gingival recession around implants: a 1-year longitudinal prospective study. Int J Oral Maxillofac Implants. 2000;15(4):527–32.
110. Kan JY, Kois JC. Predictable peri-implant gingival aesthetics: surgical and prosthodontic rationales. Pract Proced Aesthet Dent. 2001;13(9):691–8.
111. Block MS, Mercante DE, Lirette D, Mohamed W, Ryser M, Castellon P. Prospective evaluation of immediate and delayed provisional single tooth restorations. J Oral Maxillofac Surg. 2009;67(11):89–107.
112. Conserva E, Fadda M, Ferrari V, Consolo U. Predictability of a new orthodontic extrusion technique for implant site development: a retrospective consecutive case-series study. Sci World J. 2020.
113. Brown IS. The effect of orthodontic therapy on certain types of periodontal defects I—clinical findings. J Periodontol. 1973;44(12):742–56.
114. Amato F, Mirabella D, Macca U, Tarnow DP. Implant site development by orthodontic forced extraction: a preliminary study. Int J Oral Maxillofac Implants. 2012;27(2).
115. Heithersay GS. Combined endodontic-orthodontic treatment of transverse root fractures in the region of the alveolar crest. Oral Surg Oral Med Oral Pathol. 1973;36(3):404–15.
116. Ingber JS. Forced eruption: Part I. A method of treating isolated one and two wall infrabony osseous defects-rationale and case report. J Periodontol. 1974;45(4):199–206.
117. Ingber JS. Forced eruption: part II. A method of treating nonrestorable teeth – periodontal and restorative considerations. J Periodontol. 1976;47(4):203–16.
118. Potashnick SR, Rosenberg ES. Forced eruption: principles in periodontics and restorative dentistry. J Prosthet Dent. 1982;48(2):141–8.
119. Chambrone L, Chambrone LA. Forced orthodontic eruption of fractured teeth before implant placement: case report. J Can Dent Assoc. 2005;71(4):257–61.
120. Brindis MA, Block MS. Orthodontic tooth extrusion to enhance soft tissue implant esthetics. J Oral Maxillofac Surg. 2009;67(11):49–59.
121. Sun H, Wang Y, Sun C, Ye Q, Dai W, Wang X, Xu Q, Pan S, Hu R. Root morphology and development of labial inversely impacted maxillary central incisors in the mixed dentition: a retrospective cone-beam computed tomography study. Am J Orthod Dentofac Orthop. 2014;146(6):709–16.
122. da Silva VC, de Molon RS, Martins RP, Ribeiro FS, Pontes AE, Zandim-Barcelos DL, Leite FR, Neto CB, Marcantonio RA, Cirelli JA. Effects of orthodontic tooth extrusion produced by different techniques, on the periodontal tissues: a histological study in dogs. Arch Oral Biol. 2020;20:104768.
123. Schwartz O, Frederiksen K, Klausen B. Allotransplantation of human teeth. A restrospective study of 73 transplantations over a period of 28 years. Int J Oral Maxillofac Surg. 1987;16(3):285–301.
124. Schröder A, Käppler P, Nazet U, Jantsch J, Proff P, Cieplik F, Deschner J, Kirschneck C. Effects of compressive and tensile strain on macrophages during simulated orthodontic tooth movement. Mediat Inflamm. 2020;28:2020.
125. Yang CY, Jeon HH, Alshabab A, Lee YJ, Chung CH, Graves DT. RANKL deletion in periodontal ligament and bone lining cells blocks orthodontic tooth movement. Int J Oral Sci. 2018;10(1):1–9.
126. Park WY, Kim MS, Kim MS, Oh MH, Lee SY, Kim SH, Cho JH. Effects of pre-applied orthodontic force on the regeneration of periodontal tissues in tooth replantation. Korean J Orthod. 2019;49(5):299–309.
127. Suzaki Y, Matsumoto Y, Kanno Z, Soma K. Preapplication of orthodontic forces to the donor teeth affects periodontal healing of transplanted teeth. Angle Orthod. 2008;78(3):495–501.
128. Yang S, Jung BY, Pang NS. Outcomes of autotransplanted teeth and prognostic factors: a 10-year retrospective study. Clin Oral Investig. 2019;23(1):87–98.
129. Carvalho CV, Saraiva L, Bauer FP, Kimura RY, Souto ML, Bernardo CC, Pannuti CM, Romito GA, Pustiglioni FE. Orthodontic treatment in patients with aggressive periodontitis. Am J Orthod Dentofac Orthop. 2018;153(4):550–7.
130. Re S, Cardaropoli D, Abundo R, Corrente G. Reduction of gingival recession following orthodontic intrusion in periodontally compromised patients. Orthod Craniofac Res. 2004;7(1):35–9.
131. Zasciurinskiene E, Lindsten R, Slotte C, Bjerklin K. Orthodontic treatment in periodontitis-susceptible subjects: a systematic literature review. J Clin Exp Dent. 2016;2(2):162–73.

132. Cosendey VL, Frossard W, Feu D. Treatment of chronic adult periodontitis in a patient with negative overjet and multiple tooth loss. J Clin Orthod. 2016;50(4):239–49.
133. Murphy MK, MacBarb RF, Wong ME, Athanasiou KA. Temporomandibular joint disorders: a review of etiology, clinical management, and tissue engineering strategies. Int J Oral Max Impl. 2013;28:e393.
134. List T, Jensen RH. Temporomandibular disorders: old ideas and new concepts. Cephalalgia. 2017;37:692–704.
135. Willard VP, Zhang L, Athanasiou KA. Tissue engineering of the temporomandibular joint. In: Ducheyne P, Healy K, Hutmacher DW, et al., editors. Comprehensive biomaterials, vol. 5. London: Elsevier; 2011. p. 221–35.
136. Aryaei A, Vapniarsky N, Hu JC, Athanasiou KA. Recent tissue engineering advances for the treatment of temporomandibular joint disorders. Curr Osteoporos Rep. 2016;14:269–79.
137. Salash JR, Hossameldin RH, Almarza AJ, Chou JC, McCain JP, Mercuri LG, Wolford LM, Detamore MS. Potential indications for tissue engineering in temporomandibular joint surgery. J Oral Maxillofac Surg. 2016;74:705–11.
138. Acri TM, Shin K, Seol D, Laird NZ, Song I, Geary SM, Chakka JL, Martin JA, Salem AK. Tissue engineering for the temporomandibular joint. Rev Adv Healthc Mater. 2019;8:e1801236.
139. Donahue RP, Hu JC, Athanasiou KA. Remaining hurdles for tissue-engineering the temporomandibular joint disc. Rev Trends Mol Med. 2019;25:241–56.
140. Brady MA, Sivananthan S, Mudera V, Liu Q, Wiltfang J, Warnke PH. The primordium of a biological joint replacement: coupling of two stem cell pathways in biphasic ultrarapid compressed gel niches. J Cranio Maxill Surg. 2011;39:380–6.
141. Barry F, Murphy M. Mesenchymal stem cells in joint disease and repair. Nat Rev Rheumatol. 2013;9:584.
142. Chin AR, Gao J, Wang Y, Taboas JM, Almarza AJ. Regenerative potential of various soft polymeric scaffolds in the temporomandibular joint condyle. J Oral Maxillofac Surg. 2018;76:2019–26.
143. Helgeland E, Shanbhag S, Pedersen TO, Mustafa K, Rosen A. Scaffold-based temporomandibular joint tissue regeneration in experimental animal models: a systematic review. Tissue Eng Part B Rev. 2018;24:300–16.
144. Cheng B, Tu T, Shi X, Liu Y, Zhao Y, Zhao Y, Li Y, Chen H, Chen Y, Zhang M. A novel construct with biomechanical flexibility for articular cartilage regeneration. Stem Cell Res Ther. 2019;10:298.
145. Kinoshita Y, Maeda H. Recent developments of functional scaffolds for craniomaxillofacial bone tissue engineering applications. Sci World J. 2013;2013:863157.
146. Xia T, Yu F, Zhang K, Wu Z, Shi D, Teng H, Shen J, Yang X, Jiang Q. The effectiveness of allogeneic mesenchymal stem cells therapy for knee osteoarthritis in pigs. Ann Transl Med. 2018;6:404.
147. Iijima H, Isho T, Kuroki H, Takahashi M, Aoyama T. Effectiveness of mesenchymal stem cells for treating patients with knee osteoarthritis: a meta-analysis toward the establishment of effective regenerative rehabilitation. NPJ Regen Med. 2018;3:15.
148. Lee YH, Park HK, Auh QS, Nah H, Lee JS, Moon HJ, Heo DN, Kim IS, Kwon IK. Emerging potential of exosomes in regenerative medicine for temporomandibular joint osteoarthritis. Int J Mol Sci. 2020;21:E1541.

Regenerative Approaches in Oral and Maxillofacial Surgery

Seied Omid Keyhan, Hamid Reza Fallahi, Behzad Cheshmi, and Shohreh Ghasemi

1 Introduction

Bones in the craniofacial region are stress-bearing and each has their specific characteristics, however, they have similar mechanisms of turnover and for the repair of damage as other bones in the human skeleton [1]. Considering this similarity, most of the techniques and materials are used for the reconstruction of skeletal bone may be utilized in the field of craniofacial reconstruction as well. Besides, due to the unique characteristics of the craniofacial bones, it is necessary to pay attention to some supplementary considerations.

S. O. Keyhan (✉)
Oral and Maxillofacial Surgeon, Founder and Director of Maxillofacial Surgery and Implantology Research Foundation, Isfahan, Iran

H. R. Fallahi
Dental Research Center, Research Institute of Dental Sciences, Shahid Beheshti University of Medical Science, Tehran, Iran

Oral and Maxillofacial Surgeon, Founder and Director of Maxillofacial Surgery and Implantology Research Foundation, Ahvaz, Iran

B. Cheshmi
Private Practitioner, Professional Member of Maxillofacial Surgery and Implantology Research Foundation, Boroujerd, Iran

S. Ghasemi
Adjunct Assistant Professor of Oral and Maxillofacial Surgery of DCG, Georgia, USA

The main difference between craniofacial bones and other skeletal bones is their embryonic origin. Branchial arch-derived craniofacial bones and cartilages originate from the cranial neural crest. On the other hand, the axial skeleton and the limb skeleton develop from the lateral plate mesoderm. The mesenchymal tissue transforms to calcified bones through osteogenesis. The two main suggested pathways for the formation of bone include intramembranous ossification and endochondral ossification. Direct formation of the bone from mesenchymal tissue is 'Intramembranous ossification' (IO). Differentiation of mesenchymal cells to form cartilaginous templates that are subsequently remodelled and become bone is endochondral ossification (EO).

Cranial bones primarily are made via IO, however, EO is comparatively effective in this process [2]. For example, Meckel's cartilage has an important effect in mandibular morphogenesis. It forms at the proximal end of the lower jaw and then disappears once the ossification of the mandible is complete [3, 4].

During the IO of the skull, the proliferation of neural crest-derived mesenchymal cells occurs. These cells will differentiate into capillaries and osteoblasts, leading to the formation of compact condensations. Osteoblasts create collagen–proteoglycan matrix, which serves as an environment for binding calcium salts and forming bone spicules. Periosteum-derived mesenchymal cells are surrounded by calcified spicules. Also, the

© Springer Nature Switzerland AG 2021

S. Hosseinpour et al. (eds.), *Regenerative Approaches in Dentistry*,
https://doi.org/10.1007/978-3-030-59809-9_9

osteoid matrix is secreted from osteoblasts that are derived from cells located on the periosteal interior surface, and which produce several layers of bone.

Various pathways of homeostasis occur in craniofacial and non-craniofacial bones. It has been demonstrated in the literature that bone grafts from the craniofacial skeleton give better results than grafts from the iliac crest, rib, and tibia. This advantage in terms of survival and volumetric maintenance is important for maximizing clinical outcomes [5, 6].

Buchman and Ozaki showed that maintenance of bone volume depends on bone graft microarchitecture [7, 8]. Some bone disorders, such as medication-related osteonecrosis of the jaw and hyper-parathyroid jaw tumor syndrome, involve the jawbones [9, 10]. Even though the differentiation of osteoblasts in both craniofacial and skeletal bones is dependent on similar regulation factors, including the transcription factors Runx2 and osterix, the reasons for differences between craniofacial bones and skeletal bones remain controversial [11–13]. Multiple growth factors, receptors, and signalling pathways have been suggested as contributing factors affecting the behaviour of the craniofacial bones [14–16].

The purpose of regenerative medicine in oral and maxillofacial surgery (OMFS) is to exploit and accelerate the regenerative potential of human cells, to treat bone defects. Different aetiologic factors such as malignancies, trauma, and congenital anomalies can lead to minor or major bone defects. Regenerative medicine had been applied in rehabilitative procedures to treat minor bone defects, mandibular/maxillary resorption, defects from minor dental trauma, and in dental implantology.

Due to clinical limitations in the field of regenerative medicine, at the present time, oral and maxillofacial surgeons still struggle with side effects of allografts when these are used for major bone grafts. As well, the lack of development of regenerative medicine in the fields of regeneration of other types of tissues, including soft tissues, muscles, and nerves, is another significant limitation of this discipline. The current chapter reviews the history of techniques and materials, the basis of regenerative OMFS, in vitro advances, in vivo evidence, limitations and barriers, as well as future trends in this field.

2 Current Treatment and Clinical Evidence for Treatment of Defects Caused by Malignancy and Trauma

Ancient historical evidence of repair of bone defects utilizing metals can be found in the Edwin Smith Papyrus, dated 1501 BC [17]. In 1668, using a xenograft, Van Meekrenon placed the first bone graft into a patient who suffered from a skull defect [18]. Over the years, a variety of techniques and materials with different combinations, such as autogenous, allogeneic, and prosthetic materials, have been used for the reconstruction of bone defects.

Comparing all modalities, autogenous bone grafts have shown the most favourable results [19–21]. Considering advances in understanding the science and basic biology of autogenous bone grafting, there remains much to investigate and learn about regarding the inherent potential of human cells. Such investigations will allow scientists and clinicians to obtain more predictable results, and to better apply reconstruction techniques in clinical practice. Following a bone graft, the formation of a haematoma and an inflammatory reaction adjacent to the graft site that lasts for about 5–7 days following bone graft placement are the likely events. As this happens, an organized condensed fibrovascular stroma disassociates the graft from the haematoma and the surrounding tissue. After 10–14 days, angiogenesis occurs with new blood vessels growing toward the graft area, which aids the induction of cells with osteogenic abilities. Infiltrating host cells differentiate into osteoblasts and osteoclasts, which deposit and resorb bone, respectively. Also, these cells facilitate vascular tissue penetration into the bone graft.

Revascularization of cortical grafts that are modified for use in onlay augmentation occurs superficially (10–21 days) and then centrally (8–16 weeks) [22, 23]. Since vascularization of the initial graft will not occur for several days, incomplete revascularization of the necrotic region will be sealed off from the viable region. Dead bone that is trapped will only be resorbed if proper revascularization occurs.

The efficiency of angiogenesis and the revascularization process is the single major critical factor in bone graft survival. Irradiation, impaired immune responses, and also the problems at the recipient site (including the presence of necrotic/infectious tissue, or scarring) may affect the revascularization process [24]. Pre-existing healthy bone can deposit a healthy bony matrix directly.

Histologically, there are two types of cancellous and cortical bone. Cancellous bone graft revascularization is faster and more extensive than for cortical bone grafts [25]. Revascularization of cortical bone grafts often proceeds gradually, and occasionally may be incomplete [26, 27]. The penetration of new blood vessels is limited to existing pores and proceeds from the exterior surfaces into the interior segments of the graft. Since blood vessel ingrowth will be constrained by the dense cortical bone lamellar structures (including Haversian canals and Volkmann's canals), osteoclastic enlargement should occur prior to penetration of blood vessels into the cortical bone graft [28]. Revascularization of cancellous bone grafts consequently stimulates the formation of new bone via osteoinduction, to address 'bone gaps'. Alternatively, since cortical bone grafts can survive with the least resorption rate and may maintain a certain level of mechanical stability for a longer period, they are indicated for use in bone volume deficiencies [29, 30].

The presence of pre-existing cells in autogenous bone grafts makes them superior to other bone substitutes [31, 32]. Osteogenic cells originating from the periosteum, endosteum, marrow, and intracortical elements give viability to the graft and induce the production of osteoid [28]. A periosteum enriches a bone graft by providing a blood supply, while osteoprogenitor cells facilitate the further formation of bone and its vascularization [33, 34]. Unfortunately, the mature periosteum fails to retain its osteogenic capacity following detachment from the bone surface [35, 36].

A large portion of the osteocytes within a bone graft undergoes necrosis immediately following the grafting procedure. An initial implantation of large bone grafts into well-vascularized muscle tissue can accelerate the inward vascularization process, and provide a vascular pedicle that can be later used for microsurgical anastomosis. This is a recommended approach to improve the survival rate of the bone graft [37, 38].

Because of the morbidity of the donor site and the limited amount of harvestable bone, bone grafting methods have significant limitations [28]. To address these problems, stem cell-based tissue engineering of bone has now been proposed as a possible promising alternative. In 1968, Friedenstein et al. found bone marrow-derived fibroblast-like cells called mechanocytes, which are chondrogenic and adipogenic. These are now called mesenchymal stem cells (MSCs). These cells can differentiate to form haematopoietic cells, and they also can undergo osteogenic differentiation [39–42]. Besides being found in the bone marrow (BM), MSC-like cells have also been detected in other tissues such as adipose, dental pulp, and skeletal muscle [43–45]. MSC-mediated bone reconstruction has been evaluated extensively in various trials. These show that the local delivery of MSCs can improve bone regeneration [46].

The optimal method for the reconstruction of a craniofacial bony defect is an autologous bone graft (Figs. 1, 2, 3 and 4) [47, 48]. The advantages of these grafts, which mainly are harvested from the ribs and iliac crest, are ease of access, and minimal impact on the host. In contrast, source limitations and morbidity of the donor site are the main disadvantages of autogenous bone grafts. Also, these autografts do not have jawbone characteristics.

The usual complication of grafts is their lack of vascularization, which leads to resorption and deformity [5]. Preservation of the periosteum can improve the survival rate of grafts for craniofacial bones [49], by facilitating early revascularization [50]. Implant site status can also influence the rate and extent of revascularization in the graft area [51]. Muscle coverage can enhance the revascularization of grafts [52].

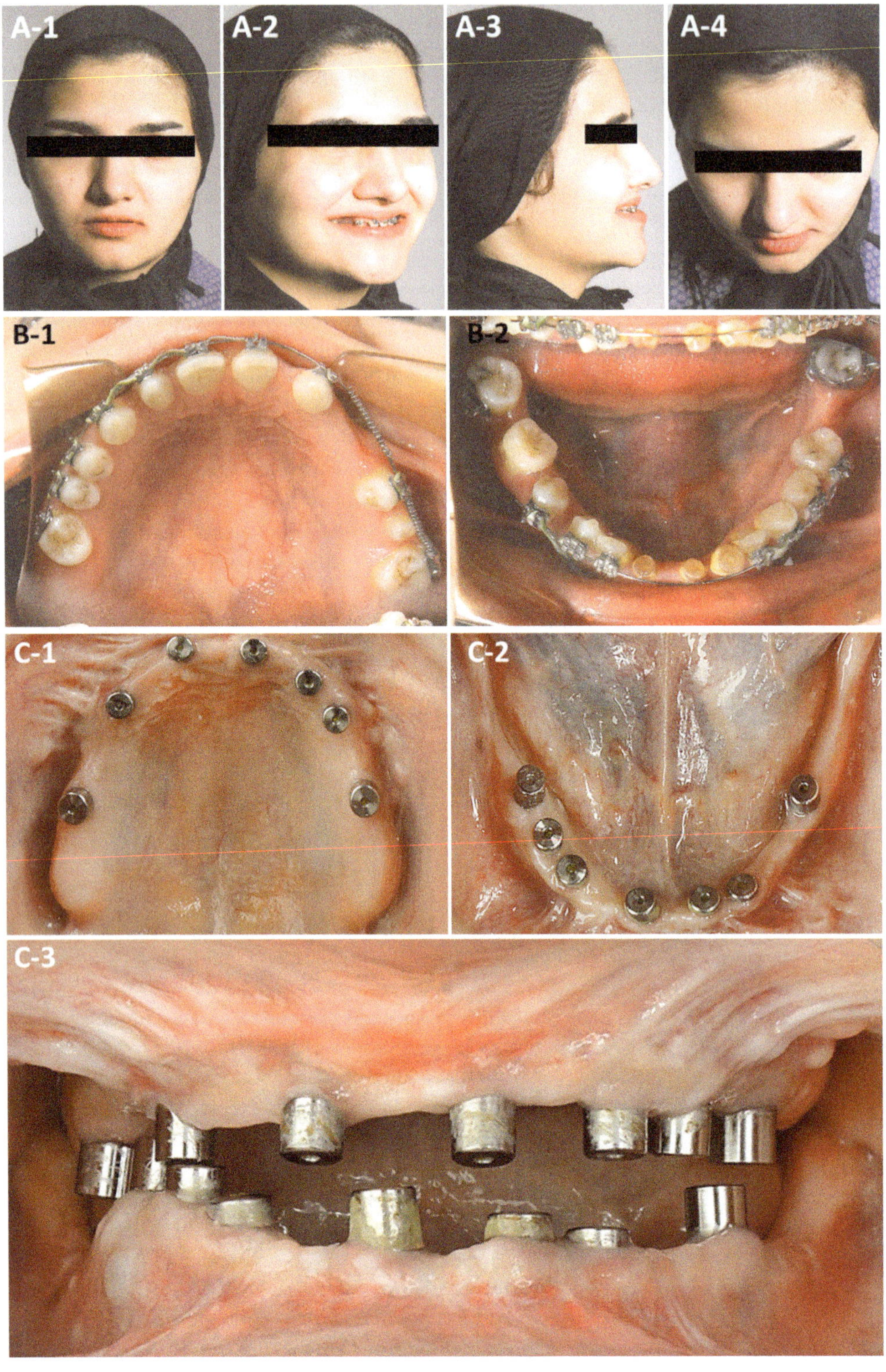

Fig. 1 (**A-1 to -4**) and (**B-1 to -2**) A 28 years old female patient who was diagnosed with anodontia; the deciduous teeth had undergone severe root resorption following orthodontic treatment and were considered hopeless and were extracted. (**C-1 to -3**) Postoperative intraoral views of the jaws after placement of implants

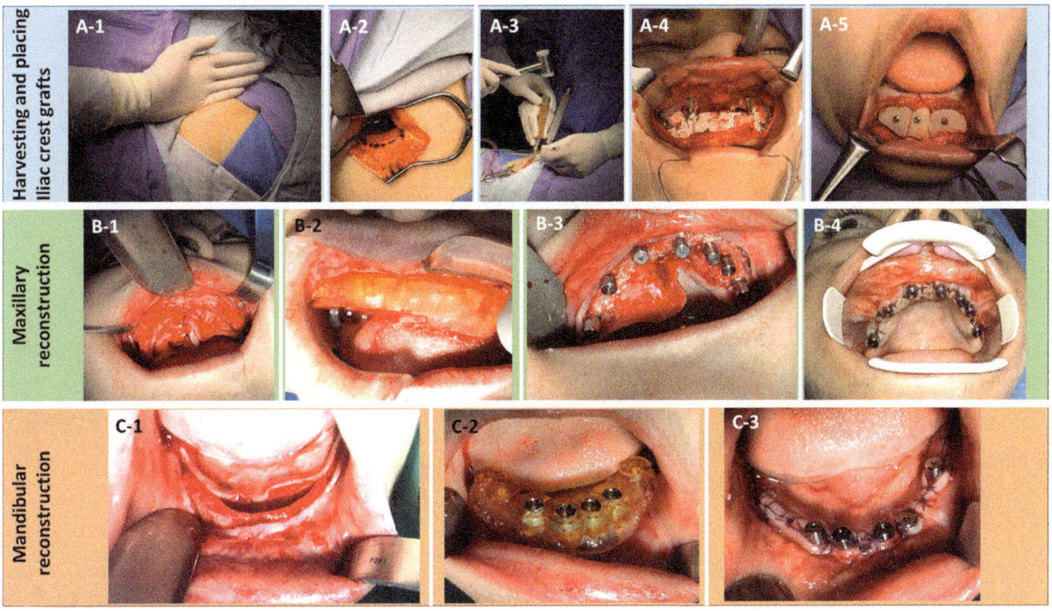

Fig. 2 An autologous iliac bone graft used in combination with a xenograft to reconstruct extensive maxillofacial bone defects in both jaws of a 28-year-old female patient diagnosed with anodontia. At first, the autogenous bone graft was harvested and combined with a xenograft, and then fixed into the defect sites with fixation screws (**A-1** to **A-5**). After 6 months, bone tissue filled the gap, and dental implants were placed to rehabilitate the dentition (**B-1** to **B-4** and **C-1** to **C-3**)

Previous studies have revealed that bony defects of reasonable size (4–6 cm) can efficiently be treated by using non-vascularized cortical-cancellous bone grafts, while for massive defects, vascularized grafts are essential [53, 54]. Another issue is the complex 3D shapes of maxillofacial bones compared to long bones in other parts of the body. Therefore, selection and reshaping of a vascularized autologous iliac crest, fibula, or rib graft for precise fit into a craniofacial bony defect is very laborious and challenging.

The initial healing of bone and accelerated revascularization can increase the retention of rigidly fixed grafts, particularly in bones with a higher motor function such as the femur [55, 56]. The mandible can be considered the only bone that undergoes significant movement, amongst the craniofacial bones. It may seem that rigid fixation does not efficiently increase the survival rate of fixed craniofacial bone grafts, however a survey of 363 patients who underwent nasal reconstruction demonstrated that by using rigid interosseous fixation, exceptional bone graft survival can be achieved [57].

The inability of bone substitutes to induce new bone in a predictable way can be considered as the main limitation of these materials. Lack of customized grafts for individuals is another important disadvantage. The extent and the severity of the bone defect may limit the use of autologous grafts, and this drives the search for external sources for the replacement of lost tissues [40]. Allogeneic or xenogeneic grafts cause numerous complications, including immediate or delayed host immunological reactions.

3 Current Treatment and Clinical Evidence for Congenital Defects

Failure in the fusion of the nasal process and palatal shelves leads to cleft lip and cleft palate (CL/CP) [58–60]. Facial deformity in CL/

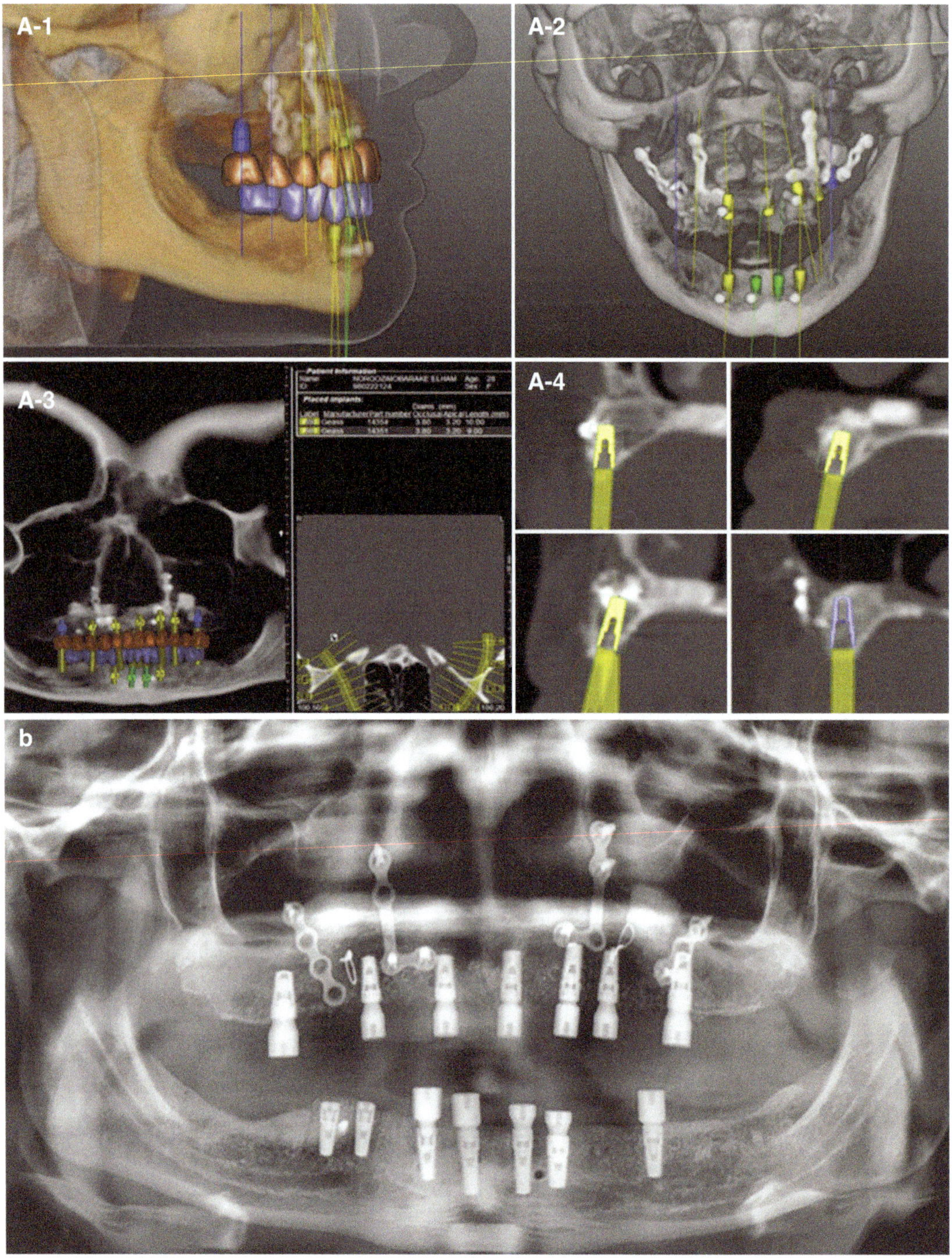

Fig. 3 (**A-1** to **A-4**) Digital design of implant treatment plan using computer software and CBCT images for the patient whose defect reconstruction is shown in Fig. 1. (**B**) The 4-month postoperative panoramic view of the patient whose defect reconstruction and digital treatment planning are shown in Fig. 1

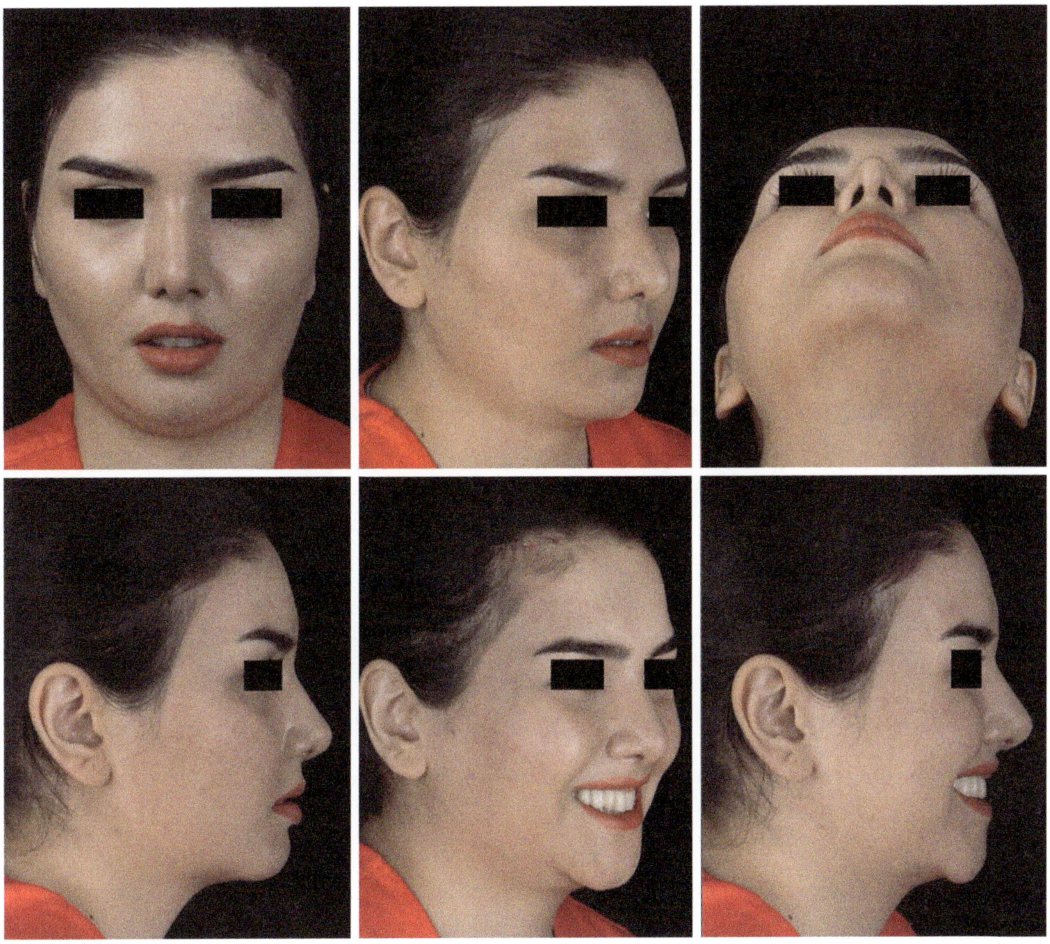

Fig. 4 Extraoral postoperative views of a 28 years old female patient who underwent bimaxillary reconstruction, rhinoplasty, and brow lift

CP patients usually includes dentoalveolar bone defects, which may be associated with missing teeth and/or a maxillary deficiency [61]. Orofacial clefts cause a severe loss of multiple tissues. Preferably, they should be reconstructed with the aid of autologous grafts. Recommended autogenous graft sources for the reconstruction of alveolar cleft deformities include the iliac crest, tibia, rib, and the cranium. The literature reports high success rates of autologous grafts of up to 88% [60–65].

The use of iliac crest bone, the most frequent graft source, results in 43.1% bone resorption during the first 12 months postoperatively [61]. Complications from these grafts and disadvantages include pain, the morbidity of the donor site, unsatisfactory bone reconstruction, additional expenses, and long operation times. Such disadvantages have led researchers and clinicians to look towards the application of regenerative options and stem cell-based tissue engineering [60, 62, 66, 67].

The treatment of cleft lip and palate is complex. Due to a child's physiological situation, the treatment of deformities has to be considered with caution, achieving a delicate balance between surgical intervention and allowing for normal growth. Considering the similarity of the

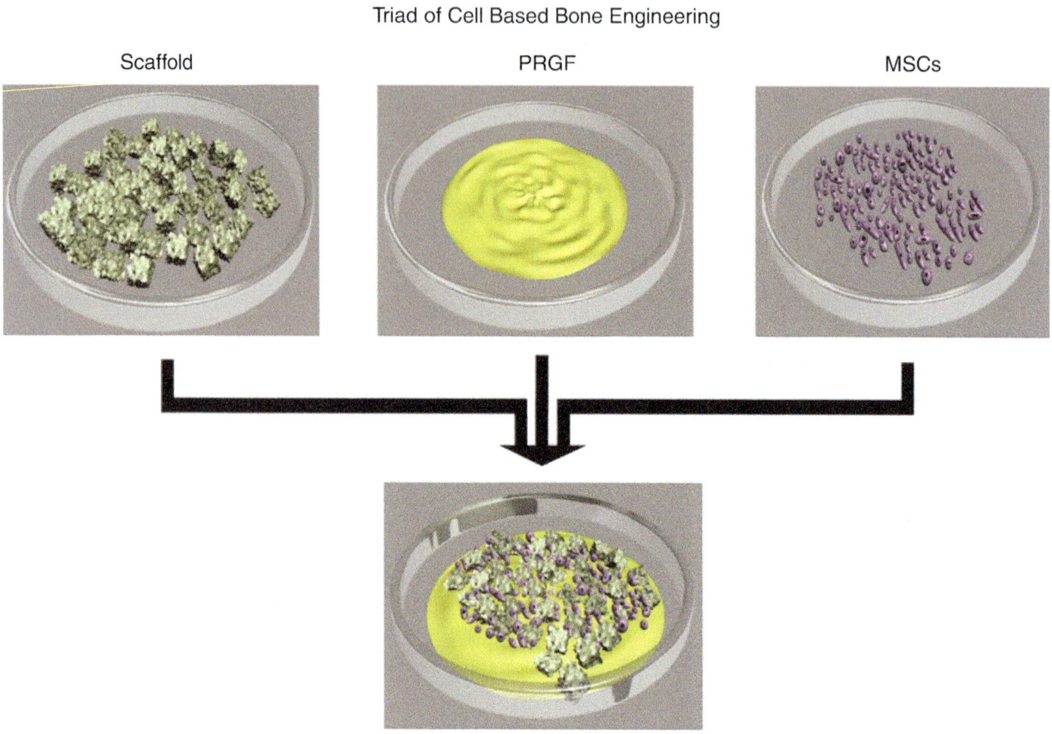

Fig. 5 The components of triad suggested by Behnia et al. [69] for the reconstruction of a bone defect in a patient with a cleft

grafts used by clinicians for treatment of defects caused by trauma or malignancies, methods to treat congenital also vary in terms of soft tissue repair. Different surgical techniques have been introduced by surgeons over time, to achieve favourable aesthetic results and to optimize the volume of soft tissues.

A major complication with such surgical approaches is wound dehiscence. This may happen as a result of poor tissue quality and excessive wound tension, resulting in an oro-nasal fistula. A palatal fistula occurs in around 4.9% of cases [68]. Major predictive factors for a palatal fistula are the type of cleft and surgical repair method, as well as wound tension, dead space, and rapid maxillary arch expansion. Local flaps, tongue flaps, and (rarely) microvascular free flaps are the treatment plans of choice if the palatal fistula is very large.

The management of a patient with cleft palate is complex. No universal agreement has been reached on an integrated and comprehensive strategy. The suggestion of foetal microsurgery for cleft repair has been rejected due to its high risk. Although a range of tissue engineering approaches exists, all have issues with expense, accessibility, and practicality (Figs. 5 and 6).

3.1 Principles and Advantages of Stem Cell-Based Therapies

Cells, scaffolds, and growth factors are the mainstays of tissue engineering. The combination of cells located in an appropriate scaffold can be used to deliver the proper biochemical signalling to induce bone growth. The appropriate design of the scaffold and its related mechanical signals is essential. The scaffold can be temporary or permanent, and natural or artificial. Nevertheless, the definitive characteristic of all scaffolds is that they must be biocompatible [70–72].

Scaffold environments improve the migration and differentiation of progenitor cells [73].

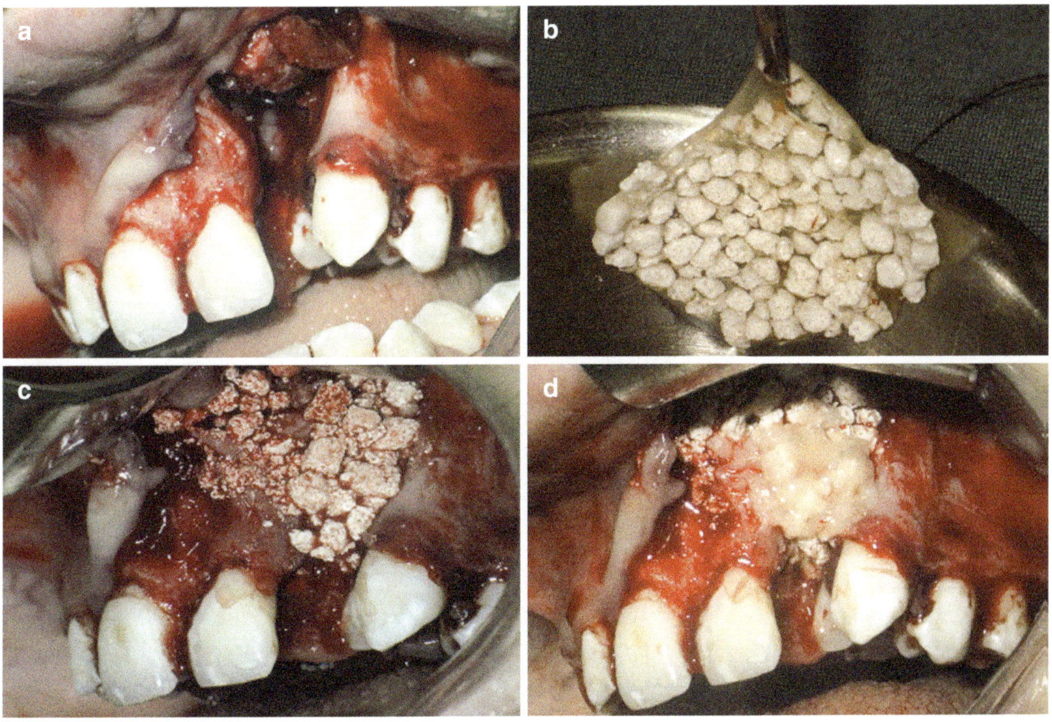

Fig. 6 (**a**) Cleft exposure and nasal floor suturing. (**b**) A biphasic hydroxyapatite/tricalcium phosphate scaffold loaded with hMSCs combined with patient-derived PDGF. (**c**) Overfilling of the defect with the triad. (**d**) Placement of a platelet fibrin membrane over the graft. Adapted from Behnia et al. J Craniomaxillofac Surg (2012)

Physical features of scaffolds, such as their biodegradability, porosity, stiffness, and resistance influence directly cell proliferation (osteoconduction), cell adhesion, and cell migration, through the action of chemical signals. The main challenge for the maxillofacial surgeon is the precise three-dimensional design of the scaffold to make it match the craniofacial defect. There have been promising results from research using computers in the design and manufacture of biomimetic scaffolds [74].

In addition to cells, scaffolds, and growth factors, which are considered the three mainstays of tissue engineering, a second group of factors is important, such as the availability of cells, whether than can be differentiated, their immunostimulatory properties, and tumourigenicity [75]. One of the most challenging issues is the selection of the best cell line for use in tissue engineering. The utilization of stem cells, and gene therapy using viral vectors for expression of growth factors in cultured cell lines, are among the newest horizons in tissue engineering [76–78]. Such approaches could enable clinicians to create living tissues, and this will expand treatment options in the future.

A major obstacle for the efficient management of tissue function in vitro is the lack of a comprehensive understanding of cell behaviour under specific physical and chemical conditions [79, 80]. Bioreactors are among the devices that can be utilized to provide a quasi-natural environment, by imitating physiological conditions. Advances in tissue engineering have now led to the wider use of three-dimensional scaffolds that can host multiple layers of cells. The spatial structure of these scaffolds needs to be designed based on the fluid flow rate, so that the scaffold maintains the ability to provide oxygen and nutrients and to remove waste. The scaffold design must also consider external mechanical forces, which can be used to induce osteoblastic proliferation.

An alternative option is in vivo graft cultivation using animal models (or humans) as a bioreactor to stimulate cellular growth [80].

3.2 Hard Tissue Repair

Stem cells have the potential to differentiate into many cell types, and because of this, they can be used in the regeneration of various tissues and organs. This aspect is important because of the complexity and variety of tissues that are needed to repair and reconstruct defects in the craniofacial region (Fig. 7).

MSCs are multipotent progenitor cells that can be detected in mature BM, adipose tissue, skin, umbilical cord, and placenta. BM-derived MSCs can be used for the treatment of bone deficiencies. Aspiration of bone marrow (BM) is a painful and invasive process for the donor, and is a technique-sensitive procedure for the operator. MSCs comprise multiple cell types, and their proliferation and differentiation can be affected by the patient's gender, age, and general health [81].

Cells with stem cell characteristics have been detected in various regions within the dental tissues, such as in the dental pulp complex, the periodontal ligament, and the dental follicle. Dental pulp stem cells (DPSCs) [82] and stem cells from human exfoliated deciduous teeth (SHED) [83] have MSC-like characteristics and have the capacity to generate osseous tissue. Since SHED have a higher rate of proliferation and are easier to collect than BM-derived MSCs, SHED may be a more suitable source of autologous stem cells for regeneration procedures in dentistry. Due to the heterogeneity of MSCs, the successful regeneration of bone requires a comprehensive understanding of now osteoblastic differentiation is regulated by growth factors and cytokines.

The main features of osteoblasts include the production of bone-related extracellular matrix proteins, and the intense enzymatic activity of alkaline phosphatase (ALP) in response to osteotropic hormones and cytokines [84, 85]. Various studies have revealed that osteoblasts, chondrocytes, adipocytes, myoblasts, tendon cells, and fibroblasts are differentiated from common precursors of BM-derived MSCs and that various transcription factors determine the cell lineage (Table 1). Runx2 steers the multipotent mesenchymal cell population toward the osteoblastic lineage, while β-catenin, Osterix, and Runx2

Table 1 Regulation of lineage differentiation by different families of transcription factors

Transcription factors	Regulated lineage
Runx2, Osterix, or β-catenin	Osteoblasts
Sox family (Sox9, Sox5, and Sox6)	Chondrocytes
MyoDs (MyoD, Myf5, and Myogenin)	Myogenics
C/EBP family (C/EBPβ, C/EBPδ, and C/EBPα) and PPARγ	Adipocytes

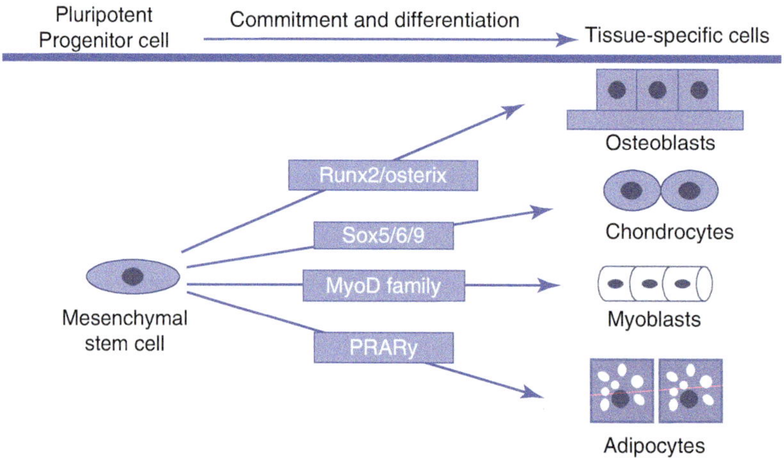

Fig. 7 Transcriptional factors including Runx2/Osterix, Sox5/6/9, MyoD family, and PPARγ can differentiate MSCs into osteoblasts, chondrocytes, myoblasts, and adipocytes, respectively

steer them to become adult osteoblasts following differentiation to preosteoblasts [84–86].

A number of hormones and cytokines, such as bone morphogenetic proteins (BMP), transforming growth factor-beta (TGF-β), Wnt signalling pathways, hedgehog, basic fibroblast growth factor (bFGF), and oestrogen, contributed to the regulation of mesenchymal cell differentiation through stimulating various intracellular signalling pathways. BMP-2 is an important cytokine that induces the formation of ectopic bone, and it effectively stimulates the differentiation of mesenchymal cells into osteoblasts.

BMPs and the mediators of the TGF-β family induce the formation of bone and cartilage, providing pivotal signals for morphogenesis, to orchestrate the development of tissue architecture [87]. These factors are also involved in skeletal morphogenesis [84, 86]. Through both the Smad pathway and non-Smad pathways, receptors on the cell membrane transduce BMP signals to the nucleus. Several extra- and intracellular molecules interact with BMPs and their signalling pathway components, to regulate signals from BMPs. The bone-osteoinductive ability of BMPs can be beneficial for regenerative treatments, however, to date BMPs have failed to provide an adequate or acceptable clinical response. The main reason may be the inhibition of BMP-mediated bone formation and osteoblast differentiation by inflammatory cytokines. For instance, TNF-α inhibits osteoblast differentiation. This has been shown in various models such as foetal calvaria, BM stromal cells, and osteoblastic cells [88–90].

Inflammatory cytokines such as TNF-α and IL-1 drive local and systemic bone loss in autoinflammatory diseases of bones (such as rheumatoid arthritis) [91]. Overproduction of TNF-α and other cytokines by inflammatory cells that infiltrate into the synovial membrane of inflamed joints leads to rheumatoid arthritis. Anti-TNF medications, such as infliximab, can reduce the signs and symptoms of the disease, and reduce joint destruction [92].

In line with clinical and animal investigations, it has been shown that TNF-α and IL-1β inhibit bone formation in a neonatal rat calvarial organ culture system [84]. TNFα and IL-1β can inhibit spontaneous and BMP-induced osteoblast differentiation, as shown by measurements of changes in the BMP2-induced expression of osteocalcin and Runx and a dose-dependent alteration in ALP activity. Various signal pathways such as mitogen-activated protein kinase, extracellular signal-regulated kinase, c-Jun N-terminal kinase, p38 kinase, and NF-κB mediate these responses. NF-κB is involved in the development of inflammation, hence inhibition of NF-κB can suppress inflammatory bone loss, through inhibition of osteoclastogenesis. This makes NF-κB a crucial target in autoinflammatory bone diseases [93].

A dominant-negative form of IKKβ inhibits NF-κB in adult rat osteoblasts, while dominant-negative IKKβ expression may lead to an increase in BMD and bone volume because of increased osteoblastic activity [94]. Overexpression of the dominant-negative form of IκBα (IκBαDN) can inhibit NF-κB, and lead to osteoblast differentiation [95]. The cell-permeable NF-κB activation antagonist TAT-NBD inhibits NF-κB activity from TNFα, and can prevent TNFα suppressing TGFβ-stimulated Smad luciferase, expression of BMP2-induced Runx2 mRNA, and osteoblast differentiation in MC3T3-E1 cells [96]. Moreover, inhibition of NF-κB increases bone formation and ameliorates osteopenia in ovariectomized rats [97]. TNFα inhibits differentiation of BMP-induced osteoblast stimulation via activation of NF-κB through Smad DNA binding inhibition [98] (Fig. 8).

NF-κB inhibition promotes BMP-induced bone regeneration [84, 93, 99]. BAY11-7082 is a selective inhibitor of NF-κB, and can enhance BMP2-induced ectopic formation of bone in rats. Considering treatments by BAY11-7082 versus BMP2, the ectopic bone induced by BMP2 has greater radiopacity. MicroCT images of BMP2-induced ectopic bone with BAY11-7082 has a thickened outer layer of bone that is filled with trabecula. In the presence of BAY11-7082, the bone mineral density of ectopic bone is enhanced.

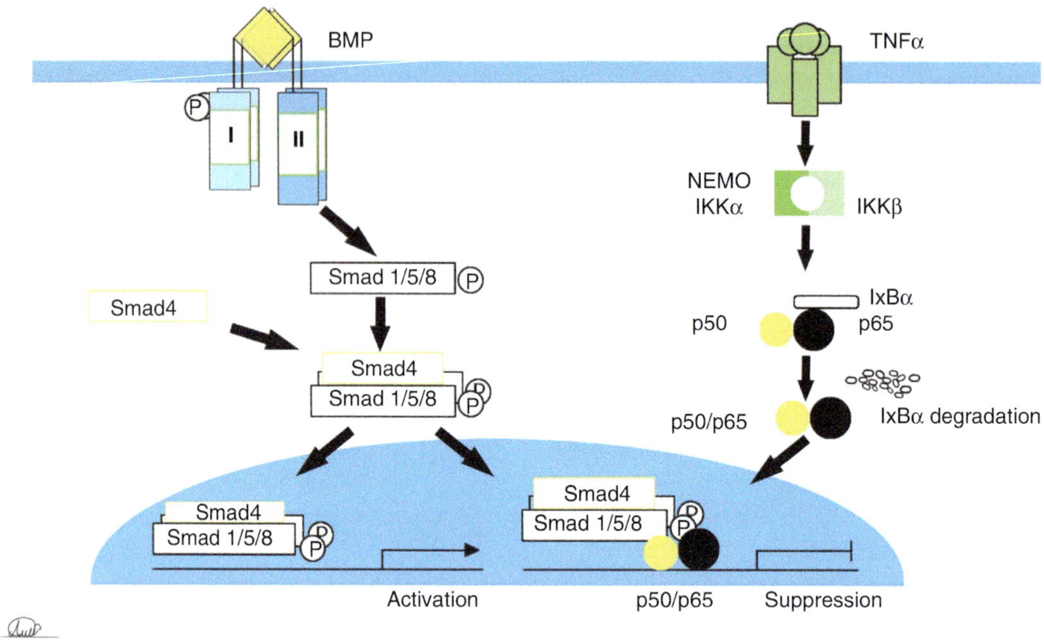

Fig. 8 NF-κB-mediated binding activity inhibition of BMP/Smad-mediated DNA

3.3 Soft Tissue Repair

The first major advances in skin engineering were made in 1975. Over the next 10 years, the Washington National Science Foundation used the term engineering to describe the applications of this specialty. In 1993, Langer and Vacanti introduced the term tissue engineering [100, 101]. The first major experimental activities in the field of tissue engineering were performed by two American teams. Rheinwald and colleagues were the first to focus on tissue engineering [102]. Using epithelial cells in a small skin sample, they created using cell culture an elementary skin graft [103]. Subsequent advances have been known as 'skin epidermis tissue engineering'.

Yannas and colleagues have explored the properties of scleroproteins and the mechanism of degradation [104]. Their research in 1975 led to the concept of 'skin dermis tissue engineering', which underpins the latest generation of artificial dermal substitutes [105]. O'Connor was among the first to use cultured autologous epithelial sheets to graft skin burns [106, 107]. Cultured epidermal autografts (CEA) are autologous cultured sheets that can effectively provide full-thickness coverage for burns in paediatric patients [108].

Burke et al. showed that artificial dermis can provide a functionally and physiologically justifiable dermis for the treatment of large full-thickness burns [109]. Heimbach et al. introduced a novel artificial dermis called Integra™ Dermal Regeneration Template, as an artificial dermis for the treatment of large burns [110]. This dermal substitute has been considered a 'gold standard' in the management of full-thickness burns [111–113].

Deficiencies of soft tissue in the craniomaxillofacial region can be caused by trauma, malignancies, or congenital anomalies. When treating these, the complex neuromuscular structure of the facial soft tissues, especially in specific areas such as the lips, eyelids, and nose, makes the reconstruction remarkably difficult for the surgeon. Preservation and restoration of aesthetics must be paramount.

The first study using tissue-engineered human ex vivo Produced Oral Mucosal Equivalent (EVPOME) was conducted by Izumi et al. [114].

EVPOME can be used in the oral cavity and has the ability for extending and reinforcing keratinized gingiva. More research is needed to examine other aspects of tissue engineering using EVPOME [114].

Free gingival grafting (FGG) uses oral connective tissues to increase the height of the attached gingiva [115]. The minimum strip width to be cut from the donor site, which must be keratinized gingiva, is 2 mm [116, 117]. Investigations of the application of autologous and allogeneic materials reveal several obstacles for FGG, such as donor site morbidity and the limitations of available tissue [118].

Nevins and McGuire [119] have stated that sufficient keratinized tissue can be attained by using bi-layered cell therapy. However, the amount of keratinized tissue that can be obtained is not much greater than what can be obtained through FGG. However, compared to FGG, bi-layered cell therapy gives more acceptable outcomes in terms of tissue texture and colour. 'Biologic dressing' refers to the concept of encompassing cells by using such approaches on wounds. A biodegradable scaffold enriched with cultured gingival autologous fibroblasts could be considered as a suitable alternative [120].

Since oral mucosal keratinocytes can be engineered more easily in an in vitro environment than epidermal keratinocytes from the skin, investing in the former for future regenerative medicine should be beneficial [121]. Regeneration of facial skin, the eyelids, the nose, and in situ tissue substitution of the urethra and conjunctiva are some of the extraoral applications that can be imagined for this platform [122, 123].

Worldwide demographic data show an increase in the proportion of the elderly in many communities, with an increase in the prevalence of systemic conditions such as diabetes and hypertension as a consequence. A growing number of untreated chronic wounds, in addition to other conditions, causes a significant economic burden, which is estimated to be 4% of the total global health budget [124, 125].

Wound healing is a complex multi-dimensional process that requires careful coordination of the various internal and external factors that are at work, and which can be interrupted by infections and mechanical irritation. The presence of any disruptive agent can weaken the local capacity for regeneration, leading to a failure of healing and the formation of a chronic wound. One of the most well-known factors that disrupt wound healing vitamin C deficiency, which can lead to non-healing wounds [124, 126]. Vitamin C is the main collagen crosslinking co-factor and reduces oxidative stresses [124, 127].

While minor changes in the balance of metabolism can disrupt the process of tissue healing, MSCs can be used for the treatment of complex wounds, such as those caused by radiotherapy, ischemia, or diabetic pathology. Potential obstacles such as age or systemic conditions can limit the clinical application of MSCs for improving wound healing [128, 129]. Fortunately, the use of stem cell-based therapies is not limited to the use of BM-derived or adipose-derived MSCs, but can also draw on peripheral blood cells (PBCs), as these also can secrete angiogenetic factors including VEGF and HIF-1 [130]. These various options for accelerating wound healing can be used in an implantable device or may be injected [130, 131].

As well, valuable advances have been made in the field of autologous cell therapy to treat salivary gland (SG) dysfunction. However, the use of autologous stem cell therapy for the treatment of SG diseases in the elderly is likely to be less successful because elderly patients have fewer stem cells. Although the exact number of cells that is needed for functional regeneration of salivary glands is unknown, recent studies on increasing the number of KIT + cells using growth factors have reported promising results. As well, some work supports the use of non-SG cells [132].

One obstacle in the development of SG cell therapy is the short lifetime of cultured cells from biopsies. Cryopreservation is one of the proposed strategies to preserve biopsied progenitor cells. Neumann et al. have introduced a cryopreservation-based banking method for SG CD49f+ CD29+ cells to store them for around 3 years without the cells suffering from genetic or functional deficits [133]. Hence, cryopreservation can be an option in the field of stem cell therapy.

Since adequate autologous SG cells may not be available, non-SG cells may also be used in the treatment of SG disease. There are several studies that support the application of non-SG cells for the treatment of irradiated salivary glands (Table 2).

Recently, a variety of studies using in mouse and rat models have been conducted into the cultivation of properly sized engineered tissues that can connect properly to remaining cells in the transplant area [148, 149]. HA gels containing primary human cells appear capable of responding to neurotransmitters following integration into resected parotid glands in immunocompromised rats [150]. Another study suggested that epithelial and mesenchymal foetal SG cells grown on a three-dimensional collagen environment can be transplanted into resected salivary glands. A suture thread can be used to identify the initial route of reconnection of the duct to the oral cavity. Future studies should investigate the administration of differentiated cells, to assess the possibility of reconnection with the remaining duct. Table 3 summarizes recent studies conducted in the field of human cell-based models, that show potential for translation.

Table 2 Use of non-salivary gland cells for potential treatment of salivary gland disease

Cell type	References
BM-derived cells	[134–136]
BM-derived MSCs	[137]
Human adipose-derived MSCs	[138–141]
SG-derived MSC-like cells	[142, 143]
Amniotic cells	[144, 145]
Embryonic stem cells	[146]
Induced-pluripotent stem cells	[147]

Table 3 Human cell-based therapy models utilized to develop three-dimensional salivary gland organoids

Study	Characteristics	Results	Limitations
Joraku et al. (2007) [151]	hSG* primary cells in three-dimensional collagen; Matrigel containing matrices	In vitro generation of differentiated salivary structures comprising amylase-secreting acinar-like cells and ductal components	• No in vivo component • Xenogeneic materials were not qualified for clinical procedures • No assessment of salivary flow rate
Feng et al. (2009) [152]	hSG progenitor cells in three-dimensional Matrigel-based matrices	• hSG progenitors differentiated into epithelial-like acinar and ductal cell types in vitro • In vitro longstanding self-renewal capacity	• No in vivo component • Xenogeneic materials were not qualified for clinical procedures • No assessment of salivary flow rate
Maria et al. (2011) [153]	hSG primary cells in Matrigel-coated dishes in serum-free conditions	• In vitro differentiation of hSG cells into salivary cells with amylase-secreting acinar structures	• No in vivo component • Xenogeneic materials were not qualified for clinical procedures • No assessment of salivary flow rate
Pradhan-Bhatt et al. (2013) [148]	hSG primary cells in a three-dimensional HA hydrogel	• HA hydrogel supported in vivo lumen establishment • Salivary phenotypic properties of hSG progenitors preserved in longstanding cultures	• No evaluation of salivary flow rate
Pringle et al. (2016) [154]	hSG primary cells in s three-dimensional matrix comprising collagen and Matrigel.	• Matrigel supports in vitro development in longstanding culture • Three-dimensional xenogeneic matrices support primary cell differentiation • Injection of hSG primary cells (>500/gland) induces functional rescue	• No in vivo component • Xenogeneic materials were not qualified for clinical procedures

HA hydroxyapatite, *hSG* human salivary gland

4 Current Clinical Evidence for Stem Cell-Based Therapies

The use of cells for the treatment of organ disorders is a traditional goal in regenerative medicine. Culturing cells in the laboratory environment, and then differentiating them to specific cell lineages to either produce tissue for use in the laboratory or for tissue healing in situ has always been an ultimate goal for researchers. The problems around tissue formation outside the human body include maintaining the nutritional supply of the cells and collecting their waste materials. Bioreactors are now starting to be used in tissue engineering [155–157].

There has been considerable diversity amongst study protocols, in terms of the methods used for cell cultivation and differentiation, and cell seeding on scaffolds. The incubation period and delivery methods can affect the efficacy of bone regeneration. Synthetic scaffolds such as tricalcium phosphate (TCP) have given better results than natural bovine bone mineral (NBBM) when used with MSCs in 6 mm rat calvarial defects. The same result was obtained when PRP was added to the scaffold [158, 159].

Biphasic synthetic HA/TCP gas given good results in the regeneration of new bone in the dog for 10 mm mandibular defects, being superior to NBBM. HA/TCP also showed superior levels of cell attachment when compared to NBBM [159, 160]. In two human studies where alveolar cleft defects were treated with MSCs, stem cells co-transplanted with HA/TCP caused more bone formation than stem cells combined with DBM/CaSO4 [69, 155].

Freeze-dried bone allograft (FDBA) has also been used. This gives less bone formation than nanoHA with silica gel when used in 8 mm rabbit calvarial defects. FDBA and demineralized freeze-dried bone allograft (DFDBA) show promising results for cell attachment when investigated using SEM, but in vivo results have not been as positive. DFDBA loaded with MSCs when used around dental implants placed in rabbit tibia gave 2.1 mm of vertical bone growth, while no new bone formation occurred in the control group which was treated with DFDBA only but with no stem cells. FDBA when used as a block for the treatment of posterior mandibular supra-crestal defects in dogs was less effective at inducing new bone formation than a PCL-TCP scaffold [161, 162]. While in vitro experiments show proper cell attachment to various scaffolds, the ultimate proof comes from in vivo application, which has demonstrated better results when MSCs are delivered by synthetic substitutes (Table 4).

Tooth sockets were used by Pelegrine et al. as a model to evaluate the potential of a BM graft to be used for the preservation of alveolar bone after tooth extraction [175]. Effective craniofacial reconstruction using a BM-derived graft is limited due to the lack of cell-type characterization [176, 177]. Meijer et al., who evaluated the application of BM grafts in craniofacial regeneration noted the same restriction [165].

Through the cell processing for the generation of 'tissue repair cells' (CD90+ cell enrichment) (TRCs), characterization of cell surface markers has revealed a moderate level of homogeneity in the TRCs produced from different persons [178, 179]. Preclinical investigations have indicated that a high CD90+ cell concentration in the TRC population is associated with the formation of ectopic bone [180]. Furthermore, there is a correlation between the in vitro osteogenic capacity of TRCs and the clinical bone mineral density (BMD) and bone volume fraction (BVF) measurements. Although the primary findings seem to be promising, in vitro and in vivo studies must be conducted on larger TRC populations to explain these effects.

The healing period suggested for grafting procedures used for alveolar bone reconstruction is 3–6 months [170]. However, an important confounding factor in the healing process for the sockets of extracted teeth is limited local progenitor cell recruitment [181, 182]. Currently, suggested practical mechanisms to address this include using transplanted cells to jump-start regeneration [183, 184].

Table 4 Results of the selected human studies aim to investigate the application of MSCs for the management of craniofacial bone deficiencies

Author(s) and year	Sample	Scaffold	Stem cell	Growth factor	Sample size	Defect size and location	Tests	Results
Filho Cerruti et al. (2007) [163]	Human	TC	BMMSCs	PRP	32	Onlay graft	– RG – HMMA – CT	– 94.7% success rate – In the anterior maxilla 6–14 mm in width and 10 mm in height – In the posterior maxilla at least 9 mm height
Shayesteh et al. (2008) [164]	Human	HA/TCP	BMMSCs	–	7	Sinus lift	– HMMA – RG	– 41.34% new bone formation (3 mo) – Preoperative bone height (BH): 2.25 mm Immediate BH after sinus graft: 12.08 mm BH after 1 year: 10.83 mm
Meijer et al. (2008) [165]	Human	HA	BMMSCs	–	6	Sinus lift	– RG – HMMA	– 100% bone formation (6 week) – The mean of additive height was 5.54 mm (3 mo), 4 mm (6 mo), 5.09 mm (9 mo), 4.72 mm (12 mo), and 4.63 mm (15 mo).
Yamada et al. (2008) [166]	Human	PRP	BMMSCs	–	12	Sinus lift	– OPG – CT – HMMA	– Mineralized tissue formation increased 8.8 ± 1.6 mm
Ueda et al. (2008) [167]	Human	–	BMMSCs	PRP	14	6 Sinus lift 8 Onlay graft	– Clinical – RG	– The mean of additive height was 5.0 mm (6 mo) – The mean of additive height was 8.7 mm (6 mo) in sinus lift cases
Behnia et al. (2009) [155]	Human	DBM/$CaSO_4$	BMMSCs	–	2	Unilateral alveolar cleft	– OPG – CT – Intraoral inspection and palpation	– The integrity of the nasal floor – *34.5% the mean patient I postoperative defect (4 mo) *25.6% the mean patient II postoperative defect (4 mo)

Table 4 (continued)

Author(s) and year	Sample	Scaffold	Stem cell	Growth factor	Sample size	Defect size and location	Tests	Results
Behnia et al. (2011) [69]	Human	HA/TCP	BMMSCs	PDGF	4	Anterior maxillary cleft	- CBCT	- 51.3% mean filling of the bone defect (3 mo)
Rickert et al. (2011) [168]	Human	BioOsss and autogenousstem cells [93]; BioOsssmixed with autogenousbone [169]	BMMSCs	–	12	Sinus lift	- HMMA	- Formation of mineralized tissue; bone formation in the test group was more than the control (3–4 mo)
Kaigler et al. (2013) [170]	Human	Absorbable gelatin sponge	BMMSCs		12 + 12		- Clinical - HMMA - RG - CT	- Formation of mineralized tissue; alveolar bone reconstruction was accelerated following the treatment (1 yr)
Rickert et al. (2014) [171]	Human	BioOss + autogenous bone/ BioOss + MSCs and	BMMSCs	–	12	Sinus lift	- Clinical - RG	- Formation of mineralized tissue; osseointegration failed in three implants (91%). (1 yr)
Wildburger et al. (2014) [172]	Human	BioOss	BMMSCs	–	7	Sinus lift	- HMMA - CT	- Formation of mineralized tissue (13.5%) (3–6 mo)
Bertolai et al. (2015) [173]	Human	PRP/ cortico-cancellous freeze-dried bone chips	BMMSCs		20		- Clinical - HMMA	- Formation of mineralized tissue (3 mo)
Kaigler et al. (2015) [174]	Human	TCP	BMMSCs	–	13 + 13	Sinus lift	- Clinical - HMMA - RG	- Mineralized tissue formed (12.2% ±3.3) (1 year)

BMMSC Bone marrow mesenchymal stem cell, *CT* computerized tomography, *CBCT* Cone beam computed tomography, *DBM* Demineralized bone mineral, *HA* hydroxyapatite, *HMMA* Histomorphometric analysis, *M* month, *RG* radiography, *TCP* Tricalcium phosphate, *PDGF* platelet-derived growth factor, *PRP* Platelet-Rich Plasma, *TC* Thrombin-calcium

5 Future Prospects

Considering the complexities and disadvantages of traditional tissue engineering methods, bioprinting seems to be an appropriate alternative method for the fabrication of organs and tissues, as it tries to mimic the proper anatomical configuration of the tissue. Currently, several types of tissues, such as blood vessels, skin, cartilages, bones, cardiac tissues, and liver are being investigated using this approach [185, 186]. So far, the most well-evolved commercial applications of bioengineered tissues are those used for high throughput screening and pharmacological studies [187, 188]. Achieving proper vascularization of artificial organs remains a problem that limits or indeed prohibits their clinical application. Bioprinting should be able to address this and many other issues in regenerative medicine [189–191].

Natural metabolic processes need a system for supplying nutrients and disposing of wastes, and in living organs, this is normally accomplished through blood vessels. Vascularization exerts a significant impact on the extent and complexity of bioengineered tissues, because any that cause a shortage of nutrition or the accumulation may lead to necrosis of the bioengineered tissues [192].

So far, some efforts have been made to develop vascular engineering. Vascular network printing should be possible using a customized thermal inkjet printing system, especially using the hollow network technique to produce vascular networks using sacrificial filaments [185, 193]. Vascular networks have been produced using 3D printed sacrificial carbohydrate glass filaments. These can then be embedded into a hydrogel containing encapsulated cells. Kolesky et al. have introduced a bioprinting system to generate vascularized structures containing various cell types, however the functionality of these products needs further study [194–196].

Because cartilage is the only tissue that requires less direct blood supply than other tissues, it is currently the most suitable tissue for engineering. To date, tissue engineering techniques have failed to produce cartilage tissue that is integrated fully with host tissue [197]. Any bioprinted cartilaginous tissue must resemble the outcome anatomically, be stable, have structural integrity, and interface with host tissues [198]. Engineered tissue should have a compressive modulus similar to the host's natural tissue and had a stable structure [199]. An example of a production pathway is shown in Fig. 9.

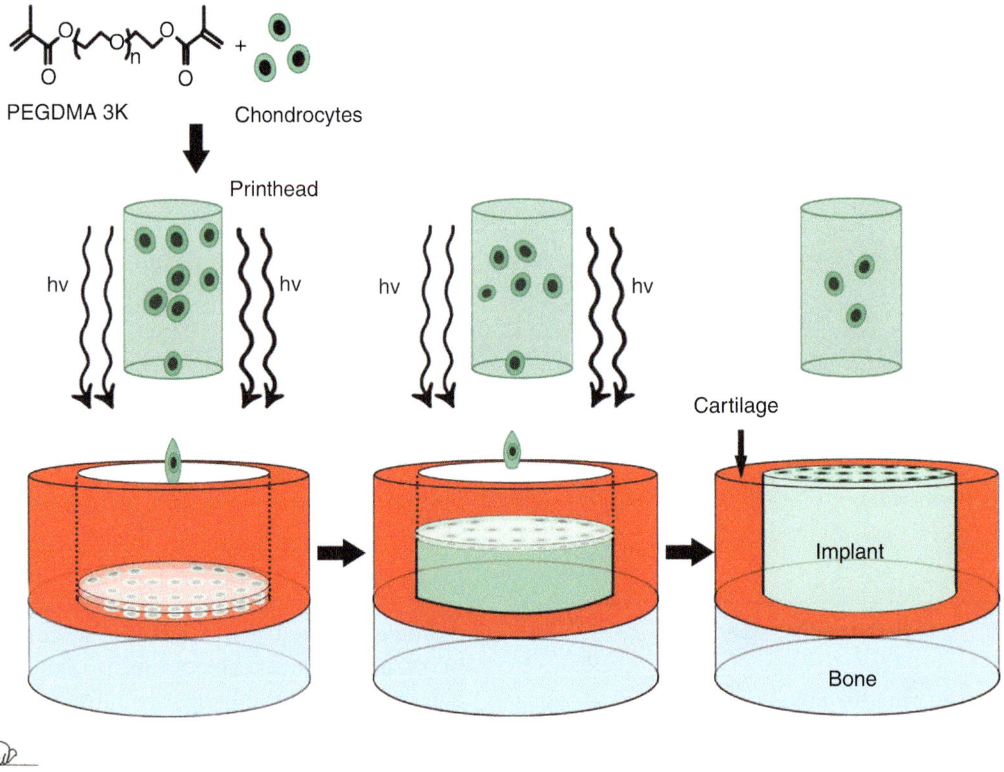

Fig. 9 Schematic illustration of the bioprinting and photopolymerization process for bioprinted polyethylene glycol dimethacrylate (PEGDA) hydrogel with human chondrocytes

Encapsulated chondrocytes in polycaprolactone (PCL) and alginate hydrogel have been used for bioprinting to produce cartilaginous tissue. PCL can provide extra stability to the printed cartilages. The addition of TGF-β can enhance the deposition of extracellular matrix [200]. Due to the possibility of a delay at the beginning of the response to a foreign body, the impact of an embedded polymeric biomaterial on the formation of the articular cartilage remains a serious obstacle to the goal of regenerating cartilaginous tissues [201]. It is not enough to modify and personalize industrial 3D printing technology in order to use it in the field of tissue reconstruction. There also needs to be a thorough understanding of the scientific principles that underpin the characteristics of the respective organ or tissue. A multidisciplinary synergy of biology, engineering, chemistry, optics, robotics, material science, and medicine is needed, to empower further innovations and advances in tissue engineering and bioprinting technologies [99, 185].

The use of bioprinting technology in clinical practice can only be effective and justifiable if it leads to the construction of products that are simple and easy to use during surgery. In situ bioprinting is a useful technique that may effectively repair scanned defects with minimal operator involvement. In situ bioprinting provides accurate deposition of cells, genes, and other factors. This method can be used in treatment procedures such as the reconstruction of craniofacial defects, soft tissue regeneration, and composite tissue reconstruction. Multi-arm bioprinters can be utilized for direct ink deposition, placing cells, biomaterials, and bio-factors into the defect area. Since this technology is still emerging, it needs further testing before clinical implementation. It appears that the use of prefabricated scaffolds is not desirable or recommended due to the possibility of deformation of these, or alterations in shape due to swelling or contraction.

6 Conclusion

Regenerative medicine has made major strides in the quest to replace tissues and organs. The science of bone regeneration has seen extraordinary progress, and this has impacted the field of dentistry, particularly in the specialty of oral and maxillofacial surgery. Today, thanks to cutting edge tissue regeneration science, patients presenting with traumatic lesions, malignancies, or congenital defects can be spared some of the side effects of conventional bone grafts.

Modern tissue regeneration leverages the potential internal capabilities of the patient's own cells for regeneration and exploits developments in 3D-bioprinting. Progress in regenerative approaches should expedite the process and benefit both patients and surgeons. Regenerative medicine and tissue engineering, especially the use of cell-based therapies, will continue to evolve and progress. A major limitation ahead of these advances in technology is the regulatory environment. The high costs of equipment are also an issue. Considering the explosive growth of science in tissue engineering, one can expect to see these technologies reach more and more from the laboratory bench to the bedside, for the benefit of patients.

References

1. Leucht P, Kim J-B, Amasha R, James AW, Girod S, Helms JA. Embryonic origin and Hox status determine progenitor cell fate during adult bone regeneration. Development. 2008;135(17):2845–54.
2. Bilezikian JP, Raisz LG, Martin TJ. Principles of bone biology. San Diego: Academic; 2008.
3. Ramaesh T, Bard JB. The growth and morphogenesis of the early mouse mandible: a quantitative analysis. J Anat. 2003;203(2):213–22.
4. Tsutsui T, Riminucci M, Holmbeck K, Bianco P, Robey P. Development of craniofacial structures in transgenic mice with constitutively active PTH/PTHrP receptor. Bone. 2008;42(2):321–31.
5. Peer LA. The fate of autogenous human bone grafts. Plast Reconstr Surg. 1951;8(1):80.
6. Sullivan WG, Szwajkun PR. Revascularization of cranial versus iliac crest bone grafts in the rat. Plast Reconstr Surg. 1991;87(6):1105–9.
7. Buchman SR, Ozaki W. The ultrastructure and resorptive pattern of cancellous onlay bone grafts in the craniofacial skeleton. Ann Plast Surg. 1999;43(1):49–56.
8. Mavropoulos A, Rizzoli R, Ammann P. Different responsiveness of alveolar and tibial bone to bone loss stimuli. J Bone Miner Res. 2007;22(3):403–10.
9. Price N, Lipton A, Jain VK, Ruggiero S. Prevention and management of osteonecrosis of the jaw associ-

ated with bisphosphonate therapy. Support Cancer Ther. 2004;2(1):14–7.
10. Simonds WF, James-Newton LA, Agarwal SK, Yang B, Skarulis MC, Hendy GN, et al. Familial isolated hyperparathyroidism: clinical and genetic characteristics of 36 kindreds. Medicine. 2002;81(1):1–26.
11. Ducy P, Zhang R, Geoffroy V, Ridall AL, Karsenty G. Osf2/Cbfa1: a transcriptional activator of osteoblast differentiation. Cell. 1997;89(5):747–54.
12. Otto F, Thornell AP, Crompton T, Denzel A, Gilmour KC, Rosewell IR, et al. Cbfa1, a candidate gene for cleidocranial dysplasia syndrome, is essential for osteoblast differentiation and bone development. Cell. 1997;89(5):765–71.
13. Nakashima M, Mizunuma K, Murakami T, Akamine A. Induction of dental pulp stem cell differentiation into odontoblasts by electroporation-mediated gene delivery of growth/differentiation factor 11 (Gdf11). Gene Ther. 2002;9(12):814–8.
14. Abzhanov A, Rodda SJ, McMahon AP, Tabin CJ. Regulation of skeletogenic differentiation in cranial dermal bone. Development. 2007;134(17):3133–44.
15. De Coster P, Mortier G, Marks L, Martens L. Cranial suture biology and dental development: genetic and clinical perspectives. J Oral Pathol Med. 2007;36(8):447–55.
16. Kimmel CB, Walker MB, Miller CT. Morphing the hyomandibular skeleton in development and evolution. J Exp Zool B Mol Dev Evol. 2007;308(5):609–24.
17. Frommelt H. Polymers for medical applications. Makromol Chem Macromol Symp. 1987;12:281–301.
18. van Meekeren J. Observationes medico-chirurgicae. Amsterdam: Ex officina Henrici & viduae Theodori Boom; 1982.
19. Hokugo A, Kubo Y, Takahashi Y, Fukuda A, Horiuchi K, Mushimoto K, et al. Prefabrication of vascularized bone graft using guided bone regeneration. Tissue Eng. 2004;10(7–8):978–86.
20. Vögelin E, Jones N, Huang J, Brekke J, Lieberman J. Healing of a critical-sized defect in the rat femur with use of a vascularized periosteal flap, a biodegradable matrix, and bone morphogenetic protein. JBJS. 2005;87(6):1323–31.
21. Gimbel M, Ashley RK, Sisodia M, Gabbay JS, Wasson KL, Heller J, et al. Repair of alveolar cleft defects: reduced morbidity with bone marrow stem cells in a resorbable matrix. J Craniofac Surg. 2007;18(4):895–901.
22. Chen NT, Glowacki J, Bucky LP, Hong H-Z, Kim W-K, Yaremchuk MJ. The roles of revascularization and resorption on endurance of craniofacial onlay bone grafts in the rabbit. Plast Reconstr Surg. 1994;93(4):714–22. discussion 23-4
23. Ozaki W, Buchman SR. Volume maintenance of onlay bone grafts in the craniofacial skeleton: micro-architecture versus embryologic origin. Plast Reconstr Surg. 1998;102(2):291–9.
24. Lukash FN, Zingaro EA, Salig J. The survival of free nonvascularized bone grafts in irradiated areas by wrapping in muscle flaps. Plast Reconstr Surg. 1984;74(6):783–8.
25. Pinholt EM, Solheim E, Talsnes O, Larsen TB, Bang G, Kirkeby OJ. Revascularization of calvarial, mandibular, tibial, and iliac bone grafts in rats. Ann Plast Surg. 1994;33(2):193–7.
26. Teng M, Liang X, Yuan Q, Nie J, Ye J, Cheng Q, et al. The inlay osteotome sinus augmentation technique for placing short implants simultaneously with reduced crestal bone height. A short-term follow-up. Clin Implant Dent Relat Res. 2013;15(6):918–26.
27. Martuscelli R, Toti P, Sbordone L, Guidetti F, Ramaglia L, Sbordone C. Five-year outcome of bone remodelling around implants in the maxillary sinus: assessment of differences between implants placed in autogenous inlay bone blocks and in ungrafted maxilla. Int J Oral Maxillofac Surg. 2014;43(9):1117–26.
28. Burchardt H. The biology of bone graft repair. Clin Orthop Relat Res. 1983;174:28–42.
29. Neu BR. Segmental bone and cartilage reconstruction of major nasal dorsal defects. Plast Reconstr Surg. 2000;106(1):160–70.
30. Uhm KI, Hwang SH, Choi BG. Cleft lip nose correction with onlay calvarial bone graft and suture suspension in oriental patients. Plast Reconstr Surg. 2000;105(2):499–503.
31. Mulliken JB, Kaban LB, Glowacki J. Induced osteogenesis—the biological principle and clinical applications. J Surg Res. 1984;37(6):487–96.
32. Oklund SA, Prolo DJ, Gutierrez RV, King S. Quantitative comparisons of healing in cranial fresh autografts, frozen autografts and processed autografts, and allografts in canine skull defects. Clin Orthop Relat Res. 1986;(205):269–91.
33. Eyre-Brook AL. The periosteum: its function reassessed. Clin Orthop Relat Res. 1984;(189):300–7.
34. Skawina A, Gorczyca W. The role of nutrient and periosteal blood vessels in the vascularization of the cortex of shafts of the long bones in human fetuses. Folia Morphol (Warsz). 1984;43(2):159–64.
35. Melcher A, Accursi G. Osteogenic capacity of periosteal and osteoperiosteal flaps elevated from the parietal bone of the rat. Arch Oral Biol. 1971;16(6):573–IN3.
36. Weng D, Hürzeler MB, Quiñones CR, Ohlms A, Caffesse RG. Contribution of the periosteum to bone formation in guided bone regeneration: a study in monkeys. Clin Oral Implants Res. 2000;11(6):546–54.
37. Warnke P, Springer I, Wiltfang J, Acil Y, Eufinger H, Wehmöller M, et al. Growth and transplantation of a custom vascularised bone graft in a man. Lancet. 2004;364(9436):766–70.
38. Mesimäki K, Lindroos B, Törnwall J, Mauno J, Lindqvist C, Kontio R, et al. Novel maxillary reconstruction with ectopic bone formation by GMP

adipose stem cells. Int J Oral Maxillofac Surg. 2009;38(3):201–9.
39. Friedenstein AJ. Precursor cells of mechanocytes. Int Rev Cytol. 1976;47(327):59.
40. Friedenstein AJ, Petrakova KV, Kurolesova AI, Frolova GP. Heterotopic of bone marrow. Analysis of precursor cells for osteogenic and hematopoietic tissues. Transplantation. 1968;6:230–47.
41. Owen M, Friedenstein A. Stromal stem cells: marrow-derived osteogenic precursors. Ciba Found Symp. 1988;136:42–60.
42. Nardi NB, da Silva Meirelles L. Mesenchymal stem cells: isolation, in vitro expansion and characterization. Stem cells. New York: Springer; 2008. p. 249–82.
43. Zuk PA, Zhu M, Mizuno H, Huang J, Futrell JW, Katz AJ, et al. Multilineage cells from human adipose tissue: implications for cell-based therapies. Tissue Eng. 2001;7(2):211–28.
44. Qu-Petersen Z, Deasy B, Jankowski R, Ikezawa M, Cummins J, Pruchnic R, et al. Identification of a novel population of muscle stem cells in mice: potential for muscle regeneration. J Cell Biol. 2002;157(5):851–64.
45. Feisst V, Brooks AE, Chen C-JJ, Dunbar PR. Characterization of mesenchymal progenitor cell populations directly derived from human dermis. Stem Cells Dev. 2014;23(6):631–42.
46. Grayson WL, Bunnell BA, Martin E, Frazier T, Hung BP, Gimble JM. Stromal cells and stem cells in clinical bone regeneration. Nat Rev Endocrinol. 2015;11(3):140.
47. Gruss JS, Mackinnon SE, Kassel EE, Cooper PW. The role of primary bone grafting in complex craniomaxillofacial trauma. Plast Reconstr Surg. 1985;75(1):17–24.
48. Manson PN, Crawley WA, Yaremchuk MJ, Rochman GM, Hoopes JE, French JJ. Midface fractures: advantages of immediate extended open reduction and bone grafting. Plast Reconstr Surg. 1985;76(1):1–12.
49. THOMPSON N, CASSON JA. Experimental onlay bone grafts to the jaws: a preliminary study in dogs. Plast Reconstr Surg. 1970;46(4):341–9.
50. KNIZE DM. The influence of periosteum and calcttonin on onlay bone graft survival a roentgenographic study. Plast Reconstr Surg. 1974;53(2):190–9.
51. Zins JE, Whitaker LA. Membranous versus endochondral bone: implications for craniofacial reconstruction. Plast Reconstr Surg. 1983;72(6):778–85.
52. Ermis I, Poole M. The effects of soft tissue coverage on bone graft resorption in the craniofacial region. Br J Plast Surg. 1992;45(1):26–9.
53. Hidalgo DA. Fibula free flap: a new method of mandible reconstruction. Plast Reconstr Surg. 1989;84(1):71–9.
54. Pogrel M, Podlesh S, Anthony JP, Alexander J. A comparison of vascularized and nonvascularized bone grafts for reconstruction of mandibular continuity defects. J Oral Maxillofac Surg. 1997;55(11):1200–6.
55. Phillips JH, Rahn BA. Fixation effects on membranous and endochondral onlay bone-graft resorption. Plast Reconstr Surg. 1988;82(5):872–7.
56. Lin KY, Bartlett SP, Yaremchuk MJ, Fallon M, Grossman RF, Whitaker LA. The effect of rigid fixation on the survival of onlay bone grafts: an experimental study. Plast Reconstr Surg. 1990;86(3):449–56.
57. Jackson I, Choi H, Clay R, Bevilacqua R, TerKonda S, Celik M, et al. Long-term follow-up of cranial bone graft in dorsal nasal augmentation. Plast Reconstr Surg. 1998;102(6):1869–73.
58. Pradel W, Tausche E, Gollogly J, Lauer G. Spontaneous tooth eruption after alveolar cleft osteoplasty using tissue-engineered bone: a case report. Oral Surg Oral Med Oral Pathol Oral Radiol Endod. 2008;105(4):440–4.
59. Vecchiatini R, Mobilio N, Raimondi F, Catapano S, Calura G. Implant-prosthetic rehabilitation for a patient with monolateral cleft lip and palate: a clinical report. Quintessence Int. 2009;40(6).
60. Luaces-Rey R, Arenaz-Búa J, López-Cedrún-Cembranos J-L, Herrero-Patiño S, Sironvalle-Soliva S, Iglesias-Candal E, et al. Is PRP useful in alveolar cleft reconstruction? Platelet-rich plasma in secondary alveoloplasty. Med Oral Patol Oral Cir Bucal. 2010;15(4):e619–23.
61. Le BT, Woo I. Alveolar cleft repair in adults using guided bone regeneration with mineralized allograft for dental implant site development: a report of 2 cases. J Oral Maxillofac Surg. 2009;67(8):1716–22.
62. Hibi H, Yamada Y, Ueda M, Endo Y. Alveolar cleft osteoplasty using tissue-engineered osteogenic material. Int J Oral Maxillofac Surg. 2006;35(6):551–5.
63. Marukawa E, Oshina H, Iino G, Morita K, Omura K. Reduction of bone resorption by the application of platelet-rich plasma (PRP) in bone grafting of the alveolar cleft. J Craniomaxillofac Surg. 2011;39(4):278–83.
64. Behnia H, Khoshzaban A, Zarinfar M, Khojasteh A. Histological evaluation of regeneration in rabbit calvarial bone defects using demineralized bone matrix, mesenchymal stem cells and platelet rich in growth factors. J Dent Sch. 2012;30(3):143–54.
65. Janssen NG, Weijs WL, Koole R, Rosenberg AJ, Meijer GJ. Tissue engineering strategies for alveolar cleft reconstruction: a systematic review of the literature. Clin Oral Investig. 2014;18(1):219–26.
66. Herford AS, Boyne PJ, Rawson R, Williams RP. Bone morphogenetic protein-induced repair of the premaxillary cleft. J Oral Maxillofac Surg. 2007;65(11):2136–41.
67. Lee C, Nishihara K, Okawachi T, Iwashita Y, Majima HJ, Nakamura N. A quantitative radiological assessment of outcomes of autogenous bone graft combined with platelet-rich plasma in the alveolar cleft. Int J Oral Maxillofac Surg. 2009;38(2):117–25.

68. Bykowski MR, Naran S, Winger DG, Losee JE. The rate of oronasal fistula following primary cleft palate surgery: a meta-analysis. Cleft Palate-Cran J. 2015;52(4):81–7.
69. Behnia H, Khojasteh A, Soleimani M, Tehranchi A, Atashi A. Repair of alveolar cleft defect with mesenchymal stem cells and platelet derived growth factors: a preliminary report. J Craniomaxillofac Surg. 2012;40(1):2–7.
70. Hosseinpour S, Ahsaie MG, Rad MR, taghi Baghani M, Motamedian SR, Khojasteh A. Application of selected scaffolds for bone tissue engineering: a systematic review. Oral Maxillofac Surg. 2017;21(2):109–29.
71. Motamedian SR, Hosseinpour S, Ahsaie MG, Khojasteh A. Smart scaffolds in bone tissue engineering: a systematic review of literature. World J Stem Cells. 2015;7(3):657.
72. Salih V. Biodegradable scaffolds for tissue engineering. Cellular response to biomaterials. Boca Raton: CRC Press; 2008. p. 185–211.
73. Mihaylova Z, Mitev V, Stanimirov P, Isaeva A, Gateva N, Ishkitiev N. Use of platelet concentrates in oral and maxillofacial surgery: an overview. Acta Odontol Scand. 2017;75(1):1–11.
74. Liu Y-f, Zhu F-d, Dong X-t, Peng W. Digital design of scaffold for mandibular defect repair based on tissue engineering. J Zhejiang Univ Sci B. 2011;12(9):769.
75. Seo S, Na K. Mesenchymal stem cell-based tissue engineering for chondrogenesis. Biomed Res Int. 2011.
76. Khojasteh A, Hosseinpour S, Dehghan M, Mashhadiabbas F, Rezai Rad M, Ansari S, et al. Antibody-mediated osseous regeneration for bone tissue engineering in canine segmental defects. Biomed Res Int. 2018.
77. Khojasteh A, Hosseinpour S, Nazeman P, Dehghan M. The effect of a platelet-rich fibrin conduit on neurosensory recovery following inferior alveolar nerve lateralization: a preliminary clinical study. Int J Oral Maxillofac Surg. 2016;45(10):1303–8.
78. Betz VM, Betz OB, Harris MB, Vrahas MS, Evans CH. Bone tissue engineering and repair by gene therapy. Front Biosci. 2008;13(13):833–41.
79. Depprich R, Handschel J, Wiesmann H-P, Jäsche-Meyer J, Meyer U. Use of bioreactors in maxillofacial tissue engineering. Br J Oral Maxillofac Surg. 2008;46(5):349–54.
80. Deb S, Mandegaran R, Di Silvio L. A porous scaffold for bone tissue engineering/45S5 bioglass® derived porous scaffolds for co-culturing osteoblasts and endothelial cells. J Mater Sci Mater Med. 2010;21(3):893–905.
81. Kern S, Eichler H, Stoeve J, Klüter H, Bieback K. Comparative analysis of mesenchymal stem cells from bone marrow, umbilical cord blood, or adipose tissue. Stem Cells. 2006;24(5):1294–301.
82. Miura M, Gronthos S, Zhao M, Lu B, Fisher LW, Robey PG, et al. SHED: stem cells from human exfoliated deciduous teeth. Proc Natl Acad Sci USA. 2003;100(10):5807–12.
83. Yamada Y, Fujimoto A, Ito A, Yoshimi R, Ueda M. Cluster analysis and gene expression profiles: a cDNA microarray system-based comparison between human dental pulp stem cells (hDPSCs) and human mesenchymal stem cells (hMSCs) for tissue engineering cell therapy. Biomaterials. 2006;27(20):3766–81.
84. Jimi E, Hirata S, Shin M, Yamazaki M, Fukushima H. Molecular mechanisms of BMP-induced bone formation: cross-talk between BMP and NF-κB signaling pathways in osteoblastogenesis. Jpn Dent Sci Rev. 2010;46(1):33–42.
85. Katagiri T, Takahashi N. Regulatory mechanisms of osteoblast and osteoclast differentiation. Oral Dis. 2002;8(3):147–59.
86. Katagiri T, Watabe T. Bone morphogenetic proteins. Cold Spring Harb Perspect Biol. 2016;8(6):a021899.
87. Bleuming SA, He XC, Kodach LL, Hardwick JC, Koopman FA, Fiebo J, et al. Bone morphogenetic protein signaling suppresses tumorigenesis at gastric epithelial transition zones in mice. Cancer Res. 2007;67(17):8149–55.
88. Canalis E. Effects of tumor necrosis factor on bone formation in vitro. Endocrinology. 1987;121(5):1596–604.
89. Nakase T, Takaoka K, Masuhara K, Shimizu K, Yoshikawa H, Ochi T. Interleukin-1β enhance and tumor necrosis factor-α inhibits bone morphogenetic protein-2-induced alkaline phosphatase activity in MC3T3-E1 osteoblastic cells. Bone. 1997;21(1):17–21.
90. Nanes MS. Tumor necrosis factor-α: molecular and cellular mechanisms in skeletal pathology. Gene. 2003;321:1–15.
91. Pacifici R, Brown C, Puscheck E, Friedrich E, Slatopolsky E, Maggio D, et al. Effect of surgical menopause and estrogen replacement on cytokine release from human blood mononuclear cells. Proc Natl Acad Sci USA. 1991;88(12):5134–8.
92. Ranganathan P. An update on pharmacogenomics in rheumatoid arthritis with a focus on TNF-blocking agents. Curr Opin Mol Ther. 2008;10(6):562–7.
93. Jimi E, Aoki K, Saito H, D'Acquisto F, May MJ, Nakamura I, et al. Selective inhibition of NF-κB blocks osteoclastogenesis and prevents inflammatory bone destruction in vivo. Nat Med. 2004;10(6):617–24.
94. Chang J, Wang Z, Tang E, Fan Z, McCauley L, Franceschi R, et al. Inhibition of osteoblastic bone formation by nuclear factor-κB. Nat Med. 2009;15(6):682.
95. Eliseev RA, Schwarz EM, Zuscik MJ, O'Keefe RJ, Drissi H, Rosier RN. Smad7 mediates inhibition of Saos2 osteosarcoma cell differentiation by NFκB. Exp Cell Res. 2006;312(1):40–50.
96. Li Y, Li A, Strait K, Zhang H, Nanes MS, Weitzmann MN. Endogenous TNFα lowers maxi-

mum peak bone mass and inhibits osteoblastic Smad activation through NF-κB. J Bone Miner Res. 2007;22(5):646–55.
97. Alles N, Soysa NS, Hayashi J, Khan M, Shimoda A, Shimokawa H, et al. Suppression of NF-κB increases bone formation and ameliorates osteopenia in ovariectomized mice. Endocrinology. 2010;151(10):4626–34.
98. Yamazaki M, Fukushima H, Shin M, Katagiri T, Doi T, Takahashi T, et al. Tumor necrosis factor α represses bone morphogenetic protein (BMP) signaling by interfering with the DNA binding of Smads through the activation of NF-κB. J Biol Chem. 2009;284(51):35987–95.
99. Jimi E, Hirata S, Osawa K, Terashita M, Kitamura C, Fukushima H. The current and future therapies of bone regeneration to repair bone defects. Int J Dent. 2012.
100. Nerem R. Tissue engineering in the USA. Med Biol Eng Comput. 1992;30(4):CE8–CE12.
101. Lanza R, Langer R, Vacanti JP. Principles of tissue engineering. San Diego: Academic; 2011.
102. Rheinwatd JG, Green H. Seria cultivation of strains of human epidemal keratinocytes: the formation keratinizin colonies from single cell is. Cell. 1975;6(3):331–43.
103. Green H, Kehinde O, Thomas J. Growth of cultured human epidermal cells into multiple epithelia suitable for grafting. Proc Natl Acad Sci USA. 1979;76(11):5665–8.
104. Yannas I, Burke J, Huang C, Gordon P. Correlation of in vivo collagen degradation rate with in vitro measurements. J Biomed Mater Res. 1975;9(6):623–8.
105. Yannas I, Burke JF. Design of an artificial skin. I. Basic design principles. J Biomed Mater Res. 1980;14(1):65–81.
106. O'Connor N, Mulliken J, Banks-Schlegel S, Kehinde O, Green H. Grafting of burns with cultured epithelium prepared from autologous epidermal cells. Lancet. 1981;317(8211):75–8.
107. Green H. The birth of therapy with cultured cells. BioEssays. 2008;30(9):897–903.
108. Gallico GG III, O'Connor NE, Compton CC, Kehinde O, Green H. Permanent coverage of large burn wounds with autologous cultured human epithelium. N Engl J Med. 1984;311(7):448–51.
109. Burke JF, Yannas IV, Quinby WC Jr, Bondoc CC, Jung WK. Successful use of a physiologically acceptable artificial skin in the treatment of extensive burn injury. Ann Surg. 1981;194(4):413.
110. Heimbach D, Luterman A, Burke J, Cram A, Herndon D, Hunt J, et al. Artificial dermis for major burns. A multi-center randomized clinical trial. Ann Surg. 1988;208(3):313.
111. Heimbach DM, Warden GD, Luterman A, Jordan MH, Ozobia N, Ryan CM, et al. Multicenter post-approval clinical trial of Integra® dermal regeneration template for burn treatment. J Burn Care Res. 2003;24(1):42–8.
112. Heitland A, Piatkowski A, Noah E, Pallua N. Update on the use of collagen/glycosaminoglycate skin substitute—six years of experiences with artificial skin in 15 German burn centers. Burns. 2004;30(5):471–5.
113. Shevchenko RV, James SL, James SE. A review of tissue-engineered skin bioconstructs available for skin reconstruction. J R Soc Interface. 2009:rsif20090403.
114. Izumi K, Neiva RF, Feinberg SE. Intraoral grafting of tissue-engineered human oral mucosa. Int J Oral Maxillofac Implants. 2013;28(5):e295.
115. Wennström JL, Zucchelli G. Increased gingival dimensions. A significant factor for successful outcome of root coverage procedures? J Clin Periodontol. 1996;23(8):770–7.
116. Crisciani S. Efficacia Clinica Degli Innesti di Collagene Xenogeno Nella Terapia Chirurgica delle Recessioni Gengivali multiple: Clinical Trial Randomizzato. Graduate thesis LC5. Universita Di Pisa. 2012.
117. Nevins M. Attached gingiva—mucogingival therapy and restorative dentistry. Int J Periodontics Restorative Dent. 1985;6(4):9–27.
118. McGuire MK, Scheyer ET, Nunn ME, Lavin PT. A pilot study to evaluate a tissue-engineered bilayered cell therapy as an alternative to tissue from the palate. J Periodontol. 2008;79(10):1847–56.
119. Nevins ML. Tissue-engineered bilayered cell therapy for the treatment of oral mucosal defects: a case series. Int J Periodontics Restorative Dent. 2010;30(1):31–9.
120. Prato GPP, Rotundo R, Mognoni C, Soranzo C. Tissue engineering technology for gingival augmentation procedures: a case report. Int J Periodontics Restorative Dent. 2000;20(6):552–9.
121. Pomahač B, Svensjö T, Yao F, Brown H, Eriksson E. Tissue engineering of skin. Crit Rev Oral Biol Med. 1998;9(3):333–44.
122. C-y T, Ueda M, K-i H, Horie K, Hibino Y, Sugimura Y, et al. Clinical results of cultured epithelial cell grafting in the oral and maxillofacial region. J Craniomaxillofac Surg. 1997;25(1):4–8.
123. Yoshizawa M, Feinberg SE, Marcelo CL, Elner VM. Ex vivo produced human conjunctiva and oral mucosa equivalents grown in a serum-free culture system. J Oral Maxillofac Surg. 2004;62(8):980–8.
124. Lanman TH, Ingalls TH. Vitamin C deficiency and wound healing: an experimental and clinical study. Ann Surg. 1937;105(4):616.
125. Lindholm C, Searle R. Wound management for the 21st century: combining effectiveness and efficiency. Int Wound J. 2016;13:5–15.
126. Aitzetmüller MM, Brett EA, Sauter M, Duscher D. Basic principles and current approach for soft tissue regeneration. In: Duscher D, Shiffman MA, editors. Regenerative medicine and plastic surgery. Cham: Springer; 2019. p. 7–15.
127. Tiesler V, Coppa A, Zabala P, Cucina A. Scurvy-related morbidity and death among Christopher

Columbus' Crew at La Isabela, the first European town in the New World (1494–1498): an assessment of the skeletal and historical information. Int J Osteoarchaeol. 2016;26(2):191–202.
128. Rennert RC, Sorkin M, Januszyk M, Duscher D, Kosaraju R, Chung MT, et al. Diabetes impairs the angiogenic potential of adipose-derived stem cells by selectively depleting cellular subpopulations. Stem Cell Res Ther. 2014;5(3):79.
129. Duscher D, Rennert RC, Januszyk M, Anghel E, Maan ZN, Whittam AJ, et al. Aging disrupts cell subpopulation dynamics and diminishes the function of mesenchymal stem cells. Sci Rep. 2014;4:7144.
130. Hadjipanayi E, Schilling AF. Regeneration through autologous hypoxia preconditioned plasma. Organogenesis. 2014;10(2):164–9.
131. Hadjipanayi E, Bauer A, Moog P, Salgin B, Kuekrek H, Fersch B, et al. Cell-free carrier system for localized delivery of peripheral blood cell-derived engineered factor signaling: towards development of a one-step device for autologous angiogenic therapy. J Control Release. 2013;169(1–2):91–102.
132. Lombaert I, Movahednia MM, Adine C, Ferreira JN. Concise review: salivary gland regeneration: therapeutic approaches from stem cells to tissue organoids. Stem Cells. 2017;35(1):97–105.
133. Neumann Y, David R, Stiubea-Cohen R, Orbach Y, Aframian DJ, Palmon A. Long-term cryopreservation model of rat salivary gland stem cells for future therapy in irradiated head and neck cancer patients. Tissue Eng Part C. 2012;18(9):710–8.
134. Lombaert IM, Wierenga PK, Kok T, Kampinga HH, dehaan G, Coppes RP. Mobilization of bone marrow stem cells by granulocyte colony-stimulating factor ameliorates radiation-induced damage to salivary glands. Clin Cancer Res. 2006;12(6):1804–12.
135. Sumita Y, Liu Y, Khalili S, Maria OM, Xia D, Key S, et al. Bone marrow-derived cells rescue salivary gland function in mice with head and neck irradiation. Int J Biochem Cell Biol. 2011;43(1):80–7.
136. Lin C-Y, Chang F-H, Chen C-Y, Huang C-Y, Hu F-C, Huang W-K, et al. Cell therapy for salivary gland regeneration. J Dent Res. 2011;90(3):341–6.
137. Lim J-Y, Yi T, Choi J-S, Jang YH, Lee S, Kim HJ, et al. Intraglandular transplantation of bone marrow-derived clonal mesenchymal stem cells for amelioration of post-irradiation salivary gland damage. Oral Oncol. 2013;49(2):136–43.
138. Kojima T, Si K, Hirano S, Tateya I, Ohno S, Nakamura T, et al. Regeneration of radiation damaged salivary glands with adipose-derived stromal cells. Laryngoscope. 2011;121(9):1864–9.
139. Lee J, Park S, Roh S. Transdifferentiation of mouse adipose-derived stromal cells into acinar cells of the submandibular gland using a co-culture system. Exp Cell Res. 2015;334(1):160–72.
140. Lim J-Y, Ra JC, Shin IS, Jang YH, An H-Y, Choi J-S, et al. Systemic transplantation of human adipose tissue-derived mesenchymal stem cells for the regeneration of irradiation-induced salivary gland damage. PLoS One. 2013;8(8):e71167.
141. Xiong X, Shi X, Chen F. Human adipose tissue-derived stem cells alleviate radiation-induced xerostomia. Int J Mol Med. 2014;34(3):749–55.
142. Jeong J, Baek H, Kim Y-J, Choi Y, Lee H, Lee E, et al. Human salivary gland stem cells ameliorate hyposalivation of radiation-damaged rat salivary glands. Exp Mol Med. 2013;45(11):e58-e.
143. Lim J-Y, Yi T, Lee S, Kim J, S-n K, Song SU, et al. Establishment and characterization of mesenchymal stem cell-like clonal stem cells from mouse salivary glands. Tissue Eng Part C. 2015;21(5):447–57.
144. Zhang N-N, Huang G-L, Han Q-B, Hu X, Yi J, Yao L, et al. Functional regeneration of irradiated salivary glands with human amniotic epithelial cells transplantation. Int J Clin Exp Patho. 2013;6(10):2039.
145. Huang G-L, Zhang N-N, Wang J-S, Yao L, Zhao Y-J, Wang Y-Y. Transdifferentiation of human amniotic epithelial cells into acinar cells using a double-chamber system. Cell Reprogram. 2012;14(4):377–83.
146. Kawakami M, Ishikawa H, Tachibana T, Tanaka A, Mataga I. Functional transplantation of salivary gland cells differentiated from mouse early ES cells in vitro. Hum Cell. 2013;26(2):80–90.
147. Ono H, Obana A, Usami Y, Sakai M, Nohara K, Egusa H, et al. Regenerating salivary glands in the microenvironment of induced pluripotent stem cells. Biomed Res Int. 2015;2015.
148. Pradhan-Bhatt S, Harrington DA, Duncan RL, Jia X, Witt RL, Farach-Carson MC. Implantable three-dimensional salivary spheroid assemblies demonstrate fluid and protein secretory responses to neurotransmitters. Tissue Eng Part A. 2013;19(13–14):1610–20.
149. Ogawa M, Oshima M, Imamura A, Sekine Y, Ishida K, Yamashita K, et al. Functional salivary gland regeneration by transplantation of a bioengineered organ germ. Nat Commun. 2013;4(1):1–10.
150. Pradhan-Bhatt S, Harrington DA, Duncan RL, Farach-Carson MC, Jia X, Witt RL. A novel in vivo model for evaluating functional restoration of a tissue-engineered salivary gland. Laryngoscope. 2014;124(2):456–61.
151. Joraku A, Sullivan CA, Yoo J, Atala A. In-vitro reconstitution of three-dimensional human salivary gland tissue structures. Differentiation. 2007;75(4):318–24.
152. Feng J, van der Zwaag M, Stokman MA, van Os R, Coppes RP. Isolation and characterization of human salivary gland cells for stem cell transplantation to reduce radiation-induced hyposalivation. Radiother Oncol. 2009;92(3):466–71.
153. Maria OM, Zeitouni A, Gologan O, Tran SD. Matrigel improves functional properties of primary human salivary gland cells. Tissue Eng Part A. 2011;17(9–10):1229–38.

154. Pringle S, Maimets M, van der Zwaag M, Stokman MA, van Gosliga D, Zwart E, et al. Human salivary gland stem cells functionally restore radiation damaged salivary glands. Stem Cells. 2016;34(3):640–52.
155. Behnia H, Khojasteh A, Soleimani M, Tehranchi A, Khoshzaban A, Keshel SH, et al. Secondary repair of alveolar clefts using human mesenchymal stem cells. Oral Surg Oral Med Oral Pathol Oral Radiol Endod. 2009;108(2):e1–6.
156. Khojasteh A, Behnia H, Dashti SG, Stevens M. Current trends in mesenchymal stem cell application in bone augmentation: a review of the literature. J Oral Maxillofac Surg. 2012;70(4):972–82.
157. Keyhan SO, Fallahi H, Jahangirnia A, SMR M, Khosravi MH, Amirzade-Iranaq MH. Tissue engineering applications in maxillofacial surgery. Stem cells in clinical practice and tissue engineering. London: IntechOpen; 2017.
158. Dohan DM, Choukroun J, Diss A, Dohan SL, Dohan AJ, Mouhyi J, et al. Platelet-rich fibrin (PRF): a second-generation platelet concentrate. Part III: leucocyte activation: a new feature for platelet concentrates? Oral Surg Oral Med Oral Pathol Oral Radiol Endod. 2006;101(3):e51–e5.
159. Khojasteh A, Eslaminejad MB, Nazarian H. Mesenchymal stem cells enhance bone regeneration in rat calvarial critical size defects more than platelete-rich plasma. Oral Surg Oral Med Oral Pathol Oral Radiol Endod. 2008;106(3):356–62.
160. Kirker-Head CA. Potential applications and delivery strategies for bone morphogenetic proteins. Adv Drug Deliv Rev. 2000;43(1):65–92.
161. Khojasteh A, Dashti SG, Dehghan MM, Behnia H, Abbasnia P, Morad G. The osteoregenerative effects of platelet-derived growth factor BB cotransplanted with mesenchymal stem cells, loaded on freeze-dried mineral bone block: a pilot study in dog mandible. J Biomed Mater Res Part B. 2014;102(8):1771–8.
162. Khojasteh A, Eslaminejad MB, Nazarian H, Morad G, Dashti SG, Behnia H, et al. Vertical bone augmentation with simultaneous implant placement using particulate mineralized bone and mesenchymal stem cells: a preliminary study in rabbit. J Oral Implantol. 2013;39(1):3–13.
163. Cerruti HF, Kerkis I, Kerkis A, Tatsui NH, da Costa Neves A, Bueno DF, et al. Allogenous bone grafts improved by bone marrow stem cells and platelet growth factors: clinical case reports. Artif Organs. 2007;31(4):268–73.
164. Shayesteh YS, Khojasteh A, Soleimani M, Alikhasi M, Khoshzaban A, Ahmadbeigi N. Sinus augmentation using human mesenchymal stem cells loaded into a β-tricalcium phosphate/hydroxyapatite scaffold. Oral Surg Oral Med Oral Pathol Oral Radiol Endod. 2008;106(2):203–9.
165. Meijer GJ, de Bruijn JD, Koole R, van Blitterswijk CA. Cell based bone tissue engineering in jaw defects. Biomaterials. 2008;29(21):3053–61.
166. Yamada Y, Nakamura S, Ito K, Kohgo T, Hibi H, Nagasaka T, et al. Injectable tissue-engineered bone using autogenous bone marrow–derived stromal cells for maxillary sinus augmentation: clinical application report from a 2–6-year follow-up. Tissue Eng Part A. 2008;14(10):1699–707.
167. Ueda M, Yamada Y, Kagami H, Hibi H. Injectable bone applied for ridge augmentation and dental implant placement: human progress study. Implant Dent. 2008;17(1):82–90.
168. Rickert D, Sauerbier S, Nagursky H, Menne D, Vissink A, Raghoebar G. Maxillary sinus floor elevation with bovine bone mineral combined with either autogenous bone or autogenous stem cells: a prospective randomized clinical trial. Clin Oral Implants Res. 2011;22(3):251–8.
169. Lin Z-Y, Duan Z-X, Guo X-D, Li J-F, Lu H-W, Zheng Q-X, et al. Bone induction by biomimetic PLGA-(PEG-ASP) n copolymer loaded with a novel synthetic BMP-2-related peptide in vitro and in vivo. J Control Release. 2010;144(2):190–5.
170. Kaigler D, Pagni G, Park CH, Braun TM, Holman LA, Yi E, et al. Stem cell therapy for craniofacial bone regeneration: a randomized, controlled feasibility trial. Cell Transplant. 2013;22(5):767–77.
171. Rickert D, Vissink A, Slot W, Sauerbier S, Meijer H, Raghoebar G. Maxillary sinus floor elevation surgery with BioOss® mixed with a bone marrow concentrate or autogenous bone: test of principle on implant survival and clinical performance. Int J Oral Maxillofac Surg. 2014;43(2):243–7.
172. Wildburger A, Payer M, Jakse N, Strunk D, Etchard-Liechtenstein N, Sauerbier S. Impact of autogenous concentrated bone marrow aspirate on bone regeneration after sinus floor augmentation with a bovine bone substitute–a split-mouth pilot study. Clin Oral Implants Res. 2014;25(10):1175–81.
173. Bertolai R, Catelani C, Aversa A, Rossi A, Giannini D, Bani D. Bone graft and mesenchimal stem cells: clinical observations and histological analysis. Clin Cases Miner Bone Metab. 2015;12(2):183.
174. Kaigler D, Avila-Ortiz G, Travan S, Taut AD, Padial-Molina M, Rudek I, et al. Bone engineering of maxillary sinus bone deficiencies using enriched CD90+ stem cell therapy: a randomized clinical trial. J Bone Miner Res. 2015;30(7):1206–16.
175. Pelegrine AA, Da Costa CES, Correa MEP, Marques JFC Jr. Clinical and histomorphometric evaluation of extraction sockets treated with an autologous bone marrow graft. Clin Oral Implants Res. 2010;21(5):535–42.
176. Marcacci M, Kon E, Moukhachev V, Lavroukov A, Kutepov S, Quarto R, et al. Stem cells associated with macroporous bioceramics for long bone repair: 6-to 7-year outcome of a pilot clinical study. Tissue Eng. 2007;13(5):947–55.
177. Soltan M, Smiler D, Soltan C, Prasad HS, Rohrer MD. Bone grafting by means of a tunnel dissection: predictable results using stem cells and matrix. Implant Dent. 2010;19(4):280–7.
178. Gastens MH, Goltry K, Prohaska W, Tschöpe D, Stratmann B, Lammers D, et al. Good manufactur-

ing practice-compliant expansion of marrow-derived stem and progenitor cells for cell therapy. Cell Trasplant. 2007;16(7):685–96.
179. Kaigler D, Avila G, Wisner-Lynch L, Nevins ML, Nevins M, Rasperini G, et al. Platelet-derived growth factor applications in periodontal and peri-implant bone regeneration. Expert Opin Biol Ther. 2011;11(3):375–85.
180. Dennis JE, Esterly K, Awadallah A, Parrish CR, Poynter GM, Goltry KL. Clinical-scale expansion of a mixed population of bone marrow-derived stem and progenitor cells for potential use in bone tissue regeneration. Stem Cells. 2007;25(10):2575–82.
181. Cardaropoli G, Araújo M, Hayacibara R, Sukekava F, Lindhe J. Healing of extraction sockets and surgically produced–augmented and non-augmented–defects in the alveolar ridge. An experimental study in the dog. J Clin Periodontol. 2005;32(5):435–40.
182. Trombelli L, Farina R, Marzola A, Bozzi L, Liljenberg B, Lindhe J. Modeling and remodeling of human extraction sockets. J Clin Periodontol. 2008;35(7):630–9.
183. Jensen SS, Terheyden H. Bone augmentation procedures in localized defects in the alveolar ridge: clinical results with different bone grafts and bone-substitute materials. Database of abstracts of reviews of effects (DARE): quality-assessed reviews [Internet]. Centre for Reviews and Dissemination (UK); 2009.
184. Caplan AI. Adult mesenchymal stem cells for tissue engineering versus regenerative medicine. J Cell Physiol. 2007;213(2):341–7.
185. Gao G, Huang Y, Schilling AF, Hubbell K, Cui X. Organ bioprinting: are we there yet? Adv Healthc Mater. 2018;7(1):1701018.
186. Peng W, Datta P, Ayan B, Ozbolat V, Sosnoski D, Ozbolat IT. 3D bioprinting for drug discovery and development in pharmaceutics. Acta Biomater. 2017;57:26–46.
187. Zhu W, Holmes B, Glazer RI, Zhang LG. 3D printed nanocomposite matrix for the study of breast cancer bone metastasis. Nanomedicine. 2016;12(1):69–79.
188. Ozbolat IT, Yu Y. Bioprinting toward organ fabrication: challenges and future trends. IEEE Trans Biomed Eng. 2013;60(3):691–9.
189. Guillemot F, Mironov V, Nakamura M. Bioprinting is coming of age: report from the international conference on bioprinting and biofabrication in Bordeaux (3B'09). Biofabrication. 2010;2(1):010201.
190. Keyhan SO, Fallahi H, Jahangirnia A, Amirzade-Iranaq MT, Amirzade-Iranaq MH. Application of 3-D printing for tissue regeneration in Oral and maxillofacial surgery: what is upcoming? Biomaterials in Regenerative Medicine. London: IntechOpen; 2017.
191. Keyhan SO, Amirzade-Iranaq MH. The use of 3D-printing technology in rhinoplasty: change horizons, change principles, change future. Glob J Otolaryngol. 2017;11(1).
192. Cui X, Boland T. Human microvasculature fabrication using thermal inkjet printing technology. Biomaterials. 2009;30(31):6221–7.
193. Miller JS, Stevens KR, Yang MT, Baker BM, Nguyen D-HT, Cohen DM, et al. Rapid casting of patterned vascular networks for perfusable engineered three-dimensional tissues. Nat Mater. 2012;11(9):768–74.
194. Kolesky DB, Homan KA, Skylar-Scott MA, Lewis JA. Three-dimensional bioprinting of thick vascularized tissues. Proc Natl Acad Sci USA. 2016;113(12):3179–84.
195. Kolesky DB, Truby RL, Gladman AS, Busbee TA, Homan KA, Lewis JA. Bioprinting: 3D bioprinting of vascularized, heterogeneous cell-laden tissue constructs. Adv Mater. 2014;26(19):2966.
196. Keyhan SO, Ghanean S, Navabazam A, Khojasteh A, Amirzade-Iranaq MH. Three-dimensional printing: a novel Technology for use in Oral and maxillofacial operations. A textbook of advanced oral and maxillofacial surgery. London: InTechOpen; 2016.
197. Cui X, Gao G, Yonezawa T, Dai G. Human cartilage tissue fabrication using three-dimensional inkjet printing technology. J Vis Exp. 2014;88:e51294.
198. Kundu J, Shim JH, Jang J, Kim SW, Cho DW. An additive manufacturing-based PCL–alginate–chondrocyte bioprinted scaffold for cartilage tissue engineering. J Tissue Eng Regen Med. 2015;9(11):1286–97.
199. Makris EA, Gomoll AH, Malizos KN, Hu JC, Athanasiou KA. Repair and tissue engineering techniques for articular cartilage. Nat Rev Rheumatol. 2015;11(1):21.
200. Mertz L. Dream it, design it, print it in 3-D: what can 3-D printing do for you? IEEE Pulse. 2013;4(6):15–21.
201. Ozbolat IT, Peng W, Ozbolat V. Application areas of 3D bioprinting. Drug Discov Today. 2016;21(8):1257–71.

Regenerative Approaches in Oral Medicine

Camile S. Farah, Antonio Celentano, Giuseppe Pantaleo, Kate Shearston, Simon Fox, Naisana Seyedasli, and Munira Xaymardan

1 Introduction and Concepts

1.1 Stem Cells

Over the last few decades, cells with regenerative potential have been identified in a variety of tissue types. These so-called stem cells are often defined by their ability to generate cells of their own kind, a property known as "self-renewal", and their capacity to differentiate into a range of tissue types [1]. The breadth of differentiation potential (potency) in stem cells is known to be inversely correlated to the developmental stage of their tissue of residence, meaning tissues at higher levels of developmental maturity, like the majority of tissues in adult organisms, harbour stem cells with more limited differentiation capacity (multipotent) compared to their embryonic equivalents. Despite this narrowing of differentiation potential in adult stem cells, however, their capacity for tissue turnover and homeostasis, disease remodelling, and regeneration have been repeatedly demonstrated [1].

In the oral cavity, stem cells reside in different anatomical locations [2]. Oral epithelial stem cells located within mucosal epithelium are the

C. S. Farah (✉)
Australian Centre for Oral Oncology Research and Education, Nedlands, WA, Australia

Australian Clinical Labs, Subiaco, WA, Australia
e-mail: camile@oralmedpath.com.au

A. Celentano
Melbourne Dental School, The University of Melbourne, Carlton, VIC, Australia
e-mail: antonio.celentano@unimelb.edu.au

G. Pantaleo
Oral Surgery, Department of Medicine and Surgery, University of Salerno, Fisciano, Italy
e-mail: gpantaleo@unisa.it

K. Shearston
UWA Dental School, University of Western Australia, Nedlands, WA, Australia
e-mail: kate.shearston@uwa.edu.au

S. Fox
Systems Biology and Genomics, Harry Perkins Institute of Medical Research, Nedlands, WA, Australia
e-mail: simon.fox@uwa.edu.au

N. Seyedasli
Faculty of Medicine and Health, University of Sydney, and Centre for Cancer Research, Westmead Institute for Medical Research, Westmead, NSW, Australia
e-mail: naisana.seyedasli@sydney.edu.au

M. Xaymardan
Faculty of Medicine and Health, Oral Biosciences, Sydney Dental School, University of Sydney, Westmead, NSW, Australia
e-mail: munira.xaymardan@sydney.edu.au

© Springer Nature Switzerland AG 2021
S. Hosseinpour et al. (eds.), *Regenerative Approaches in Dentistry*,
https://doi.org/10.1007/978-3-030-59809-9_10

most studied among all oral stem cells [3, 4]. Approaches based on isolation of cells with slow cycling behaviour that specifically capture cells retaining incorporated modified nucleotides or specific genetic tags after pulse-chase tracking have highlighted the presence of the so-called label-retaining stem cells in the gingival epithelium, the buccal mucosa, the palate, and the tongue with the majority of cells located at the basal layer of the epithelium [5–7].

Oral epithelial stem cells are by far the only cells within the oro–dental structures capable of epithelialization, an expression of the ability to generate a fully stratified oral epithelial equivalent [3, 8]. This capacity was evident more clearly in marker-based stem cell isolation strategies. The integrin $\alpha6\beta4^{pos}$ $CD71^{neg}$ keratinocyte stem cells from gingivae, for example, show full capacity to form stratified epithelium while retaining other stem-like characters including colony formation and expression of additional stem cell markers [9]. Likewise, epithelial stem cells expressing the neurotrophin receptor p75 located in gingivae and buccal mucosa, could also generate fully stratified epithelium ex vivo [10].

A second class of cells with stem-like properties known as mesenchymal stem cells (MSC), have also been reported in multiple oro–dental structures [11]. Originally identified in the bone marrow [12] and soon after detected in almost all adult organs, MSC/-like cells are defined as multipotent, clonogenic, and self-renewing cells with the potential to at least differentiate to osteocytes, chondrocytes, and adipocytes although broader lineages have also been reported [13]. In the oral cavity, MSCs are found in the dental pulp [14], exfoliated deciduous teeth [15], apical papilla [16, 17], and the periodontal ligament [18]. Although different studies have suggested a variety of differentiation lineages for these MSCs, it is commonly agreed that they contribute to the generation of odontoblasts and the dental pulp although with varying capacities depending on MSC subtype [19–21].

The oral mucosa and the ventral tongue also contain stromal stem cells primarily derived from the neural crest that are marked by the expression of LGR5 [22]. $LGR5^{pos}$ cells are involved in stromal homeostasis and have been shown to efficiently reconstitute the entire stroma in transplantation experiments [22]. Oral MSCs have shown osteo-regenerative potential in the context of orofacial bone defects. Biocomplexes containing autologous dental pulp stem cells (DPSC) with a collagen matrix, implanted in defective maxillary alveolar bone at the site of an extracted third molar, could efficiently regenerate bone and periodontal defects proximal to the transplantation site [23]. Other studies have also confirmed the potential of DPSCs in the regeneration of alveolar bones in tooth extraction sockets [24], and in the expansion of maxillary bones during sinus lift procedure [25]. Their osteo-regenerative potential has also been reported in periodontal ligament-derived stem cells (PDLSC) where regeneration of deep intrabony defects in extraction sockets and the periodontal region have been observed [26, 27]. Tissue engineered bone grafts composed of alveolar stem cells have proven to have promising regenerative potential in the repair of alveolar clefts [28, 29] and mandibular defects caused by cyst enucleation [30].

Tissue-resident stem cells in the oral cavity have been implicated in local regenerative and repair processes including the regeneration of dentine, dental pulp, and alveolar bones both during normal homeostasis and in the context of oro–facial disease and malignancies [31]. Cells with stem-like properties, also known as cancer stem cells (CSC) have reportedly been isolated and characterized from a range of human cancers and premalignant lesions [32]. CSCs are known for their capacity to repopulate and rebuild the entire tumour in isolation, and as such are regarded as the most potent within the tumour's heterogeneous nature [32, 33]. It is notable that although a strong link exists between endogenous stem cells and CSCs, studies also suggest cancer stem-like phenotypes to be induced during post-therapy tumour remodelling [33–35].

Regardless of their origin, CSCs play key roles in multiple stages of the tumour life cycle, from initiation to progression and metastasis [36, 37]. These cells are also well known to drive cancer recurrence and resistance to therapy [36]. In oral squamous cell carcinoma, the most common type of head and neck cancer, several populations of CSCs have been reported, which will be covered in detail later in this chapter [38, 39]. Given the key functions of cancer stem cells in tumorigenesis and response to treatment, therapeutic strategies have specifically targeted this cell population to more precisely result in cell death [40] ensuring effective targeting of the entire tumour.

1.2 Growth Factors

Growth factors are naturally occurring substances capable of stimulating cellular growth, proliferation, and differentiation, the latter being one of the key factors for tissue regeneration. Usually, growth factors are a protein or a steroid hormone, with unique functions according to the need of the body. Various growth factors have been tested in the field of oral surgery to regenerate defective tissues and to increase the therapeutic effect of conventional treatment. They can work as stimulators and/or growth inhibitors. They are able to either stimulate or inhibit cell migration, act as chemotactic agents, modulate the activity of cells involved in apoptosis, stimulate angiogenesis, and promote cell survival [41, 42].

Of significant importance is the release of growth factors by platelets (fundamental for primary haemostasis). There are various platelet-derived growth factors of note applicable to the field of oral medicine. Fibroblast growth factor (FGF) stimulates the proliferation of fibroblasts and osteoblast chemotaxis and increases the expression of osteocalcin to improve wound healing. Platelet-derived angiogenesis factor (PDAF) induces vascularization and stimulates vascular endothelial cells. Platelet-derived endothelial growth factor (PDEGF) promotes wound healing by stimulating the proliferation of keratinocytes and dermal fibroblasts. Epidermal growth factor (EGF) stimulates cell proliferation and turnover of the extracellular matrix, chemotaxis of fibroblasts, and differentiation of epithelial cells. Platelet Factor 4 (PF4) promotes chemotaxis of inflammatory cells (neutrophils) and the migration and mitosis of endothelial cells [43].

Other cytokines and growth factors are more relevant to bone regeneration. Platelet-derived growth factor (PDGF) is the first of the growth factors to appear after an injury. It promotes the regeneration of connective tissue through the synthesis of proteins and collagen. PDGF is a glycoprotein with a molecular weight of around 30 kD, with three isoforms; PDGF-AA, PDGF-BB, and PDGF-AB. The primary effect of PDGF in bone regeneration is to induce mitogenic activity in cells of mesodermal origin such as fibroblasts, vascular muscle cells, glial cells, and chondrocytes. PDGF also attracts and activates inflammatory cells such as neutrophils, monocytes, and fibroblasts and stimulates the synthesis of additional growth factors. PDGF activates osteoblast chemotaxis by promoting fracture healing. Periodontal defects and alveolar bone defects have been successfully regenerated using PDGF [44, 45].

Transforming growth factor-β (TGF-β) and growth factors of the same family as Bone morphogenetic proteins (BMPs) play an important role in cell division, differentiation, migration, adhesion, organization, and apoptosis. In humans, there are three TGF-β subtypes, but TGF-β1 and TGF-β2 are the most important in reference to repair and regeneration of connective and bone tissue. TGF-β can inhibit the formation of osteoclasts and the synthesis of proteases, promote the differentiation of osteoblasts, and the synthesis of the osteoid matrix thus determining positive and favourable bone apposition [46]. Among BMPs, BMP-2 and BMP-7 have been studied for clinical applications and used in combination with collagen/collagen composite scaffolds. To date, it has been possible to reconstruct mandibular defects, maxillary sinus augmentation, and cleft alveolus using BMP-2 or BMP-7 combined with collagen sponge as a carrier [42].

Another important therapeutic application of BMPs is for maxillary bone regeneration to allow replacement of lost teeth by osseointegrated dental implants. This approach involves the regeneration of peri-implant bone after implant fixation or bone height improvement in areas below the maxillary sinus [47]. Currently, the use of BMPs to treat several refractory diseases, including bisphosphonate-related osteonecrosis of the jaw (BRONJ), is being attempted in the field of oral surgery [41]. BMP-2 is well recognized as the most effective bone induction treatment. There is evidence that BMP-2 can directly influence osteoclastogenesis through osteoblast/stromal cells. Some authors have found that a lower dose of BMP-2, was able to induce osteoclast potency, indicating that at lower concentrations, BMP-2 can have osteoinductive potency [48].

Insulin growth factor (IGF-I) stimulates osteoblasts. IGF has been found to increase bone turnover in patients with low bone density [49]. Vascular endothelial growth factor (VEGF) is the main regulator of vasculogenesis and angiogenesis. It also plays an important role in the regeneration of bone tissue [50].

Connective tissue growth factor (CTGF) adheres to platelets in injured sites where it appears to be overexpressed during platelet coagulation. Inactivated platelets contain a considerable amount of CTGF. Once released, it promotes angiogenic activity, cartilage regeneration, and fibrosis. Some authors have shown that CTGF is expressed in bone marrow cells, but not in megakaryocytes [51]. This suggests that the total amount of CTGF in platelets is the result of endocytosis conducted by the extracellular environment of bone marrow [51].

In addition to platelets, leukocytes also play an active role in tissue healing by releasing cytokines and growth factors at the lesion site that can influence wound remodelling, such as transforming growth factor (TGF), interleukin-1 (IL-1), fibroblast growth factor (FGF), and tumour necrosis factor (TNF). Granulocytes and macrophages synthesize inflammatory mediators such as leukotriene B4 and platelet activating factor (PAF), which increase the permeability of blood vessels and stimulate the production of inflammatory cytokines and proteolytic enzymes. IL-1 and TNF act on blood vessel endothelial cells, causing the adhesion of neutrophils and lymphocytes and their diapedesis (migration outside the vessel). Lymphocytes also produce growth factors and contribute to tissue remodelling during the late stages of healing [50, 52]. According to some authors, activated T lymphocytes are capable of producing insulin-like growth factor I (IGF-I) to promote wound healing [53].

Despite promising clinical success of growth factors for tissue regeneration, controversies still hound their clinical use for tissue regeneration in the maxillofacial region. The greatest debate deals with the oncogenic potential of growth factors. Many tumours including malignancies of the oral and maxillofacial region are known to overexpress growth factors such as TGF-β and BMPs, which ironically are beneficial for tissue regeneration when expressed in adequate amounts in tissue. For the clinical application of growth factors, supraphysiological doses are required for effective tissue regeneration, but this increases the risk of tumour development.

Unfortunately, the optimum concentration and appropriate timing for administering growth factors are not fully established, and this is crucial for successful clinical outcomes without side effects. Biologically, growth factors usually react with each other in a very delicate and sophisticated manner, exchanging feedback from responding cells and tissues. Growth factors often have biphasic features depending on the condition of the tissue. Therefore, in figuring out the appropriate dose and timing of growth factors, more studies are required to understand the exact mechanisms of the cascades of growth factors when the target tissue is regenerated. In addition, a single dose of exogenous protein is well known not to induce an adequate biological response in compromised tissue conditions. Therefore, injudicious use of growth factors to target tissue regeneration should be avoided, and more detailed considerations should be endorsed prior to growth factor use [41].

1.3 Matrices

A matrix is a microscopic structure inserted into the body to give form to a macroscopic and organized structure [54]. Matrices are either organic or inorganic. Organic matrices include collagen type I, the most abundant protein in human and animal tissues, which is required for structural maintenance of tissues.

All inorganic matrices have common features, such as fragility, non-susceptibility to corrosion, and small fatigue resistance. The most significant property of inorganic matrices is their long resorption time in the body associated with the absence of an induced inflammatory response. However, none possess the properties of osteogenesis and bone induction [55].

Hydroxyapatite and calcium phosphates are commonly used inorganic matrices. Calcium phosphate-based inorganic matrices can be further classified as ceramic or cement matrices. Ceramic matrices are divided into bioglass and hydroxyapatite ceramics, whereas β-TCP is included in a subgroup of cement matrices [56]. Hydroxyapatite can be derived from animal bone, coral, or can be manufactured as purely synthetic hydroxyapatite. Hydroxyapatite has several disadvantages that include reduced mechanical resistance, long resorption time, and difficulty in controlling pore size [57].

Bioglass ceramics are entirely synthetic in nature [58]. Bioglass is composed of oxides of silicon, sodium, calcium, phosphorus, and boron, although the final chemical composition is extremely variable in terms of the percentage of the various elements present. It has been shown that ions released from Bioglass inserted into the body induce intra- and extracellular responses, by stimulating osteoblast differentiation and revascularization [54, 59].

Moreover, Bioglass undergoes controlled resorption in an optimal time, and demonstrates efficient bioactivity, and the ability to modulate cell migration. Bioglass can form chemical bonds with tissues. If placed in contact with a liquid medium, Bioglass forms a layer of hydroxycarbonate gel of calcium and silica, which facilitates the absorption of proteins used by osteoblasts to produce matrix [60]. Furthermore, Bioglass can support revascularization, and the adhesion, growth, and differentiation of osteoblasts [57, 61]. Despite these exceptional characteristics, Bioglass has a low resistance to load, similar to other inorganic matrices.

Tissue- or cell-derived extracellular matrix bone grafts can preserve most of the biological and biomechanical properties of mature bone. Extracellular matrix bone grafts preserve the architecture and native bioactive factors such as growth factors and cytokines. Thus, these materials show promise for fulfilling the biophysical and biochemical requirements of new bone formation or bone regeneration [62, 63].

For bone therapy, tissue-derived extracellular matrix scaffolds are generally of two types: demineralized bone matrix and decellularized bone [63, 64]. Demineralized bone matrix has a similar composition, function, and structure to that of an autograft, hence this kind of matrix has broad application in multiple bone therapeutic surgeries [64]. The creation of demineralized bone matrix was inspired by a simple hydrochloric acid extraction process for the decalcification of bone matrix, leaving a complex with osteoinductive and osteoconductive properties derived from native bone [65, 66]. Some authors have demonstrated that organic components inside the bone extracellular matrix promote marrow-directed osteogenesis [67]. In vitro studies that have demonstrated the ability of demineralized bone matrix to stimulate osteogenesis in periosteal cells and human mesenchymal stromal cells [68, 69]. Other studies have demonstrated similar results in vivo for healing of bone defects in rats [70].

Demineralized bone matrix has several advantages over autologous bone for clinical applications, as well as reducing donor site morbidity and operative time [64]. Demineralized bone matrix has been used to supplement autologous bone grafts in numerous clinical trials [66]. Nevertheless, there is as yet no clinical evidence to support the independent use of these materials.

Tissue-derived bone extracellular matrix (demineralized bone matrix and decellularized bone) from xenografts or allografts may induce immunological responses or pose a risk of transmitting infectious diseases. Hence, extracellular matrix derived from autologous cells constitutes an alternative source [71]. Extracellular matrix derived from cells, either from osteoblasts or osteoblast-like cells, contains minerals, growth factors, proteoglycans, and proteins. These components have an important role in proliferation, differentiation, cell adhesion, and migration [71,

72]. Extracellular matrix from stem cells facilitates the property preservation of stem cells, but prevents their differentiation into osteoblasts [73]. Extracellular matrix from osteoblasts or osteoblast-like cells is osteogenic, and can bind growth factors [74]. It can also modulate the local inflammatory response.

Several authors have used extracellular matrix from fibroblasts and chondrocytes to promote osteogenic differentiation of stem cells [75, 76]. The main challenge in using cell-derived extracellular matrix is finding a way to upscale the process in such a way that it can be applied clinically for human regenerative therapies [62, 77]. For bone repair, tissue-derived extracellular matrix bone grafts can promote osteogenesis. Demineralized bone matrix is produced and applied as a supplement to autograft materials. It can provide mechanical support for the repair of some tissue defects. Cell-derived extracellular matrix has no mechanical strength. It is not simple to collect in adequate quantities for bone repair, so it has no indication in treating large size defects.

The keys to the successful clinical application of extracellular matrix-derived tissues for hard tissue repair (for example, in BRONJ) are the preservation of bioactive factors and the elimination of any immunogenic substance, which remains in the source tissue during decellularization or demineralization procedures. Current acellular methods normally cause loss of bioactive factors and/or the destruction of tissue ultrastructure due to the extreme conditions experienced during their processing [78].

1.4 Scaffolds and Grafts

Scaffolds are biomaterials that can provide support as a biological platform to facilitate the restoration and repair of the physiological features of injured tissues [54, 55, 79–81]. In tissue regeneration, a biocompatible scaffold permits cell adhesion, and induces cell proliferation and differentiation without initiating an inflammatory reaction or rejection by the body [81].

An ideal bone scaffold must be osteoinductive, osteoconductive, and osteogenic. An osteogenic material can generate bone tissue, similar to that by osteoblasts. An ideal scaffold should also be an osteoconductive material that stimulates bone cells to grow on its surface [55, 82]. Nevertheless, these fundamental characteristics alone are insufficient for the successful in vivo application of a bone scaffold. To fabricate a bone scaffold, an ideal biomaterial must also be bioinert, bioactive, biocompatible, and biodegradable. The scaffold–cell interaction must ensure easy penetration, distribution, and proliferation. The biomaterial should be porous, with a suitable pore diameter to enable cells to penetrate the biomaterial, hence, securing the growth of new bone tissue and vascularization [54, 83].

Biocompatibility, nevertheless, remains the most essential property that scaffolds must possess. Biomaterials should not induce inflammatory responses in the body, or show any cytotoxicity or immunogenicity [57]. It is also fundamental that the scaffold biomaterial be efficiently resorbed along with the deposition of new bone tissue, so that new bone can replace it entirely, while maintaining its shape and thickness [84]. Craniofacial scaffolds must fill three-dimensionally complex defects and provide adequate resistance to temporary load during regeneration.

In relation to scaffolds, it is necessary to distinguish between bone grafts, matrices, polymeric materials, and combined (composite) scaffolds. Bone grafts are classified as autologous, homologous, heterologous (xenografts), or alloplastic grafts. Currently, autologous bone grafts are considered as the gold standard among all bone grafts.

In autologous bone grafting, the donor's own bone is taken from a healthy part of the donor's body and grafted in the affected region, minimizing rejection issues. The donor sites can be intraoral or extraoral [85, 86]. Allogeneic grafts or allografts are derived from donors of the same species, but lose many of their characteristics during the long process of decellularization and sterilization. These grafts can be classified as freeze-dried bone allograft (FDBA) and demineralized freeze-dried bone allograft (DFDBA) [87, 88].

Xenografts are a less expensive alternative to allografts, but undergo similar sterilization procedures. They are derived from animals, principally pigs or horses. Several studies have shown that these materials provide support and survival similar to those of autologous bone grafts, but without any osteogenic properties [89, 90]. Several alloplastic grafts have been used for bone regeneration, such as β-tricalcium phosphate (TCP), hydroxyapatite, and bioceramics. They can be manufactured at lower cost than heterologous biomaterials; however, their resorption is not always ideal.

Polymeric materials are classified into natural or synthetic polymers. Natural polymers are organic in origin. The most representative organic polymer is collagen. Other organic polymers used as scaffolds include hyaluronic acid, alginate, chitosan, and peptide hydrogel [55, 91].

Synthetic polymers are classified as absorbable or non-absorbable polymers. They include polylactic acid (PLA), polyglycolic acid (PGA), polylactic–polyglycolic acid (PLGA), polyethylene glycol (PEG), PEG with PLGA (PEG-PLGA), and polycaprolactone (PCL) [92]. The most commonly used non-absorbable polymer in bone regeneration is polytetrafluoroethylene-expanded (e-PTFE). This polymer is used in the form of a membrane to cover bone grafts. It forms a barrier between the graft material and the soft tissues of the flap to inhibit the early onset of gingiva-derived fibroblast formation [93].

PLA and PGA are synthetic polymers with excellent biomaterial characteristics. Their properties can be controlled during their synthesis [94]. Their rapid degradation is a disadvantage, as this may cause early failure of the graft. Additionally, intracellular degradation of an acid can induce an inflammatory response [94, 95]. To reduce inflammation, hybrid scaffolds have been created, combining PLA and PGA with Bioglass and calcium phosphate [96].

PLGA is a copolymer obtained by the union of lactic and glycolic acids through ester bonds. Currently, this polymeric bone substitute is used extensively for bone regeneration in dentistry, and it has been combined with growth factors to obtain good results [94]. PEG is a polyether with a high molecular weight that is very resistant to resorption. It has been used in combination with multipotent stem cells and peptides with good results. In addition, it has been used as a scaffold for neuronal regeneration in the treatment of pathologies of the central nervous system [97]. PCL is an aliphatic polyester. It has good mechanical characteristics and very long resorption times [57]. It has been combined with hydroxyapatite and chitosan to form hybrid scaffolds with high mechanical resistance [57, 98].

In some cases, it is possible to combine the materials described above, to improve their mechanical characteristics and osteoconductivity. Composite scaffolds obtained when PLA is enriched with dicalcium phosphate or PGA and PLGA can be combined with hydroxyapatite or β-tricalcium phosphate. This can increase the degradation time and improve the mechanical properties of scaffolds [99].

Scaffolds containing hydroxyapatite reinforced with collagen have been developed to overcome the mechanical strength limitations of collagen, and to stimulate the differentiation of stromal cells in vitro and in vivo [100]. Collagen can also be enriched with growth factors to induce osteogenesis, or associated with multipotent stem cells and polypeptides in order to improve cellular colonization.

Scaffolds enriched with hydroxyapatite have shown very positive outcomes in maxillofacial surgery [101]. In particular, different collagen formulations enriched with nano-hydroxyapatite have been created to increase the migration and differentiation of progenitor cells involved in bone regeneration. PLGA scaffolds have been enriched with hydroxyapatite to slow the process of graft resorption, and to enhance the mechanical properties. Surprisingly, this combination also increases cell engraftment and the amount of newly formed bone tissue [102, 103].

1.5 Bioprinting

Bioprinting is an extended application of additive manufacturing or rapid prototyping that has emerged as a promising avenue in tissue engineering and regenerative medicine. The process is defined as the printing of structures containing viable cells, biological materials, and bioactive molecules, to produce three-dimensional cell-laden microstructures mimicking native tissues [104]. Bioprinting is a process of biomimicry that has the potential to provide individualized spatial geometry, and controlled structural design and positioning of cells and molecules [105, 106]. Compared to tissue engineering processes that necessitate post-fabrication cell seeding, bioprinted scaffolds retain a more homogeneous and controlled cell distribution. This positively impacts on their integration with the host tissue, providing grafts with reduced risk of rejection and uniform tissue growth in vivo [104, 107].

Bioprinting technology is experiencing rapid growth and continuous improvement. The most common bioprinting platforms use inkjet-based, extrusion-based, or laser-based printing and fabrication technologies [104]. In inkjet-based bioprinting, cells and biomaterials are positioned into a desired pattern using droplets ejected via piezoelectric or thermal processes [108]. Although inkjet printing is widely available, affordable, and rapid, and gives reasonable cell viability, the low droplet directionality and unreliable encapsulation of cells and biomaterials in this method cause concern [108].

In extrusion-based bioprinting, the bioink is extruded through a nozzle under pneumatic or mechanical pressure and deposited in a predesigned structure [109]. This technology can print with very high cell densities. It has, however, limited resolution and so is suited mostly to hard tissues with larger size defects. As such, it has been used widely in bone and cartilage tissue engineering [110].

Laser-based bioprinting is a nozzle-free technique that uses ultraviolet (UV) or visible light to cure photosensitive polymers in sequential layers. This technology eliminates the negative impact of shear stress, and offers fast and accurate fabrication, achieving resolutions ranging from 5 to 300 μm. The main risk with this technology, however, is possible cytotoxic effects from the photoinitiators or the UV light [104, 107, 111].

Regardless of the printing technology that is used, the bioprinting material is often encapsulated in a bioink. This biomaterial ensures that cell viability is maintained. Bioink can be used to incorporate additional biomolecules [112]. The majority of cell-laden bioinks are hydrogel-based, as their high water content ensures both retention of moisture for cell survival, and protection against fabrication-induced shear forces [112, 113].

Bioinks have served as the sweet point of technological innovation in bioprinting, such as generation of personalized tissue transplants. Platelet-rich plasma, for example, is a rich source of patient-specific autologous growth factors. This has been combined with alginate-based hydrogels to generate 3D bioprinted constructs which show enhanced angiogenesis, stem cell recruitment, and tissue regeneration [114].

In addition, bioinks have been the transition point towards 4D biofabrication strategies, where a dynamic temporal shape-morphing of bioprinted material is introduced. Development of hollow self-folding bioprinted transplants to generate tubes with controlled diameters of up to 20 μm resolution is an example of such approaches [115].

Although bioinks have proven to be an essential component of 3D biofabrication, efforts towards the development of inkless or hydrogel-free printing to generate complex flat or multilayered tissues are noteworthy. As an example, fibroblasts have been positioned on a printed retinal pigmented epithelial layer as hydrogel free droplets using inkjet bioprinting [116]. Although maintaining cellular viability is still a challenge in this approach, there is hope that further tailored technologies might make this possible in the future.

3D bioprinting has major applications in preclinical studies, disease modelling, and drug screening [117–122]. The ultimate goal for the technology, however, is its use in regenerative therapy and tissue replacement in the clinical setting. As such, 3D bioprinting has also been an approach of interest in regenerative dentistry and craniofacial tissue engineering [123]. Reconstruction of the complex structure of the tooth, including dental pulp and dentine as well as innervating nerves and the vasculature together with supporting tissues such as the periodontal ligaments and alveolar bones, is the desired outcome, but an ambitious aim [124].

3D bioprinting has been successfully applied in various fields of regenerative dentistry. Light-assisted 3D printing technologies have been used in the generation of in vitro models for tooth buds [125, 126], and to mimic pulp revascularization using hydrogels encapsulating dental pulp stem cells [127]. There have also been successful attempts to reconstruct the complex structure of the periodontium, with cementum, periodontal ligament, alveolar bone, and fibre bundles in defined directions [128–131].

3D fabrication has also been used experimentally to reconstruct the dentine–pulp interface, providing a tool for regenerative endodontics to mimic the production of reparative dentine [132], and to facilitate pulp revascularization [133, 134] and neurotization [135, 136]. Furthermore, 3D reconstructed tooth germs have promising applications, including eruption through the jaw bone in preclinical in vivo transplantation experiments [137–139]. There is increasing demand, however, to achieve efficient tooth germ reconstruction while simulating the 3D anatomy of teeth, to account for the individual differences between molar and incisor teeth as an example.

3D bioprinting has also been applied to the biofabrication of alveolar bone and mandibular joints, and to reconstruction of other craniofacial structures [123, 140]. 3D printing has been applied to epithelial tissue engineering for the regeneration of defects in oral mucosal epithelium. As such, these so-called "ex vivo produced oral mucosal equivalents" have been efficiently used in pre-prosthetic procedures, such as vestibuloplasty, repair of superficial mucosal defects, and for prelamination of surgical free flaps [141–143]. 3D bioprinting has also been used to effectively recapitulate the secretory epithelium [144], and to partially regenerate salivary glands by stimulating epithelial growth in glandular tissue [145].

2 Current Applications of Stem Cells for Treating or Preventing Oral Conditions

2.1 Oral Mucosal Lesions

Oral mucosal lesions include a range of benign or potentially malignant clinical conditions affecting the oral cavity, including oral lichen planus (OLP), oral leukoplakia (OLK), and oral pigmented lesions (OPL). This section will outline current applications of stem cells for the treatment and prevention of oral potentially malignant disorders (OPMDs).

2.1.1 Oral Lichen Planus

Oral lichen planus (OLP) is a T-cell-mediated inflammatory condition characterized by white or mixed red white lesions. It is a relatively common condition that affects 0.5–2.2% of the population, and is more prevalent in females and in the 30–60 year old age group [146]. Multiple lesions are common and can be classified into a number of clinical sub-groups: erosive, reticular, plaque-like, atrophic, or bullous type, with erosive lesions being associated with a greater risk of malignant transformation [147]. Histologically, OLP is characterized by a lymphocytic inflammatory infiltrate accompanied with basal layer liquefaction, which is thought to be a result of keratinocyte apoptosis driven by activation of cytotoxic T cells. Keratinocytes are known to perform antigen presentation to helper T cells, which are stimulated to produce IL-2 and IFN-gamma [148].

Current treatments for OLP include corticosteroids and other immunosuppressive measures. Corticosteroids are used widely, either in systemic or topical form, but can be associated with side effects such as oral candidosis. Retinoids have also been used, but appear to be less effective than corticosteroids, and are associated with more severe side effects. Calcineurin inhibitors include cyclosporine and newer drugs such as tacrolimus and pimecrolimus. These inhibit cytokine release and have also been used in OLP, however, this drug class is associated with a potential increased risk of malignant transformation [149]. A recent Cochrane review found limited evidence supporting the efficacy of corticosteroids in OLP, and some evidence that tacrolimus may be more effective at pain resolution [150].

Photobiomodulation (PBM) and Low-Level Laser Therapy (LLLT) are largely overlapping terms referring to the use of low intensity laser energy to stimulate cellular regeneration and healing. They have been used, primarily in erosive/atrophic type lesions of OLP, which are more challenging to treat with conventional methods. A number of studies have investigated PBM or LLLT in the treatment of OLP with mixed results [151]. Cafaro et al., for example, found that 78% of lesions resolved completely in an average of 11 sessions, while Trehan and Taylor found that the majority of patients responded well [152, 153]. Jajarm et al. found the treatment to be as effective as corticosteroids in 10 treatments [154], while Mirza et al., and Kazancioglu and Erisen found LLLT to be less effective than alternative treatments [155, 156].

Mesenchymal stem cells (MSCs) have considerable potential as a novel treatment for OLP, but to date there is minimal evidence available. A small study in humans assessed the ability of bone marrow-derived undifferentiated pluripotent stem cells. The cells were isolated from patients undergoing surgery, and then injected into the site of OLP excision, to promote wound healing, however, the study was small in scale and the results were unclear [157].

Aside from wound healing after surgical biopsy or excision, MSCs, particularly those isolated from the dental pulp (DPSCs), have shown particular promise in their ability to modulate immune and inflammatory processes (reviewed in [158–160]). In animal models, DPSC have been used to inhibit injury-induced neural inflammation [161] and diabetic polyneuropathy [162] among others, but to date human clinical applications have been limited to pulpal and bone regeneration [23, 163].

DPSCs can secrete an array of inflammatory signalling molecules, including transforming growth factor beta (TGF-β), indoleamine 2,3-dioxygenase (IDO), prostaglandin E2 (PGE-2), and human leukocyte antigen G5 (HLA-G5), that interact with immune cells [160]. The ability of MSCs to home into sites of tissue damage and modulate immune responses has been demonstrated to involve IDO. A recent study sought to identify and isolate MSCs from OLP lesions. Zhang et al. found that MSCs were present in OLP lesions, and were in greater numbers than in normal tissue, suggesting that they were perhaps moving to sites of inflammation. OLP MSCs were, however, less effective at suppressing T-lymphocytes in vitro than MSCs from normal tissue. This raises the possibility that delivery of MSCs from normal buccal mucosa to OLP tissues could be effective at inhibiting T-lymphocyte-mediated damage [164].

An intriguing pair of studies in cats with feline chronic gingivostomatitis, a feline T-cell-mediated oral inflammatory condition with similarities to human OLP, has demonstrated potential benefit of both autologous and allogenic stem cells. Infusion of allogenic adipose derived stem cells was associated with clinical remission in 57% of cases (4/7), with a slightly higher proportion responding to the autologous equivalent (71%; 5/7) [165, 166]. Although the use of MSCs to treat OLP is not yet at the stage of clinical application, there are definite possibilities in the field.

2.1.2 Oral Leukoplakia and Epithelial Dysplasia

Oral leukoplakia (OLK) is an oral lesion defined as a white patch or plaque that cannot be characterized clinically or pathologically as any other definable lesion, having excluded other known diseases or disorders that carry no increased risk for cancer [167–169]. It is one of the most common oral potentially malignant disorders (OPMDs), with a worldwide prevalence of approximately 4.11% [170]. A subset of OLK lesions (between 0.13 and 34%) undergo malignant transformation to oral squamous cell carcinoma (OSCC). Factors such as increased lesion size, a non-homogeneous appearance, and the presence of patient risk factors such as smoking and alcohol use can increase this risk [171, 172]. Typically, OLK lesions are biopsied and submitted for histopathological analysis in order to exclude other diagnoses, for example, OLP and to determine the presence or absence and grade of oral epithelial dysplasia (OED) [173]. The main treatment for OLK lesions displaying OED is surgical excision, either by scalpel or by laser, while non-dysplastic lesions are more often reviewed for potentially sinister changes in appearance.

Excision of suspicious OLK can be challenging, particularly with large lesions. PBM/LLLT has been used in a number of in vivo and in vitro *studies* to enhance the healing of experimentally induced ulcers (punch biopsies), suggesting it could be a useful adjunct to scalpel or laser excision of suspect lesions [174–176]. This technique has also been used clinically in a small preliminary study, which found that LLLT could reduce the size of leukoplakia by approximately 50% [177]. Another small study combined cryotherapy and LLLT and found that LLLT was associated with a reduction in pain [178]. Platelet-rich fibrin (PRF) has also been proposed to have healing benefits that could be of value after the excision of OPMDs. Combinations of PRF and allogenic MSCs have been used successful in vivo to repair muscle in a rabbit model [179, 180].

Kasai et al. have developed a method for the generation of oral mucosal epithelial cell sheets isolated from a small biopsy, which could be of value in repairing large excision sites [181]. This involved collection of a small 6 mm punch biopsy, yielding a single 23-mm diameter round cell sheet and a plasma sample, which was used as a serum source for the cell culture [182]. Larger biopsies (140 mm^2) have also been trialled, and these can yield between 6 and 17 cell sheets. Enzymatically isolated cells are seeded onto temperature responsive tissue culture inserts and cultured for 14 days. When cooled to 20 °C, the cell sheets lift from the insert, and due to the maintenance of extracellular matrix proteins on the bottom surface of the sheet they will adhere to tissue without sutures. The sheets have been used to prevent oesophageal stricture after removing superficial neoplasms using endoscopic submucosal dissection, and in this setting they increased the speed of healing (3.5 vs. 8 weeks) and avoided stricture [183]. They have also been applied to corneal repair in corneal limbal epithelial stem cell deficiency, a success rate of 64% in patients [184].

Similar grafts using keratinocytes isolated from an oral mucosal biopsy then seeded onto a dermal equivalent matrix have been used for repair of cancer or premalignant lesion sites [141]. These have potential for use when large or multiple excisions of areas of OED would benefit from rapid healing or in sites where stricture or shrinkage could affect oral function. It would be critical, however, to ensure that the site of the donor cell biopsy does not include areas that are molecularly abnormal which could be challenging in some cases. Kasai et al. have used brush biopsy and immunoblotting as a strategy for assessing the "normality" of cells prior to fabrication of these cell sheets, although cytology, with or without immunostaining is another potential tool [181].

MSCs have not yet been used clinically in the treatment of OLK lesions, but there is some intriguing data in cell and animal models. Typically, MSCs can be differentiated into adipocytes, chondrocytes, and osteocytes and are highly proliferative. Li et al. compared the characteristics of MSCs isolated from normal mucosa, OLK (with mild-to-moderate OED) and OSCC, and found that OLK-MSCs underwent differentiation in the same way as OSCC-MSCs, but proliferation less when compared with OSCC-MSCs and MSCs from normal tissue [185]. Exosomes secreted by OLK-MSCs were able to enhance the proliferation and migration of dysplastic DOK and malignant SCC15 cells compared with those from normal MSCs and behaved similarly to exosomes from OSCC-MSCs. This suggests that exosomes secreted by OLK MSCs could be a potential target to block malignant transformation of OLK [185].

Another study which isolated MSCs from OLK compared MSCs derived from normal tissue (OMMSCs) and OLK tissue with a histological diagnosis of OED (OLKd-MSCs) or hyperkeratosis (OLKh-MSCs). Zhang et al. found that overall, OLK-MSCs were more proliferative and displayed enhanced colony-forming ability when compared with OMMSCs, and that this effect was augmented in OLKd-MSCs. These effects may be correlated with the process of epithelial-to-mesenchymal transition known to be critical during the development of OED and OSCC. OEDd-MSCs were less effective at osteogenic, neurogenic, and adipogenic differentiation than MSCs derived from normal or non-dysplastic tissue. Interestingly when MSCs were transplanted into mice in a complex with fibrin, OLDd-MSCs partly lost the ability to form mesenchymal tissue and maintain regeneration. This was linked with decreased production of collagen IV and enhanced expression of the matrix-metalloprotease 9 (MMP-9; degrades collagen IV), suggesting a loss of basement membrane integrity. Overall, OLK-MSCs displayed impaired regenerative capacities, differentiated less effectively and displayed proliferative abilities that increased with the grade of dysplasia, suggesting that oral mucosal MSCs may lose their protective effect in OED [186].

In animal studies, MSCs were able to reduce the proportion of hamsters undergoing MT in a 7,12-Dimethylbenz[*a*]anthracene (DMBA) carcinogenesis model, but only when they were administered at later stages of the model when dysplastic lesions and papillomas had started to appear [187]. The DMBA hamster model provides a model for oral carcinogenesis over a 13-week period, with the sequential appearance of hyperplastic lesions (weeks 4–6), dysplastic lesions (weeks 6–8) and papillomas (weeks 9–12), with extensive OSCC present by the end of the study. Lower doses of MSCs administered at dysplasia or papilloma stage resulted in a lower proportion of lesions progressing to OSCC, and for MSCs administered at papilloma stage, resulted in a reduction in lesion size. Interestingly the MSCs rarely homed to the site of pre-malignant lesions, suggesting that the reduction in OSCC burden was due to a systemic immunomodulatory action, rather than a direct effect. This is supported by a number of studies where MSCs were able to inhibit inflammation in animal models (reviewed in [158–160]) and the association of inflammation with oral malignant transformation.

MSC-derived exosomes loaded with miR-185 also can provide benefit, significantly reducing tissue-specific and systemic inflammation, as well as OPMD progression to OSCC in a hamster DMBA model. However, there is also evidence that MSCs can drive malignant transformation. In another animal chemical carcinogenesis model, MSCs were increased by 4-nitroquinoline 1-oxide (4NQO) treatment. The MSCs derived from oral cancer could block T-cell proliferation, suggesting that MSCs may be involved in oral carcinogenesis via immunosuppression [188].

Cancer stem cells (CSCs) are a small cell population within tumours with enhanced metastatic properties and resistance to radiotherapy and chemotherapy [189]. While CSCs are known to be resident within OSCC, they have also been identified in OLK lesions, and have the potential to be used to identify higher risk lesions and to provide therapeutic targets. OSCC CSCs can be characterized by a number of cell surface markers including the glycoproteins CD44, CD133 and podoplanin, ATP-binding cassette super-family G member 2 (ABCG2), Aldehyde dehydrogenase 1 (ALDH1), tyrosine-protein kinase Met (c-MET, also known as hepatocyte growth factor receptor), Polycomb complex protein BMI-1 (Bmi1), and leucine-rich repeat-containing G-protein-coupled receptor 5 (LGR5). Various transcription factors including octamer-binding transcription factor 4 (Oct4), the homeobox protein NANOG (Nanog), SRY (sex determining region Y)-box 2 protein (Sox2), Krüppel-like factor 4 (KLF4) and Myc are also important in CSCs [190, 191], and the expression/activation of a number of these markers has been investigated in OPMDs.

CD44 expression was increased in dysplastic OPMD, particularly those with moderate or severe OED, compared with normal and non-dysplastic OPMD [192]. Interestingly, this study also identified increased expression of stromal cell-derived factor 1 (SDF1) and C-X-C chemokine receptor type 4 (CXCR4) with increasing OED, which have been implicated in CSC-specific signalling in HNSCC [192].

CD133 expression has also been shown to be very low in normal tissues, but significant increases in mildly dysplastic tissue, and increases even further in moderate–severe OED [193]. CD133 has also been implicated in OLP MT in a case-control study of 110 patients, a cohort in which a rather high proportion of cases (10/110) developed OSCC. Of the patients that transformed, 80% displayed CD133 expression, versus 29% of non-transforming OLP cases [194].

Saintigny et al. identified c-MET as a potentially actionable marker for the development of oral cancer using a microarray, then demonstrated that expression of c-MET in OLK was associated with a reduced time to OSCC with immunohistochemistry. MET activation was further demonstrated in OPMD and OSCC cell lines. MET inhibition was able to reduced malignant transformation rates in a 4NQO mouse model [195]. c-MET expression in oral submucosis fibrosis was also associated with malignant transformation, with all the cases of OSMF that transformed to OSCC displaying staining [196]. ATP-binding cassette super-family G member 2 (ABCG2) and Polycomb complex protein BMI-1 (Bmi1) are additional stemness markers that are expressed in OLK lesions, and are associated with an increased risk of malignant progression, but these display variability that limit their use as a tool to identify high risk lesions [197, 198].

The expression of podoplanin in OLK has been correlated with the presence of OED. Podoplanin-positive OLK lesions are associated with an increased risk of malignant transformation [199–202]. When protein expression levels of ALDH1 were assessed along with podoplanin, 66% of patients expressing both proteins in OLK lesions developed oral cancer.

ALDH1 was assessed in another cohort of OLK patients with long-term follow-up, along with the expression of the glycoprotein CD133. CD133 and ALDH1 expressions were associated with a 2.9 and 4.2 fold increased risk in the development of oral cancer, respectively [203]. LGR-5 expression has also been investigated in OED and found to be increased significantly in moderately-severely dysplastic OLK compared with normal tissue. This was associated with a shift from basal layer staining to more extensive staining throughout the tissue [204].

The evidence for the influence of CSC-associated transcription factors in OED is less extensive to date, but there have been some studies using human FFPE tissue. Sox-2 expression when assessed by IHC was enhanced in the suprabasal layer of OLK samples compared with normal samples [205]. In a smaller initial study, Sox-2 and Oct-4 expressions were also demonstrated in the majority of cases in a mixed group of OPMD including OLK and OLP (18/20 and 14/20, respectively) [206]. Oct-4 appears to be less consistently expressed in OED than Oct-4, with a recent study demonstrating only 9% positivity in OED samples, compared with 59% positivity for Sox-2 [207]. In a mouse 4NQO carcinogenesis model, knockdown of KLF-4 was associated with an increase in the severity of dysplastic lesions [208]. In a DMBA carcinogenesis hamster model, Myc staining reached a peak in hyperplastic oral tissues then remained steady throughout OED and OSCC. This suggests that Myc overexpression was an early event in oral carcinogenesis [209]. Myc expression was also demonstrated in human tissue in OED and OSCC [210, 211]. Cytoplasmic expression of Nanog, while only present in a relatively small subset of OED samples (16.4%), has been associated with an increased risk of malignant transformation [212].

Overall these studies demonstrate that subgroups of CSCs are present at various stages in the progression of oral carcinogenesis, and in many cases CSC markers are able to identify potentially malignant lesions at higher risk of undergoing malignant transformation. These provide both putative predictive markers for OSCC progression to shape management but also the opportunity for novel chemotherapeutic treatment to block tumour development.

2.2 Oral Malignancies

Head and neck malignancies, of which oral squamous cell carcinoma is the most common, are globally significant cancers and represent a serious health burden. This impact is seen not only in mortality but also upon quality of life even following successful therapy, due to the longer term consequences of the interventions employed. Stem cell technologies have so far had limited impact upon therapies for these cancers, however, there are a number of promising avenues by which they may potentially impact both directly and indirectly based upon current research. These include the application of tissue engineering to the restoration of structures compromised by therapeutic intervention, the development of in vitro models using stem cells for investigation of anticancer therapies, and the leveraging of knowledge of normal stem cell biology to target cancer cells with stem-like properties that are frequently termed "cancer stem cells".

2.2.1 Head and Neck Squamous Cell Carcinoma

In the clinic, surgery remains a mainstay of head and neck cancer therapy. Surgery may also be accompanied by localized radiotherapy and/or chemotherapy, depending on the staging of the tumour, the anatomical location and the involvement of lymph nodes, and metastasis. Surgical intervention can require subsequent reconstructive surgery, particularly in the case of more advanced disease undergoing extensive resection. Such reconstructive surgery in the head and neck region is particularly complicated and challenging due to the complexity of the tissue and anatomical structures. Thus, regenerative medical therapies based upon stem cells and/or biomimetic scaffolds offer promise for the repair of oncology-related tissue damage, to restore function and enhance outcomes.

Restorative therapies required after oncological surgery frequently extend beyond regenerative therapies in the oral cavity and surrounding tissue to include structures of the neck. A number of these anatomical structures have been the target of recent progress in tissue engineering. In particular, efforts to develop methods for the restoration and repair of vocal cords have demonstrated great promise in experimental models [213, 214]. Even tissues with complex 3D structures such as the oesophagus and trachea have been the subject of significant experimental advances using combinations of 3D printed biomimetic structures and complex combinations of cell populations including stem cells, with demonstrated feasibility in animal models [215–217]. These studies have integrated a number of different techniques to produce complex structures by first generating multicellular spheroids grown in culture from mesenchymal stem cells, fibroblasts, endothelial cells and smooth muscle cells, then robotically harvesting the spheroids and arraying them in a supported tubular structure [217, 218]. These structures were cultured until they formed sufficient cellular structure and matrix to be self-supporting, whereupon the artificial supports could be removed, and the grafts then implanted in vivo [217]. These ambitious efforts demonstrate the feasibility of 3D bioprinting tissue engineering technologies (reviewed in [219]), and are encouraging for the future application of stem cell-based tissue engineering applications in head and neck cancer restorative therapy.

Stem and tissue engineering technologies are also making a significant impact in the development of improved in vitro models for the investigation of cancer biology and for the testing of novel therapeutics. Reviews of clinical trials of new therapeutics have demonstrated alarmingly low success rates for cancer therapies even in comparison with the relative poor success rates of drugs targeting other conditions [220]. One of the most common reasons for failure of these trials is lack of efficacy at the phase II and III stages. It is recognized widely that improvements in assessing compounds at the preclinical stage will be essential to addressing this. These findings are driving efforts to develop experimental models with greater fidelity [221], which can robustly predict clinical efficacy at earlier stages in the development process because they are more physiologically and clinically relevant.

Historically, in vitro cancer models have involved 2D culture of cancer cell lines, however, over the past decade there has been considerable development of greatly improved preclinical cancer models. The most useful of these new models include patient-derived xenografts (PDX), in which a piece of a patient's tumour is implanted in immune-deficient mice, and patient-derived 3D organoid cultures of cells, which grow in a spatially organized manner in a biological matrix [222]. The advantages of these models are that they recapitulate more faithfully the heterogeneity and 3D architecture found in clinical tumours.

PDX models are particularly useful because they allow the exploration of tumour–stroma interactions and immunological therapies; however, they can be difficult to establish, and are not suitable for high throughput screening (HTS) [221]. In contrast, 3D organoid models are amenable to HTS, while effectively modelling key characteristics of tumour heterogeneity and physiological features. The object here is not to review in detail the development of 3D organoid models of head and neck cancers and how they are being applied, but rather to discuss how advances in stem cell technology are contributing to the development of improved tumour models.

Since the original report of the establishment of 3D organoids from mouse gut stem cells which replicate in vitro the cellular architecture and functional features of that tissue [223], organoids have been grown from many different normal epithelia including, most recently, oral mucosa [224, 225]. Thus, tissue-specific stem cells from adult tissues when cultured in 3D conditions with a suitable matrix will proliferate and give rise to differentiated cells, which will self-organize given suitable culture conditions [226]. Since cancers also proliferate and give rise to specific tissue organization, this technology has also been applied to cells isolated from primary patient tumour samples to establish tumour organoids that model tumour behaviour in vitro.

Using approaches developed for oral mucosal organoids from normal tissues, organoids have been developed which display morphological and genetic features of HNSCC [224]. These organoids maintain these morphological and genetic features even after long-term expansion, and they retain their tumorigenic capability in animal models [224]. As yet, such patient-derived tumour organoid models have not been applied to head and neck cancers to the same extent as they have for other cancers, however, in several studies such models have demonstrated promise for the prediction of therapeutic responses to cisplatin, docetaxel, and EGFR-targeted therapies [227, 228]. Thus, HNSCC organoids generated from patient tumour samples displayed similar in vitro drug sensitivity to the corresponding in vivo response to drug therapy [227] indicating the feasibility of this approach as significantly less costly and time consuming to personalized therapies than PDX models. The success rate for organoid establishment was relatively low (30%); however, other studies of HNSCC organoids have reported higher rates (60%) [224], which support further investigation of this platform.

Tumorigenesis is characterized by the development of highly heterogeneous cell populations, which exhibit phenotypic, genetic and epigenetic diversity. Different models have been developed to explain the heterogeneity found within tumours. The cancer stem cell (CSC) theory proposes that there exists within tumours a subpopulation of tumour propagating cells with stem cell like properties which can self-renew and generate different progenies (reviewed in [229]).

The CSC hypothesis also suggests that these cells are particularly resilient, being resistant to chemo- and radiotherapy, as well as the agents of metastases. The CSC model has been a topic of intense interest in cancer biology because it offers an explanation for clinical observations regarding the refractoriness of cancers to therapy, and also inspires strategies to circumvent this resistance. CSC theory has been the subject of debate, and recent major developments have led to modifications in the CSC model [229]. Advances in the understanding of adult stem cells, such as the inherent plasticity of epithelial populations, have also driven similar findings in tumour cells. For example, ablation of basal stem cells in the epithelia of mouse trachea resulted in the reversion of differentiated cells, to produce functional stem cells [230]. Similar phenomena have been described in cancer cells, which also undergo transitions between non-stem and stem-like states [231, 232].

What these studies suggest is that for many cancers, and likely for HNSCC, therapies which target CSCs alone may not be as effective as anticipated, since other tumour cells may prove capable of dedifferentiation to a stem-like state. Furthermore, there is evidence that tumour cells which do not express conventional CSC markers are capable of tumour initiation in animal models [233] that previously were considered a defining phenotype of CSC, thereby introducing additional ambiguity.

Many studies have sought to identify CSC populations in HNSCC in order to understand the role of these cells in disease progression and resistance. The search for such CSC is based upon finding presumed CSC markers which are derived from knowledge of adult stem cell biology. In HNSCC, a number of proposed CSC markers, including CD44 [234], CD133 [235, 236], CD98 [237], CD10 [238], and ALDH1 [239], have been identified as indicators of subpopulations of tumour cells that show reduced sensitivity to chemo- or radiotherapy and enhanced tumour forming potential. The investigation of stem cell markers in HNSCC has recently been reviewed [240].

Many early studies of CSC isolation in HNSCC did not conduct the in vivo xenograft analyses, which are preferred as a means of confirmation [241]. The most commonly used marker was CD44, which is used widely as a CSC marker in many cancers [242] and has been shown to be associated with prognosis and therapeutic responses in HNSCC [243, 244]. However, there is some ambiguity regarding the application of CD44 as a CSC marker in HNSCC based upon the expression pattern in normal tissue and in various experimental conditions [190, 245]. These earlier studies of CSC and their biomarkers in HNSCC need to be viewed through the prism of recent advances in stem cell and CSC biology, which demonstrate clearly that CSC themselves are a heterogeneous population, and that the expression profile of markers of stem-like properties vary both within a tumour and between patients and tumour types [229, 246]. For clinical applications and therapeutic targeting, an approach based upon personalized therapy is indicated. As well, CSC identification and characterization needs to be performed based upon multiple markers and phenotypic studies if possible.

Recognition that CSC are key players in the refractoriness of HNSCC to chemo- and radiotherapy has given impetus to the search for therapies which specifically target this cell population. While in other cancers such efforts are more advanced [247], in HNSCC there have been a small number of preclinical studies investigating CSC targeting strategies. For example, one avenue being explored is the inhibition of aldehyde dehydrogenase 1 A1 (ALDH1) to reduce CSC capacity and enhance sensitivity to cisplatin [248].

Notably, a number of studies have reported promising preclinical data using the strategy of targeting cells expressing the stem cell marker BMI1, which is associated with malignant progression in the oral cavity [197], demonstrating growth inhibition and drug sensitization in both in vitro and in vivo models of HNSCC [249–251]. These findings are promising, however, the recent evidence that non-CSC may undergo dedifferentiation and change their phenotype acquiring stem-like properties and tumorigenicity provides a complicating factor for therapeutic designs targeting CSC since one time elimination of these cells may be insufficient.

Lessons from stem cell biology in healthy tissues may provide some alternative strategies to circumvent this barrier [252]. It is well recognized that the specialized microenvironment is key to the maintenance and function of adult stem cells, and that this finding also extends to CSC. There is evidence in HNSCC for the role of the microenvironment and signalling by stromal cells in the maintenance of the stem-like state in CSC. Notably, EGF secreted by endothelial cells has been shown to support the stemness of OSCC CSC populations [253, 254] while fibroblast-derived Wnt signalling has similarly been implicated in the maintenance of the CSC niche [255] As a consequence, disruption of this microenvironment rather than direct targeting CSC has already been explored in some HNSCC preclinical models, to provide insight into the potential of this approach, including by application of small molecule inhibitors of Wnt signalling in PDX tumour models and EGFR targeted therapy in in vitro models [255–258]. These studies, while still in their infancy, provide optimism that even though there is increased complexity to the CSC targeting strategy than previously envisaged, these difficulties are not insurmountable. The rapid pace of advances in our understanding of stem cell biology in healthy tissues and cancers will undoubtedly continue to drive this in the future.

2.3 Salivary Gland Disorders

Human saliva is a biological fluid with unique properties and diagnostic capabilities. It is an exocrine solution consisting of 99% water. The remaining 1% contains many biological molecules, including nucleic acids, proteins, metabolites, and the microbiota [259]. Saliva plays a key role in maintaining oral health, helping to build and maintain the health of soft and hard tissues. Salivary gland dysfunction includes three broad pathological alterations: hyposalivation, hypersalivation, and changes in saliva composition. The consequences of all these three events can be severe and greatly impact a patient's quality of life.

The most common salivary gland disorder is xerostomia, which is the subjective feeling of dryness throughout the mouth. This may be reversible, such as in medication-induced salivary gland dysfunction, but there are also some permanent causes of xerostomia such as congenital abnormalities, autoimmune disorders such as Sjogren's syndrome, and head and neck irradiation, chemotherapy, or salivary gland tumour resection among others. Consequences of hyposalivation include rampant dental caries, periodontitis, dysgeusia, difficulty with mastication and swallowing, and decreased quality of life.

Treatments for hyposalivation are limited, and current therapies for xerostomia focus on oral lubricants and saliva substitutes, [260] and on medications to stimulate salivation from residual glands to attempt to reduce patient symptoms. These treatments are not sufficient to restore gland secretory function, and symptoms can only be relieved temporarily with current strategies. The regeneration of functional salivary gland tissue is an important therapeutic goal in the field of regenerative medicine, and several biological therapies have been proposed at preclinical stages, predominantly in the last decade. Examples of biotechnology strategies that have been attempted so far include:

- Use of salivary gland stem/progenitor cells implanted as salispheres into salivary glands to replace the functionally damaged cells [261, 262].
- Transplantation of adult stem cells, such as mesenchymal stem cells, with directed differentiation in mono- or co-culture systems [263, 264].
- Tissue engineering techniques combining cells with or without environmental cues in 3D biomaterial constructs [265, 266].
- 3D organotypic spheroid cultures assembled from epithelial cells with capabilities to grow and mature into a secretory organoid [267].

However, all these proposed strategies are affected by lack of clinical evidence for salivary gland repair or regeneration [268]. In addition, there are several issues that seem to delay the advancement of this field of research and its translation to clinical applicability. In fact, compared to other epithelial cells, salivary acinar cells are very fragile, hence specialized techniques are needed to isolate and culture them. To achieve the level of a functioning, implantable reassembled gland, human-based matrices are also needed to culture these cells in an appropriate 3D tissue architecture [269]. However, even when this becomes feasible, more obstacles arise such as considerable regulatory hurdles for tissue engineered systems.

Even if the above-cited models can use autologous cells, regulations from regulatory agencies such as the US FDA deny the use of most animal-based products for human transplantation as Class III devices, in order to prevent an immune response against the implant [269]. Current cell culture media for epithelial adherent cells commonly necessitate the addition of foetal bovine serum as a crucial factor to allow cell adhesion and expansion in tissue culture plates, and this presents a significant obstacle. To obtain approval for human use from regulatory agencies an alternative medium without animal products should be used for culture of salivary gland cells. However, so far, this has proved difficult to implement.

2.3.1 Local Obstructive, Inflammatory, or Reactive Disorders

Sialolithiasis

Sialolithiasis refers to the formation of salivary calculus or stones within the salivary gland parenchyma and/or ducts. It is the most frequent cause of inflammation of the major salivary glands. The submandibular gland is the most frequently affected (87% of cases), followed by the parotid gland (10%) and the sublingual gland (3% of cases) [270]. The treatment of choice for this condition is removal of the sialolith, and in most cases sialoliths can be surgically removed by an intraoral approach. Removal can be performed even during acute inflammatory stages, if the sialolith is accessible for the operator and as long as the infected gland is simultaneously drained.

In rare cases, when small sialoliths are located near the duct orifice, they may be removed with a lacrimal probe by performing a widening of the orifices. When sialoliths are associated with the minor salivary glands instead, they are usually excised with the associated gland. Conversely, for deeply situated sialoliths, the priority is given to antibiotic therapy to control the infection, after which removal of the sialolith should follow.

During the last decade, the most established therapeutic option for intraglandular sialoliths has been the submandibular sialadenectomy or partial parotidectomy, a technique with a low rate of complication [271]. Specifically, the low-risk standard treatment for sialolithiasis of the middle or distal portion of the Wharton's duct has been resection, under local anaesthesia, with or without partial removal of the sublingual gland. The treatment of sialoliths that are located proximally in the hilum requires more invasive and complex approaches. However, with accumulated experience and the development of new diagnostic techniques, sialadenectomy procedures have been progressively reserved, as a first therapeutic option, exclusively for tumour pathology.

The traditional management of obstructive salivary disorders has been replaced by minimally invasive gland-preserving techniques including shock wave lithotripsy, sialendoscopy, interventional radiology, and endoscopically video-assisted trans-oral and cervical stone retrieval. Of these, sialendoscopy is considered to be the method of first choice [272]. Lithiasis, stenosis, and chronic inflammatory conditions can benefit from minimally invasive procedures associated with sialadenoscopy, and many authors have demonstrated that the conservation of the salivary gland is characterized by functional and anatomical recovery after the period of obstruction [273].

Larger sialoliths (>5 mm in diameter) can be removed using extracorporeal shock wave lithotripsy externally or intraductally [274]. In addition, multiple studies have shown promising results using lasers as adjuvant tools to fragment and remove these calcifications [270]. Laser therapy has been demonstrated to be a relatively simple and particularly safe method to manage larger sialoliths. The different types of laser investigated in the literature include pulsed excimer, pulsed dye, carbon dioxide, Holmium:YAG, Erbium:YAG, diode, and Thulium:YAG, with reported treatment success rates ranging from 81% to 100% [270]. The use of lasers in this field has also been reported in the literature for surgical removal of sialoliths from the salivary glands. The types of laser that have been used for this purpose include diode lasers [275, 276] and the CO_2 laser [277–281].

Notably, in the literature to date in the field of sialolithiasis there is a complete lack of support for regenerative procedures such as those involving stem cells. Therefore, the only hint of regenerative medicine in the field of obstructive salivary gland disorders remains the use of laser therapy, which is considered to be superior to conventional surgical treatment due to its photobiomodulation abilities (leading to increased production of collagenase and enhanced wound-healing quality), safety and minimal invasiveness of the procedure, resulting in reduced morbidity. Nevertheless, increased operative time and the high cost of equipment represent limitations to be considered [282].

Mucoceles and Ranulas

Mucoceles are common fluid-filled lesions (salivary gland mucous extravasation cysts or mucous retention cysts), which typically present as a swelling. They commonly appear in the lip, floor of the mouth, or soft palate. This group of lesions also includes Blandin and Nuhn mucoceles, rare benign lesions, which develop on the ventral side of the tongue of patients, including newborns and paediatric patients. Effective treatment strategies include surgical removal by complete excision of the lesion and the associated minor salivary gland [283]. Also, both open and closed cryotherapy approaches have been described as effective treatment [284, 285].

To date, there is a lack of regenerative medicine approaches reported in the literature for mucoceles. However, the use of laser therapy in this field has been proposed. In the past, the use of CO_2 lasers, either in a pure vaporization or mixed excision-vaporization mode, has been reported for successful ablation of ranulas [286, 287]. More recently, diode laser surgery, using lasers with wavelengths of 930 nm (continuous mode, power setting = 1.8 W) [288] and 800 ± 10 nm (continuous mode, power setting = 1.5 W) [289] has been proposed as an alternative to conventional surgery to promote minimal discomfort and pain, particularly in children, and to reduce scarring. Laser diode procedures appear to be particularly preferable in newborns and uncooperative children, with increased acceptance of the procedure by parents.

2.3.2 Salivary Gland Tumours

Salivary gland tumours are an uncommon entity. They comprise only 3% to 6% of all head and neck tumours, and 0.6% of all tumours of human body. Some 80% of cases occur in the parotid gland [290]. A small percentage of salivary gland tumours occur in the submandibular and sublingual glands, while about 20% occur in minor salivary glands [291]. Most salivary gland tumours are benign (80%), and they are almost never life threatening. There are many types of benign salivary gland tumours, including pleomorphic adenoma, oncocytoma, and Warthin tumour. Benign tumours are treated mostly by surgery and very rarely progress to cancer if left untreated or if they are not completely removed and recur.

Salivary gland malignancies are remarkably diverse and heterogeneous. This is mainly due to the fact that normal salivary glands are composed of several different cell types, and tumours can originate from any of these. It is beyond the scope of this chapter to discuss all salivary gland tumours. A detailed description of salivary gland tumours can be found in other comprehensive resources [292].

The management of salivary gland tumours has changed over the last few decades. In the case of parotid gland tumours, local excision and enucleation are no longer recommended, due to high rates of recurrence, facial nerve injury, and malignant progression [290, 293–295]. Nowadays, the treatment of choice for both benign and malignant tumours is partial or total parotidectomy with preservation of the facial nerve, based on the extent of the neoplasm [296]. Facial nerve removal is mandatory in patients with direct invasion. Additionally, radiotherapy may be useful in malignancies as adjuvant therapy, while chemotherapy is rarely used [297].

Only one study has reported the use of stem cells in salivary gland tumour management [298]. In this study, the authors investigated the use of induced pluripotent stem cells for treating induced salivary gland cancer and for restoring salivary gland function both in vitro and in vivo using male albino rats. They showed that cases treated with induced pluripotent stem cells exhibited regeneration of salivary glands, although minor degenerative and vascularization changes remained [298].

One of the few regenerative approaches used in the field of salivary gland tumours is the use of the buccal fat pad. This technique has been used since 1977 [299]. The buccal pad is used commonly as a graft to cover minor and medium intraoral defects, including those caused by the excision of benign tumours [300, 301]. Several advantages and characteristics of the buccal fat pad make it desirable for grafting. It is autologous, readily available, within the same field of dissection (thus averting further incisions), easily dissected, with low risk of damaging neighbouring anatomical structures, richly vascularized, and sufficient in size for many surgical scenarios (especially in children and in young females). Grafting is associated with low morbidity in the donor site, and low complication rates, and grafts undergo epithelialization between 3 and 6 months.

Moreover, besides being a convenient and versatile graft, the buccal fat pad it is also known to be a desirable source of stem cells. In fact, even though bone marrow is the most common source of stem cells for clinical application, adipose tissue from the buccal pad represents a valid alternative, particularly for oral and maxillofacial surgeons, for harvesting a high number of adipose-derived mesenchymal stem cells.

Broccaioli and colleagues were the first to characterize the adipose-derived mesenchymal stem cells from the buccal fat pad [302]. Stem cells isolated from a small amount of buccal fat pad tissue were found to have clonogenic capabilities, a typical mesenchymal stem cell immunophenotype, and increased multipotentiality under the influence of autocrine growth factors. To date, some knowledge accumulated around the characteristics and the potential uses of cells from the buccal fat pad, including their ability to differentiate to adipocytes [302], osteocytes, and chondrocytes [303]. These cells also adhere to biological supports, such as alveolar bone and periodontal ligament, and to synthetic scaffolds such as collagen membrane [302], and ceramic-coated scaffolds [304]. Many examples of regenerative medicine through the use of the buccal fat pad have been reported, including its successful use in:

- The treatment of maxillary atrophy, as a regenerative aid to increase the osteogenic capacity of non-vascularized iliac bone [305].
- A mouse model for Parkinson's disease, where the cells differentiate into neural cells [306].
- The prophylaxis of Frey's syndrome following parotidectomy [304].
- Cases of pleomorphic adenoma for closure of palatal defects [307–309].
- Two cases of mucoepidermoid carcinoma for bone defect reconstruction [310, 311].

Despite the multitude of studies reported above, to date the regenerative properties of the buccal fat pad derived stem cells have not been explored clinically for regenerating damaged or hypofunctioning salivary gland tissue. There is only one animal study reported in the literature that explored this topic. Kawakami and colleagues induced the differentiation of salivary gland cells by co-culturing human adipose-derived stem cells isolated from buccal fat pads with human salivary gland-derived fibroblasts, and transplanting them into the submandibular gland of SCID mice [312]. Using a three-dimensional (3D) culture system, they demonstrated that these induced cells reconstituted glandular structures in the 3D culture system, expressed salivary gland-related markers, and generated new tissues following transplantation in vivo.

In summary, for most parotid gland diseases including pleomorphic adenoma, a partial parotidectomy is a reliable surgical method. However, radiotherapy should be offered to patients with parotid malignant disease, as it is associated with lower recurrence rates [290]. Laser use for the management of salivary gland tumours has rarely been reported in the literature. For example, carbon dioxide laser surgery has been effectively used to treat a case of papillary cystadenoma in the lower lip [313]. Despite the large potential of transplanted salivary gland cells differentiated from stem cells in culture for regenerating salivary gland tissue, to date there are currently no established regenerative therapies for atrophied and hypofunctioning salivary glands.

2.3.3 Immune-Mediated Salivary Disorders

Another group of salivary gland disorders is immune-mediated diseases. These are heterogenous group of conditions that include Sjögren's syndrome, rheumatoid arthritis, systemic lupus erythematosus, sarcoidosis, scleroderma, mixed connective tissue disease, and IgG4-related disease. Most of these conditions can be managed initially with medical treatment. Regardless of their local or systemic involvement, all these conditions have been associated with salivary hypofunction [283].

Sjögren's syndrome is the condition that has been studied the most in the field of regenerative medicine over the last two decades, and will be used as a model for discussion in this section.

Sjögren's syndrome is a systemic autoimmune disease that leads to progressive dryness of the eyes and mouth because of tissue injury resulting from lymphocyte infiltration into the lacrimal and minor salivary glands, respectively. In the salivary glands, acinar cells are involved frequently. Sjögren's syndrome can occur as a primary disorder, or in connection with other connective tissue diseases.

Current therapies for Sjögren's syndrome are aimed largely at symptomatic relief, and typically involve stimulation of the residual salivary cells with cholinergic medications. These approaches are effective only if enough secretory cells remain. In addition, oral lubricants and salivary substitutes are often prescribed. When Sjögren's syndrome is related to other connective tissue diseases, therapy with systemic immunosuppressive agents or B cell-directed therapies are treatment options [314–316]. These therapies are aimed at reducing the autoimmune attack responsible for exocrine tissue destruction, but not at promoting regeneration of already destroyed acinar cells. The effectiveness of such treatments is controversial [314–316]. For example, a recent study by Pringle et al. confirmed that antirheumatic drugs used to treat primary Sjögren's syndrome are not likely to restore saliva production, and should be supplemented with fresh salivary gland stem cells to recover saliva production [317]. Therefore, therapies aiming at regenerating the secretory cells are needed for patients with Sjögren's syndrome.

Laser phototherapy has been reported to be an effective treatment for Sjögren's syndrome in one study. Specifically, Simões et al. reported the use of a diode laser (wavelength of 780 nm, 3.8 J/cm^2, power setting = 15 mW) directed at the parotid, submandibular, and sublingual glands of one patient, when used three times per week, for a period of 8 months [318]. The authors measured salivary flow rate and xerostomia symptoms of patients before, during, and after laser treatment. They reported an overall improvement of dry mouth symptoms during treatment, and resolution of parotid salivary gland pain and swelling after treatment.

A promising field of investigation is the potential therapeutic role of mesenchymal stem cells (MSC) in the treatment of primary Sjögren's syndrome. To date, evidence has accumulated that transfusion of MSCs can suppress autoimmunity and restore salivary gland secretory function in both mouse models and in patients with primary Sjögren's syndrome. This effect is mediated by inducing regulatory T cells, suppressing Th1, Th17, and T follicular helper cell responses.

Moreover, there is evidence that MSCs can differentiate into salivary epithelial cells, representing a suitable alternative treatment for this condition. Non-obese diabetic mice (NOD) are the most common Sjögren's syndrome animal model, as they exhibit an autoimmune Sjögren-like disease. Xu et al. investigated the function of MSCs in NOD mice and in human Sjögren's syndrome patients [319]. They were the first to demonstrate that the immunoregulatory functions and biologic properties of MSCs in Sjögren's syndrome patients were significantly impaired. In addition, they showed that treatment with allogeneic bone marrow MSCs could alleviate experimental Sjögren's syndrome-like disease in mice. Importantly, they showed that infusion with allogenic umbilical cord MSCs improved salivary gland function and suppressed disease activity and autoimmunity in patients with Sjögren's syndrome. Of note, no adverse events during or after MSCs infusion were reported, and all patients tolerated the treatment. All 11 patients with xerostomia showed an increased unstimulated salivary flow 2 weeks after MSC infusion, with a twofold increase 1 month after infusion [319].

Chen et al. treated 24 Sjögren's syndrome patients (including 11 with xerostomia) with an intravenous injection of allogeneic MSCs, and concluded that this method was safe and reliable [320]. Further, Khalili et al. found that intravenous injection of bone marrow MSCs prevented the loss of salivary flow, and reduced the extent of lymphocytic infiltration of the salivary glands, the influx of T and B cells, the frequency of FoxP3þ (Treg) cells, and inflammation in mice at the initial stages of Sjögren-like disease [321]. In a subsequent study, the same authors investigated the treatment for salivary gland hypofunction using two cell-based therapies at both initial and advanced stages of Sjögren-like disease in NOD mice [322]. They compared bone marrow versus spleen cell therapy, and monitored the mice for 1 year. Spleen cell therapy was superior to bone marrow cell therapy, but both cell therapies were effective in preserving normal salivary gland functions when transplanted during the initial or advanced phases of Sjögren-like disease. Both cell therapies downregulated inflammatory and upregulated salivary tissue homeostasis/maintenance genes [322].

Recently, Liu et al. explored the feasibility and mechanisms of umbilical cord MSC therapy in the treatment of NOD mice and Sjögren's syndrome patients, with a particular emphasis on regulatory T cells [323]. Umbilical cord MSCs conferred potent immunosuppressive effects by inducing regulatory T cells both in NOD mice and in Sjögren's syndrome patients. NOD mice that received cell therapy showed inhibited production of Sjögren's syndrome-related antibodies and decreased lymphocytic infiltration. These features resulted in increased salivary flow in the treated animals compared to controls [323].

Despite the optimism from these gains, there are still many challenges to overcome for clinical application of MSCs in the treatment of Sjögren's syndrome. Firstly, there is a lack of standardized protocols for MSC isolation and culture. MSC delivery methods, transfusion frequency, and doses also vary across the reported studies. Secondly, there is a lack of definition of quality control procedures for the different types of MSCs isolated and used. Thirdly, an additional restriction for clinical application of MSCs is related to potential biosafety issues around tumorigenicity [324, 325]. Fourthly, all studies undertaken thus far suffer from a limited sample size, so are not sufficient to provide reliable conclusions.

Moreover, a major obstacle in stem/progenitor cell therapies is the limited lifespan of the cells obtained from in vitro cultivation systems, hence needing to be used within a short time window. Finally, it is still unknown if MSC transfusion will have a positive long-term effect in primary Sjögren's syndrome patients, given the limited follow-up periods reported in the few published studies.

In summary, MSCs can be obtained from different sources (such as adipose tissue and dental pulp). These stem cells possess potent immunomodulatory functions, and are capable of repairing damaged tissues. Currently, there is no curative clinical treatment for primary Sjögren's syndrome. MSC-based therapies show great potential for this, due to their ability to significantly suppress inflammation and preserve salivary function in Sjögren's syndrome animal models. However, evidence is currently lacking regarding the use of MSC-based therapies in clinical practice. Future randomized controlled trials of the therapeutic use of MSCs in Sjögren's syndrome patients will be of considerable interest.

2.3.4 Radiation-Induced Xerostomia

Post Radiation Xerostomia

Xerostomia is defined as the subjective feeling of dry mouth, which may or may not be associated with objective evidence of hyposalivation. Dry mouth is a debilitating adverse effect of head and neck radiotherapy. Salivary gland hypofunction has been observed in more than 60% of patients with head and neck cancer undergoing radiotherapy [326–328]. Therapeutic irradiation for head and neck cancer leads to irreversible damage to acinar cells. Consequent loss of salivary parenchyma causes irreversible salivary gland hypofunction. Radiation doses between 26 and 39 Gy have been considered sufficient to induce a significant reduction in saliva production [329–332]. The disruptive effects of radiation on the salivary glands usually commence during the course of radiotherapy [333], with tissue damage usually peaking at 6 days following irradiation [334]. Such patients cannot produce adequate levels of saliva, and this leads to considerable morbidity.

To manage radiation-induced hyposalivation, various therapeutic strategies have been applied in patients. These include the use of cytokines, or medications such as pilocarpine or amifostine to increase the secretion of saliva [335, 336]. However, treatment with pilocarpine is associated with a plethora of side effects, such as dizziness, flushing, tremor, and diarrhoea, among others.

Therapeutic strategies developed to date to rescue the salivary glands from radiation damage include surgical salivary gland transfer. The aim of this technique is to preserve a salivary gland by surgically transferring it into a different anatomical space and shielding it from the full dose of radiation, for example, transferring the submandibular gland to the submental space, or the submandibular gland to the parotid region [337]. These procedures are proven to reduce the rate of radiation-induced xerostomia. The disadvantages include possible delays in cancer treatment, even if this delay does not negatively affect patient survival [338].

A recent multicentre phase III trial found that submandibular salivary gland transfer was superior to pilocarpine in management of radiation-induced xerostomia [339]. However, in some cases, salivary gland transfer surgery can be contraindicated. Regenerative medicine approaches may represent a valid alternative to overcome these side effects and limitations.

Nguyen et al. utilized growth factor systemic administration in mice, and ex vivo sphere formation cell culture to assess the self-renewal capacity of cells derived from irradiated parotid glands [340]. Their study showed a significant reduction in sphere formation from irradiated parotid glands, and that treatment with growth factors such as Insulin-like growth factor 1 (IGF-1) can improve salisphere generation, restoring self-renewal properties to levels similar to unirradiated controls [340].

A recent in vivo murine study investigated the effect of Alpha-Lipoic Acid on radiation-induced salivary gland dysfunction [341]. Alpha-Lipoic Acid could rescue radiation-induced hyposalivation by preserving parasympathetic innervation and regenerative potentials. Alpha-Lipoic Acid is involved in enhanced parasympathetic protection via the rescue of GFRα2 and AchE expression, and in preserving levels of the nerve growth factors BDNF (brain-derived neurotrophic factor) and neurturin in salivary glands. Alpha-Lipoic Acid is also involved in the Hedgehog signal pathway, which is known to be activated during the functional regeneration of adult salivary glands [341].

An additional treatment strategy that has been reported as being able to repair irradiated-injured salivary glands is the use of a bone marrow cell extract (also known as "bone marrow soup") [263, 342, 343]. However, an important limitation of this approach is the need to harvest bone marrow cells, either allogeneic or autologous, which is an invasive procedure.

The exact mechanisms underlying cell-based therapy are still enigmatic. However current studies suggest that the transplanted cells participate in the tissue repair and regeneration processes by releasing several angiogenesis-related growth factors and cytokines, such as CD26, FGF, HGF, MMP-8, MMP-9, OPN, PF4, SDF-1, IL-1ra, IL-16, and some additional paracrine factors [342]. Of note, bone marrow soup is a cell-free therapy, as it contains only the soluble intracellular contents. It should be less tumorigenic and immunogenic than injected intact cells [342]. Recently, to overcome the instability of the above-cited proteins, a lyophilized version has been developed, and demonstrated to be effective [344].

In regenerative medicine, stem cell therapy is an attractive option for the long-term treatment of xerostomia induced by radiation. Adult stem cell therapy has also been investigated over the last two decades. Several studies have attempted to isolate salivary gland progenitor cell populations from adult human salivary glands, most of them using biomarkers that were identified in previous murine animal studies or cell culture studies [345]. Jeong et al. isolated progenitor cell populations without the use of specific markers and developed a human salivary gland progenitor culture system that exhibited MSC-like characteristics and could rescue acinar structure and hyposalivation in rats that received radiotherapy [346].

The use of MSCs is a very promising regenerative treatment option for increasing saliva flow rates and thus relieving xerostomia. However, the majority of studies using MSCs and radiation-induced xerostomia include only preclinical models. Li et al. investigated the effect of the systemic administration of adipose tissue-derived MSCs into C57BL/6 mice, immediately after radiation at a dose of 18 Gy [347]. Their functional evaluations, conducted 8 weeks after radiation, in addition to morphological and microscopic parameters, concluded that systemic administration of adipose tissue-derived MSCs exerted protective effects against external radiation-induced salivary gland damage.

Lim et al. studied radiation-induced salivary gland dysfunction and xerostomia using mice irradiated with 15 Gy. They measured the stimulated salivary flow before and 12 weeks after radiation [264]. Their study group was treated with a local injection of bone marrow-derived clonal MSCs 24 h after radiation (directly into the submandibular gland), while the control group received an injection of saline only. They observed significantly increased salivary secretion (41%) and an increased weight of the glands in the study group, compared with the saline-treated control. Morphological, histological, and immunofluorescent analysis also showed more functional acini, less apoptotic cells, and an increased micro-vessel density in the study group.

Using a similar model, but injecting adipose tissue-derived MSCs, Kojima et al. found improved salivary flow, angiogenesis, and other paracrine effects in their study group compared with controls [348]. Similarly, in a model of radiation-induced xerostomia, Lin et al. found that transplantation of mesenchymal stem cells increased salivary flow, salivary gland weight, and body weight of the treatment group compared with the control group [349]. More recently, Serrano Martinez et al. developed a murine parotid gland organoid model where unlimited expansion of multipotent stem cells could be achieved, to explore the radiation response of stem cells in vitro [350].

To date, there are several in vivo animal studies that have investigated stem cell therapies for salivary gland disorders. Nevertheless, murine animal studies with appropriate long term follow-up and a relevant time of treatment are still required.

Other additional crucial factors that prevent possible translation of these models to the clinic include the discrepancy between the radiotherapy modalities used in murine models and those used for head and neck cancer patients. In clinical practice, radiotherapy regimens are fractionated and spread across several weeks, and target the glands in a variable patient-specific manner. In contrast, murine studies utilize radiation exposure as a single standardized dose.

The timing of stem cell treatment following radiotherapy is also variable. In fact, during the acute phase of radiotherapy for head and neck cancer, an inflammatory infiltrate is found in salivary glands. It has been demonstrated that mesenchymal stem cells are able to decrease the inflammatory responses in many different scenarios, therefore potentially reducing immune-mediated destruction of gland parenchyma [351]. Finally, there is the potential risk that these cells could stimulate cancer growth, as some in vitro studies have suggested [324, 325].

To date, only a single study has assessed the safety and efficacy of stem cell therapy for radiation-induced xerostomia in humans. This was a randomized, placebo-controlled phase 1/2 trial that included 30 patients, who received transplantation of adipose tissue-derived MSCs or a placebo to the submandibular glands [352]. The patients had oropharyngeal squamous cell carcinoma and had previously received radiotherapy. Unstimulated whole salivary flow rates were significantly increased in the stem cell treated group at 1 month (by 33%; $p = 0.048$) and at 4 months (by 50%; $p = 0.003$), but no changes were seen in the placebo group, compared to baseline. Their findings suggest that therapy for radiation-induced hypofunction with adipose tissue-derived MSCs is safe, and significantly improves salivary gland function and patient-reported outcomes.

In summary, the majority of studies in the field of radiation-induced xerostomia describe preclinical models. Until transplantation studies are performed, there remains a dearth of information about the ability of these cells to contribute to the salivary gland regeneration in vivo.

2.3.5 Radioiodine-Induced Xerostomia

A parallel field of research in this area relates to treatment of salivary gland disorders caused by radioiodine therapy. In this case, salivary gland dysfunction is caused by irreversible damage to acinar, ductal, and endothelial cells. Radioiodine therapy is a nuclear medicine treatment mainly used to treat hyperthyroidism. A small dose of an isotope of iodine (radioactive iodine I-131) is administered orally, is absorbed into the bloodstream and concentrates within the thyroid gland, where it destroys cells due to emitted radiation. It is also the treatment of choice for differentiated thyroid carcinoma, and in this case the ablative radioiodine therapy is performed following conventional surgery. The mechanism that allows gland tissue destruction is associated with radiation damage through the sodium iodide symporter [353].

Unfortunately, the expression of the sodium iodide symporter gene is also observed in salivary glands, mammary glands, hair, and stomach. Therefore, during the transport of the radioiodine through the bloodstream, there is a significant dose-dependent uptake of isotope by salivary gland tissue, with a concentration in the salivary gland increasing up to 100-fold higher than in serum [354]. Salivary gland dysfunction is the most common chronic complication after radioiodine therapy, and more than 50% of all patients experience xerostomia or sialadenitis in the post-treatment period [355].

Several researchers have attempted to develop effective protective and therapeutic strategies in the field of regenerative medicine to manage radioiodine-induced salivary dysfunction. These include salivary gland autologous cell transplantation [356], the implantation of engineered artificial salivary gland cells [357], and stem cell therapy [358]. Using structural and functional analyses, Saylam et al. showed that systemic adipose tissue-derived MSC administration ameliorated radioiodine-induced histologic changes and salivary dysfunction in a rat model [359]. As an alternative, Kim et al. showed that a local injection of adipose tissue-derived MSCs into the submandibular gland (4 weeks after radioiodine treatment) enhanced salivary gland secretory function after radioiodine treatment by attenuating or preventing isotope-induced damage [360].

2.3.6 Post-chemotherapy Xerostomia

There is currently minimal literature describing regenerative medicine approaches addressing chemotherapy-induced salivary gland dysfunction. Chemotherapy regimens during cancer treatment have been associated with temporary salivary gland hypofunction, however, it remains unclear if this hypofunction is caused by the chemotherapeutic agents themselves or by other concomitant medications, such as antiemetics or other factors [361].

Some studies have indicated that cancer patients may experience hyposalivation preceding the administration of a chemotherapeutic agent [362, 363], highlighting the value of salivary evaluation of subjects prior to chemotherapy treatment. However, chemotherapy itself can induce compositional changes in saliva, including immunological composition [362, 364, 365].

There is still no gold standard for the treatment of xerostomia and hyposalivation. To be able to choose the best treatment for patients, it is crucial to understand the mechanism of action of the treatment modalities for salivary hypofunction, as well as their adverse effects. None of the regenerative treatment approaches described in the articles included in this chapter had an effect on re-establishing glandular function in a fashion that could be translated to clinic. There are still no controlled clinical studies supporting this evidence. The use of stem cells derived from adipose tissue has shown promise; however, it requires complex and expensive procedures. Further research on stem cell therapy is required to move to extend our knowledge of managing hyposalivation.

2.4 Craniofacial Bone Pathology

Successful regeneration of bone requires the concomitant processes of osteogenesis and neovascularization. Current methods of repair and reconstruction include rigid fixation, grafting, and free tissue transfer [366]. However, these methods carry innate complications, including plate extrusion, non-union, graft/flap failure, and donor site morbidity.

Recent research efforts have focused on using stem cells and synthetic scaffolds to heal critical-sized bone defects similar to those sustained from traumatic injury or ablative oncologic surgery. Growth factors can be used to augment both osteogenesis and neovascularization across these defects. Many different growth factor delivery techniques and scaffold compositions have been explored, yet none have emerged as the universally accepted standard [367, 368].

Traditional means of repair of bony defects of the craniofacial skeleton include bone grafting, rigid fixation, and microvascular free tissue transfer for larger defects. While these current methods work well for smaller fractures and defects, the methods for larger reconstructive problems carry significant morbidities, and are not always successful [369]. Osteoconduction, osteogenesis, and osteoinduction are the three mechanisms that need to act together to regenerate osseous defects. Efficacious bone tissue engineering requires some combination of a sound osteoconductive scaffold, vascularization, appropriate intercellular signalling, and the presence of osteoblastic cells. Stem cells are the primary source for osteoblastic cells, and require activation by osteoinductive factors for new bone formation [370, 371]. Stem cells may be harvested and induced to differentiate into osteoblasts either in vitro or in vivo [372]. Proangiogenic and osteogenic growth factors can be loaded in biosynthetic scaffolds before implantation into bony defects inducing native stem cells to differentiate into osteoblasts. Stem cells have powerful osteogenic potential given their ability to differentiate into osteoblasts. Stem cells can be used in a variety of ways to supplement osteogenesis [368]. They can be implanted into living tissue and allowed to differentiate into the surrounding tissue type.

Mesenchymal stem cells derived from adult bone marrow are potentially useful for craniofacial tissue engineering of bone, adipose, muscle, and cartilage [373]. Kaigler et al. found that implanted cells led to an increase in alveolar bone regeneration and a reduced need for secondary bone grafting compared with conventional guided bone regeneration [374]. Yang et al. demonstrated the viability of adipose-derived stem cell osteogenic capabilities by engineering biomimetic scaffolds cross-linked with rabbit adipose-derived stem cells along with collagen into critical-sized defects in rabbit radii. Complete repair of the defect was achieved in 12 weeks, suggesting a role for the use of adipose-derived stem cells in osteogenesis [375].

While having the appropriate scaffold, cell type, and blood supply is important for new bone growth, stimulated cells must also receive the appropriate signals to grow in a regulated, organized fashion. These signals include mechanical, chemical, and even electrical stimuli. These signals must then be communicated between cells for the coordinated response needed to guide osteogenesis [368]. Further research is needed in several aspects of the field of implantable osteogenic constructs for the craniofacial skeleton. These include finding the ideal biomaterial, exploring the efficacies of protein versus gene-based strategies of osseo-implants, and defining the optimal use of stem cells in repairing craniofacial defects.

Bone is a constantly adapting and changing tissue that plays a significant role in human health, maintaining and producing blood cells, providing a physical barrier to vital organs, and acting as a support structure for movement [376]. Similar to other parts of the body, bone is susceptible to many diseases that can cause it to react in a dynamic way [377]. There are many possible causes of bone diseases (including hereditary, infectious, metabolic, malignant, and idiopathic), and they can arise in all age groups, races, and genders [378, 379]. While osseous diseases can affect any bone in the body, in this chapter attention is focussed on the maxillofacial skeleton.

The maxilla and mandible suffer from both generalized and localized forms of skeletal pathologies, to a greater extent than other bones in the body [380]. The presence of teeth creates a unique condition for some of these diseases, making the jaws more susceptible to a variety of forces and infections, and altering the response of bone to injury. The diagnosis of bone pathologies is still based on radiographic interpretation and hard tissue biopsy evaluations by a specialist [381]. Treatment timing and appropriate management approaches have increased with advances in technology [382]. In this chapter, the discussion is limited to the most common pathologies associated with the jaws, focusing on the difficulty of diagnosis and the type of treatment.

2.4.1 Ameloblastoma

Ameloblastoma is one of the most common odontogenic tumours diagnosed. It has an epithelial origin [383]. Ameloblasts are the cells that produce and secrete the enamel on the crown of the tooth. While considered to be a benign tumour, an ameloblastoma can often be locally invasive and slow growing. There are three variants: solid/multicystic (traditional), unicystic, and peripheral [384]. Radiographically, there is a soap bubble or honeycomb appearance. Many cases involve bony expansion, and the roots of teeth near the tumour are often resorbed [385, 386]. Although these characteristics suggest the diagnosis of an ameloblastoma, definitive identification requires histological evaluation via biopsy. The histopathological characteristics of all three variants show similarities. Much of the lining is composed of islands of epithelium within a mature fibrous connective tissue stroma. Also contained within this mixture are ameloblast-like cells [387]. Enamel is the hardest substance in the body, as its inorganic component comprises a higher percentage than that found in bone. In some cases, there is a degree of osseous metaplasia within the dense fibrous tissue, giving the tumour a mineralized character [388].

The solid/multicystic (traditional) ameloblastoma tends to infiltrate the surrounding trabecular bone, and therefore, the actual margin of the tumour may not be adequately known until it has already caused significant destruction. It can be a devastating and potentially deadly neoplasm when it spreads to involve vital structures [389]. Most surgeons designate 1.0–1.5 cm margins past the tumour for the margin of resection [390]. Attention to the type of resection performed and risk of recurrence is essential to ensuring a lasting reconstruction. The use of regenerative medicine in the form of recombinant human bone morphogenetic protein (rhBMP2) has shown excellent regeneration of large mandibular defects after resection of ameloblastoma, obviating the need for autogenous bone graft [391].

2.4.2 Odontogenic Keratocyst

Radiographically, the odontogenic keratocyst can resemble a dentigerous cyst, yet clinical and histological evaluation reveals a much different picture [392]. The odontogenic keratocyst exhibits a distinct growth pattern from the dentigerous cyst [393]. While the cyst can also be associated with swelling and pain, there is no obvious bony expansion and there may be associated drainage from the area. Radiographically, larger lesions may contain multiple small areas of radiolucency [394]. There is not commonly any associated resorption of nearby roots. The diagnosis is confirmed by histopathology, as histologic features are distinctive. The cyst lumen is often filled with a cheese-like material that is composed of a mixture of keratin and oily-like keratinaceous debris [395, 396]. The lumen of the cyst appears corrugated or wavy when examined microscopically.

Much like the dentigerous cyst, the typical mode of treatment is surgical enucleation. Complete removal is difficult because the lesion is extremely fragile, necessitating its removal in several pieces [397, 398]. The recurrence rate is high when compared with a dentigerous cyst. Nevertheless, the prognosis is favourable [399]. Some authors have proposed guided tissue regeneration as treatment after removal of a keratocyst [400].

2.4.3 Osteosarcoma

The most common malignancy of osseous origin is osteosarcoma. The malignant mesenchymal cells produce a poorly formed bone matrix [401].

Although the frequency of this neoplasm is significantly less in the jaws than in the extraskeletal region, its discussion is necessary as its consequences are both severe and devastating [402]. The most common symptoms are swelling and pain of the associated area, with increased mobility of the teeth [403]. The radiographic appearance varies from dense sclerosis to mixed radiopaque/radiolucent lesions to an entirely radiolucent lesion [404]. As with many tumours of the jaws, there is marked resorption of the roots of the nearby teeth, which is demonstrated by a gradual spiking appearance. Nevertheless, the traditional sun-ray appearance often seen in osteosarcomas of the limbs is also seen in the jaws, due to osteophytic bone production on the surface of the lesion.

Histopathologically, there is a great variety among osteosarcomas, and this is primarily due to the varying amounts of osteoid produced by the defective mesenchymal cells [405]. Additionally, chondroid, a precursor to cartilage, can also be found within the fibrous connective tissue. This variability leads to the three categories of osteosarcomas: osteoblastic, chondroblastic, and fibroblastic [406]. Of the three types, chondroblastic osteosarcomas comprise the majority of those found in the jaws, with some cases demonstrating lobules of cartilage with surrounding osteoid production [407].

When compared to osteosarcomas in the extremities, those found in the jaws tend to be less aggressive. Radical surgical excision of osteosarcomas in the jaw is the mainstay of treatment, and the main challenge is incomplete surgical excision of the tumour [408]. Chemotherapy is also used in combination with radical surgery [409]. When surgeons resect well past the known margins of the tumour to ensure complete removal, this comes at a high cost for the bones of the jaw, impacting function and aesthetics, and requiring complex rehabilitation [408].

2.4.4 Cemento-Osseous Dysplasia

Considered by many to be the most common fibro-osseous lesion in clinical dentistry, cemento-osseous dysplasia remains a confusing disorder in terms of diagnosis and classification [410]. Due to similarities, both clinically and histopathologically, with other lesions, it is often either misdiagnosed or undiagnosed. Radiographically, the lesion may be entirely radiolucent, or it may be densely radiopaque with a radiolucent rim. It is now commonly agreed that the lesion usually appears as mixed radiolucent/radiopaque in nature [411]. Variations of this disorder are common, including focal, periapical, and florid variants. Radiographically, the periapical variant resembles an inflammatory or cystic lesion, and this difference can lead to drastic changes in treatment options [412].

Histologically, the lesions are composed of portions of mesenchymal tissue, consisting of fibroblasts and collagen fibres, with some vascular components interspersed. The fibrous connective tissue contains a mixture of osseous portions of woven bone and lamellar bone. The mineralized portion varies from site to site and lesion to lesion. Maturing lesions demonstrate an increased ratio of mineral to fibrous connective tissue as time progresses [413]. If surgery is undertaken, the lesion is gritty and fragments readily when removed from the affected area, which is in contrast with lesions such as ossifying fibromas, which separate cleanly from the bone [414]. Fortunately, this disorder is not sinister and rarely needs surgical excision; however, increased sclerosis of the lesion may lead to decreased vascularity and eventual necrosis. There is a possibility for exposure of the lesion in the oral cavity, and at this point surgical intervention may be indicated [412].

2.4.5 Ossifying Fibroma

Unlike cemento-osseous dysplasia, an ossifying fibroma is a true neoplasm with a high growth potential. It consists of a mixture of trabecular bone and cementum with fibrous connective tissue, and is sometimes confused with cemento-osseous dysplasia [415]. The origin of this neoplasm remains a mystery; it was once postulated that it arose from a remnant after tooth formation. This disorder can manifest as a painless swelling of the involved osseous structure, leading to facial asymmetry. Radiographically, it is a large radiopaque lesion [416]. Because of its high rate of mineralization, there is a clear delineation between the lesion and the surrounding bone [417]. The lesion has a more random pattern of ossification than fibrous dysplasia [416]. For larger, aggressive lesions, the treatment of choice is surgical resection with bone grafting, and it has a generally favourable prognosis [418].

2.4.6 Fibrous Dysplasia

In fibrous dysplasia, the normal osseous structure is replaced by an increase in cellular fibrous connective tissue with areas of irregular bony trabeculae. There is a painless swelling, which grows slowly over time. The diagnosis is often first made clinically made, and then confirmed by histopathology [419]. Radiographically, the hallmark of this lesion is a "ground glass" pattern of opacification, which is due to the disorganization of poorly calcified trabecular bone [420]. This coincides with histological findings of irregularly shaped trabecular bone with a dispersed arranged fibrous connective tissue component. These features are key to distinguishing it from other lesions, since fibrous dysplasia demonstrates a more uniform pattern as compared to the random mixture of immature woven bone.

As the lesion matures, the trabeculae of bone run parallel to each other [419, 421]. In terms of clinical management, patients with cosmetic or functional issues may require corrective surgery; however, due to the unique physiological characteristics of this disorder, there is a chance of regrowth [422].

2.4.7 Osteogenesis Imperfecta

Osteogenesis imperfecta refers to a group of disorders related to defects in collagen maturation. Because collagen is a ubiquitous protein, the effects of this disorder are seen in a variety of structures within the body, such as the bone, dentine, sclera of the eyes, the skin, and the ligaments of joints. This disorder is considered to be the most common type of inherited bone disease.

The defect in collagen maturation leads to bone formation, with thin cortical bone and even more delicate trabecular bone. If the area suffers a fracture, there will be an abundance of calluses formed. Any area of the body can be affected, including the teeth and the jawbones [423]. The dental alterations in osteogenesis imperfecta are distinctive, with a blue to brown discolouration of the dentine and greater translucency, as well as reduced strength. Extreme bone fragility may lead to a spontaneous fractures, particularly in the jaws [424].

Histologically, while osteoblasts are present in the bone, the rate of formation of bone matrix is low, which causes the bone to resemble immature bone [425], and woven bone fails to mature into lamellar bone. There are several variations within this disorder, and depending upon the type found, the prognosis can vary from good to poor [426].

2.4.8 Idiopathic Osteosclerosis

This disorder represents a unique group of entities characterized radiographically by an isolated area of increased radiodensity. The cause is unknown, and fortunately patients with idiopathic osteosclerosis experience no remarkable symptoms [427]. Histologically, the lesion consists of dense lamellar bone with some fibrous and fatty marrow, which distinguishes it from other more dangerous pathologic entities [428].

2.4.9 Paget's Disease

This remains one of the more common bone disorders, and varies in the bones which can be affected. Pain is the most common symptom, leasing it to be mistaken for arthritis when pain occurs near joints [429]. When it affects the skull, there is a progressive increase in the size of the cranium, which leads clinicians to ask whether the patient's cap size has changed. When the skull is affected, diastemas develop and the teeth become spaced apart. In patients who wear dentures, the fit of the prosthesis alters and dentures feel too tight.

In the osteoblastic phase of this disease, patchy areas of sclerotic bone form, leading to a classic radiographic appearance of cotton wool [430]. Histologically, in Paget's disease there are osteoclasts in the resorptive phase and concurrent deposition of bone by osteoblasts, with formation of areas of osteoid matrix around the bony trabeculae, and replacement of fatty bone marrow by a highly vascular fibrous connective tissue [431].

The diagnosis is based on clinical, radiographic, and supportive laboratory results. Fortunately, this condition is usually not life threatening, and patients are able to continue leading relatively normal lives with supportive therapy [429].

2.4.10 Osteonecrosis

Osteonecrosis is the death of a bone or a part of a bone, that results as a consequence of a wide variety of systemic and local factors that compromise blood flow within bone. Such factors include haemoglobinopathies, anticardiolipin antibodies, defects of the thrombotic and fibrinolytic systems, fat emboli, alcoholism, systemic lupus erythematosus, and corticosteroid administration [432].

From a clinical point of view, bone necrosis of the jaws usually appears as an exposure of avascular bone in the mandible, in the maxilla, or in both [433]. The exposed necrotic bone is infected, and the area is usually painful. Patients may complain of difficulty in eating and speaking, pain, bleeding and, when the necrosis is extensive and near to the mandibular branch of the trigeminal nerve, paraesthesia of the lower lip [434].

The main cause of bone necrosis is a defect in vascularization. In the oral cavity, bone necrosis in immunosuppressed patients is probably related to the presence of unhealthy teeth, which increase the risk of infection. This also explains why bone necrosis is usually related to tooth extraction [433, 435].

Osteonecrosis of the jaws (ONJ) has been reported in osteoporosis patients treated with oral bisphosphonates since 2004 [436]. The first largely accepted definition of bisphosphonate-related osteonecrosis of the jaws (BRONJ) was released by the American Association of Oral Maxillofacial Surgeons (AAOMS) in 2007, and the condition was defined as the presence of exposed necrotic bone in the maxillofacial region that has persisted for more than 8 weeks in patients with current or previous treatment with bisphosphonates, and no history of head and neck radiation to the jaws [437]. This definition was confirmed in 2009 by AAOMS [438].

It is important to note that recently the BRONJ name has been converted by AAOMS to medication-related osteonecrosis of the jaw (MRONJ) to integrate the growing number of osteonecrosis cases involving the maxilla and mandible associated with other antiresorptive drugs (such as denosumab) and targeted biological therapies [439]. Tooth extraction, dental trauma, radio- and chemotherapy, infectious disease, and concomitant therapy with corticosteroids are factors that may increase the risk of BRONJ developing [440].

Clinical symptoms of BRONJ may vary from an asymptomatic process to local pain, development of fistulas, ulceration, inflammatory reactions of the soft tissues, tooth mobility, and exposure of necrotic bone [441, 442]. BRONJ may also cause sinus pain related to the thinning of the sinus walls and infection [443].

Due to difficulties in treating BRONJ as a complication of bisphosphonate therapy or the use of antiresorptive medications, it seems reasonable to apply vigorous prevention, such as ensuring proper oral hygiene before commencing bisphosphonate therapy [444]. Teeth that do not qualify for conservative, endodontic, or prosthetic treatment should be extracted. If possible, the antiresorptive treatment should be postponed until the extraction wound is fully healed. Surgical procedures require separate prevention methods, such as atraumatic extraction. Stopping bisphosphonate therapy just before a surgical procedure does not decrease the risk of BRONJ [445, 446], because the concentration of bisphosphonates in the bone remains high for many years after the therapy has been stopped. The exact duration of the half-life period of bisphosphonates is not well defined due to technical difficulties in marking their concentration in the urine and blood. It is estimated that alendronate can remain for 10 years, even after just a single dose [447]. If it is necessary to perform a surgical procedure in the maxillofacial area on a patient using bisphosphonates, the patient should always be informed about the potential risk of bone necrosis.

Based on the probable correlation between jaw osteonecrosis and bisphosphonates, both before and during bisphosphonates therapy, in patients with cancer on intravenous high dose bisphosphonate therapy, careful evaluation of the patient and strong collaboration between the dentists and the oncologists is essential. Before the beginning of bisphosphonate treatment, it is necessary to evaluate the patient's oral health to establish good oral hygiene and plaque control, and to undertake appropriate restorative, endodontic, periodontal, and surgical treatments in order to prevent or eliminate possible causes of infection, which may reach the bone. It is necessary to avoid invasive oral interventions such as dental extractions or periodontal surgical treatments in the near and intermediate future [448]. Antibiotic prophylaxis is used in oral surgery in such high risk cases to prevent infections. It may also be considered in other patients whose general health conditions or specific medical procedures that they have undergone make them more susceptible to contracting surgical infections [449]. Patients treated with bisphosphonates show suppressed bone remodelling, inhibited angiogenesis, and delayed healing [450]. If antibiotics are used, penicillin still remains a drug of choice in antibiotic prophylaxis, but the decision of which medicine should be used depends on the pathogens present as well as on the patient's tolerance or allergies [451]. According to the literature, a longer application period of antibiotics may have a positive influence on the healing process [452]. Some research does not support the effectiveness of antibiotic usage in preventing BRONJ [453]; however it is difficult to define blanket rules on using prophylaxis to prevent osteonecrosis of the jaws from bisphosphonate therapy, so each case should be considered individually.

In terms of regenerative medicine approaches, several regeneration protocols have been proposed. In order to limit the development of BRONJ, Kaibuchi et al. transplanted a sheet of mesenchymal stem cells onto bone that was exposed after tooth extraction in a BRONJ-like rat model [454]. After 2 weeks, healing was significantly improved over the control group where the animals received sham surgery. Neovascularization was found in the transplant group through the secretion of angiogenic factors from transplanted mesenchymal cells.

Currently, treatments for MRONJ have been limited, and results have been inconclusive. Hyperbaric oxygen (HBO) has been proposed, but it is not recommended as a sole treatment [455]. Recently, the use of platelet-rich plasma has been reported to have potential for the prevention and treatment of MRONJ [456, 457]. The mechanism of action of plasma rich growth factors (PRGF) in tissue regeneration has been studied widely [456, 458]. In the case of BRONJ, bisphosphonates inhibit bone resorption and angiogenesis by blocking the action of vascular endothelial growth factor (VEGF). The application of PRGF provides proteins and growth factors to the local milieu, including angiogenic factors (VEGF and angiopoietin), and factors that promote osteogenic differentiation, which can activate and accelerate the regeneration of involved tissues [457]. Moreover, PRGF has recently been demonstrated to exert a cytoprotective role in zoledronic acid-treated oral alveolar osteoblasts and gingival fibroblasts [459].

2.4.11 Post-trauma

Craniofacial trauma is one of the most challenging injuries that confront health care workers, including those involved in emergency care as well as those in office-based clinical practice [460]. These injuries can not only be life-threatening, but also cause disfigurement [461], and be aesthetically significant. There are many causes of craniofacial trauma, ranging from road traffic accidents to physical assaults. The demographic patterns vary widely in different regions of the world [462, 463]. Leading causes of facial injuries include motor vehicle accidents, pedestrian collisions, stumbling, sports injuries, industrial accidents, assaults, and warfare.

Facial injuries can be categorized into soft tissue injuries, skeletal injuries, and a combination of both. Like any other discipline of medicine, clinical evaluation start with detailed history taking [464]. The details which need to be included are the mechanism of injury, any history of loss of consciousness, and any history of previous medical or surgical disease. The history should document the circumstances under which the injury has occurred, and the direction of the impact. The physical examination should include careful but gentle palpation, assessment of the mobility of bone fragments, and for signs and symptoms such as numbness or paresthesia, occlusal disharmony, displacement of bony segments, and changes in vision [465]. Craniofacial injuries can be associated with many other concomitants life-threatening major injuries namely cranial trauma, limb fractures, chest injuries, spinal injuries, and orbital injuries. Evaluation of patients must be systematic so that no injuries are overlooked [464, 466].

Plain radiographs such as skull radiographs and panoramic radiographs have a limited role in the radiological evaluation of bony injuries. With significant technical advancements in computed tomography, multislice CT imaging has become the primary modality of imaging. A CT scan will detect craniofacial injuries in detail, and can help to exclude intracranial haemorrhages, as well as being used to assess, identify, and classify associated bone injuries. A CT scan can also help differentiate fractures in anatomically difficult areas which cannot be seen on conventional radiographs. A CT scan also has the capacity to evaluate the facial skeleton in both the axial and coronal planes [467]. Magnetic Resonance Imaging (MRI) may be used to investigate diffuse axonal injuries and to evaluate complications and soft tissue trauma [468].

The initial management of any patient with the clinical suspicion of craniofacial injury who comes to the emergency room follows a standard protocol, namely management of the airway, breathing, and circulation. Once the general condition of the patient is stabilized, further investigations and appropriate surgical intervention can be planned [464]. Maintenance of the airway is of primary importance. A chin lift or jaw thrust is used to prevent the tongue from obstructing the airway. Endotracheal intubation or tracheotomy may be needed to secure the airway [469].

Before planning for surgery, it is necessary to identify the extent of injuries, namely whether these are injuries just to soft tissues or the facial skeleton, or if it is a combination of both. The purpose of surgical intervention is to regain function with a good aesthetic outcome. Fractures will need rigid fixation of the fragments to achieve perfect reduction. An adequate soft tissue cover is needed followed by sufficient time for bone healing.

2.4.12 Maxillofacial Infections

Maxillofacial infections are an important public health problem due to their great potential to spread to involve important and vital anatomical structures, such as the respiratory system and mediastinum, thus increasing the risk of septicemia and death [470]. While most infections are limited in nature and easily treated, there are cases of fatal outcomes due to airway obstruction.

These infections can progress rapidly if not adequately treated. Maxillofacial infections are characterized as polymicrobial in nature. The pathogens are endogenous and opportunistic [471]. The literature shows that maxillofacial infections more often affect male patients, both in child and in adult populations.

Odontogenic infections have multiple aetiological factors [472]. They can be related to dental caries, periapical and/or periodontal abscess, and pericoronitis. Most dentoalveolar or odontogenic infections arise from one of two sources—the periapical region (due to pulp necrosis and subsequent bacterial invasion) and the periodontium (as a result of periodontal disease) that allows inoculation of bacteria into deep tissues [473]. Under normal circumstances, inflammation is self-limited [474]. However, in the presence of a substrate, microbial colonization may result in formation of a biofilm that provides a sanctuary for resident flora, and is difficult to eliminate [475]. Biofilms are involved in many human infections. For example, in periodontitis, bacteria produce a variety of products that elicit a host response involving the expression of various signalling molecules and mediators, and the recruitment of inflammatory cells [476]. This process may culminate in tissue destruction and interfere with tissue regeneration and repair. Removal of the offending agent allows for resolution of the inflammatory response. If the microbial challenge cannot be eliminated, the inflammatory response will persist and become chronic, leading to tissue destruction, as in periodontitis [477]. Given the association between infection, inflammation, and tissue destruction, it is not surprising that inflammation may interfere with the process of bone healing and regeneration [474]. Interventions that interfere with microbial colonization of the surgical site are likely to promote regeneration. Some of these are simple, such as proper aseptic technique during surgery.

Major clinical objectives of treating these infections are pain relief, recovery of function, preservation of vital structures, preventing flares or relapses, and limiting the period of incapacity [473, 478]. If an antibiotic is used, the process of choice must follow four criteria. First, the antibiotic must be effective against all the microorganisms that are usually responsible for the infection. Second, the medication should have as limited a spectrum as possible, in order not to interfere with the normal microbiota of the patient. Third, the antibiotic must be low in toxicity. Fourth, it should be bactericidal, because a maxillofacial infection can have serious consequences, and bacteriostatic antibiotics will mean a slower recovery [473, 479]. Inflammation, repair, and regeneration are carefully choreographed processes that occur in a specific temporal and physical contexts. A better understanding of these associations will allow for the identification of new therapeutic targets and the development of novel interventions to promote bone regeneration.

2.5 Temporomandibular Joint Disorders

Temporomandibular joint disorders (TMD) are a heterogeneous collection of conditions involving the temporomandibular joint (TMJ). The TMJ has a unique structure, with two synovial joint cavities, each lined by a synovial membrane. TMJ articulation is formed between the mandibular condyle and the temporal bone, and the articulation surfaces are covered by fibrocartilage. The temporal articular surface is large, and consists of the mandibular fossa and the articular tubercle. Along this large articular temporal surface, each mandibular condyle has a wide range of motion, with both rotation and translation being possible, allowing the complex movements necessary for mastication and speech. The fibrocartilaginous disc cushions mechanical stresses that exist between the temporal and the mandibular articular surfaces. The high collagen content of this disc provides great rigidity and durability. The TMJ disc has no direct vascularization or innervation. However, its posterior attachment, known as the retrodiscal tissue, features many vessels and nerves which are crucial during physiopathological processes [480].

The TMJ also has a unique embryonic origin. In embryos, the condyle head and cartilage develop through endochondral ossification at week 7, with neural crest mesenchymal stem cells (MSCs) providing cellular components of these tissues, while the main mandibular bodies undergo intramembranous ossification mediated by cells derived from mesodermal derived MSCs [481].

The possible causes of TMD range from minor self-resolving disruption of muscular and neuromuscular harmony, to degenerative diseases of the condyle and/or the disc structures of the TMJ that lead to debilitating pain, jaw function limitation, and culminate in malformation of the joint and facial structures. Degenerative disorders include non-inflammatory diseases such as osteoarthritis and congenital TMJ and condyle disorders. Inflammatory diseases include rheumatoid arthritis, psoriatic arthritis, capsulitis, synovitis, and ankylosing spondylitis.

The reported incidence of TMD in the general population is around 18–35%, and around 10% in juveniles, indicating an early onset of these diseases. A high frequency of TMD has been found in autopsy reports. Degenerative TMJ disease has been found in 28% of individuals in the 16–39-year-old age group, and in up to 50% of people over 50 years of age. The TMJ articular cartilage has a limited potential for repair due to a lack of vascularity and limited cellularity [482, 483], and as a consequence, cartilage injuries can result in chronic disability.

Osteoarthritis (OA) of the TMJ is a non-inflammatory, degenerative joint disease, and is the most common form of TMJ pathology. It is characterized by breakdown of the articular cartilage, architectural changes in bone, and degeneration of the synovial tissues, causing pain and/or dysfunction in movements of the jaw [484]. In TMD patients, once joint breakdown commences, OA can be crippling, leading to morphological deformity and functional obstruction [484]. The TMJ can be the first joint to develop osteoarthritis, whereas in rheumatoid arthritis, it is the last joint to be affected [485]. This high incidence of TMJ OA is related to the nature of the human TMJ, which is load-bearing under function [486].

Although the disc of the TMJ plays a very important role in stress distribution [487], it is at high risk of displacement [488], which leaves the condylar cartilage in friction during function, and leads to wear and tear of articular surfaces. Around 70% of patients may start with a TMJ disc displacement with or without reduction. Progression of the disease leads to severe TMJ osteoarthritis, with damaged cartilage surface and malformed subchondral bone.

In up to 67–70% of TMD cases, especially in OA patients, TMD is accompanied by malpositioning of the TMJ disc, termed "internal derangement" (ID) [489, 490]. Magnetic resonance imaging (MRI) studies have found that in asymptomatic patients, the discs are identified in the "normal" anatomical position and show minimal morphological changes of the condyle and articular surfaces. In symptomatic patients, however, substantial osseous change is observed with ID, including disc perforation [491].

Rheumatoid arthritis (RA) of TMJ is a chronic, systemic, autoimmune inflammatory disorder that is characterized by joint inflammation, erosion of the joint, and symmetric multiple joint involvements. The TMJs are rarely affected in the early phase of RA. However, arthroscopically, high frequencies of synovitis, degenerative changes, and fibrosis leading to ankylosis are observed in RA patients compared to OA, despite a shorter duration of TMJ symptoms. A correlation has been noted between lateral joint tenderness and pronounced synovitis in RA patients [492].

Congenital deformities of the TMJ complex can present as a heterogeneous continuum of growth disturbances of the mandibular condyle, articular eminence, and temporal bone. Some present as components of syndromes with congenital condylar deformity, including mandibulofacial dysostosis (Treacher Collins syndrome), hemifacial microsomia, oculoauriculovertebral syndrome, oculomandibulodyscephaly (Hallermann-Streiff syndrome), and Nager syndrome [493]. Additionally, isolated TMJ hyperplasia, hypoplasia, and bifidity have been reported [494].

A range of therapeutic options are available for TMD, which include non-surgical modalities such as control of contributory factors, occlusal appliances, pharmacological interventions (including corticosteroids delivered locally or systemically, NSAIDs, anxiolytics, and antidepressants), as well as physiotherapy and low-level laser therapy. Surgical treatment options include intraarticular injections, arthrocentesis as well as attempts at repair or replacement of portions of the TMJ. The majority of treatments aim to reduce pain and prevent disease progression without final resolution of disease pathology [495–497].

Development of tissue engineering protocols may create possibilities for reconstruction of condylar bone and cartilage using a bioprosthesis, exploiting combinations of scaffolds and stem cells to provide better functional recovery. To this end, three strategies have been developed: (1) intraarticular cell injection; (2) cell-free scaffold implantation to induce local stem cell migration and differentiation; and (3) Ex vivo seeding of scaffold prior to implantation.

Mesenchymal Stem Cells (MSCs) are the most frequently used cell types in bone and cartilage regeneration, with varied efficiencies reported, presumably due to the heterogeneous nature of MSCs. The term MSC is vague, and not all MSCs are equal in the sense that they are of different embryonic origins, and of different potencies for differentiation and marker identification.

In general, MSCs are characterized for their abilities to adhere to plastic culture dishes to form colonies. They can be cultured long term, and are able to differentiate into chondrocytes, adipogenic, and osteogenic cell types in vitro. Two classes of MSCs can be defined from their embryonic origin. Cranial Neural Crest MSCs give rise to cranial bones (excluding maxillary and main body of the mandible), dentine, pulp, and periodontal tissues as well as condylar bone and cartilage in addition to neural tissues in embryos. Trunk MSCs include bone marrow, adipose, and other organ derived mesenchymal population.

The markers of adult MSCs include stromal markers such as CD90, CD44, CD13, CD19, and CD79; and pericyte markers such as CD146. In the mouse, MSCs are marked within a broader stromal population of the Stem Cell Antigen 1 (Sca-1) positive cohorts [498]. In humans, Stro-1 and CD34 are described as MSC markers that can be used to isolate CFU-F forming cells in vitro, although these markers are expressed by other cell types and are thus not specific [498]. Cranial neural crest MSCs derived from dental origin, such as dental pulp stem cells (DPSCs), have been shown to express embryonic cell markers that are not currently identified in other adult MSCs. These markers include Oct-4, Nanog, SSEA-3, SSEA-4, TRA-1-60, and TRA-1-81 [499, 500], and their presence may signify higher potency and enhanced stem cell activity for DSPCs compared to other adult tissue-derived stem cells. In vivo correlation of cultured cells is not well studied.

Platelet-derived growth factor receptor alpha (PDGFRa) is used to enrich MSC colony-forming ability [501, 502] and its use leads to higher efficiency of ectopic bone formation, but it has not been shown to increase the efficiency of chondrocyte differentiation.

Bone marrow MSCs provide an autologous transplantation opportunity. They have been largely trialled in the larger joints of the body; direct injection into the TMJ has been investigated. This intraarticular route has led to improved cartilage repair in early-stage OA [503]. However, the effect is largely attributed to the trophic effects of injected MSCs that change the inflammatory environment without structural integration [503, 504] to regenerate the joints. Similarly, human umbilical cord-derived MSCs display repair benefits which are also thought to be from a paracrine effect, by suppressing inflammation in cases of TMJ OA [505].

Human exfoliated deciduous MSCs have been injected intravenously into TMJ OA mice, resulting in a decrease of IL-1β, iNOS, and MMP-13 expressing chondrocytes, and an increase in proliferation of chondrocytes. This indicates their role in slowing the progression of OA via cytokine secretion and immunoregulation [506]. Their anti-inflammatory effects appear to be more potent than bone marrow MSCs for treating RA in large joints [507].

DPSCs have also been shown to have colony-forming capacity, in addition to odonto/osteogenic and adipogenic differentiation capacity. Chondrogenic differentiation is evidenced by type II collagen and aggrecan immunostaining. After in vivo transplantation in mice, these cells appear able to generate new cartilage-like tissues [508].

Although "mesenchymal cells" may give rise to many of the embryonic connective tissues, adult MSCs do not display a promising potential for differentiation into functional parenchymal cells. This may partly be due to the elusive identity of MSCs. It is possible that adult MSCs are too far downstream of their embryonic namesakes. As well, complexity of finding the "real" stem cells in adult tissues has yet to be mastered. However, the immune-modulatory effects of MSCs seem to withstand the test of time, as demonstrated in cell-based therapies in TMD, where the trophic effect seems to play a major role in the short term due to the reduction in inflammation. A lack of engraftment and articular regeneration discourages however any expectations of full anatomical and functional recovery of the TMJ.

Another type of stem cell is the so-called "Induced Pluripotent Stem Cells" (iPSCs), which are fast becoming a focus of stem cell research. iPSCs potentially can provide an abundant and immune compatible cell source. They are pluripotent in nature, and may provide functional cell types for chondrocyte differentiation [509, 510], with the advantage of being able to produce their own matrix, enabling scaffold-free tissue regeneration. However, cell generation is time consuming and expensive, and uncontrolled de-differentiation poses risks for teratogenicity.

The current focus of iPSC research is around cardiovascular and neuronal diseases, with a lack of reliable literature on their possible use for TMJ regeneration. More research is required before a clinically useful modality becomes available.

Lastly, primary cultures of fibrocartilage stem cells (FCSCs) or chondrocytes have been used for joint regeneration. These include meniscus disc chondrocytes and costal cartilage chondrocytes. This approach has shown some promising results both in vitro and in vivo in animal models [511–514]. Perhaps one of the most elegant studies in this area is that performed by Embree et al. [515]. They identified FCSCs residing within the superficial zone niche in the TMJ condyle, utilizing the BrdU pulse-chase method. BrdU was injected into pregnant mice when the foetuses were 16.5 days old, and the label-retaining cells were chased in the postnatal offspring at 16 weeks of age. Single FCSCs isolated from the proliferation zone were able to spontaneously generate a cartilage anlage, remodel into bone, organize a haematopoietic microenvironment, and form structurally healthy bone and cartilage when transplanted into adult mice [515]. This study highlights the importance of an in-depth understanding of developmental biology for optimizing stem cell therapeutic modalities [516].

Traditional alloplastic joint replacement treatments have been used in clinical practice since 1961 where they were first employed in hip joint prostheses for the treatment of severe hip lesions [517, 518]. A TMJ prosthesis has also been applied in the field of craniomaxillofacial surgery, with various rates of success [519, 520], and more recently using 3D printed precision joints, allowing customized fabrication of prostheses [521].

The idea of using cell conducive scaffolds is to provide structural support and morphological guidance, and to permit local stem cells to migrate and differentiate into the desired cell types that assume normal anatomical morphology and function. A range of materials have been used, including synthetic materials such as polyethylene glycol (PEG)-based hydrogels [522], polylactic acid, and poly(vinyl alcohol) (PVA), alone and in combination with hydroxyapatite [523]. These hydrogels can be structurally and mechanically similar to cartilage, and allow efficient load transfer [524, 525].

PVA hydrogels have been researched extensively. These display the best mechano-weight bearing properties for the TMJ, but their main disadvantages are non-integration with surrounding tissues, and compression failure over time [525]. In addition to structural support, these gels are also used as drug delivery systems for the gradual release of medications such as glucocorticoids and antibiotics [526].

Naturally derived polymers, such as agarose, alginate, chitosan, hyaluronan, collagen, fibrin, and polysaccharides, are attractive biomaterials because they are biochemically similar to cartilage, and can be degraded by cell-secreted enzymes. Many of these are available commercially (for review, see [527]). As cell-seeded tissue engineering scaffolds, hydrogels are also extremely useful, as they promote chondrocyte attachment in a manner that is similar to cartilage ECM [528]. They maintain the chondrocyte phenotype in a way that is impossible in monolayer culture [529–531]. Their viscoelastic nature permits effective transfer of loads to chondrocytes, which depend on mechanical signals for survival [532, 533]. Perhaps the success of this approach will depend on the health of the cells in situ. At present, it is not clear how the inflammatory milieu influences the local stem cells. Thus, the potential of the local cell reaction is unclear, and the prognosis of the treatment may not be predictable [534, 535].

Bioactive materials such as porcine de-cellularized tissue combined with host stem cells may provide a time efficient and cost-effective alternative. Recent studies have shown that porcine cartilage provides type II collagen that causes only minor rejection reactions in the human body [536, 537]. Using α-galactosidase to remove α-gal from porcine cartilage can further reduce the likelihood of rejection [512]. Indeed, studies have shown that the porcine de-cellularized cartilage can enhance cartilage regeneration, proliferation, and matrix synthesis of TMJ disc-derived cells [538]. As yet, its use in condylar cartilage regeneration has not been reported.

Tissue engineering approaches also include combinations of scaffolds with ex vivo cell seeding before transplantation. The scaffold provides a TMJ-shaped mould for cells to migrate into, and then differentiate into anatomically and functionally correct structures. Both synthetic and natural scaffolds have been used to host appropriate cell types, and are often used in the presence of growth factors and cytokines, and sometimes under mechanical inductive conditions [539]. TMJ-shaped moulds treated with transforming growth factor-β1 have been reported to synergistically increase Young's modulus and to promote collagen fibril alignment, akin to native tissue [511]. PVA meshes seeded with TMJ disc cells, using a spinner flask, produce thick collagen in tissue engineering for the TMJ disc [540].

Autologous bone marrow MSCs combining natural hydrogels seem to provide functional benefits and histological improvement of fibrocartilage in large joints [541, 542], however, data on TMJ repair are absent. Similarly, human DPSCs combined with an alginate hydrogel co-culture and implanted in a rabbit model of cartilage damage showed significant cartilage regeneration [543]. A PVA scaffold with cultured DPSCs combined with exogenous FGF9 was able to simultaneously promote the chondrogenesis of DPSCs, and could partially inhibit their mineralization by enhancing the phosphorylation of ERK1/2 in DPSCs [513]. The combined method involving DPSCs in a variety of animal models including mice, rats, rabbits, and miniature pigs, shows promising results in a combined scaffold approach [544, 545], with two studies focused on preclinical models of cartilage tissue engineering. Again, the data derives from large joints and does not specifically address the TMJ.

Growth factors that enhance chondrogenic pathways such as TGF-β1 show some positive effects on cellular proliferation and on the production of extracellular matrix in TMJ disc implants [546], as well as increased collagen synthesis, increased Young's modulus and compressive stiffness in co-culture of articular chondrocytes and fibrochondrocytes [546]. Additionally, a significant increase in collagen and matrix deposition inside the engineered cartilage is seen [547]. GAG synthesis is also significantly stimulated by bFGF [511]. PDGF increases GAG synthesis [511]. Further, PDGF significantly increases the proliferation rate of TMJ-disc derived cells, and collagen and hyaluronic acid synthesis in engineered TMJ discs. It upregulates RNA levels of type I and II collagens, matrix metalloproteinases (MMPs), and their specific tissue inhibitors (TIMPs) [548].

Overall, joint and disc reconstruction and regeneration for the TMJ poses a formidable challenge, but significant experience has accumulated by studying large joints. Following a successful clinical trial of a knee joint approach [549], full regeneration of bilateral condyles has been reported in a patient with TMD. Specifically, the patient received transplantation of autologous nasal septum derived cells expanded in vitro and co-transplanted with glycosaminoglycan [550]. The authors speculated that the neural crest origin of the nasal septum may have contributed to the success of the transplanted cells [550].

Preclinical studies in larger animals such as dogs and mini pigs also provide valuable guides towards successful human studies. In one study in canines which involved bilateral resection of the meniscus, one side received bladder collagen matrix and achieved near perfect disc regeneration, while the control side showed degeneration in the fossae [538, 551]. In mini pigs, the use of allogeneic costal chondrocytes has been investigated for the development of transplantable, cell-based, scaffold-free TMJ implants. Implants were found to be well integrated, and the mechanical strength of the defects was more robust in the implanted meniscus than in untreated defects [552]. Combining the accumulated knowledge and multidisciplinary approaches including surgery, materials science, and stem cell science [514] may provide new therapeutic modalities for TMJ repair.

2.6 Orofacial Pain

Pain in the oral and craniofacial region affects approximately 5–12% of the population. This severely affects the life quality of sufferers, and is costly for society. The normal sensory function of the orofacial region is carried out by the trigeminal or fifth cranial nerve, which contains three sensory branches that innervate the oral cavity and facial area. The trigeminal nerve has a large representation in the central sensory region of the brain, representing 50% of the sensory cortex, and it is the only nerve that possesses an intracranial distal root ganglion. This over representation is partly due to evolutionary needs of small animals to sense their locations and environment [553] (e.g., whiskers of rodents [554]).

The trigeminal sensory ganglions develop early in mammalian embryos from the pharyngeal placodes, which have unique structures and functional roles in processing orofacial nociception as well as non-noxious sensations in comparison to the spinal nerve system. The trigeminal nerve is distributed to the oral mucosa, tongue, teeth, temporomandibular joints, and other orofacial structures by small diameter light myelinated A delta and unmyelinated C fibres that process orofacial nociception [555].

In humans, heightened sensory nerve function may have protective effects under normal physiological conditions. However, in pathology, this heightened sensitivity seems to bring especially unbearable pain, which is also prolonged for the sufferer due to trigeminal neurons becoming hyperactive for a considerable time following damage or inflammation [556].

Orofacial pain has been defined as "relatively localized syndromes of the head and neck" with complex aetiologies that include many disorders [557]. Most pain conditions in the orofacial region are caused by odontogenic sources, or factors such as infection and inflammation. Other sources include masticatory myalgia, and debilitating conditions such as neuropathic pain, whether idiopathic or iatrogenic [558].

2.6.1 Inflammatory Orofacial Pain

Inflammatory pain usually results from nociceptive stimuli, such as infection, trauma, or underlying systemic diseases, and it plays a pivotal role in generation of an effective defence to injury. Depending on the condition, neutrophils are followed typically by macrophages into the wound space, and release cytokines and chemokines that induce an immune response and the accompanying inflammatory process. Cytokines such as TNF-α, IL-1, and IL-6 are initially involved in the development of inflammation [559]. Further, chemokines that regulate bone metabolism via the CCR1, CCR2, CXCR3, and CXCR4 receptors expressed on osteoclast precursors, mature osteoclasts, and osteoblasts, can drive osteoclastic activity, and thus bone resorption can occur. This is one of the common sequelae of oral inflammation that can lead to chronic inflammatory pain.

The key inflammatory factors responsible for pain mediation are prostaglandins, which additionally cause vasodilation, increased blood flow, and the formation of inflammatory exudates. The majority of oral inflammatory pain can be attributed to pulpal inflammation originating from dental caries and its consequent dento-alveolar lesions. Oral mucosal lesions including ulcers, infective lesions, and malignancies can also cause inflammatory pain.

Inflammatory factors result in sensitization of nociceptive nerve endings. Peripheral nerve injury or orofacial inflammation often causes changes in the excitability of trigeminal neurons, thereby resulting in pain hypersensitivities such as allodynia and hyperalgesia [560]. Moreover, miscommunication in the trigeminal ganglion may occur due to dysfunctions of peptide signalling, which alter nitric oxide and nerve growth factors (NGF) pathways. This has been found in TMJ inflammation [560]. Glial cell abnormalities post injury or inflammation of peripheral nerves [555] can also contribute.

2.6.2 Neuropathic Orofacial Pain

Neuropathic orofacial pain (NOP), including trigeminal neuralgia, is a group of chronic pain conditions that affect the trigeminal nerve. These result from a dysfunction or primary lesion of the nerve [561]. NOP is heterogeneous in nature and has a range of aetiological factors. Neuropathic pain can be primary, with no recognized pathological process, or pain can be secondary, with an underlying pathology, as in painful post-traumatic neuropathies. Based on presenting symptoms, NOP can be broadly categorized as follows [558]:

(a) Paroxysmal neuropathy, e.g. trigeminal neuralgia, which is characterized by short electrical or sharp pain.
(b) Continuous, characterized by the long span of painful experience, as seen in post-herpetic neuralgia. It is sometimes of a burning quality, and is a characteristic feature of post-traumatic neuropathy.

Trigeminal neuropathic pain may occur due to either a decreased pain threshold (allodynia) or an increased pain intensity (hyperalgesia), or both. Allodynic neuropathic pain patients complain of severe pain normally following non-noxious mechanical or thermal stimulation of the extensive area innervated by the injured nerve. Hyperalgesic patients complain of stronger pain compared to that of nociceptive pain [562].

Neuropathic pain has peripheral and central hypersensitization of the trigeminal nervous system, albeit at much-heightened levels compared to nociceptive pain. Dysfunction of the micro-environment facilitates ectopic firing of peripheral nerves, with heightened perception of pain. Additionally, products associated with Wallerian degeneration that are released near uninjured nerves may trigger changes in ion channels and in receptor expression in uninjured nerves, contributing to neuropathic pain [562]. Mechanisms involved include the dysregulation of microglial ATP activities that mediate P2 receptors, and brain-derived nerve factor pathways which cause disinhibition of the normally inhibitory actions of GABA to become excitatory in lamina 1 neurons [563].

2.6.3 Masticatory Myalgia

Myalgia or muscular pain is the most common non-dental pain in the orofacial region. It is categorized as a deep somatic pain that is usually constant, dull, and aching, and may be accompanied by occasional exacerbations of sharp pain that are typically difficult to localize [564]. Although the mechanisms of myalgia are not well understood, it is generally classified as a nociceptive pain due to mechanical overloading and inflammatory stimuli. Endogenous inflammatory mediators including bradykinin (BK), serotonin, and prostaglandin activate the nociceptors that send pain signals to the central nervous system. Similar to inflammatory pain, the trigeminal nervous system becomes hypersensitized through NGF, leading to further modulation of GABA disinhibition [565].

Currently, orofacial pain is difficult to treat. The existing treatment modalities are typically removal of the cause if identifiable, as well as pharmacotherapeutic agents such as anticonvulsants, antidepressants, non-steroidal anti-inflammatory drugs, and opioids. These medications have strong side effects, and they only control the symptoms of the condition without altering its pathological root cause. Patients often require long-term treatment with minimal benefit, and can go through many episodes of relapse during or after therapy.

New therapies and treatment modalities are required urgently to address the pathology of orofacial pain and to repair damaged nerves. Using stem cells may provide opportunities for controlling and healing orofacial pain by (1) modulation of factors to alter neuronal pathophysiology and produce a disease-modifying effect; and (2) regeneration and repair of the damaged nerve to restore the anatomy and function of the affected nerve.

Inflammatory factors released by injury or infection cause changes in the microenvironment leading to orofacial pain, which further heightens the spontaneous activity of nerve fibres, while products associated with Wallerian degeneration released near uninjured nerves may contribute to neuropathic pain. Therefore, it is of utmost importance for the surrounding environment to protect the nerve fibres from degeneration as this exacerbates neuropathic pain. Neurotrophic growth factors are known to promote neuron development and survival, and are critical to providing a protective microenvironment. They also maintain functional integrity, promote regeneration, regulate neuronal plasticity, and aid in repairing damaged nerves [566].

Neurotrophic factors include various factors such as NGF, BDNF, neurotrophin 3 (NT-3), NT-4/5, insulin-like growth factor 1 and 2 (IGF 1/2), glial cell line-derived neurotrophic factor (GDNF), neurturin (NRTN), persephin, and ciliary neurotrophic factor (CNTF). Various neurotrophic factors affect different cell populations within the peripheral and central nervous system.

Mesenchymal stem cells (MSCs) produce a large variety of trophic factors, and they do not trigger immune rejection [567]. This makes them suitable for use in a range of conditions. They can significantly reduce both mechanical allodynia and thermal hyperalgesia in mouse and rat models. Bone marrow MSCs (BMMSCs) are thought to promote peripheral nerve regeneration. This happens not only via their direct release of neurotrophic factors, but also through indirect modulation of the behaviours of stem cells. This has been shown in a rat sciatic nerve regeneration model [568]. The effects of BMMSCs have been reported to reduce pain behaviours in mouse and rat models when injected intravenously or at the site [569, 570].

Adipose-derived stem cells have become a popular source for MSCs due to the abundance of lipoaspirates. These cells express a range of neurotrophic factors, namely NGF, BDNF, GDNF, and NT-4 [571]. They may also increase angiogenesis [572]. Clinical trials of autologous adipose-derived cells injected into female patients with neuropathic trigeminal pain lasting 4 months to 6 years showed that 6/9 patients had reduced pain scores at 6 months after treatment [573].

The secretion of neurotrophic factors by different stem cell populations suggests that no matter the source of the stem cell, there is a possible use for them in treating neuropathic pain, by either providing neuroprotection or through neuro-regenerative effects.

Recently, matured human induced pluripotent stem cell (iPSC)-derived GABAergic neurons (iGABAergic neurons) were differentiated ex vivo and then transplanted into the dorsal horn to alleviate pain elicited by spinal nerve injury (SNI) as a model of persistent neuropathic pain in mice. iGABAergic neurons were able to survive in the SNI-injured spinal cord, and were able to alleviate pain long term with a single treatment [574] similar to iPSC induced GABAergic progenitor transplants from rats and mice [575].

Human iPSCs have been used to model human sensory nociceptors, providing a new rationale for voltage-gated sodium channel (Na_V) 1.7 function, which also promises to be a valuable translational tool to profile and develop more efficacious clinical analgesics [576]. It is interesting that patient-derived iPSCs can also be used as "drug testing" platforms to identify patient-specific changes in nociceptor excitability in peripheral neuropathy, to deliver more effective treatments [577].

Human MSCs from Wharton's jelly (HMSCs) are able to differentiate into neuroglial-like cells associated with poly (DL-lactide-ε-caprolactone) membranes in vitro, however, they failed to repair the sciatic nerve injury in a rat model [578]. Adipose-derived MSCs have been induced to differentiate into neuron-like cells, as evidenced by neuronal morphology and the presence of neuronal markers including microtubule-associated protein 2, neuronal nuclear antigen, and β-tubulin III [579].

Regeneration of damaged nerves has wide implications in a range of neurological diseases and conditions such as traumatic injury, as well as vascular, degenerative, and inflammatory conditions of the central and peripheral nervous systems. Investigations into stem and progenitor cell therapies will likely remain a hot topic of research.

Dental pulp stem cells (DPSCs) derived from a cranial neural crest lineage, retain a remarkable potential for neuronal differentiation, and additionally they express multiple factors that are linked to neuronal and axonal regenerations [580, 581]. DPSCs express neural cell markers such as oligodendrocyte and glial fibrillary acidic protein. Osteocytes have been grown from dental pulp MSCs in vitro [582]. DPSCs filled in a degradable poly-DL-lactide-co-glycolide (PLGA) tube were able to effectively repair gaps in the facial nerves with myelinated fibres of rats [583]. Transplantation of human DPSCs into the completely transected adult rat spinal cord resulted in the marked recovery of hind limb locomotor functions [584], while BMMSCs or skin-derived fibroblasts led to substantially less recovery of locomotor function [585].

A number of published studies have described the successful differentiation of NSCs/NPCs from iPSCs, and their subsequent engraftment into spinal cord animal models, followed by functional recovery from injury [586, 587]. Similarly, peripheral nerve rebuilding by regeneration of both neuron and supportive cells has been reported in sciatic nerve injury models [588, 589]. In the neuropathic pain arena, iPSC induced GABAergic progenitors were able to survive long term in the injured spinal cord, with evidence of synaptic integration. This provides a proof-of-concept for possible therapeutic approaches in humans [575].

The regeneration of nervous tissues is a long-term goal. There still is a long way to go before efficient human therapeutic modalities are available. Issues that need to be sorted include selection of the correct stem cell types, finding appropriate cues for differentiation pathways, facilitating the engraftment of transplanted cells, and avoiding potential undesired outcomes such as teratogenicity of iPSCs.

Stem cells offer an exciting therapeutic potential for treating neuropathic pain. Although the mechanism of action of stem cells is not completely understood, studies demonstrate that they have the potential to arrest degenerative processes, inhibit apoptotic pathways, and augment the survival/recovery pathways of both injured and uninjured nerves. Coupled with the neurotrophic factor-releasing nature of stem cells, the ability of stem cells to modify cellular processes provides for both a protective and a restorative microenvironment that can potentially fully reverse the onset of neuropathic pain. More excitingly, future stem cell treatments promise regeneration and repair of damaged or degenerated nerves to restore the function of neurons and neural tissues.

3 Conclusion and Future Directions

Stem cells and regenerative medicine is an evolving science with significant and exciting clinical implications. Despite the substantial advances made in these areas, however, the applications in terms of diseases affecting the oral and maxillofacial complex have been modest at best. The greater majority of advancement has been made in terms of craniofacial bone biology and associated defects. Work on soft tissue applications to tackle problems related to oral mucosal diseases, head and neck cancers, salivary disorders, and temporomandibular and orofacial pain conditions has been less impressive in its progress and outcomes.

Bioprinting is an exciting yet challenging approach in regenerative medicine, and has the potential to change the manner in which many oral mucosal, salivary, and neuropathic conditions may be managed or treated. This area of investigation, which couples together biology and engineering, remains a major hope for the future of clinical oral medicine. Homogeneity of stem cells, methods of delivery, quality of regenerated tissues, and their potential integration into the host are still problems, however, requiring more research to solve.

Obstacles to clinical stem cell therapy include a deficiency in generating sufficient cells, and determining how to direct their differentiation into functional cells and tissues. Although numerous applications have been reported in isolated cases or in laboratory settings, clinical applicability is still hindered by the lack of robust, controllable, and routinely reproducible approaches on a mass scale. Until these problems can be resolved, the full potential of regenerative medicine in clinical oral medicine remains elusive.

References

1. Prentice DA. Adult stem cells. Circ Res. 2019; 124(6):837–9. https://doi.org/10.1161/CIRCRESAHA.118.313664.
2. Paz AG, Maghaireh H, Mangano FG. Stem cells in dentistry: types of intra- and extraoral tissue-derived stem cells and clinical applications. Stem Cells Int. 2018;2018:4313610. https://doi.org/10.1155/2018/4313610.
3. Jones KB, Klein OD. Oral epithelial stem cells in tissue maintenance and disease: the first steps in a long journey. Int J Oral Sci. 2013;5(3):121–9. https://doi.org/10.1038/ijos.2013.46.

4. Papagerakis S, Pannone G, Zheng L, About I, Taqi N, Nguyen NP, Matossian M, McAlpin B, Santoro A, McHugh J, Prince ME, Papagerakis P. Oral epithelial stem cells – implications in normal development and cancer metastasis. Exp Cell Res. 2014;325(2):111–29. https://doi.org/10.1016/j.yexcr.2014.04.021.
5. Asaka T, Akiyama M, Kitagawa Y, Shimizu H. Higher density of label-retaining cells in gingival epithelium. J Dermatol Sci. 2009;55(2):132–4. https://doi.org/10.1016/j.jdermsci.2009.03.006.
6. Bickenbach JR. Identification and behavior of label-retaining cells in oral mucosa and skin. J Dent Res. 1981;60:1611–20. https://doi.org/10.1177/00220345 8106000311011.
7. Bickenbach JR, Mackenzie IC. Identification and localization of label-retaining cells in hamster epithelia. J Invest Dermatol. 1984;82(6):618–22. https://doi.org/10.1111/1523-1747.ep12261460.
8. DiPietro LA. Oral stem cells: the fountain of youth for epithelialization and wound therapy? Adv Wound Care (New Rochelle). 2014;3(7):465–7. https://doi.org/10.1089/wound.2012.0421.
9. Calenic B, Ishkitiev N, Yaegaki K, Imai T, Costache M, Tovaru M, Tovaru S, Parlatescu I. Characterization of oral keratinocyte stem cells and prospects of its differentiation to oral epithelial equivalents. Romanian J Morphol Embryol. 2010;51(4):641–5.
10. Nakamura T, Endo K, Kinoshita S. Identification of human oral keratinocyte stem/progenitor cells by neurotrophin receptor p75 and the role of neurotrophin/p75 signaling. Stem Cells. 2007;25(3):628–38. https://doi.org/10.1634/stemcells.2006-0494.
11. Bansal R, Jain A. Current overview on dental stem cells applications in regenerative dentistry. J Nat Sci Biol Med. 2015;6(1):29–34. https://doi.org/10.4103/0976-9668.149074.
12. Friedenstein AJ, Chailakhjan RK, Lalykina KS. The development of fibroblast colonies in monolayer cultures of guinea-pig bone marrow and spleen cells. Cell Tissue Kinet. 1970;3(4):393–403. https://doi.org/10.1111/j.1365-2184.1970.tb00347.x.
13. Pittenger MF, Discher DE, Peault BM, Phinney DG, Hare JM, Caplan AI. Mesenchymal stem cell perspective: cell biology to clinical progress. NPJ Regen Med. 2019;4:22. https://doi.org/10.1038/s41536-019-0083-6.
14. Gronthos S, Mankani M, Brahim J, Robey PG, Shi S. Postnatal human dental pulp stem cells (DPSCs) *in vitro* and *in vivo*. Proc Natl Acad Sci USA. 2000;97(25):13625–30. https://doi.org/10.1073/pnas.240309797.
15. Miura M, Gronthos S, Zhao M, Lu B, Fisher LW, Robey PG, Shi S. SHED: stem cells from human exfoliated deciduous teeth. Proc Natl Acad Sci USA. 2003;100(10):5807–12. https://doi.org/10.1073/pnas.0937635100.
16. Sonoyama W, Liu Y, Fang D, Yamaza T, Seo BM, Zhang C, Liu H, Gronthos S, Wang CY, Wang S, Shi S. Mesenchymal stem cell-mediated functional tooth regeneration in swine. PLoS One. 2006;1:e79. https://doi.org/10.1371/journal.pone.0000079.
17. Sonoyama W, Liu Y, Yamaza T, Tuan RS, Wang S, Shi S, Huang GT. Characterization of the apical papilla and its residing stem cells from human immature permanent teeth: a pilot study. J Endod. 2008;34(2):166–71. https://doi.org/10.1016/j.joen.2007.11.021.
18. Seo BM, Miura M, Gronthos S, Bartold PM, Batouli S, Brahim J, Young M, Robey PG, Wang CY, Shi S. Investigation of multipotent postnatal stem cells from human periodontal ligament. Lancet. 2004;364(9429):149–55. https://doi.org/10.1016/S0140-6736(04)16627-0.
19. An Z, Sabalic M, Bloomquist RF, Fowler TE, Streelman T, Sharpe PT. A quiescent cell population replenishes mesenchymal stem cells to drive accelerated growth in mouse incisors. Nat Commun. 2018;9(1):378. https://doi.org/10.1038/s41467-017-02785-6.
20. Kang KJ, Ryu CJ, Jang YJ. Identification of dentinogenic cell-specific surface antigens in odontoblast-like cells derived from adult dental pulp. Stem Cell Res Ther. 2019;10(1):128. https://doi.org/10.1186/s13287-019-1232-y.
21. Yang X, Li L, Xiao L, Zhang D. Recycle the dental fairy's package: overview of dental pulp stem cells. Stem Cell Res Ther. 2018;9(1):347. https://doi.org/10.1186/s13287-018-1094-8.
22. Boddupally K, Wang G, Chen Y, Kobielak A. Lgr5 Marks neural crest derived multipotent oral stromal stem cells. Stem Cells. 2016;34(3):720–31. https://doi.org/10.1002/stem.2314.
23. d'Aquino R, De Rosa A, Lanza V, Tirino V, Laino L, Graziano A, Desiderio V, Laino G, Papaccio G. Human mandible bone defect repair by the grafting of dental pulp stem/progenitor cells and collagen sponge biocomplexes. Eur Cell Mater. 2009;18:75–83. https://doi.org/10.22203/ecm.v018a07.
24. Monti M, Graziano A, Rizzo S, Perotti C, Del Fante C, d'Aquino R, Redi CA, Rodriguez YBR. *In vitro* and *In vivo* differentiation of progenitor stem cells obtained after mechanical digestion of human dental pulp. J Cell Physiol. 2017;232(3):548–55. https://doi.org/10.1002/jcp.25452.
25. Sybil D, Jain V, Mohanty S, Husain SA. Oral stem cells in intraoral bone formation. J Oral Biosci. 2020;62(1):36–43. https://doi.org/10.1016/j.job.2019.12.001.
26. Feng F, Akiyama K, Liu Y, Yamaza T, Wang TM, Chen JH, Wang BB, Huang GT, Wang S, Shi S. Utility of PDL progenitors for *in vivo* tissue regeneration: a report of 3 cases. Oral Dis. 2010;16(1):20–8. https://doi.org/10.1111/j.1601-0825.2009.01593.x.
27. Shalini HS, Vandana KL. Direct application of autologous periodontal ligament stem cell niche in treatment of periodontal osseous defects: a randomized controlled trial. J Indian Soc Periodontol. 2018;22(6):503–12. https://doi.org/10.4103/jisp.jisp_92_18.

28. Pradel W, Tausche E, Gollogly J, Lauer G. Spontaneous tooth eruption after alveolar cleft osteoplasty using tissue-engineered bone: a case report. Oral Surg Oral Med Oral Pathol Oral Radiol Endod. 2008;105(4):440–4. https://doi.org/10.1016/j.tripleo.2007.07.042.
29. Pradel W, Lauer G. Tissue-engineered bone grafts for osteoplasty in patients with cleft alveolus. Ann Anat. 2012;194(6):545–8. https://doi.org/10.1016/j.aanat.2012.06.002.
30. Pradel W, Eckelt U, Lauer G. Bone regeneration after enucleation of mandibular cysts: comparing autogenous grafts from tissue-engineered bone and iliac bone. Oral Surg Oral Med Oral Pathol Oral Radiol Endod. 2006;101(3):285–90. https://doi.org/10.1016/j.tripleo.2005.06.001.
31. Zheng C, Chen J, Liu S, Jin Y. Stem cell-based bone and dental regeneration: a view of microenvironmental modulation. Int J Oral Sci. 2019;11(3):23. https://doi.org/10.1038/s41368-019-0060-3.
32. Capp JP. Cancer stem cells: from historical roots to a new perspective. J Oncol. 2019;2019:5189232. https://doi.org/10.1155/2019/5189232.
33. Beck B, Blanpain C. Unravelling cancer stem cell potential. Nat Rev Cancer. 2013;13(10):727–38. https://doi.org/10.1038/nrc3597.
34. Martins-Neves SR, Cleton-Jansen AM, Gomes CMF. Therapy-induced enrichment of cancer stem-like cells in solid human tumors: where do we stand? Pharmacol Res. 2018;137:193–204. https://doi.org/10.1016/j.phrs.2018.10.011.
35. Freitas DP, Teixeira CA, Santos-Silva F, Vasconcelos MH, Almeida GM. Therapy-induced enrichment of putative lung cancer stem-like cells. Int J Cancer. 2014;134(6):1270–8. https://doi.org/10.1002/ijc.28478.
36. Peitzsch C, Tyutyunnykova A, Pantel K, Dubrovska A. Cancer stem cells: the root of tumor recurrence and metastases. Semin Cancer Biol. 2017;44:10–24. https://doi.org/10.1016/j.semcancer.2017.02.011.
37. Phi LTH, Sari IN, Yang YG, Lee SH, Jun N, Kim KS, Lee YK, Kwon HY. Cancer stem cells (CSCs) in drug resistance and their therapeutic implications in cancer treatment. Stem Cells Int. 2018;2018:5416923. https://doi.org/10.1155/2018/5416923.
38. Shin KH, Kim RH. An updated review of oral cancer stem cells and their stemness regulation. Crit Rev Oncog. 2018;23(3–4):189–200. https://doi.org/10.1615/CritRevOncog.2018027501.
39. Kyrodimos E, Chrysovergis A, Tsiambas E, Papanikolaou V. Cancer stem cells in oral squamous cell carcinoma. J BUON. 2018;23(5):1558–9.
40. Shibata M, Hoque MO. Targeting cancer stem cells: a strategy for effective eradication of cancer. Cancers (Basel). 2019;11(5):732. https://doi.org/10.3390/cancers11050732.
41. Lee BK. Growth factors in oral and maxillofacial surgery: potentials and challenges. J Korean Assoc Oral Maxillofac Surg. 2013;39(6):255–6. https://doi.org/10.5125/jkaoms.2013.39.6.255.
42. Cicciù M, Herford AS, Juodžbalys G, Stoffella E. Recombinant human bone morphogenetic protein type 2 application for a possible treatment of bisphosphonates-related osteonecrosis of the jaw. J Craniofac Surg. 2012;23(3):784–8. https://doi.org/10.1097/SCS.0b013e31824dbdd4.
43. Werner S, Grose R. Regulation of wound healing by growth factors and cytokines. Physiol Rev. 2003;83(3):835–70. https://doi.org/10.1152/physrev.2003.83.3.835.
44. Ross R, Raines EW, Bowen-Pope DF. The biology of platelet-derived growth factor. Cell. 1986;46(2):155–69. https://doi.org/10.1016/0092-8674(86)90733-6.
45. Alvarez RH, Kantarjian HM, Cortes JE. Biology of platelet-derived growth factor and its involvement in disease. Mayo Clin Proc. 2006;81(9):1241–57. https://doi.org/10.4065/81.9.1241.
46. Massagué J, Gomis RR. The logic of TGFbeta signaling. FEBS Lett. 2006;580(12):2811–20. https://doi.org/10.1016/j.febslet.2006.04.033.
47. Kaigler D, Cirelli JA, Giannobile WV. Growth factor delivery for oral and periodontal tissue engineering. Expert Opin Drug Deliv. 2006;3(5):647–62. https://doi.org/10.1517/17425247.3.5.647.
48. Kwon TK, Song JM, Kim IR, Park BS, Kim CH, Cheong IK, Shin SH. Effect of recombinant human bone morphogenetic protein-2 on bisphosphonate-treated osteoblasts. J Korean Assoc Oral Maxillofac Surg. 2014;40(6):291–6. https://doi.org/10.5125/jkaoms.2014.40.6.291.
49. Conover CA. *In vitro* studies of insulin-like growth factor I and bone. Growth Horm IGF Res. 2000;10(Suppl B):S107–10. https://doi.org/10.1016/s1096-6374(00)80020-9.
50. Dohan Ehrenfest DM, Bielecki T, Jimbo R, Barbé G, Del Corso M, Inchingolo F, Sammartino G. Do the fibrin architecture and leukocyte content influence the growth factor release of platelet concentrates? An evidence-based answer comparing a pure platelet-rich plasma (P-PRP) gel and a leukocyte- and platelet-rich fibrin (L-PRF). Curr Pharm Biotechnol. 2012;13(7):1145–52. https://doi.org/10.2174/138920112800624382.
51. Kubota S, Kawata K, Yanagita T, Doi H, Kitoh T, Takigawa M. Abundant retention and release of connective tissue growth factor (CTGF/CCN2) by platelets. J Biochem. 2004;136(3):279–82. https://doi.org/10.1093/jb/mvh126.
52. Ohki Y, Heissig B, Sato Y, Akiyama H, Zhu Z, Hicklin DJ, Shimada K, Ogawa H, Daida H, Hattori K, Ohsaka A. Granulocyte colony-stimulating factor promotes neovascularization by releasing vascular endothelial growth factor from neutrophils. FASEB J. 2005;19(14):2005–7. https://doi.org/10.1096/fj.04-3496fje.
53. Toulon A, Breton L, Taylor KR, Tenenhaus M, Bhavsar D, Lanigan C, Rudolph R, Jameson J, Havran WL. A role for human skin-resident T cells in wound healing. J Exp Med. 2009;206(4):743–50. https://doi.org/10.1084/jem.20081787.

54. Ceccarelli G, Presta R, Benedetti L, Cusella De Angelis MG, Lupi SM, Rodriguez YBR. Emerging perspectives in scaffold for tissue engineering in oral surgery. Stem Cells Int. 2017;2017:4585401. https://doi.org/10.1155/2017/4585401.
55. Polo-Corrales L, Latorre-Esteves M, Ramirez-Vick JE. Scaffold design for bone regeneration. J Nanosci Nanotechnol. 2014;14(1):15–56. https://doi.org/10.1166/jnn.2014.9127.
56. Lei X, Gao J, Xing F, Zhang Y, Ma Y, Zhang G. Comparative evaluation of the physicochemical properties of nano-hydroxyapatite/collagen and natural bone ceramic/collagen scaffolds and their osteogenesis-promoting effect on MC3T3-E1 cells. Regen Biomater. 2019;6(6):361–71. https://doi.org/10.1093/rb/rbz026.
57. Rezwan K, Chen QZ, Blaker JJ, Boccaccini AR. Biodegradable and bioactive porous polymer/inorganic composite scaffolds for bone tissue engineering. Biomaterials. 2006;27(18):3413–31. https://doi.org/10.1016/j.biomaterials.2006.01.039.
58. Fernandes JS, Gentile P, Pires RA, Reis RL, Hatton PV. Multifunctional bioactive glass and glass-ceramic biomaterials with antibacterial properties for repair and regeneration of bone tissue. Acta Biomater. 2017;59:2–11. https://doi.org/10.1016/j.actbio.2017.06.046.
59. El-Rashidy AA, Roether JA, Harhaus L, Kneser U, Boccaccini AR. Regenerating bone with bioactive glass scaffolds: a review of *in vivo* studies in bone defect models. Acta Biomater. 2017;62:1–28. https://doi.org/10.1016/j.actbio.2017.08.030.
60. Souza MT, Tansaz S, Zanotto ED, Boccaccini AR. Bioactive glass Fiber-reinforced PGS matrix composites for cartilage regeneration. Materials (Basel). 2017;10(1):83. https://doi.org/10.3390/ma10010083.
61. Aguilar-Pérez FJ, Vargas-Coronado RF, Cervantes-Uc JM, Cauich-Rodríguez JV, Covarrubias C, Pedram-Yazdani M. Preparation and bioactive properties of nano bioactive glass and segmented polyurethane composites. J Biomater Appl. 2016;30(9):1362–72. https://doi.org/10.1177/0885328215626361.
62. Benders KE, van Weeren PR, Badylak SF, Saris DB, Dhert WJ, Malda J. Extracellular matrix scaffolds for cartilage and bone regeneration. Trends Biotechnol. 2013;31(3):169–76. https://doi.org/10.1016/j.tibtech.2012.12.004.
63. Papadimitropoulos A, Scotti C, Bourgine P, Scherberich A, Martin I. Engineered decellularized matrices to instruct bone regeneration processes. Bone. 2015;70:66–72. https://doi.org/10.1016/j.bone.2014.09.007.
64. Cheng CW, Solorio LD, Alsberg E. Decellularized tissue and cell-derived extracellular matrices as scaffolds for orthopaedic tissue engineering. Biotechnol Adv. 2014;32(2):462–84. https://doi.org/10.1016/j.biotechadv.2013.12.012.
65. Urist MR. Bone: formation by autoinduction. Science. 1965;150(3698):893–9. https://doi.org/10.1126/science.150.3698.893.
66. Yin H, Wang Y, Sun Z, Sun X, Xu Y, Li P, Meng H, Yu X, Xiao B, Fan T, Wang Y, Xu W, Wang A, Guo Q, Peng J, Lu S. Induction of mesenchymal stem cell chondrogenic differentiation and functional cartilage microtissue formation for *in vivo* cartilage regeneration by cartilage extracellular matrix-derived particles. Acta Biomater. 2016;33:96–109. https://doi.org/10.1016/j.actbio.2016.01.024.
67. Bahar H, Yaffe A, Boskey A, Binderman I. Influence of bone-derived matrices on generation of bone in an ectopic rat model. J Orthop Res. 2010;28(5):664–70. https://doi.org/10.1002/jor.21017.
68. Becerra J, Andrades JA, Ertl DC, Sorgente N, Nimni ME. Demineralized bone matrix mediates differentiation of bone marrow stromal cells *in vitro*: effect of age of cell donor. J Bone Miner Res. 1996;11(11):1703–14. https://doi.org/10.1002/jbmr.5650111114.
69. Honsawek S, Bumrungpanichthaworn P, Thitiset T, Wolfinbarger L Jr. Gene expression analysis of demineralized bone matrix-induced osteogenesis in human periosteal cells using cDNA array technology. Genet Mol Res. 2011;10(3):2093–103. https://doi.org/10.4238/vol10-3gmr1329.
70. Gepstein R, Weiss RE, Hallel T. Bridging large defects in bone by demineralized bone matrix in the form of a powder. A radiographic, histological, and radioisotope-uptake study in rats. J Bone Joint Surg Am. 1987;69(7):984–92.
71. Yang Y, Lin H, Shen H, Wang B, Lei G, Tuan RS. Mesenchymal stem cell-derived extracellular matrix enhances chondrogenic phenotype of and cartilage formation by encapsulated chondrocytes *in vitro* and *in vivo*. Acta Biomater. 2018;69:71–82. https://doi.org/10.1016/j.actbio.2017.12.043.
72. Ragelle H, Naba A, Larson BL, Zhou F, Prijić M, Whittaker CA, Del Rosario A, Langer R, Hynes RO, Anderson DG. Comprehensive proteomic characterization of stem cell-derived extracellular matrices. Biomaterials. 2017;128:147–59. https://doi.org/10.1016/j.biomaterials.2017.03.008.
73. Chen XD, Dusevich V, Feng JQ, Manolagas SC, Jilka RL. Extracellular matrix made by bone marrow cells facilitates expansion of marrow-derived mesenchymal progenitor cells and prevents their differentiation into osteoblasts. J Bone Miner Res. 2007;22(12):1943–56. https://doi.org/10.1359/jbmr.070725.
74. Bhat A, Boyadjiev SA, Senders CW, Leach JK. Differential growth factor adsorption to calvarial osteoblast-secreted extracellular matrices instructs osteoblastic behavior. PLoS One. 2011;6(10):e25990. https://doi.org/10.1371/journal.pone.0025990.
75. Lau TT, Lee LQ, Vo BN, Su K, Wang DA. Inducing ossification in an engineered 3D scaffold-free living cartilage template. Biomaterials. 2012;33(33):8406–17. https://doi.org/10.1016/j.biomaterials.2012.08.025.

76. Tour G, Wendel M, Tcacencu I. Bone marrow stromal cells enhance the osteogenic properties of hydroxyapatite scaffolds by modulating the foreign body reaction. J Tissue Eng Regen Med. 2014;8(11):841–9. https://doi.org/10.1002/term.1574.
77. Xue JX, Gong YY, Zhou GD, Liu W, Cao Y, Zhang WJ. Chondrogenic differentiation of bone marrow-derived mesenchymal stem cells induced by acellular cartilage sheets. Biomaterials. 2012;33(24):5832–40. https://doi.org/10.1016/j.biomaterials.2012.04.054.
78. Zimmermann G, Moghaddam A. Allograft bone matrix versus synthetic bone graft substitutes. Injury. 2011;42(Suppl 2):S16–21. https://doi.org/10.1016/j.injury.2011.06.199.
79. Barabaschi GD, Manoharan V, Li Q, Bertassoni LE. Engineering pre-vascularized scaffolds for bone regeneration. Adv Exp Med Biol. 2015;881:79–94. https://doi.org/10.1007/978-3-319-22345-2_5.
80. Holzapfel BM, Rudert M, Hutmacher DW. Scaffold-based bone tissue engineering. Orthopade. 2017;46(8):701–10. https://doi.org/10.1007/s00132-017-3444-0.
81. Mobasseri R, Tian L, Soleimani M, Ramakrishna S, Naderi-Manesh H. Peptide modified nanofibrous scaffold promotes human mesenchymal stem cell proliferation and long-term passaging. Mater Sci Eng C Mater Biol Appl. 2018;84:80–9. https://doi.org/10.1016/j.msec.2017.11.017.
82. Alford AI, Kozloff KM, Hankenson KD. Extracellular matrix networks in bone remodeling. Int J Biochem Cell Biol. 2015;65:20–31. https://doi.org/10.1016/j.biocel.2015.05.008.
83. Jeong HJ, Gwak SJ, Seo KD, Lee S, Yun JH, Cho YS, Lee SJ. Fabrication of three-dimensional composite scaffold for simultaneous alveolar bone regeneration in dental implant installation. Int J Mol Sci. 2020;21(5):1863. https://doi.org/10.3390/ijms21051863.
84. Engstrand T. Biomaterials and biologics in craniofacial reconstruction. J Craniofac Surg. 2012;23(1):239–42. https://doi.org/10.1097/SCS.0b013e318241c0f4.
85. Azi ML, Aprato A, Santi I, Kfuri M Jr, Masse A, Joeris A. Autologous bone graft in the treatment of post-traumatic bone defects: a systematic review and meta-analysis. BMC Musculoskelet Disord. 2016;17(1):465. https://doi.org/10.1186/s12891-016-1312-4.
86. Moura LB, Carvalho PH, Xavier CB, Post LK, Torriani MA, Santagata M, Chagas Júnior OL. Autogenous non-vascularized bone graft in segmental mandibular reconstruction: a systematic review. Int J Oral Maxillofac Surg. 2016;45(11):1388–94. https://doi.org/10.1016/j.ijom.2016.05.004.
87. Demetter RS, Calahan BG, Mealey BL. Histologic evaluation of wound healing after ridge preservation with cortical, cancellous, and combined cortico-cancellous freeze-dried bone allograft: a randomized controlled clinical trial. J Periodontol. 2017;88(9):860–8. https://doi.org/10.1902/jop.2017.170155.
88. Lang NP, Zitzmann NU. Clinical research in implant dentistry: evaluation of implant-supported restorations, aesthetic and patient-reported outcomes. J Clin Periodontol. 2012;39(Suppl 12):133–8. https://doi.org/10.1111/j.1600-051X.2011.01842.x.
89. Bracey DN, Jinnah AH, Willey JS, Seyler TM, Hutchinson ID, Whitlock PW, Smith TL, Danelson KA, Emory CL, Kerr BA. Investigating the osteoinductive potential of a decellularized xenograft bone substitute. Cells Tissues Organs. 2019;207(2):97–113. https://doi.org/10.1159/000503280.
90. Manfro R, Fonseca FS, Bortoluzzi MC, Sendyk WR. Comparative, histological and histomorphometric analysis of three anorganic bovine xenogenous bone substitutes: bio-oss, bone-fill and gen-ox anorganic. J Maxillofac Oral Surg. 2014; 13(4):464–70. https://doi.org/10.1007/s12663-013-0554-z.
91. Song R, Murphy M, Li C, Ting K, Soo C, Zheng Z. Current development of biodegradable polymeric materials for biomedical applications. Drug Des Devel Ther. 2018;12:3117–45. https://doi.org/10.2147/dddt.S165440.
92. Pilipchuk SP, Plonka AB, Monje A, Taut AD, Lanis A, Kang B, Giannobile WV. Tissue engineering for bone regeneration and osseointegration in the oral cavity. Dent Mater. 2015;31(4):317–38. https://doi.org/10.1016/j.dental.2015.01.006.
93. Naung NY, Shehata E, Van Sickels JE. Resorbable versus nonresorbable membranes: when and why? Dent Clin North Am. 2019;63(3):419–31. https://doi.org/10.1016/j.cden.2019.02.008.
94. Gentile P, Chiono V, Carmagnola I, Hatton PV. An overview of poly(lactic-co-glycolic) acid (PLGA)-based biomaterials for bone tissue engineering. Int J Mol Sci. 2014;15(3):3640–59. https://doi.org/10.3390/ijms15033640.
95. Jain RA. The manufacturing techniques of various drug loaded biodegradable poly(lactide-co-glycolide) (PLGA) devices. Biomaterials. 2000;21(23):2475–90. https://doi.org/10.1016/s0142-9612(00)00115-0.
96. Cao H, Kuboyama N. A biodegradable porous composite scaffold of PGA/beta-TCP for bone tissue engineering. Bone. 2010;46(2):386–95. https://doi.org/10.1016/j.bone.2009.09.031.
97. Li X, Wang Y, Wang Z, Qi Y, Li L, Zhang P, Chen X, Huang Y. Composite PLA/PEG/nHA/dexamethasone scaffold prepared by 3D printing for bone regeneration. Macromol Biosci. 2018;18(6):e1800068. https://doi.org/10.1002/mabi.201800068.
98. Thuaksuban N, Nuntanaranont T, Pattanachot W, Suttapreyasri S, Cheung LK. Biodegradable polycaprolactone-chitosan three-dimensional scaffolds fabricated by melt stretching and multilayer deposition for bone tissue engineering: assessment of the physical properties and cellular response. Biomed Mater. 2011;6(1):015009. https://doi.org/10.1088/1748-6041/6/1/015009.

99. Gandolfi MG, Zamparini F, Degli Esposti M, Chiellini F, Aparicio C, Fava F, Fabbri P, Taddei P, Prati C. Polylactic acid-based porous scaffolds doped with calcium silicate and dicalcium phosphate dihydrate designed for biomedical application. Mater Sci Eng C Mater Biol Appl. 2018;82:163–81. https://doi.org/10.1016/j.msec.2017.08.040.
100. Prati C, Gandolfi MG. Calcium silicate bioactive cements: biological perspectives and clinical applications. Dent Mater. 2015;31(4):351–70. https://doi.org/10.1016/j.dental.2015.01.004.
101. de Almeida AD, Leite FG, Chaud MV, Rebelo MA, Borges L, Viroel FJM, Hataka A, Grotto D. Safety and efficacy of hydroxyapatite scaffold in the prevention of jaw osteonecrosis *in vivo*. J Biomed Mater Res B Appl Biomater. 2018;106(5):1799–808. https://doi.org/10.1002/jbm.b.33995.
102. Huang YX, Ren J, Chen C, Ren TB, Zhou XY. Preparation and properties of poly(lactide-co-glycolide) (PLGA)/ nano-hydroxyapatite (NHA) scaffolds by thermally induced phase separation and rabbit MSCs culture on scaffolds. J Biomater Appl. 2008;22(5):409–32. https://doi.org/10.1177/0885328207077632.
103. Kim SS, Sun Park M, Jeon O, Yong Choi C, Kim BS. Poly(lactide-co-glycolide)/hydroxyapatite composite scaffolds for bone tissue engineering. Biomaterials. 2006;27(8):1399–409. https://doi.org/10.1016/j.biomaterials.2005.08.016.
104. Kacarevic ZP, Rider PM, Alkildani S, Retnasingh S, Smeets R, Jung O, Ivanisevic Z, Barbeck M. An introduction to 3D bioprinting: possibilities, challenges and future aspects. Materials (Basel). 2018;11(11):2199. https://doi.org/10.3390/ma11112199.
105. Ayan B, Heo DN, Zhang Z, Dey M, Povilianskas A, Drapaca C, Ozbolat IT. Aspiration-assisted bioprinting for precise positioning of biologics. Sci Adv. 2020;6(10):eaaw5111. https://doi.org/10.1126/sciadv.aaw5111.
106. Wust S, Muller R, Hofmann S. Controlled positioning of cells in biomaterials-approaches towards 3D tissue printing. J Funct Biomater. 2011;2(3):119–54. https://doi.org/10.3390/jfb2030119.
107. Rider P, Kacarevic ZP, Alkildani S, Retnasingh S, Barbeck M. Bioprinting of tissue engineering scaffolds. J Tissue Eng. 2018;9:2041731418802090. https://doi.org/10.1177/2041731418802090.
108. Xia Z, Jin S, Ye K. Tissue and organ 3D bioprinting. SLAS Technol. 2018;23(4):301–14. https://doi.org/10.1177/2472630318760515.
109. Kyle S, Jessop ZM, Al-Sabah A, Whitaker IS. 'Printability' of candidate biomaterials for extrusion based 3D printing: state-of-the-art. Adv Healthc Mater. 2017;6(16). https://doi.org/10.1002/adhm.201700264.
110. Dhawan A, Kennedy PM, Rizk EB, Ozbolat IT. Three-dimensional bioprinting for bone and cartilage restoration in orthopaedic surgery. J Am Acad Orthop Surg. 2019;27(5):e215–26. https://doi.org/10.5435/JAAOS-D-17-00632.
111. Keriquel V, Oliveira H, Remy M, Ziane S, Delmond S, Rousseau B, Rey S, Catros S, Amedee J, Guillemot F, Fricain JC. In situ printing of mesenchymal stromal cells, by laser-assisted bioprinting, for *in vivo* bone regeneration applications. Sci Rep. 2017;7(1):1778. https://doi.org/10.1038/s41598-017-01914-x.
112. Gungor-Ozkerim PS, Inci I, Zhang YS, Khademhosseini A, Dokmeci MR. Bioinks for 3D bioprinting: an overview. Biomater Sci. 2018;6(5):915–46. https://doi.org/10.1039/c7bm00765e.
113. Malda J, Visser J, Melchels FP, Jungst T, Hennink WE, Dhert WJ, Groll J, Hutmacher DW. 25th anniversary article: engineering hydrogels for biofabrication. Adv Mater. 2013;25(36):5011–28. https://doi.org/10.1002/adma.201302042.
114. Faramarzi N, Yazdi IK, Nabavinia M, Gemma A, Fanelli A, Caizzone A, Ptaszek LM, Sinha I, Khademhosseini A, Ruskin JN, Tamayol A. Patient-specific bioinks for 3D bioprinting of tissue engineering scaffolds. Adv Healthc Mater. 2018;7(11):e1701347. https://doi.org/10.1002/adhm.201701347.
115. Kirillova A, Maxson R, Stoychev G, Gomillion CT, Ionov L. 4D biofabrication using shape-morphing hydrogels. Adv Mater. 2017;29(46):1703443. https://doi.org/10.1002/adma.201703443.
116. Masaeli E, Marquette C. Direct-write bioprinting approach to construct multilayer cellular tissues. Front Bioeng Biotechnol. 2019;7:478. https://doi.org/10.3389/fbioe.2019.00478.
117. Meng F, Meyer CM, Joung D, Vallera DA, McAlpine MC, Panoskaltsis-Mortari A. 3D bioprinted *In vitro* metastatic models via reconstruction of tumor microenvironments. Adv Mater. 2019;31(10):e1806899. https://doi.org/10.1002/adma.201806899.
118. Hou S, Tiriac H, Sridharan BP, Scampavia L, Madoux F, Seldin J, Souza GR, Watson D, Tuveson D, Spicer TP. Advanced development of primary pancreatic organoid tumor models for high-throughput phenotypic drug screening. SLAS Discov. 2018;23(6):574–84. https://doi.org/10.1177/2472555218766842.
119. Ma Y, Ji Y, Huang G, Ling K, Zhang X, Xu F. Bioprinting 3D cell-laden hydrogel microarray for screening human periodontal ligament stem cell response to extracellular matrix. Biofabrication. 2015;7(4):044105. https://doi.org/10.1088/1758-5090/7/4/044105.
120. Hou X, Liu S, Wang M, Wiraja C, Huang W, Chan P, Tan T, Xu C. Layer-by-layer 3D constructs of fibroblasts in hydrogel for examining transdermal penetration capability of nanoparticles. SLAS Technol. 2017;22(4):447–53. https://doi.org/10.1177/2211068216655753.
121. Mittal R, Woo FW, Castro CS, Cohen MA, Karanxha J, Mittal J, Chhibber T, Jhaveri VM. Organ-on-chip

models: implications in drug discovery and clinical applications. J Cell Physiol. 2019;234(6):8352–80. https://doi.org/10.1002/jcp.27729.
122. Liu J, He J, Liu J, Ma X, Chen Q, Lawrence N, Zhu W, Xu Y, Chen S. Rapid 3D bioprinting of *in vitro* cardiac tissue models using human embryonic stem cell-derived cardiomyocytes. Bioprinting. 2019;13. https://doi.org/10.1016/j.bprint.2019.e00040.
123. Obregon F, Vaquette C, Ivanovski S, Hutmacher DW, Bertassoni LE. Three-dimensional bioprinting for regenerative dentistry and craniofacial tissue engineering. J Dent Res. 2015;94(9 Suppl):143S–52S. https://doi.org/10.1177/0022034515588885.
124. Ma Y, Xie L, Yang B, Tian W. Three-dimensional printing biotechnology for the regeneration of the tooth and tooth-supporting tissues. Biotechnol Bioeng. 2019;116(2):452–68. https://doi.org/10.1002/bit.26882.
125. Smith EE, Zhang W, Schiele NR, Khademhosseini A, Kuo CK, Yelick PC. Developing a biomimetic tooth bud model. J Tissue Eng Regen Med. 2017;11(12):3326–36. https://doi.org/10.1002/term.2246.
126. Monteiro N, Smith EE, Angstadt S, Zhang W, Khademhosseini A, Yelick PC. Dental cell sheet biomimetic tooth bud model. Biomaterials. 2016; 106:167–79. https://doi.org/10.1016/j.biomaterials.2016.08.024.
127. Khayat A, Monteiro N, Smith EE, Pagni S, Zhang W, Khademhosseini A, Yelick PC. GelMA-encapsulated hDPSCs and HUVECs for dental pulp regeneration. J Dent Res. 2017;96(2):192–9. https://doi.org/10.1177/0022034516682005.
128. Staples RJ, Ivanovski S, Vaquette C. Fibre guiding scaffolds for periodontal tissue engineering. J Periodontal Res. 2020;55(3):331–41. https://doi.org/10.1111/jre.12729.
129. Ivanovski S, Vaquette C, Gronthos S, Hutmacher DW, Bartold PM. Multiphasic scaffolds for periodontal tissue engineering. J Dent Res. 2014;93(12):1212–21. https://doi.org/10.1177/0022034514544301.
130. Vaquette C, Saifzadeh S, Farag A, Hutmacher DW, Ivanovski S. Periodontal tissue engineering with a multiphasic construct and cell sheets. J Dent Res. 2019;98(6):673–81. https://doi.org/10.1177/0022034519837967.
131. Vaquette C, Pilipchuk SP, Bartold PM, Hutmacher DW, Giannobile WV, Ivanovski S. Tissue engineered constructs for periodontal regeneration: current status and future perspectives. Adv Healthc Mater. 2018;7(21):e1800457. https://doi.org/10.1002/adhm.201800457.
132. Athirasala A, Tahayeri A, Thrivikraman G, Franca CM, Monteiro N, Tran V, Ferracane J, Bertassoni LE. A dentin-derived hydrogel bioink for 3D bioprinting of cell laden scaffolds for regenerative dentistry. Biofabrication. 2018;10(2):024101. https://doi.org/10.1088/1758-5090/aa9b4e.
133. Sooppan R, Paulsen SJ, Han J, Ta AH, Dinh P, Gaffey AC, Venkataraman C, Trubelja A, Hung G, Miller JS, Atluri P. *In vivo* anastomosis and perfusion of a three-dimensionally-printed construct containing microchannel networks. Tissue Eng Part C Methods. 2016;22(1):1–7. https://doi.org/10.1089/ten.TEC.2015.0239.
134. Miller JS, Stevens KR, Yang MT, Baker BM, Nguyen DH, Cohen DM, Toro E, Chen AA, Galie PA, Yu X, Chaturvedi R, Bhatia SN, Chen CS. Rapid casting of patterned vascular networks for perfusable engineered three-dimensional tissues. Nat Mater. 2012;11(9):768–74. https://doi.org/10.1038/nmat3357.
135. Sachdeva GS, Sachdeva LT, Goel M, Bala S. Regenerative endodontic treatment of an immature tooth with a necrotic pulp and apical periodontitis using platelet-rich plasma (PRP) and mineral trioxide aggregate (MTA): a case report. Int Endod J. 2015;48(9):902–10. https://doi.org/10.1111/iej.12407.
136. Dhillon H, Kaushik M, Sharma R. Regenerative endodontics—creating new horizons. J Biomed Mater Res B Appl Biomater. 2016;104(4):676–85. https://doi.org/10.1002/jbm.b.33587.
137. Ono M, Oshima M, Ogawa M, Sonoyama W, Hara ES, Oida Y, Shinkawa S, Nakajima R, Mine A, Hayano S, Fukumoto S, Kasugai S, Yamaguchi A, Tsuji T, Kuboki T. Practical whole-tooth restoration utilizing autologous bioengineered tooth germ transplantation in a postnatal canine model. Sci Rep. 2017;7:44522. https://doi.org/10.1038/srep44522.
138. Oshima M, Mizuno M, Imamura A, Ogawa M, Yasukawa M, Yamazaki H, Morita R, Ikeda E, Nakao K, Takano-Yamamoto T, Kasugai S, Saito M, Tsuji T. Functional tooth regeneration using a bioengineered tooth unit as a mature organ replacement regenerative therapy. PLoS One. 2011;6(7):e21531. https://doi.org/10.1371/journal.pone.0021531.
139. Oshima M, Tsuji T. Functional tooth regenerative therapy: tooth tissue regeneration and whole-tooth replacement. Odontology. 2014;102(2):123–36. https://doi.org/10.1007/s10266-014-0168-z.
140. Tao O, Kort-Mascort J, Lin Y, Pham HM, Charbonneau AM, ElKashty OA, Kinsella JM, Tran SD. The applications of 3D printing for craniofacial tissue engineering. Micromachines (Basel). 2019;10(7):140. https://doi.org/10.3390/mi10070480.
141. Izumi K, Song J, Feinberg SE. Development of a tissue-engineered human oral mucosa: from the bench to the bed side. Cells Tissues Organs. 2004;176(1–3):134–52. https://doi.org/10.1159/000075034.
142. Feinberg SE, Aghaloo TL, Cunningham LL Jr. Role of tissue engineering in oral and maxillofacial reconstruction: findings of the 2005 AAOMS research summit. J Oral Maxillofac Surg. 2005;63(10):1418–25. https://doi.org/10.1016/j.joms.2005.07.004.

143. Hotta T, Yokoo S, Terashi H, Komori T. Clinical and histopathological analysis of healing process of intraoral reconstruction with ex vivo produced oral mucosa equivalent. Kobe J Med Sci. 2007;53(1–2):1–14.
144. Urkasemsin G, Rungarunlert S, Ferreira JN. Bioprinting strategies for secretory epithelial organoids. Methods Mol Biol. 2020;2140:243–9. https://doi.org/10.1007/978-1-0716-0520-2_16.
145. Adine C, Ng KK, Rungarunlert S, Souza GR, Ferreira JN. Engineering innervated secretory epithelial organoids by magnetic three-dimensional bioprinting for stimulating epithelial growth in salivary glands. Biomaterials. 2018;180:52–66. https://doi.org/10.1016/j.biomaterials.2018.06.011.
146. Al-Hashimi I, Schifter M, Lockhart PB, Wray D, Brennan M, Migliorati CA, Axéll T, Bruce AJ, Carpenter W, Eisenberg E, Epstein JB, Holmstrup P, Jontell M, Lozada-Nur F, Nair R, Silverman B, Thongprasom K, Thornhill M, Warnakulasuriya S, van der Waal I. Oral lichen planus and Oral lichenoid lesions: diagnostic and therapeutic considerations. Oral Surg Oral Med Oral Pathol Oral Radiol Endodont. 2007;103:S25.e21–12. https://doi.org/10.1016/j.tripleo.2006.11.001.
147. Müller S. Oral lichenoid lesions: distinguishing the benign from the deadly. Mod Pathol. 2017; 30(s1):S54–67. https://doi.org/10.1038/modpathol.2016.121.
148. Sugerman PB, Savage NW, Walsh LJ, Zhao ZZ, Zhou XJ, Khan A, Seymour GJ, Bigby M. The pathogenesis of oral lichen planus. Crit Rev Oral Biol Med. 2002;13(4):350–65. https://doi.org/10.1177/154411130201300405.
149. Al-Hashimi I, Schifter M, Lockhart PB, Wray D, Brennan M, Migliorati CA, Axéll T, Bruce AJ, Carpenter W, Eisenberg E, Epstein JB, Holmstrup P, Jontell M, Lozada-Nur F, Nair R, Silverman B, Thongprasom K, Thornhill M, Warnakulasuriya S, van der Waal I. Oral lichen planus and oral lichenoid lesions: diagnostic and therapeutic considerations. Oral Surg Oral Med Oral Pathol Oral Radiol Endodontol. 2007;103:S25.e21–12. https://doi.org/10.1016/j.tripleo.2006.11.001.
150. Lodi G, Manfredi M, Mercadante V, Murphy R, Carrozzo M. Interventions for treating oral lichen planus: corticosteroid therapies. Cochrane Database Syst Rev. 2020;2:CD001168. https://doi.org/10.1002/14651858.CD001168.pub3.
151. Kalhori KAM, Vahdatinia F, Jamalpour MR, Vescovi P, Fornaini C, Merigo E, Fekrazad R. Photobiomodulation in oral medicine. Photobiomodul Photomed Laser Surg. 2019;37(12):837–61. https://doi.org/10.1089/photob.2019.4706.
152. Cafaro A, Arduino PG, Massolini G, Romagnoli E, Broccoletti R. Clinical evaluation of the efficiency of low-level laser therapy for oral lichen planus: a prospective case series. Lasers Med Sci. 2014;29(1):185–90. https://doi.org/10.1007/s10103-013-1313-6.
153. Trehan M, Taylor CR. Low-dose excimer 308-nm laser for the treatment of oral lichen planus. Arch Dermatol. 2004;140(4):415–20. https://doi.org/10.1001/archderm.140.4.415.
154. Jajarm HH, Falaki F, Mahdavi O. A comparative pilot study of low intensity laser versus topical corticosteroids in the treatment of erosive-atrophic oral lichen planus. Photomed Laser Surg. 2011;29(6):421–5. https://doi.org/10.1089/pho.2010.2876.
155. Kazancioglu HO, Erisen M. Comparison of low-level laser therapy versus ozone therapy in the treatment of oral lichen Planus. Ann Dermatol. 2015;27(5):485–91. https://doi.org/10.5021/ad.2015.27.5.485.
156. Mirza S, Rehman N, Alrahlah A, Alamri WR, Vohra F. Efficacy of photodynamic therapy or low level laser therapy against steroid therapy in the treatment of erosive-atrophic oral lichen planus. Photodiagn Photodyn Ther. 2018;21:404–8. https://doi.org/10.1016/j.pdpdt.2018.02.001.
157. Pavan Kumar B, Ram Mohan S, Mohan AP, Jeevan Kumar KA, Yashwanth Yadav B. Versatility of pleuripotent undifferentiated stem cells aspirated from bone marrow and its applications in oral and maxillofacial surgery. J Maxillofac Oral Surg. 2016;15(1):1–11. https://doi.org/10.1007/s12663-015-0793-2.
158. Yamada Y, Nakamura-Yamada S, Konoki R, Baba S. Promising advances in clinical trials of dental tissue-derived cell-based regenerative medicine. Stem Cell Res Ther. 2020;11(1):175. https://doi.org/10.1186/s13287-020-01683-x.
159. Yamada Y, Nakamura-Yamada S, Kusano K, Baba S. Clinical potential and current progress of dental pulp stem cells for various systemic diseases in regenerative medicine: a concise review. IJMS. 2019;20(5):1132. https://doi.org/10.3390/ijms20051132.
160. Li Z, Jiang CM, An S, Cheng Q, Huang YF, Wang YT, Gou YC, Xiao L, Yu WJ, Wang J. Immunomodulatory properties of dental tissue-derived mesenchymal stem cells. Oral Dis. 2014;20(1):25–34. https://doi.org/10.1111/odi.12086.
161. Ullah I, Choe Y-h, Khan M, Bharti D, Shivakumar SB, Lee H-J, Son Y-B, Shin Y, Lee S-L, Park B-W, Ock S-A, Rho G-J. Dental pulp-derived stem cells can counterbalance peripheral nerve injury-induced oxidative stress and supraspinal neuro-inflammation in rat brain. Science. 2018;8(1):15795. https://doi.org/10.1038/s41598-018-34151-x.
162. Omi M, Hata M, Nakamura N, Miyabe M, Kobayashi Y, Kamiya H, Nakamura J, Ozawa S, Tanaka Y, Takebe J, Matsubara T, Naruse K. Transplantation of dental pulp stem cells suppressed inflammation in sciatic nerves by promoting macrophage polarization towards anti-inflammation phenotypes and ameliorated diabetic polyneuropathy. J Diabet Investig. 2016;7(4):485–96. https://doi.org/10.1111/jdi.12452.
163. Nakashima M, Iohara K, Murakami M, Nakamura H, Sato Y, Ariji Y, Matsushita K. Pulp regeneration by transplantation of dental pulp stem cells

in pulpitis: a pilot clinical study. Stem Cell Res Ther. 2017;8(1):61. https://doi.org/10.1186/s13287-017-0506-5.
164. Zhang Z, Han Y, Song J, Luo R, Jin X, Mu D, Su S, Ji X, Ren Y-F, Liu H. Interferon-γ regulates the function of mesenchymal stem cells from oral lichen planus via indoleamine 2,3-dioxygenase activity. J Oral Pathol Med. 2015;44(1):15–27. https://doi.org/10.1111/jop.12224.
165. Arzi B, Clark KC, Sundaram A, Spriet M, Verstraete FJM, Walker NJ, Loscar MR, Fazel N, Murphy WJ, Vapniarsky N, Borjesson DL. Therapeutic efficacy of fresh, allogeneic mesenchymal stem cells for severe refractory feline chronic gingivostomatitis. Stem Cells Transl Med. 2017;6(8):1710–22. https://doi.org/10.1002/sctm.17-0035.
166. Arzi B, Mills-Ko E, Verstraete FJ, Kol A, Walker NJ, Badgley MR, Fazel N, Murphy WJ, Vapniarsky N, Borjesson DL. Therapeutic efficacy of fresh, autologous mesenchymal stem cells for severe refractory gingivostomatitis in cats. Stem Cells Transl Med. 2016;5(1):75–86. https://doi.org/10.5966/sctm.2015-0127.
167. Kramer IR, Lucas RB, Pindborg JJ, Sobin LH. Definition of leukoplakia and related lesions: an aid to studies on oral precancer. Oral Surg Oral Med Oral Pathol. 1978;46(4):518–39.
168. Pindborg JJRP, Smith CJ, van der Waal I. World Health Organization International histological classification of tumours. Histological typing of cancer and precancer of the oral mucosa. 2nd ed. Heidelberg: Springer; 1997. p. 1–85.
169. Warnakulasuriya S, Johnson NW, van der Waal I. Nomenclature and classification of potentially malignant disorders of the oral mucosa. J Oral Pathol Med. 2007;36(10):575–80. https://doi.org/10.1111/j.1600-0714.2007.00582.x.
170. Mello FW, Miguel AFP, Dutra KL, Porporatti AL, Warnakulasuriya S, Guerra ENS, Rivero ERC. Prevalence of oral potentially malignant disorders: a systematic review and meta-analysis. J Oral Pathol Med. 2018;47(7):633–40. https://doi.org/10.1111/jop.12726.
171. Speight PM, Khurram SA, Kujan O. Oral potentially malignant disorders: risk of progression to malignancy. Oral Surg Oral Med Oral Pathol Oral Radiol. 2018;125(6):612–27. https://doi.org/10.1016/j.oooo.2017.12.011.
172. Dost F, Le Cao KA, Ford PJ, Farah CS. A retrospective analysis of clinical features of oral malignant and potentially malignant disorders with and without oral epithelial dysplasia. Oral Surg Oral Med Oral Pathol Oral Radiol. 2013;116(6):725–33. https://doi.org/10.1016/j.oooo.2013.08.005.
173. El-Naggar AK, Chan JKC, Takata T, Grandis JR, Slootweg PJ. The fourth edition of the head and neck World Health Organization blue book: editors' perspectives. Hum Pathol. 2017;66:10–2. https://doi.org/10.1016/j.humpath.2017.05.014.
174. de Farias Gabriel A, Wagner VP, Correa C, Webber LP, Pilar EFS, Curra M, Carrard VC, Martins MAT, Martins MD. Photobiomodulation therapy modulates epigenetic events and NF-κB expression in oral epithelial wound healing. Lasers Med Sci. 2019;34(7):1465–72. https://doi.org/10.1007/s10103-019-02745-0.
175. Pellicioli AC, Martins MD, Dillenburg CS, Marques MM, Squarize CH, Castilho RM. Laser phototherapy accelerates oral keratinocyte migration through the modulation of the mammalian target of rapamycin signaling pathway. J Biomed Opt. 2014;19(2):028002. https://doi.org/10.1117/1.JBO.19.2.028002.
176. Wagner VP, Curra M, Webber LP, Nor C, Matte U, Meurer L, Martins MD. Photobiomodulation regulates cytokine release and new blood vessel formation during oral wound healing in rats. Lasers Med Sci. 2016;31(4):665–71. https://doi.org/10.1007/s10103-016-1904-0.
177. Jagtap B, Bhate K, Santhoshkumar SN. Low level laser therapy reduces oral leukoplakia lesion size: results from a preliminary study. Oral Oncol. 2018;85:108–9. https://doi.org/10.1016/j.oraloncology.2018.08.002.
178. Ribeiro AS, de Aguiar MCF, do Carmo MAV, de Abreu MHNG, Silva TA, Mesquita RA. 660 AsGaAl laser to alleviate pain caused by cryosurgical treatment of oral leukoplakia: a preliminary study. Photomed Laser Surg. 2011;29(5):345–50. https://doi.org/10.1089/pho.2010.2824.
179. Utomo DN, Mahyudin F, Hernugrahanto KD, Suroto H, Chilmi MZ, Rantam FA. Implantation of platelet rich fibrin and allogenic mesenchymal stem cells facilitate the healing of muscle injury: an experimental study on animal. Int J Surg Open. 2018;11:4–9. https://doi.org/10.1016/j.ijso.2018.03.001.
180. Mahajan M, Gupta MK, Bande C, Meshram V. Comparative evaluation of healing pattern after surgical excision of oral mucosal lesions by using platelet-rich fibrin (PRF) membrane and collagen membrane as grafting materials—a randomized clinical trial. J Oral Maxillofac Surg. 2018;76(7):1469.e1461–9. https://doi.org/10.1016/j.joms.2018.02.031.
181. Kasai Y, Takagi R, Kobayashi S, Owaki T, Yamaguchi N, Fukuda H, Sakai Y, Sumita Y, Kanai N, Isomoto H, Kanetaka K, Ohki T, Asahina I, Nagai K, Nakao K, Takeda N, Okano T, Eguchi S, Yamato M. A stable protocol for the fabrication of transplantable human oral mucosal epithelial cell sheets for clinical application. Regenerat Therapy. 2020;14:87–94. https://doi.org/10.1016/j.reth.2019.11.007.
182. Ohki T, Yamato M, Murakami D, Takagi R, Yang J, Namiki H, Okano T, Takasaki K. Treatment of oesophageal ulcerations using endoscopic transplantation of tissue-engineered autologous oral mucosal epithelial cell sheets in a canine model.

Gut. 2006;55(12):1704–10. https://doi.org/10.1136/gut.2005.088518.
183. Ohki T, Yamato M, Ota M, Takagi R, Murakami D, Kondo M, Sasaki R, Namiki H, Okano T, Yamamoto M. Prevention of esophageal stricture after endoscopic submucosal dissection using tissue-engineered cell sheets. Gastroenterology. 2012;143(3):582–588.e582. https://doi.org/10.1053/j.gastro.2012.04.050.
184. Burillon C, Huot L, Justin V, Nataf S, Chapuis F, Decullier E, Damour O. Cultured autologous oral mucosal epithelial cell sheet (CAOMECS) transplantation for the treatment of corneal limbal epithelial stem cell deficiency. Invest Ophthalmol Vis Sci. 2012;53(3):1325–31. https://doi.org/10.1167/iovs.11-7744.
185. Li W, Han Y, Zhao Z, Ji X, Wang X, Jin J, Wang Q, Guo X, Cheng Z, Lu M, Wang G, Wang Y, Liu H. Oral mucosal mesenchymal stem cell-derived exosomes: a potential therapeutic target in oral premalignant lesions. Int J Oncol. 2019;54(5):1567–78. https://doi.org/10.3892/ijo.2019.4756.
186. Zhang Z, Song J, Han Y, Mu D, Su S, Ji X, Liu H. Impairment of mesenchymal stem cells derived from oral leukoplakia. Int J Clin Exp Pathol. 2015;8(9):10026–37.
187. Bruna F, Plaza A, Arango M, Espinoza I, Conget P. Systemically administered allogeneic mesenchymal stem cells do not aggravate the progression of precancerous lesions: a new biosafety insight. Stem Cell Res Ther. 2018;9(1):137. https://doi.org/10.1186/s13287-018-0878-1.
188. Chen Y, Wang X, Fang J, Song J, Ma D, Luo L, He B, Xia J, Lui VWY, Cheng B, Wang Z. Mesenchymal stem cells participate in oral mucosa carcinogenesis by regulating T cell proliferation. Clin Immunol. 2019;198:46–53. https://doi.org/10.1016/j.clim.2018.12.001.
189. Bakhshinyan D, Adile AA, Qazi MA, Singh M, Kameda-Smith MM, Yelle N, Chokshi C, Venugopal C, Singh SK. Introduction to cancer stem cells: past, present, and future. Methods Mol Biol. 2018;1692:1–16. https://doi.org/10.1007/978-1-4939-7401-6_1.
190. Major AG, Pitty LP, Farah CS. Cancer stem cell markers in head and neck squamous cell carcinoma. Stem Cells Int. 2013;2013:319489. https://doi.org/10.1155/2013/319489.
191. Hadjimichael C, Chanoumidou K, Papadopoulou N, Arampatzi P, Papamatheakis J, Kretsovali A. Common stemness regulators of embryonic and cancer stem cells. World J Stem Cells. 2015;7(9):1150–84. https://doi.org/10.4252/wjsc.v7.i9.1150.
192. Surendran S, Siddappa G, Mohan A, Hicks W Jr, Jayaprakash V, Mimikos C, Mahri M, Almarzouki F, Morrell K, Ravi R, Govindan S, Sushma CN, Raghavan N, Birur P, Ilayaraja J, Merzianu M, Reid M, Suresh A, Kuriakose MA. Cancer stem cell and its niche in malignant progression of oral potentially malignant disorders. Oral Oncol. 2017;75:140–7. https://doi.org/10.1016/j.oraloncology.2017.11.003.
193. Ravindran G, Devaraj H. Aberrant expression of CD133 and musashi-1 in preneoplastic and neoplastic human oral squamous epithelium and their correlation with clinicopathological factors. Head Neck. 2012;34(8):1129–35. https://doi.org/10.1002/hed.21896.
194. Sun L, Feng J, Ma L, Liu W, Zhou Z. CD133 expression in oral lichen planus correlated with the risk for progression to oral squamous cell carcinoma. Ann Diagn Pathol. 2013;17(6):486–9. https://doi.org/10.1016/j.anndiagpath.2013.06.004.
195. Saintigny P, William WN Jr, Foy JP, Papadimitrakopoulou V, Lang W, Zhang L, Fan YH, Feng L, Kim ES, El-Naggar AK, Lee JJ, Mao L, Hong WK, Lingen MW, Lippman SM. Met receptor tyrosine kinase and chemoprevention of oral cancer. J Natl Cancer Inst. 2018;110(3):250–7. https://doi.org/10.1093/jnci/djx186.
196. Bazarsad S, Zhang X, Kim KY, Illeperuma R, Jayasinghe RD, Tilakaratne WM, Kim J. Identification of a combined biomarker for malignant transformation in oral submucous fibrosis. J Oral Pathol Med. 2017;46(6):431–8. https://doi.org/10.1111/jop.12483.
197. Dalley AJ, Pitty LP, Major AG, Abdulmajeed AA, Farah CS. Expression of ABCG2 and Bmi-1 in oral potentially malignant lesions and oral squamous cell carcinoma. Cancer Med. 2014;3(2):273–83. https://doi.org/10.1002/cam4.182.
198. Liu W, Feng JQ, Shen XM, Wang HY, Liu Y, Zhou ZT. Two stem cell markers, ATP-binding cassette, G2 subfamily (ABCG2) and BMI-1, predict the transformation of oral leukoplakia to cancer: a long-term follow-up study. Cancer. 2012;118(6):1693–700. https://doi.org/10.1002/cncr.26483.
199. de Vicente JC, Rodrigo JP, Rodriguez-Santamarta T, Lequerica-Fernandez P, Allonca E, Garcia-Pedrero JM. Podoplanin expression in oral leukoplakia: tumorigenic role. Oral Oncol. 2013;49(6):598–603. https://doi.org/10.1016/j.oraloncology.2013.02.008.
200. Gissi DB, Gabusi A, Tarsitano A, Luccarini L, Morandi L, Montebugnoli L. Podoplanin expression as a predictive marker of dysplasia in oral leukoplakia. J Craniomaxillofac Surg. 2018;46(5):759–64. https://doi.org/10.1016/j.jcms.2018.02.016.
201. Habiba U, Hida K, Kitamura T, Matsuda AY, Higashino F, Ito YM, Ohiro Y, Totsuka Y, Shindoh M. ALDH1 and podoplanin expression patterns predict the risk of malignant transformation in oral leukoplakia. Oncol Lett. 2017;13(1):321–8. https://doi.org/10.3892/ol.2016.5379.
202. Kreppel M, Kreppel B, Drebber U, Wedemayer I, Rothamel D, Zoller JE, Scheer M. Podoplanin expression in oral leukoplakia: prognostic value and clinicopathological implications. Oral Dis. 2012;18(7):692–9. https://doi.org/10.1111/j.1601-0825.2012.01927.x.
203. Liu W, Wu L, Shen XM, Shi LJ, Zhang CP, Xu LQ, Zhou ZT. Expression patterns of cancer stem cell markers ALDH1 and CD133 correlate with a high

risk of malignant transformation of oral leukoplakia. Int J Cancer. 2013;132(4):868–74. https://doi.org/10.1002/ijc.27720.
204. Dalley AJ, Abdul Majeed AA, Pitty LP, Major AG, Farah CS. LGR5 expression in oral epithelial dysplasia and oral squamous cell carcinoma. Oral Surg Oral Med Oral Pathol Oral Radiol. 2015;119(4):436–40.e431. https://doi.org/10.1016/j.oooo.2014.11.014.
205. Luiz ST, Modolo F, Mozzer I, Dos Santos EC, Nagashima S, Camargo Martins AP, de Azevedo MLV, Azevedo Alanis LR, Hardy A, de Moraes RS, Aguiar MCF, Ignacio SA, Jham BC, Noronha L, Johann A. Immunoexpression of SOX-2 in oral leukoplakia. Oral Dis. 2018;24(8):1449–57. https://doi.org/10.1111/odi.12922.
206. Qiao B, He B, Cai J, Yang W. The expression profile of Oct4 and Sox2 in the carcinogenesis of oral mucosa. Int J Clin Exp Pathol. 2014;7(1):28–37.
207. Vijayakumar G, Narwal A, Kamboj M, Sen R. Association of SOX2, OCT4 and WNT5A expression in oral epithelial dysplasia and oral squamous cell carcinoma: an immunohistochemical study. Head Neck Pathol. 2020;14:749–57. https://doi.org/10.1007/s12105-019-01114-1.
208. Paparella ML, Abrigo M, Bal de Kier Joffe E, Raimondi AR. Oral-specific ablation of Klf4 disrupts epithelial terminal differentiation and increases premalignant lesions and carcinomas upon chemical carcinogenesis. J Oral Pathol Med. 2015;44(10):801–9. https://doi.org/10.1111/jop.12307.
209. Papakosta V, Vairaktaris E, Vylliotis A, Derka S, Nkenke E, Vassiliou S, Lazaris A, Mourouzis C, Rallis G, Spyridonidou S, Anagnostopoulou S, Perrea D, Donta I, Yapijakis C, Patsouris E. The co-expression of c-myc and p53 increases and reaches a plateau early in oral oncogenesis. Anticancer Res. 2006;26(4B):2957–62.
210. Gokulan R, Halagowder D. Expression pattern of notch intracellular domain (NICD) and Hes-1 in preneoplastic and neoplastic human Oral squamous epithelium: their correlation with c-Myc, clinicopathological factors and prognosis in oral cancer. Med Oncol. 2014;31(8):126. https://doi.org/10.1007/s12032-014-0126-1.
211. Pallavi N, Nalabolu GRK, Hiremath SKS. Bcl-2 and c-Myc expression in oral dysplasia and oral squamous cell carcinoma: An immunohistochemical study to assess tumor progression. J Oral Maxillofac Pathol. 2018;22(3):325–31. https://doi.org/10.4103/jomfp.JOMFP_197_18.
212. de Vicente JC, Rodriguez-Santamarta T, Rodrigo JP, Allonca E, Vallina A, Singhania A, Donate-Perez Del Molino P, Garcia-Pedrero JM. The emerging role of NANOG as an early cancer risk biomarker in patients with oral potentially malignant disorders. J Clin Med. 2019;8(9):1376. https://doi.org/10.3390/jcm8091376.
213. Ling C, Li Q, Brown ME, Kishimoto Y, Toya Y, Devine EE, Choi K-O, Nishimoto K, Norman IG, Tsegyal T, Jiang JJ, Burlingham WJ, Gunasekaran S, Smith LM, Frey BL, Welham NV. Bioengineered vocal fold mucosa for voice restoration. Sci Transl Med. 2015;7(314):314ra187. https://doi.org/10.1126/scitranslmed.aab4014.
214. Fukahori M, Chitose S-I, Sato K, Sueyoshi S, Kurita T, Umeno H, Monden Y, Yamakawa R. Regeneration of vocal fold mucosa using tissue-engineered structures with oral mucosal cells. PLoS One. 2016;11(1):e0146151. https://doi.org/10.1371/journal.pone.0146151.
215. Machino R, Matsumoto K, Taniguchi D, Tsuchiya T, Takeoka Y, Taura Y, Moriyama M, Tetsuo T, Oyama S, Takagi K, Miyazaki T, Hatachi G, Doi R, Shimoyama K, Matsuo N, Yamasaki N, Nakayama K, Nagayasu T. Replacement of rat tracheas by layered, trachea-like, scaffold-free structures of human cells using a bio-3D printing system. Adv Healthc Mater. 2019;8(7):e1800983. https://doi.org/10.1002/adhm.201800983.
216. Taniguchi D, Matsumoto K, Tsuchiya T, Machino R, Takeoka Y, Elgalad A, Gunge K, Takagi K, Taura Y, Hatachi G, Matsuo N, Yamasaki N, Nakayama K, Nagayasu T. Scaffold-free trachea regeneration by tissue engineering with bio-3D printing. Interact Cardiovasc Thorac Surg. 2018;26(5):745–52. https://doi.org/10.1093/icvts/ivx444.
217. Takeoka Y, Matsumoto K, Taniguchi D, Tsuchiya T, Maachino R, Moriyama M, Oyama S, Tetsuo T, Taura Y, Takagi K, Yoshida T, Elgalad A, Matsuo N, Kunizaki M, Tobinaga S, Nonaka T, Hidaka S, Yamasaki N, Nakayama K, Nagayasu T. Regeneration of esophagus using a scaffold-free biomimetic structure created with bio-three-dimensional printing. PLoS One. 2019;14(3):e0211339. https://doi.org/10.1371/journal.pone.0211339.
218. Fennema E, Rivron N, Rouwkema J, van Blitterswijk C, de Boer J. Spheroid culture as a tool for creating 3D complex tissues. Trends Biotechnol. 2013;31(2):108–15. https://doi.org/10.1016/j.tibtech.2012.12.003.
219. Gu BK, Choi DJ, Park SJ, Kim YJ, Kim CH. 3D bioprinting technologies for tissue engineering applications. Adv Exp Med Biol. 2018;1078:15–28. https://doi.org/10.1007/978-981-13-0950-2_2.
220. Harrison RK. Phase II and phase III failures: 2013-2015. Nat Rev Drug Discov. 2016;15(12):817–8. https://doi.org/10.1038/nrd.2016.184.
221. Bleijs M, van de Wetering M, Clevers H, Drost J. Xenograft and organoid model systems in cancer research. EMBO J. 2019;38(15):e101654. https://doi.org/10.15252/embj.2019101654.
222. Yang L, Yang S, Li X, Li B, Li Y, Zhang X, Ma Y, Peng X, Jin H, Fan Q, Wei S, Liu J, Li H. Tumor organoids: from inception to future in cancer research. Cancer Lett. 2019;454:120–33. https://doi.org/10.1016/j.canlet.2019.04.005.
223. Sato T, Vries RG, Snippert HJ, van de Wetering M, Barker N, Stange DE, van Es JH, Abo A, Kujala P, Peters PJ, Clevers H. Single Lgr5 stem cells build crypt-villus structures *in vitro* without a mesenchy-

mal niche. Nature. 2009;459(7244):262–5. https://doi.org/10.1038/nature07935.

224. Driehuis E, Kolders S, Spelier S, Lõhmussaar K, Willems SM, Devriese LA, de Bree R, de Ruiter EJ, Korving J, Begthel H, van Es JH, Geurts V, He G-W, van Jaarsveld RH, Oka R, Muraro MJ, Vivié J, Zandvliet MMJM, Hendrickx APA, Iakobachvili N, Sridevi P, Kranenburg O, van Boxtel R, Kops GJPL, Tuveson DA, Peters PJ, van Oudenaarden A, Clevers H. Oral mucosal organoids as a potential platform for personalized Cancer therapy. Cancer Discov. 2019;9(7):852–71. https://doi.org/10.1158/2159-8290.CD-18-1522.

225. Driehuis E, Oosterom N, Heil SG, Muller IB, Lin M, Kolders S, Jansen G, de Jonge R, Pieters R, Clevers H, van den Heuvel-Eibrink MM. Patient-derived oral mucosa organoids as an *in vitro* model for methotrexate induced toxicity in pediatric acute lymphoblastic leukemia. PLoS One. 2020;15(5):e0237488. https://doi.org/10.1371/journal.pone.0231588.

226. Sasai Y, Eiraku M, Suga H. *In vitro* organogenesis in three dimensions: self-organising stem cells. Development. 2012;139(22):4111–21. https://doi.org/10.1242/dev.079590.

227. Tanaka N, Osman AA, Takahashi Y, Lindemann A, Patel AA, Zhao M, Takahashi H, Myers JN. Head and neck cancer organoids established by modification of the CTOS method can be used to predict *in vivo* drug sensitivity. Oral Oncol. 2018;87:49–57. https://doi.org/10.1016/j.oraloncology.2018.10.018.

228. Driehuis E, Spelier S, Beltrán Hernández I, de Bree R, M Willems S, Clevers H, Oliveira S. Patient-derived head and neck cancer organoids recapitulate EGFR expression levels of respective tissues and are responsive to EGFR-targeted photodynamic therapy. J Clin Med. 2019;8(11):1880. https://doi.org/10.3390/jcm8111880.

229. Batlle E, Clevers H. Cancer stem cells revisited. Nat Med. 2017;23(10):1124–34. https://doi.org/10.1038/nm.4409.

230. Tata PR, Mou H, Pardo-Saganta A, Zhao R, Prabhu M, Law BM, Vinarsky V, Cho JL, Breton S, Sahay A, Medoff BD, Rajagopal J. Dedifferentiation of committed epithelial cells into stem cells *in vivo*. Nature. 2013;503(7475):218–23. https://doi.org/10.1038/nature12777.

231. Chaffer CL, Brueckmann I, Scheel C, Kaestli AJ, Wiggins PA, Rodrigues LO, Brooks M, Reinhardt F, Su Y, Polyak K, Arendt LM, Kuperwasser C, Bierie B, Weinberg RA. Normal and neoplastic nonstem cells can spontaneously convert to a stem-like state. PNAS. 2011;108(19):7950–5. https://doi.org/10.1073/pnas.1102454108.

232. Ischenko I, Zhi J, Moll UM, Nemajerova A, Petrenko O. Direct reprogramming by oncogenic Ras and Myc. PNAS. 2013;110(10):3937–42. https://doi.org/10.1073/pnas.1219592110.

233. Huang SD, Yuan Y, Tang H, Liu XH, Fu CG, Cheng HZ, Bi JW, Yu YW, Gong DJ, Zhang W, Chen J, Xu ZY. Tumor cells positive and negative for the common cancer stem cell markers are capable of initiating tumor growth and generating both progenies. PLoS One. 2013;8(1):e54579. https://doi.org/10.1371/journal.pone.0054579.

234. Prince ME, Sivanandan R, Kaczorowski A, Wolf GT, Kaplan MJ, Dalerba P, Weissman IL, Clarke MF, Ailles LE. Identification of a subpopulation of cells with cancer stem cell properties in head and neck squamous cell carcinoma. PNAS. 2007;104(3):973–8. https://doi.org/10.1073/pnas.0610117104.

235. Lee J, Park M, Ko Y, Kim B, Kim O, Hyun H, Kim D, Sohn H, Moon YL, Lim W. Ectopic overexpression of CD133 in HNSCC makes it resistant to commonly used chemotherapeutics. Tumour Biol. 2017;39(4):1010428317695534. https://doi.org/10.1177/1010428317695534.

236. Wu Y, Zhang Y, Niu M, Shi Y, Liu H, Yang D, Li F, Lu Y, Bo Y, Zhang R, Li Z, Luo H, Cui J, Sang J, Xiang C, Gao W, Wen S. Whole-transcriptome analysis of CD133+CD144+ cancer stem cells derived from human laryngeal squamous cell carcinoma cells. Cell Physiol Biochem. 2018;47(4):1696–710. https://doi.org/10.1159/000490992.

237. Martens-de Kemp SR, Brink A, Stigter-van Walsum M, Damen JMA, Rustenburg F, Wu T, van Wieringen WN, Schuurhuis GJ, Braakhuis BJM, Slijper M, Brakenhoff RH. CD98 marks a subpopulation of head and neck squamous cell carcinoma cells with stem cell properties. Stem Cell Res. 2013;10(3):477–88. https://doi.org/10.1016/j.scr.2013.02.004.

238. Fukusumi T, Ishii H, Konno M, Yasui T, Nakahara S, Takenaka Y, Yamamoto Y, Nishikawa S, Kano Y, Ogawa H, Hasegawa S, Hamabe A, Haraguchi N, Doki Y, Mori M, Inohara H. CD10 as a novel marker of therapeutic resistance and cancer stem cells in head and neck squamous cell carcinoma. Br J Cancer. 2014;111(3):506–14. https://doi.org/10.1038/bjc.2014.289.

239. Clay MR, Tabor M, Owen JH, Carey TE, Bradford CR, Wolf GT, Wicha MS, Prince ME. Single-marker identification of head and neck squamous cell carcinoma cancer stem cells with aldehyde dehydrogenase. Head Neck. 2010;32(9):1195–201. https://doi.org/10.1002/hed.21315.

240. Curtarelli RB, Gonçalves JM, Dos Santos LGP, Savi MG, Nör JE, Mezzomo LAM, Rodríguez Cordeiro MM. Expression of cancer stem cell biomarkers in human head and neck carcinomas: a systematic review. Stem Cell Rev Rep. 2018;14(6):769–84. https://doi.org/10.1007/s12015-018-9839-4.

241. Nassar D, Blanpain C. Cancer stem cells: basic concepts and therapeutic implications. Annu Rev Pathol. 2016;11:47–76. https://doi.org/10.1146/annurev-pathol-012615-044438.

242. Yan Y, Zuo X, Wei D. Concise review: emerging role of CD44 in cancer stem cells: a promising biomarker and therapeutic target. Stem Cells Transl Med. 2015;4(9):1033–43. https://doi.org/10.5966/sctm.2015-0048.
243. Lin J-T, Chang T-H, Chang C-S, Wang W-H, Su B-W, Lee K-D, Chang P-J. Prognostic value of pretreatment CD44 mRNA in peripheral blood of patients with locally advanced head and neck cancer. Oral Oncol. 2010;46(5):e29–33. https://doi.org/10.1016/j.oraloncology.2010.02.011.
244. Kokko L-L, Hurme S, Maula S-M, Alanen K, Grénman R, Kinnunen I, Ventelä S. Significance of site-specific prognosis of cancer stem cell marker CD44 in head and neck squamous-cell carcinoma. Oral Oncol. 2011;47(6):510–6. https://doi.org/10.1016/j.oraloncology.2011.03.026.
245. Chen J, Zhou J, Lu J, Xiong H, Shi X, Gong L. Significance of CD44 expression in head and neck cancer: a systemic review and meta-analysis. BMC Cancer. 2014;14:15. https://doi.org/10.1186/1471-2407-14-15.
246. Qian X, Nie X, Wollenberg B, Sudhoff H, Kaufmann AM, Albers AE. Heterogeneity of head and neck squamous cell carcinoma stem cells. Adv Exp Med Biol. 2019;1139:23–40. https://doi.org/10.1007/978-3-030-14366-4_2.
247. Brower V. Cancer stem cell hypothesis evolves with emerging research. JNCI. 2016;108(5):djw139. https://doi.org/10.1093/jnci/djw139.
248. Kulsum S, Sudheendra HV, Pandian R, Ravindra DR, Siddappa G, R N, Chevour P, Ramachandran B, Sagar M, Jayaprakash A, Mehta A, Kekatpure V, Hedne N, Kuriakose MA, Suresh A. Cancer stem cell mediated acquired chemoresistance in head and neck cancer can be abrogated by aldehyde dehydrogenase 1 A1 inhibition. Mol Carcinog. 2017;56(2):694–711. https://doi.org/10.1002/mc.22526.
249. Chen D, Wu M, Li Y, Chang I, Yuan Q, Ekimyan-Salvo M, Deng P, Yu B, Yu Y, Dong J, Szymanski JM, Ramadoss S, Li J, Wang C-Y. Targeting BMI1+ cancer stem cells overcomes chemoresistance and inhibits metastases in squamous cell carcinoma. Cell Stem Cell. 2017;20(5):621–634.e626. https://doi.org/10.1016/j.stem.2017.02.003.
250. Hu J, Mirshahidi S, Simental A, Lee SC, De Andrade Filho PA, Peterson NR, Duerksen-Hughes P, Yuan X. Cancer stem cell self-renewal as a therapeutic target in human oral cancer. Oncogene. 2019;38(27):5440–56. https://doi.org/10.1038/s41388-019-0800-z.
251. Wang J, Ji H, Zhu Q, Yu X, Du J, Jiang Z. Co-inhibition of BMI1 and Mel18 enhances chemosensitivity of esophageal squamous cell carcinoma *in vitro* and *in vivo*. Oncol Lett. 2019;17(6):5012–22. https://doi.org/10.3892/ol.2019.10160.
252. Sun H-R, Wang S, Yan S-C, Zhang Y, Nelson PJ, Jia H-L, Qin L-X, Dong Q-Z. Therapeutic strategies targeting cancer stem cells and their microenvironment. Front Oncol. 2019;9:1104. https://doi.org/10.3389/fonc.2019.01104.
253. Zhang Z, Dong Z, Lauxen IS, Filho MS, Nor JE. Endothelial cell-secreted EGF induces epithelial to mesenchymal transition and endows head and neck cancer cells with stem-like phenotype. Cancer Res. 2014;74(10):2869–81. https://doi.org/10.1158/0008-5472.CAN-13-2032.
254. Xu Q, Zhang Q, Ishida Y, Hajjar S, Tang X, Shi H, Dang CV, Le AD. EGF induces epithelial-mesenchymal transition and cancer stem-like cell properties in human oral cancer cells via promoting Warburg effect. Oncotarget. 2017;8(6):9557–71. https://doi.org/10.18632/oncotarget.13771.
255. Le PN, Keysar SB, Miller B, Eagles JR, Chimed T-S, Reisinger J, Gomez KE, Nieto C, Jackson BC, Somerset HL, Morton JJ, Wang X-J, Jimeno A. Wnt signaling dynamics in head and neck squamous cell cancer tumor-stroma interactions. Mol Carcinog. 2019;58(3):398–410. https://doi.org/10.1002/mc.22937.
256. Lee SH, Koo BS, Kim JM, Huang S, Rho YS, Bae WJ, Kang HJ, Kim YS, Moon JH, Lim YC. Wnt/β--catenin signalling maintains self-renewal and tumourigenicity of head and neck squamous cell carcinoma stem-like cells by activating Oct4. J Pathol. 2014;234(1):99–107. https://doi.org/10.1002/path.4383.
257. Yu B, Wu K, Wang X, Zhang J, Wang L, Jiang Y, Zhu X, Chen W, Yan M. Periostin secreted by cancer-associated fibroblasts promotes cancer stemness in head and neck cancer by activating protein tyrosine kinase 7. Cell Death Dis. 2018;9(11):1082. https://doi.org/10.1038/s41419-018-1116-6.
258. Leong HS, Chong FT, Sew PH, Lau DP, Wong BH, Teh BT, Tan DS, Iyer NG. Targeting cancer stem cell plasticity through modulation of epidermal growth factor and insulin-like growth factor receptor signaling in head and neck squamous cell cancer. Stem Cells Transl Med. 2014;3(9):1055–65. https://doi.org/10.5966/sctm.2013-0214.
259. de Almeida Pdel V, Grégio AM, Machado MA, de Lima AA, Azevedo LR. Saliva composition and functions: a comprehensive review. J Contemp Dent Pract. 2008;9(3):72–80.
260. Dost F, Farah CS. Stimulating the discussion on saliva substitutes: a clinical perspective. Aust Dent J. 2013;58(1):11–7. https://doi.org/10.1111/adj.12023.
261. Lombaert IM, Brunsting JF, Wierenga PK, Faber H, Stokman MA, Kok T, Visser WH, Kampinga HH, de Haan G, Coppes RP. Rescue of salivary gland function after stem cell transplantation in irradiated glands. PLoS One. 2008;3(4):e2063. https://doi.org/10.1371/journal.pone.0002063.

262. Coppes RP, Stokman MA. Stem cells and the repair of radiation-induced salivary gland damage. Oral Dis. 2011;17(2):143–53. https://doi.org/10.1111/j.1601-0825.2010.01723.x.
263. Tran SD, Liu Y, Xia D, Maria OM, Khalili S, Wang RW, Quan VH, Hu S, Seuntjens J. Paracrine effects of bone marrow soup restore organ function, regeneration, and repair in salivary glands damaged by irradiation. PLoS One. 2013;8(4):e61632. https://doi.org/10.1371/journal.pone.0061632.
264. Lim JY, Yi T, Choi JS, Jang YH, Lee S, Kim HJ, Song SU, Kim YM. Intraglandular transplantation of bone marrow-derived clonal mesenchymal stem cells for amelioration of post-irradiation salivary gland damage. Oral Oncol. 2013;49(2):136–43. https://doi.org/10.1016/j.oraloncology.2012.08.010.
265. Aframian DJ, Palmon A, Nahlieli O. Future therapy strategies for salivary gland impairment. Refuat Hapeh Vehashinayim. 2004;21(3):43–50. 93
266. Cantara SI, Soscia DA, Sequeira SJ, Jean-Gilles RP, Castracane J, Larsen M. Selective functionalization of nanofiber scaffolds to regulate salivary gland epithelial cell proliferation and polarity. Biomaterials. 2012;33(33):8372–82. https://doi.org/10.1016/j.biomaterials.2012.08.021.
267. Pradhan-Bhatt S, Harrington DA, Duncan RL, Jia X, Witt RL, Farach-Carson MC. Implantable three-dimensional salivary spheroid assemblies demonstrate fluid and protein secretory responses to neurotransmitters. Tissue Eng Part A. 2013;19(13–14):1610–20. https://doi.org/10.1089/ten.TEA.2012.0301.
268. Urkasemsin G, Ferreira JN. Unveiling stem cell heterogeneity toward the development of salivary gland regenerative strategies. Adv Exp Med Biol. 2019;1123:151–64. https://doi.org/10.1007/978-3-030-11096-3_9.
269. Wu D, Chapela P, Farach-Carson MC. Reassembly of functional human stem/progenitor cells in 3D culture. Methods Mol Biol. 2018;1817:19–32. https://doi.org/10.1007/978-1-4939-8600-2_3.
270. Guenzel T, Hoch S, Heinze N, Wilhelm T, Gueldner C, Franzen A, Coordes A, Lieder A, Wiegand S. Sialendoscopy plus laser lithotripsy in sialolithiasis of the submandibular gland in 64 patients: a simple and safe procedure. Auris Nasus Larynx. 2019;46(5):797–802. https://doi.org/10.1016/j.anl.2019.01.009.
271. Hernando M, Echarri RM, Taha M, Martin-Fragueiro L, Hernando A, Mayor GP. Surgical complications of submandibular gland excision. Acta Otorrinolaringol Esp. 2012;63(1):42–6. https://doi.org/10.1016/j.otorri.2011.08.001.
272. Capaccio P, Torretta S, Pignataro L, Koch M. Salivary lithotripsy in the era of sialendoscopy. Acta Otorhinolaryngol Ital. 2017;37(2):113–21. https://doi.org/10.14639/0392-100X-1600.
273. Saga-Gutierrez C, Chiesa-Estomba CM, Larruscain E, Gonzalez-Garcia JA, Sistiaga JA, Altuna X. Transoral sialolitectomy as an alternative to submaxilectomy in the treatment of submaxillary sialolithiasis. Ear Nose Throat J. 2019;98(5):287–90. https://doi.org/10.1177/0145561319841268.
274. Kopeć T, Szyfter W, Wierzbicka M. Sialoendoscopy and combined approach for the management of salivary gland stones. Eur Arch Otorhinolaryngol. 2013;270(1):219–23. https://doi.org/10.1007/s00405-012-2145-x.
275. Liberati A, Altman DG, Tetzlaff J, Mulrow C, Gøtzsche PC, Ioannidis JP, Clarke M, Devereaux PJ, Kleijnen J, Moher D. The PRISMA statement for reporting systematic reviews and meta-analyses of studies that evaluate health care interventions: explanation and elaboration. J Clin Epidemiol. 2009;62(10):e1–34. https://doi.org/10.1016/j.jclinepi.2009.06.006.
276. Kılınç Y, Çetiner S. Surgical removal of a giant sialolith by diode laser. Open J Stomato. 2014;4(10):484–8.
277. Azaz B, Regev E, Casap N, Chicin R. Sialolithectomy done with a CO2 laser: clinical and scintigraphic results. J Oral Maxillofac Surg. 1996;54(6):685–8.; discussion 689. https://doi.org/10.1016/s0278-2391(96)90681-3.
278. Barak S, Horowitz I, Katz J, Kaplan I. Experiences with the CO_2 laser in the surgical treatment of intraoral salivary gland pathology. J Clin Laser Med Surg. 1991;9(4):295–9. https://doi.org/10.1089/clm.1991.9.295.
279. Barak S, Katz J, Mintz S. Use of the carbon dioxide laser to locate small sialoliths. J Oral Maxillofac Surg. 1993;51(4):379–81. https://doi.org/10.1016/s0278-2391(10)80349-0.
280. López Alvarez-Buhilla P, Blanco Bruned JL, Torres Piedra C, Alfonso Sánchez L. CO(2) laser treatment of sialolithiasis. An Esp Pediatr. 2000;53(1):62–3.
281. Yang SW, Chen TA. Transoral carbon dioxide laser sialolithectomy with topical anaesthesia. A simple, effective, and minimally invasive method. Int J Oral Maxillofac Surg. 2011;40(2):169–72. https://doi.org/10.1016/j.ijom.2010.09.020.
282. Haas OL Jr, Scolari N, da Silva Meirelles L, Favoretto AX, de Oliveira RB. Sialolith removal in the submandibular region using surgical diode laser: report of two cases and literature review. Oral Maxillofac Surg. 2018;22(1):105–11. https://doi.org/10.1007/s10006-018-0674-1.
283. Jensen SB, Vissink A, Firth N. Salivary gland disorders and diseases. In: Farah CS, Balasubramaniam R, McCullough MJ, editors. Contemporary Oral medicine: a comprehensive approach to clinical practice. Cham: Springer International Publishing; 2018. p. 1–85. https://doi.org/10.1007/978-3-319-28100-1_19-1.
284. Farah CS, Savage NW. Cryotherapy for treatment of oral lesions. Aust Dent J. 2006;51(1):2–5. https://doi.org/10.1111/j.1834-7819.2006.tb00392.x.
285. Farah CS, Koelmeyer N, Kaney A, Simanovic B. Nitrous oxide cryotherapy for the management

of benign lesions of the oral cavity. J Oral Pathol Med. 2019;48(7):611–8. https://doi.org/10.1111/jop.12912.
286. Mintz S, Barak S, Horowitz I. Carbon dioxide laser excision and vaporization of nonplunging ranulas: a comparison of two treatment protocols. J Oral Maxillofac Surg. 1994;52(4):370–2.
287. Hanna R, Parker S. The advantages of carbon dioxide laser applications in paediatric oral surgery. A prospective cohort study. Lasers Med Sci. 2016;31(8):1527–36. https://doi.org/10.1007/s10103-016-1978-8.
288. Bagher SM, Sulimany AM, Kaplan M, Loo CY. Treating mucocele in pediatric patients using a diode laser: three case reports. Dent J (Basel). 2018;6(2):09. https://doi.org/10.3390/dj6020013.
289. De Falco D, Di Venere D, Maiorano E. Diode laser excision of Blandin-Nuhn Mucocele. Cureus. 2020;12(3):e7441. https://doi.org/10.7759/cureus.7441.
290. Correia-Sa IB, Correia-Sa M, Costa-Ferreira P, Silva A, Marques M. Eleven years of parotid gland surgery in a plastic and reconstructive department. J Craniofac Surg. 2016;27(1):e26–33. https://doi.org/10.1097/SCS.0000000000002299.
291. Waldron CA, el-Mofty SK, Gnepp DR. Tumors of the intraoral minor salivary glands: a demographic and histologic study of 426 cases. Oral Surg Oral Med Oral Pathol. 1988;66(3):323–33. https://doi.org/10.1016/0030-4220(88)90240-x.
292. Farah CS, Balasubramaniam R, McCullough MJ. Contemporary oral medicine: a comprehensive approach to clinical practice. Cham: Springer; 2019.
293. Buchman C, Stringer SP, Mendenhall WM, Parsons JT, Jordan JR, Cassisi NJ. Pleomorphic adenoma: effect of tumor spill and inadequate resection on tumor recurrence. Laryngoscope. 1994;104(10):1231–4.
294. McGurk M, Renehan A, Gleave EN, Hancock BD. Clinical significance of the tumour capsule in the treatment of parotid pleomorphic adenomas. Br J Surg. 1996;83(12):1747–9. https://doi.org/10.1002/bjs.1800831227.
295. Stennert E, Wittekindt C, Klussmann JP, Guntinas-Lichius O. New aspects in parotid gland surgery. Otolaryngol Pol. 2004;58(1):109–14.
296. Johnson JT, Ferlito A, Fagan JJ, Bradley PJ, Rinaldo A. Role of limited parotidectomy in management of pleomorphic adenoma. J Laryngol Otol. 2007;121(12):1126–8. https://doi.org/10.1017/s0022215107000345.
297. Yu GY, Ma DQ. Carcinoma of the salivary gland: a clinicopathologic study of 405 cases. Semin Surg Oncol. 1987;3(4):240–4. https://doi.org/10.1002/ssu.2980030405.
298. Alaa El-Din Y, Sabry D, Abdelrahman AH, Fathy S. Potential therapeutic effects of induced pluripotent stem cells on induced salivary gland cancer in experimental rats. Biotech Histochem. 2019;94(2):92–9. https://doi.org/10.1080/10520295.2018.1508747.
299. Egyedi P. Utilization of the buccal fat pad for closure of oro-antral and/or oro-nasal communications. J Maxillofac Surg. 1977;5(4):241–4. https://doi.org/10.1016/s0301-0503(77)80117-3.
300. Kim MK, Han W, Kim SG. The use of the buccal fat pad flap for oral reconstruction. Maxillofac Plast Reconstr Surg. 2017;39(1):5. https://doi.org/10.1186/s40902-017-0105-5.
301. Chaudhary B, Gong Z, Lin Z, Abbas K, Ling B, Liu H. Reconstruction of intraoral maxillary defect with buccal fat pad. J Craniofac Surg. 2014;25(6):2174–7. https://doi.org/10.1097/scs.0000000000001075.
302. Broccaioli E, Niada S, Rasperini G, Ferreira LM, Arrigoni E, Yenagi V, Brini AT. Mesenchymal stem cells from Bichat's fat pad: *In vitro* comparison with adipose-derived stem cells from subcutaneous tissue. Biores Open Access. 2013;2(2):107–17. https://doi.org/10.1089/biores.2012.0291.
303. Ghaderi H, Razmkhah M, Kiany F, Chenari N, Haghshenas MR, Ghaderi A. Comparison of osteogenic and chondrogenic differentiation ability of buccal fat pad derived mesenchymal stem cells and gingival derived cells. J Dent (Shiraz). 2018;19(2):124–31.
304. Oliveira Neto JQ, Cetira Filho EL, Andrade GS, Silveira DXD, Carvalho A. Technique of the buccal fat pad flap as an alternative for the surgical defect of pleomorphic adenoma. J Craniofac Surg. 2019;30(3):798–9. https://doi.org/10.1097/SCS.0000000000004890.
305. Khojasteh A, Sadeghi N. Application of buccal fat pad-derived stem cells in combination with autogenous iliac bone graft in the treatment of maxillomandibular atrophy: a preliminary human study. Int J Oral Maxillofac Surg. 2016;45(7):864–71. https://doi.org/10.1016/j.ijom.2016.01.003.
306. Takahashi H, Ishikawa H, Tanaka A. Regenerative medicine for Parkinson's disease using differentiated nerve cells derived from human buccal fat pad stem cells. Hum Cell. 2017;30(2):60–71. https://doi.org/10.1007/s13577-017-0160-3.
307. Moon SY. Surgical management of the palatal pleomorphic adenoma. J Craniofac Surg. 2019;30(6):e580–2. https://doi.org/10.1097/scs.0000000000005608.
308. Bataineh AB, al-Dwairi ZN. Surgical management of pleomorphic adenoma of the palate. J Ir Dent Assoc. 2002;48(4):126–31.
309. Alkan A, Inal S. Closure of palatal defects following excision of palatal pleomorphic adenomas. J Contemp Dent Pract. 2008;9(6):99–107.
310. Seok H, Kim MK, Kim SG. Reconstruction of partial maxillectomy defect with a buccal fat pad flap and application of 4-hexylresorcinol: a case report. J Korean Assoc Oral Maxillofac Surg. 2016;42(6):370–4. https://doi.org/10.5125/jkaoms.2016.42.6.370.
311. Devaraju R, Gantala R, Aitha H, Gotoor SG. Mucoepidermoid carcinoma. BMJ Case Rep. 2014:bcr-2013-202776. https://doi.org/10.1136/bcr-2013-202776

312. Kawakami M, Ishikawa H, Tanaka A, Mataga I. Induction and differentiation of adipose-derived stem cells from human buccal fat pads into salivary gland cells. Hum Cell. 2016;29(3):101–10. https://doi.org/10.1007/s13577-016-0132-z.
313. Halbritter SA, Altermatt HJ, Caversaccio M, Bornstein MM. Apocrine papillary cystadenoma of a minor salivary gland on the lower lip: case presentation. Quintessence Int. 2009;40(2):167–9.
314. Mavragani CP, Moutsopoulos HM. Conventional therapy of Sjogren's syndrome. Clin Rev Allergy Immunol. 2007;32(3):284–91. https://doi.org/10.1007/s12016-007-8008-3.
315. Ramos-Casals M, Brito-Zerón P, Sisó-Almirall A, Bosch X, Tzioufas AG. Topical and systemic medications for the treatment of primary Sjögren's syndrome. Nat Rev Rheumatol. 2012;8(7):399–411. https://doi.org/10.1038/nrrheum.2012.53.
316. Mavragani CP, Moutsopoulos NM, Moutsopoulos HM. The management of Sjögren's syndrome. Nat Clin Pract Rheumatol. 2006;2(5):252–61. https://doi.org/10.1038/ncprheum0165.
317. Pringle S, Wang X, Verstappen G, Terpstra JH, Zhang CK, He A, Patel V, Jones RE, Baird DM, Spijkervet FKL, Vissink A, Bootsma H, Coppes RP, Kroese FGM. Salivary gland stem cells age prematurely in primary Sjogren's syndrome. Arthritis Rheumatol. 2019;71(1):133–42. https://doi.org/10.1002/art.40659.
318. Simoes A, Platero MD, Campos L, Aranha AC, Eduardo Cde P, Nicolau J. Laser as a therapy for dry mouth symptoms in a patient with Sjogren's syndrome: a case report. Spec Care Dentist. 2009;29(3):134–7.
319. Xu J, Wang D, Liu D, Fan Z, Zhang H, Liu O, Ding G, Gao R, Zhang C, Ding Y, Bromberg JS, Chen W, Sun L, Wang S. Allogeneic mesenchymal stem cell treatment alleviates experimental and clinical Sjogren syndrome. Blood. 2012;120(15):3142–51. https://doi.org/10.1182/blood-2011-11-391144.
320. Chen FY, Xue B, Wang H. Analysis of the clinical efficacy of yiqi fumai injection combined hydroxychloroquine sulfate tablet for treating Sjogren's syndrome. Zhongguo Zhong Xi Yi Jie He Za Zhi. 2012;32(12):1621–3.
321. Khalili S, Liu Y, Kornete M, Roescher N, Kodama S, Peterson A, Piccirillo CA, Tran SD. Mesenchymal stromal cells improve salivary function and reduce lymphocytic infiltrates in mice with Sjögren's-like disease. PLoS One. 2012;7(6):e38615. https://doi.org/10.1371/journal.pone.0038615.
322. Khalili S, Faustman DL, Liu Y, Sumita Y, Blank D, Peterson A, Kodama S, Tran SD. Treatment for salivary gland hypofunction at both initial and advanced stages of Sjogren-like disease: a comparative study of bone marrow therapy versus spleen cell therapy with a 1-year monitoring period. Cytotherapy. 2014;16(3):412–23. https://doi.org/10.1016/j.jcyt.2013.10.006.
323. Liu Y, Li C, Wang S, Guo J, Guo J, Fu J, Ren L, An Y, He J, Li Z. Human umbilical cord mesenchymal stem cells confer potent immunosuppressive effects in Sjögren's syndrome by inducing regulatory T cells. Mod Rheumatol. 2020;1–11. https://doi.org/10.1080/14397595.2019.1707996.
324. Yong KW, Choi JR, Dolbashid AS, Wan Safwani WKZ. Biosafety and bioefficacy assessment of human mesenchymal stem cells: what do we know so far? Regen Med. 2018;13(2):219–32. https://doi.org/10.2217/rme-2017-0078.
325. Jensen DH, Oliveri RS, Trojahn Kolle SF, Fischer-Nielsen A, Specht L, Bardow A, Buchwald C. Mesenchymal stem cell therapy for salivary gland dysfunction and xerostomia: a systematic review of preclinical studies. Oral Surg Oral Med Oral Pathol Oral Radiol. 2014;117(3):335–342.e331. https://doi.org/10.1016/j.oooo.2013.11.496.
326. Vissink A, Jansma J, Spijkervet FK, Burlage FR, Coppes RP. Oral sequelae of head and neck radiotherapy. Crit Rev Oral Biol Med. 2003;14(3):199–212. https://doi.org/10.1177/154411130301400305.
327. Langendijk JA, Doornaert P, Verdonck-de Leeuw IM, Leemans CR, Aaronson NK, Slotman BJ. Impact of late treatment-related toxicity on quality of life among patients with head and neck cancer treated with radiotherapy. J Clin Oncol. 2008;26(22):3770–6. https://doi.org/10.1200/jco.2007.14.6647.
328. Ho KF, Farnell DJ, Routledge JA, Burns MP, Sykes AJ, Slevin NJ, Davidson SE. Developing a CTCAEs patient questionnaire for late toxicity after head and neck radiotherapy. Eur J Cancer. 2009;45(11):1992–8. https://doi.org/10.1016/j.ejca.2009.04.010.
329. Eisbruch A, Ten Haken RK, Kim HM, Marsh LH, Ship JA. Dose, volume, and function relationships in parotid salivary glands following conformal and intensity-modulated irradiation of head and neck cancer. Int J Radiat Oncol Biol Phys. 1999;45(3):577–87. https://doi.org/10.1016/s0360-3016(99)00247-3.
330. Murdoch-Kinch CA, Kim HM, Vineberg KA, Ship JA, Eisbruch A. Dose-effect relationships for the submandibular salivary glands and implications for their sparing by intensity modulated radiotherapy. Int J Radiat Oncol Biol Phys. 2008;72(2):373–82. https://doi.org/10.1016/j.ijrobp.2007.12.033.
331. Ortholan C, Chamorey E, Benezery K, Thariat J, Dassonville O, Poissonnet G, Bozec A, Follana P, Peyrade F, Sudaka A, Gerard JP, Bensadoun RJ. Modeling of salivary production recovery after radiotherapy using mixed models: determination of optimal dose constraint for IMRT planning and construction of convenient tools to predict salivary function. Int J Radiat Oncol Biol Phys. 2009;73(1):178–86. https://doi.org/10.1016/j.ijrobp.2008.03.068.
332. Roesink JM, Moerland MA, Battermann JJ, Hordijk GJ, Terhaard CH. Quantitative dose-volume response analysis of changes in parotid gland function after radiotherapy in the head-and-neck region.

Int J Radiat Oncol Biol Phys. 2001;51(4):938–46. https://doi.org/10.1016/s0360-3016(01)01717-5.
333. Burlage FR, Coppes RP, Meertens H, Stokman MA, Vissink A. Parotid and submandibular/sublingual salivary flow during high dose radiotherapy. Radiother Oncol. 2001;61(3):271–4. https://doi.org/10.1016/s0167-8140(01)00427-3.
334. Bralic M, Muhvic-Urek M, Stemberga V, Golemac M, Jurkovic S, Borcic J, Braut A, Tomac J. Cell death and cell proliferation in mouse submandibular gland during early post-irradiation phase. Acta Med Okayama. 2005;59(4):153–9. https://doi.org/10.18926/amo/31948.
335. Lombaert IM, Brunsting JF, Wierenga PK, Kampinga HH, de Haan G, Coppes RP. Cytokine treatment improves parenchymal and vascular damage of salivary glands after irradiation. Clin Cancer Res. 2008;14(23):7741–50. https://doi.org/10.1158/1078-0432.Ccr-08-1449.
336. Burlage FR, Faber H, Kampinga HH, Langendijk JA, Vissink A, Coppes RP. Enhanced proliferation of acinar and progenitor cells by prophylactic pilocarpine treatment underlies the observed amelioration of radiation injury to parotid glands. Radiother Oncol. 2009;90(2):253–6. https://doi.org/10.1016/j.radonc.2008.11.011.
337. Marzouki HZ, Elkhalidy Y, Jha N, Scrimger R, Debenham BJ, Harris JR, O'Connell DA, Seikaly H. Modification of the submandibular gland transfer procedure. Laryngoscope. 2016;126(11):2492–6. https://doi.org/10.1002/lary.26029.
338. Morand GB, Madana J, Da Silva SD, Roskies M, Sultanem K, Black MJ, Mlynarek AM, Hier MP. Survival and quality of life in oropharyngeal cancer patients treated with primary chemoradiation after salivary gland transfer. J Laryngol Otol. 2016;130(8):755–62. https://doi.org/10.1017/S0022215116008100.
339. Jha N, Seikaly H, Harris J, Williams D, Sultanem K, Hier M, Ghosh S, Black M, Butler J, Sutherland D, Kerr P, Barnaby P. Phase III randomized study: oral pilocarpine versus submandibular salivary gland transfer protocol for the management of radiation-induced xerostomia. Head Neck. 2009;31(2):234–43. https://doi.org/10.1002/hed.20961.
340. Nguyen VT, Dawson P, Zhang Q, Harris Z, Limesand KH. Administration of growth factors promotes salisphere formation from irradiated parotid salivary glands. PLoS One. 2018;13(3):e0193942. https://doi.org/10.1371/journal.pone.0193942.
341. Kim JH, Jeong BK, Jang SJ, Yun JW, Jung MH, Kang KM, Kim TG, Woo SH. Alpha-lipoic acid ameliorates radiation-induced salivary gland injury by preserving parasympathetic innervation in rats. Int J Mol Sci. 2020;21(7):2260. https://doi.org/10.3390/ijms21072260.
342. Fang D, Hu S, Liu Y, Quan VH, Seuntjens J, Tran SD. Identification of the active components in bone marrow soup: a mitigator against irradiation-injury to salivary glands. Sci Rep. 2015;5:16017. https://doi.org/10.1038/srep16017.
343. Fang D, Shang S, Liu Y, Bakkar M, Sumita Y, Seuntjens J, Tran SD. Optimal timing and frequency of bone marrow soup therapy for functional restoration of salivary glands injured by single-dose or fractionated irradiation. J Tissue Eng Regen Med. 2018;12(2):e1195–205. https://doi.org/10.1002/term.2513.
344. Su X, Fang D, Liu Y, Ruan G, Seuntjens J, Kinsella J, Tran S. Lyophilized bone marrow cell extract functionally restores irradiation-injured salivary glands. Oral Dis. 2018;24(1–2):202–6. https://doi.org/10.1111/odi.12728.
345. Emmerson E, Knox SM. Salivary gland stem cells: a review of development, regeneration and cancer. Genesis. 2018;56(5):e23211. https://doi.org/10.1002/dvg.23211.
346. Jeong J, Baek H, Kim YJ, Choi Y, Lee H, Lee E, Kim ES, Hah JH, Kwon TK, Choi IJ, Kwon H. Human salivary gland stem cells ameliorate hyposalivation of radiation-damaged rat salivary glands. Exp Mol Med. 2013;45:e58. https://doi.org/10.1038/emm.2013.121.
347. Li Z, Wang Y, Xing H, Wang Z, Hu H, An R, Xu H, Liu Y, Liu B. Protective efficacy of intravenous transplantation of adipose-derived stem cells for the prevention of radiation-induced salivary gland damage. Arch Oral Biol. 2015;60(10):1488–96. https://doi.org/10.1016/j.archoralbio.2015.07.016.
348. Kojima T, Kanemaru S, Hirano S, Tateya I, Ohno S, Nakamura T, Ito J. Regeneration of radiation damaged salivary glands with adipose-derived stromal cells. Laryngoscope. 2011;121(9):1864–9. https://doi.org/10.1002/lary.22080.
349. Lin CY, Chang FH, Chen CY, Huang CY, Hu FC, Huang WK, Ju SS, Chen MH. Cell therapy for salivary gland regeneration. J Dent Res. 2011;90(3):341–6. https://doi.org/10.1177/0022034510386374.
350. Serrano Martinez P, Cinat D, van Luijk P, Baanstra M, de Haan G, Pringle S, Coppes RP. Mouse parotid salivary gland organoids for the *in vitro* study of stem cell radiation response. Oral Dis. n/a (n/a). https://doi.org/10.1111/odi.13475
351. Uccelli A, Moretta L, Pistoia V. Mesenchymal stem cells in health and disease. Nat Rev Immunol. 2008;8(9):726–36. https://doi.org/10.1038/nri2395.
352. Grønhøj C, Jensen DH, Vester-Glowinski P, Jensen SB, Bardow A, Oliveri RS, Fog LM, Specht L, Thomsen C, Darkner S, Jensen M, Müller V, Kiss K, Agander T, Andersen E, Fischer-Nielsen A, von Buchwald C. Safety and efficacy of mesenchymal stem cells for radiation-induced xerostomia: a randomized, placebo-controlled phase 1/2 trial (MESRIX). Int J Radiat Oncol Biol Phys. 2018;101(3):581–92. https://doi.org/10.1016/j.ijrobp.2018.02.034.
353. Alotaibi H, Tuzlakoğlu-Öztürk M, Tazebay UH. The thyroid Na+/I- symporter: molecular character-

ization and genomic regulation. Mol Imaging Radionucl Ther. 2017;26(Suppl 1):92–101. https://doi.org/10.4274/2017.26.suppl.11.
354. Jo KS, An YS, Lee SJ, Soh EY, Lee J, Chung YS, Kim DJ, Yoon SH, Lee DH, Yoon JK. Significance of salivary gland radioiodine retention on post-ablation (131)I scintigraphy as a predictor of salivary gland dysfunction in patients with differentiated thyroid carcinoma. Nucl Med Mol Imaging. 2014;48(3):203–11. https://doi.org/10.1007/s13139-014-0274-4.
355. Jeong SY, Kim HW, Lee SW, Ahn BC, Lee J. Salivary gland function 5 years after radioactive iodine ablation in patients with differentiated thyroid cancer: direct comparison of pre- and postablation scintigraphies and their relation to xerostomia symptoms. Thyroid. 2013;23(5):609–16. https://doi.org/10.1089/thy.2012.0106.
356. Shi L, Cong X, Zhang Y, Ding C, Ding QW, Fu FY, Wu LL, Yu GY. Carbachol improves secretion in the early phase after rabbit submandibular gland transplantation. Oral Dis. 2010;16(4):351–9. https://doi.org/10.1111/j.1601-0825.2009.01633.x.
357. Aframian DJ, Palmon A. Current status of the development of an artificial salivary gland. Tissue Eng Part B Rev. 2008;14(2):187–98. https://doi.org/10.1089/ten.teb.2008.0044.
358. Nicolay NH, Lopez Perez R, Saffrich R, Huber PE. Radio-resistant mesenchymal stem cells: mechanisms of resistance and potential implications for the clinic. Oncotarget. 2015;6(23):19366–80. https://doi.org/10.18632/oncotarget.4358.
359. Saylam G, Bayır Ö, Gültekin SS, Pınarlı FA, Han Ü, Korkmaz MH, Sancaktar ME, Tatar İ, Sargon MF, Tatar E. Protective/restorative role of the adipose tissue-derived mesenchymal stem cells on the radioiodine-induced salivary gland damage in rats. Radiol Oncol. 2017;51(3):307–16. https://doi.org/10.1515/raon-2017-0022.
360. Kim JW, Kim JM, Choi ME, Kim SK, Kim YM, Choi JS. Adipose-derived mesenchymal stem cells regenerate radioiodine-induced salivary gland damage in a murine model. Science. 2019;9(1):15752. https://doi.org/10.1038/s41598-019-51775-9.
361. Jensen SB, Pedersen AM, Vissink A, Andersen E, Brown CG, Davies AN, Dutilh J, Fulton JS, Jankovic L, Lopes NN, Mello AL, Muniz LV, Murdoch-Kinch CA, Nair RG, Napeñas JJ, Nogueira-Rodrigues A, Saunders D, Stirling B, von Bültzingslöwen I, Weikel DS, Elting LS, Spijkervet FK, Brennan MT. A systematic review of salivary gland hypofunction and xerostomia induced by cancer therapies: prevalence, severity and impact on quality of life. Support Care Cancer. 2010;18(8):1039–60. https://doi.org/10.1007/s00520-010-0827-8.
362. Harrison T, Bigler L, Tucci M, Pratt L, Malamud F, Thigpen JT, Streckfus C, Younger H. Salivary sIgA concentrations and stimulated whole saliva flow rates among women undergoing chemotherapy for breast cancer: an exploratory study. Spec Care Dentist. 1998;18(3):109–12. https://doi.org/10.1111/j.1754-4505.1998.tb00914.x.
363. Napeñas JJ, Miles L, Guajardo-Streckfus C, Streckfus CF. Salivary flow rates among women diagnosed with benign and malignant tumors. Spec Care Dentist. 2013;33(3):102–10. https://doi.org/10.1111/scd.12017.
364. Jensen SB, Mouridsen HT, Reibel J, Brünner N, Nauntofte B. Adjuvant chemotherapy in breast cancer patients induces temporary salivary gland hypofunction. Oral Oncol. 2008;44(2):162–73. https://doi.org/10.1016/j.oraloncology.2007.01.015.
365. Meurman JH, Laine P, Keinànen S, Pyrhönen S, Teerenhovi L, Lindqvist C. Five-year follow-up of saliva in patients treated for lymphomas. Oral Surg Oral Med Oral Pathol Oral Radiol Endod. 1997;83(4):447–52. https://doi.org/10.1016/s1079-2104(97)90143-8.
366. Luo Z, Shang X, Zhang H, Wang G, Massey PA, Barton SR, Kevil CG, Dong Y. Notch signaling in osteogenesis, osteoclastogenesis, and angiogenesis. Am J Pathol. 2019;189(8):1495–500. https://doi.org/10.1016/j.ajpath.2019.05.005.
367. Qin Y, Wang L, Gao Z, Chen G, Zhang C. Bone marrow stromal/stem cell-derived extracellular vesicles regulate osteoblast activity and differentiation *in vitro* and promote bone regeneration *in vivo*. Sci Rep. 2016;6:21961. https://doi.org/10.1038/srep21961.
368. Fishero BA, Kohli N, Das A, Christophel JJ, Cui Q. Current concepts of bone tissue engineering for craniofacial bone defect repair. Craniomaxillofac Trauma Reconstr. 2015;8(1):23–30. https://doi.org/10.1055/s-0034-1393724.
369. Marx RE. Bone and bone graft healing. Oral Maxillofac Surg Clin North Am. 2007;19(4):455–66., v. https://doi.org/10.1016/j.coms.2007.07.008.
370. Mulligan RP, Friedman JA, Mahabir RC. A nationwide review of the associations among cervical spine injuries, head injuries, and facial fractures. J Trauma. 2010;68(3):587–92. https://doi.org/10.1097/TA.0b013e3181b16bc5.
371. Wang XX, Allen RJ Jr, Tutela JP, Sailon A, Allori AC, Davidson EH, Paek GK, Saadeh PB, McCarthy JG, Warren SM. Progenitor cell mobilization enhances bone healing by means of improved neovascularization and osteogenesis. Plast Reconstr Surg. 2011;128(2):395–405. https://doi.org/10.1097/PRS.0b013e31821e6e10.
372. Zaidi N, Nixon AJ. Stem cell therapy in bone repair and regeneration. Ann NY Acad Sci. 2007;1117:62–72. https://doi.org/10.1196/annals.1402.074.
373. Fong KD, Nacamuli RP, Song HM, Warren SM, Lorenz HP, Longaker MT. New strategies for craniofacial repair and replacement: a brief review. J Craniofac Surg. 2003;14(3):333–9. https://doi.org/10.1097/00001665-200305000-00011.
374. Kaigler D, Pagni G, Park CH, Braun TM, Holman LA, Yi E, Tarle SA, Bartel RL, Giannobile WV. Stem cell therapy for craniofacial bone regeneration: a ran-

domized, controlled feasibility trial. Cell Transplant. 2013;22(5):767–77. https://doi.org/10.3727/096368912x652968.
375. Yang P, Huang X, Wang C, Dang X, Wang K. Repair of bone defects using a new biomimetic construction fabricated by adipose-derived stem cells, collagen I, and porous beta-tricalcium phosphate scaffolds. Exp Biol Med (Maywood). 2013;238(12):1331–43. https://doi.org/10.1177/1535370213505827.
376. Valderrábano RJ, Wu JY. Bone and blood interactions in human health and disease. Bone. 2019;119:65–70. https://doi.org/10.1016/j.bone.2018.02.019.
377. Piper K, Valentine G. Bone pathology. Methods Mol Biol. 2012;915:51–88. https://doi.org/10.1007/978-1-61779-977-8_4.
378. Tandon R, Herford AS. Future of bone pathology, bone grafting, and osseointegration in oral and maxillofacial surgery: how applying optical advancements can help both fields. J Biomed Opt. 2013;18(5):55006. https://doi.org/10.1117/1.Jbo.18.5.055006.
379. Chinoy A, Mughal MZ, Padidela R. Metabolic bone disease of prematurity: causes, recognition, prevention, treatment and long-term consequences. Arch Dis Child Fetal Neonatal Ed. 2019;104(5):F560–f566. https://doi.org/10.1136/archdischild-2018-316330.
380. Meng Y, Zhao YN, Zhang YQ, Liu DG, Gao Y. Three-dimensional radiographic features of ameloblastoma and cystic lesions in the maxilla. Dentomaxillofac Radiol. 2019;48(6):20190066. https://doi.org/10.1259/dmfr.20190066.
381. Miller TT. Bone tumors and tumorlike conditions: analysis with conventional radiography. Radiology. 2008;246(3):662–74. https://doi.org/10.1148/radiol.2463061038.
382. Andreu-Arasa VC, Chapman MN, Kuno H, Fujita A, Sakai O. Craniofacial manifestations of systemic disorders: CT and MR imaging findings and imaging approach. Radiographics. 2018;38(3):890–911. https://doi.org/10.1148/rg.2018170145.
383. McClary AC, West RB, McClary AC, Pollack JR, Fischbein NJ, Holsinger CF, Sunwoo J, Colevas AD, Sirjani D. Ameloblastoma: a clinical review and trends in management. Eur Arch Otorhinolaryngol. 2016;273(7):1649–61. https://doi.org/10.1007/s00405-015-3631-8.
384. Effiom OA, Ogundana OM, Akinshipo AO, Akintoye SO. Ameloblastoma: current etiopathological concepts and management. Oral Dis. 2018;24(3):307–16. https://doi.org/10.1111/odi.12646.
385. Kreppel M, Zöller J. Ameloblastoma-clinical, radiological, and therapeutic findings. Oral Dis. 2018;24(1–2):63–6. https://doi.org/10.1111/odi.12702.
386. Adeel M, Rajput MSA, Arain AA, Baloch M, Khan M. Ameloblastoma: management and outcome. Cureus. 2018;10(10):e3437. https://doi.org/10.7759/cureus.3437.
387. Milman T, Ying GS, Pan W, LiVolsi V. Ameloblastoma: 25 year experience at a single institution. Head Neck Pathol. 2016;10(4):513–20. https://doi.org/10.1007/s12105-016-0734-5.
388. Khalele BA, Al-Shiaty RA. A novel marker of ameloblastoma and systematic review of immunohistochemical findings. Ann Diagn Pathol. 2016;22:18–24. https://doi.org/10.1016/j.anndiagpath.2016.01.005.
389. Mendenhall WM, Werning JW, Fernandes R, Malyapa RS, Mendenhall NP. Ameloblastoma. Am J Clin Oncol. 2007;30(6):645–8. https://doi.org/10.1097/COC.0b013e3181573e59.
390. Hendra FN, Natsir Kalla DS, Van Cann EM, de Vet HCW, Helder MN, Forouzanfar T. Radical vs conservative treatment of intraosseous ameloblastoma: systematic review and meta-analysis. Oral Dis. 2019;25(7):1683–96. https://doi.org/10.1111/odi.13014.
391. Cicciù M, Herford AS, Cicciù D, Tandon R, Maiorana C. Recombinant human bone morphogenetic protein-2 promote and stabilize hard and soft tissue healing for large mandibular new bone reconstruction defects. J Craniofac Surg. 2014;25(3):860–2. https://doi.org/10.1097/scs.0000000000000830.
392. Chuong R, Donoff RB, Guralnick W. The odontogenic keratocyst. J Oral Maxillofac Surg. 1982;40(12):797–802. https://doi.org/10.1016/0278-2391(82)90177-x.
393. Goyault G, Moser T, Lutz JC, Neuville A, Buy X, Freitas R, Roy C, Gangi A. Odontogenic keratocyst. J Radiol. 2007;88(11 Pt 1):1733–5. https://doi.org/10.1016/s0221-0363(07)74055-6.
394. Gomes CC, Diniz MG, Gomez RS. Review of the molecular pathogenesis of the odontogenic keratocyst. Oral Oncol. 2009;45(12):1011–4. https://doi.org/10.1016/j.oraloncology.2009.08.003.
395. Preston RD, Narayana N. Peripheral odontogenic keratocyst. J Periodontol. 2005;76(12):2312–5. https://doi.org/10.1902/jop.2005.76.12.2312.
396. Polak K, Jędrusik-Pawłowska M, Drozdzowska B, Morawiec T. Odontogenic keratocyst of the mandible: a case report and literature review. Dent Med Probl. 2019;56(4):433–6. https://doi.org/10.17219/dmp/110682.
397. Slusarenko da Silva Y, Stoelinga PJW, Naclério-Homem MDG. Recurrence of nonsyndromic odontogenic keratocyst after marsupialization and delayed enucleation vs. enucleation alone: a systematic review and meta-analysis. Oral Maxillofac Surg. 2019;23(1):1–11. https://doi.org/10.1007/s10006-018-0737-3.
398. Guna TP, Rilna P, Sathyanarayanan R. Is surgical treatment based on one-step or two-step protocol effective in managing the odontogenic keratocyst? J Oral Maxillofac Surg. 2019;77(12):2367. https://doi.org/10.1016/j.joms.2019.07.022.

399. Johnson NR, Gannon OM, Savage NW, Batstone MD. Frequency of odontogenic cysts and tumors: a systematic review. J Investig Clin Dent. 2014;5(1):9–14. https://doi.org/10.1111/jicd.12044.
400. O'Neill R, Al-Hezaimi K. Identification of an odontogenic keratocyst and treatment with guided tissue regeneration: case report. J Can Dent Assoc. 2011;77:b6.
401. Ritter J, Bielack SS. Osteosarcoma. Ann Oncol. 2010;21(Suppl 7):vii320–5. https://doi.org/10.1093/annonc/mdq276.
402. ElKordy MA, ElBaradie TS, ElSebai HI, KhairAlla SM, Amin AAE. Osteosarcoma of the jaw: challenges in the diagnosis and treatment. J Egypt Natl Canc Inst. 2018;30(1):7–11. https://doi.org/10.1016/j.jnci.2018.02.001.
403. Simpson E, Brown HL. Understanding osteosarcomas. Jaapa. 2018;31(8):15–9. https://doi.org/10.1097/01.JAA.0000541477.24116.8d.
404. Biazzo A, De Paolis M. Multidisciplinary approach to osteosarcoma. Acta Orthop Belg. 2016;82(4):690–8.
405. Xie Y, Huang J, Wu M, Zhou Y. Expression of CD133 protein in osteosarcoma and its relationship with the clinicopathological features and prognosis. J Cancer Res Ther. 2018;14(4):892–5. https://doi.org/10.4103/jcrt.JCRT_461_17.
406. Shi K, Wang SL, Shen B, Yu FQ, Weng DF, Lin JH. Clinicopathological and prognostic values of fibronectin and integrin αvβ3 expression in primary osteosarcoma. World J Surg Oncol. 2019;17(1):23. https://doi.org/10.1186/s12957-019-1566-z.
407. Sarsilmaz A, Argin M, Sezak M, Altay C, Erdogan N. Primary osteosarcoma arising from subcutaneous tissue: 5-year follow-up. Clin Imaging. 2012;36(4):402–5. https://doi.org/10.1016/j.clinimag.2011.10.016.
408. Bishop MW, Janeway KA, Gorlick R. Future directions in the treatment of osteosarcoma. Curr Opin Pediatr. 2016;28(1):26–33. https://doi.org/10.1097/mop.0000000000000298.
409. Kansara M, Teng MW, Smyth MJ, Thomas DM. Translational biology of osteosarcoma. Nat Rev Cancer. 2014;14(11):722–35. https://doi.org/10.1038/nrc3838.
410. Fenerty S, Shaw W, Verma R, Syed AB, Kuklani R, Yang J, Ali S. Florid cemento-osseous dysplasia: review of an uncommon fibro-osseous lesion of the jaw with important clinical implications. Skelet Radiol. 2017;46(5):581–90. https://doi.org/10.1007/s00256-017-2590-0.
411. Brody A, Zalatnai A, Csomo K, Belik A, Dobo-Nagy C. Difficulties in the diagnosis of periapical translucencies and in the classification of cemento-osseous dysplasia. BMC Oral Health. 2019;19(1):139. https://doi.org/10.1186/s12903-019-0843-0.
412. Yang DD, Ghuman A, Stratton R, Hall M, Parthipun A. Appearance of florid cemento-osseous dysplasia on SPECT/CT. Clin Nucl Med. 2019;44(5):e357–9. https://doi.org/10.1097/rlu.0000000000002512.
413. Patel MM, Wilkey JF, Abdelsayed R, D'Silva NJ, Malchoff C, Mallya SM. Analysis of GNAS mutations in cemento-ossifying fibromas and cemento-osseous dysplasias of the jaws. Oral Surg Oral Med Oral Pathol Oral Radiol Endod. 2010;109(5):739–43. https://doi.org/10.1016/j.tripleo.2009.12.016.
414. Aiuto R, Gucciardino F, Rapetti R, Siervo S, Bianch AE. Management of symptomatic florid cemento-osseous dysplasia: literature review and a case report. J Clin Exp Dent. 2018;10(3):e291–5. https://doi.org/10.4317/jced.54577.
415. Chrcanovic BR, Gomez RS. Juvenile ossifying fibroma of the jaws and paranasal sinuses: a systematic review of the cases reported in the literature. Int J Oral Maxillofac Surg. 2020;49(1):28–37. https://doi.org/10.1016/j.ijom.2019.06.029.
416. MacDonald-Jankowski DS. Ossifying fibroma: a systematic review. Dentomaxillofac Radiol. 2009;38(8):495–513. https://doi.org/10.1259/dmfr/70933621.
417. Katti G, Khan MM, Chaubey SS, Amena M. Cemento-ossifying fibroma of the jaw. BMJ Case Rep. 2016. https://doi.org/10.1136/bcr-2015-214327.
418. Titinchi F, Morkel J. Ossifying fibroma: analysis of treatment methods and recurrence patterns. J Oral Maxillofac Surg. 2016;74(12):2409–19. https://doi.org/10.1016/j.joms.2016.05.018.
419. Kransdorf MJ, Moser RP Jr, Gilkey FW. Fibrous dysplasia. Radiographics. 1990;10(3):519–37. https://doi.org/10.1148/radiographics.10.3.2188311.
420. Burke AB, Collins MT, Boyce AM. Fibrous dysplasia of bone: craniofacial and dental implications. Oral Dis. 2017;23(6):697–708. https://doi.org/10.1111/odi.12563.
421. Ricalde P, Magliocca KR, Lee JS. Craniofacial fibrous dysplasia. Oral Maxillofac Surg Clin North Am. 2012;24(3):427–41. https://doi.org/10.1016/j.coms.2012.05.004.
422. Florez H, Peris P, Guañabens N. Fibrous dysplasia. Clinical review and therapeutic management. Med Clin (Barc). 2016;147(12):547–53. https://doi.org/10.1016/j.medcli.2016.07.030.
423. Rauch F, Glorieux FH. Osteogenesis imperfecta. Lancet. 2004;363(9418):1377–85. https://doi.org/10.1016/s0140-6736(04)16051-0.
424. Van Dijk FS, Sillence DO. Osteogenesis imperfecta: clinical diagnosis, nomenclature and severity assessment. Am J Med Genet A. 2014;164a(6):1470–81. https://doi.org/10.1002/ajmg.a.36545.
425. Bregou Bourgeois A, Aubry-Rozier B, Bonafé L, Laurent-Applegate L, Pioletti DP, Zambelli PY. Osteogenesis imperfecta: from diagnosis and multidisciplinary treatment to future perspectives. Swiss Med Wkly. 2016;146:w14322. https://doi.org/10.4414/smw.2016.14322.
426. Thomas IH, DiMeglio LA. Advances in the classification and treatment of osteogenesis imperfecta. Curr Osteoporos Rep. 2016;14(1):1–9. https://doi.org/10.1007/s11914-016-0299-y.

427. Ledesma-Montes C, Jiménez-Farfán MD, Hernández-Guerrero JC. Idiopathic osteosclerosis in the maxillomandibular area. Radiol Med. 2019;124(1):27–33. https://doi.org/10.1007/s11547-018-0944-x.
428. Sisman Y, Ertas ET, Ertas H, Sekerci AE. The frequency and distribution of idiopathic osteosclerosis of the jaw. Eur J Dent. 2011;5(4):409–14.
429. Hsu E. Paget's disease of bone: updates for clinicians. Curr Opin Endocrinol Diabetes Obes. 2019;26(6):329–34. https://doi.org/10.1097/med.0000000000000503.
430. Winn N, Lalam R, Cassar-Pullicino V. Imaging of Paget's disease of bone. Wien Med Wochenschr. 2017;167(1–2):9–17. https://doi.org/10.1007/s10354-016-0517-3.
431. Kravets I. Paget's disease of bone: diagnosis and treatment. Am J Med. 2018;131(11):1298–303. https://doi.org/10.1016/j.amjmed.2018.04.028.
432. Merigo E, Manfredi M, Meleti M, Guidotti R, Ripasarti A, Zanzucchi E, D'Aleo P, Corradi D, Corcione L, Sesenna E, Ferrari S, Poli T, Bonaninil M, Vescovi P. Bone necrosis of the jaws associated with bisphosphonate treatment: a report of twenty-nine cases. Acta Biomed. 2006;77(2):109–17.
433. Migliario M, Mergoni G, Vescovi P, Martino I, Alessio M, Benzi L, Renò F, Fusco V. Osteonecrosis of the jaw (ONJ) in osteoporosis patients: report of delayed diagnosis of a multisite case and commentary about risks coming from a restricted ONJ definition. Dent J (Basel). 2017;5(1):13. https://doi.org/10.3390/dj5010013.
434. Hess LM, Jeter JM, Benham-Hutchins M, Alberts DS. Factors associated with osteonecrosis of the jaw among bisphosphonate users. Am J Med. 2008;121(6):475–483.e473. https://doi.org/10.1016/j.amjmed.2008.01.047.
435. Hasegawa T, Kawakita A, Ueda N, Funahara R, Tachibana A, Kobayashi M, Kondou E, Takeda D, Kojima Y, Sato S, Yanamoto S, Komatsubara H, Umeda M, Kirita T, Kurita H, Shibuya Y, Komori T. A multicenter retrospective study of the risk factors associated with medication-related osteonecrosis of the jaw after tooth extraction in patients receiving oral bisphosphonate therapy: can primary wound closure and a drug holiday really prevent MRONJ? Osteoporos Int. 2017;28(8):2465–73. https://doi.org/10.1007/s00198-017-4063-7.
436. Ruggiero SL, Mehrotra B, Rosenberg TJ, Engroff SL. Osteonecrosis of the jaws associated with the use of bisphosphonates: a review of 63 cases. J Oral Maxillofac Surg. 2004;62(5):527–34. https://doi.org/10.1016/j.joms.2004.02.004.
437. American Association of Oral and Maxillofacial Surgeons position paper on bisphosphonate-related osteonecrosis of the jaws. J Oral Maxillofac Surg. 2007;65(3):369–76. https://doi.org/10.1016/j.joms.2006.11.003.
438. Ruggiero SL, Dodson TB, Assael LA, Landesberg R, Marx RE, Mehrotra B. American Association of Oral and Maxillofacial Surgeons position paper on bisphosphonate-related osteonecrosis of the jaws—2009 update. J Oral Maxillofac Surg. 2009;67(5 Suppl):2–12. https://doi.org/10.1016/j.joms.2009.01.009.
439. Yee AJ, Raje NS. Denosumab for the treatment of bone disease in solid tumors and multiple myeloma. Future Oncol. 2018;14(3):195–203. https://doi.org/10.2217/fon-2017-0403.
440. Kishimoto H, Noguchi K, Takaoka K. Novel insight into the management of bisphosphonate-related osteonecrosis of the jaw (BRONJ). Jpn Dent Sci Rev. 2019;55(1):95–102. https://doi.org/10.1016/j.jdsr.2018.09.002.
441. Endo Y, Kumamoto H, Nakamura M, Sugawara S, Takano-Yamamoto T, Sasaki K, Takahashi T. Underlying mechanisms and therapeutic strategies for bisphosphonate-related osteonecrosis of the jaw (BRONJ). Biol Pharm Bull. 2017;40(6):739–50. https://doi.org/10.1248/bpb.b16-01020.
442. Rollason V, Laverrière A, MacDonald LC, Walsh T, Tramèr MR, Vogt-Ferrier NB. Interventions for treating bisphosphonate-related osteonecrosis of the jaw (BRONJ). Cochrane Database Syst Rev. 2016;2(2):Cd008455. https://doi.org/10.1002/14651858.CD008455.pub2.
443. Fliefel R, Tröltzsch M, Kühnisch J, Ehrenfeld M, Otto S. Treatment strategies and outcomes of bisphosphonate-related osteonecrosis of the jaw (BRONJ) with characterization of patients: a systematic review. Int J Oral Maxillofac Surg. 2015;44(5):568–85. https://doi.org/10.1016/j.ijom.2015.01.026.
444. Nieckula P, Stempniewicz A, Tubaja M. Prophylaxis of osteonecrosis in the case of patients treated with bisphosphonates: a review paper. Dent Med Probl. 2018;55(4):425–9. https://doi.org/10.17219/dmp/99021.
445. Ficarra G, Beninati F. Bisphosphonate-related osteonecrosis of the jaws: an update on clinical, pathological and management aspects. Head Neck Pathol. 2007;1(2):132–40. https://doi.org/10.1007/s12105-007-0033-2.
446. Kalra S, Jain V. Dental complications and management of patients on bisphosphonate therapy: a review article. J Oral Biol Craniofac Res. 2013;3(1):25–30. https://doi.org/10.1016/j.jobcr.2012.11.001.
447. Iglesias JE, Salum FG, Figueiredo MA, Cherubini K. Important aspects concerning alendronate-related osteonecrosis of the jaws: a literature review. Gerodontology. 2015;32(3):169–78. https://doi.org/10.1111/ger.12093.
448. Kilic E, Doganay O. Current management concepts for bisphosphonate-related osteonecrosis of the jaw: a review. Gen Dent. 2018;66(6):e1–5.
449. Dodson TB. Intravenous bisphosphonate therapy and bisphosphonate-related osteonecrosis of the jaws. J Oral Maxillofac Surg. 2009;67(5 Suppl):44–52. https://doi.org/10.1016/j.joms.2008.12.004.

450. Vescovi P, Nammour S. Bisphosphonate-related osteonecrosis of the jaw (BRONJ) therapy. A critical review. Minerva Stomatol. 2010;59(4):181–203. 204–113
451. Akashi M, Kusumoto J, Takeda D, Shigeta T, Hasegawa T, Komori T. A literature review of perioperative antibiotic administration in surgery for medication-related osteonecrosis of the jaw. Oral Maxillofac Surg. 2018;22(4):369–78. https://doi.org/10.1007/s10006-018-0732-8.
452. Hoefert S, Eufinger H. Relevance of a prolonged preoperative antibiotic regime in the treatment of bisphosphonate-related osteonecrosis of the jaw. J Oral Maxillofac Surg. 2011;69(2):362–80. https://doi.org/10.1016/j.joms.2010.06.200.
453. Edwards BJ, Hellstein JW, Jacobsen PL, Kaltman S, Mariotti A, Migliorati CA. Updated recommendations for managing the care of patients receiving oral bisphosphonate therapy: an advisory statement from the American Dental Association Council on Scientific Affairs. J Am Dent Assoc. 2008;139(12):1674–7. https://doi.org/10.14219/jada.archive.2008.0110.
454. Kaibuchi N, Iwata T, Yamato M, Okano T, Ando T. Multipotent mesenchymal stromal cell sheet therapy for bisphosphonate-related osteonecrosis of the jaw in a rat model. Acta Biomater. 2016;42:400–10. https://doi.org/10.1016/j.actbio.2016.06.022.
455. Freiberger JJ, Feldmeier JJ. Evidence supporting the use of hyperbaric oxygen in the treatment of osteoradionecrosis of the jaw. J Oral Maxillofac Surg. 2010;68(8):1903–6. https://doi.org/10.1016/j.joms.2010.02.001.
456. Mozzati M, Gallesio G, Arata V, Pol R, Scoletta M. Platelet-rich therapies in the treatment of intravenous bisphosphonate-related osteonecrosis of the jaw: a report of 32 cases. Oral Oncol. 2012;48(5):469–74. https://doi.org/10.1016/j.oraloncology.2011.12.004.
457. Anitua E, Begoña L, Orive G. Treatment of hemimandibular paresthesia in a patient with bisphosphonate-related osteonecrosis of the jaw (BRONJ) by combining surgical resection and PRGF-Endoret. Br J Oral Maxillofac Surg. 2013;51(8):e272–4. https://doi.org/10.1016/j.bjoms.2012.08.018.
458. Mozzati M, Arata V, Gallesio G. Tooth extraction in patients on zoledronic acid therapy. Oral Oncol. 2012;48(9):817–21. https://doi.org/10.1016/j.oraloncology.2012.03.009.
459. Anitua E, Zalduendo M, Troya M, Orive G. PRGF exerts a cytoprotective role in zoledronic acid-treated oral cells. Clin Oral Investig. 2016;20(3):513–21. https://doi.org/10.1007/s00784-015-1528-y.
460. Trivedi B, Kesterke MJ, Bhattacharjee R, Weber W, Mynar K, Reddy LV. Craniofacial injuries seen with the introduction of bicycle-share electric scooters in an urban setting. J Oral Maxillofac Surg. 2019;77(11):2292–7. https://doi.org/10.1016/j.joms.2019.07.014.
461. Hull AM, Lowe T, Finlay PM. The psychological impact of maxillofacial trauma: an overview of reactions to trauma. Oral Surg Oral Med Oral Pathol Oral Radiol Endod. 2003;95(5):515–20. https://doi.org/10.1067/moe.2003.161.
462. Arslan ED, Solakoglu AG, Komut E, Kavalci C, Yilmaz F, Karakilic E, Durdu T, Sonmez M. Assessment of maxillofacial trauma in emergency department. World J Emerg Surg. 2014;9(1):13. https://doi.org/10.1186/1749-7922-9-13.
463. Aksoy E, Unlü E, Sensöz O. A retrospective study on epidemiology and treatment of maxillofacial fractures. J Craniofac Surg. 2002;13(6):772–5. https://doi.org/10.1097/00001665-200211000-00012.
464. Gassner R, Tuli T, Hächl O, Rudisch A, Ulmer H. Cranio-maxillofacial trauma: a 10 year review of 9,543 cases with 21,067 injuries. J Craniomaxillofac Surg. 2003;31(1):51–61. https://doi.org/10.1016/s1010-5182(02)00168-3.
465. van den Bergh B, Karagozoglu KH, Heymans MW, Forouzanfar T. Aetiology and incidence of maxillofacial trauma in Amsterdam: a retrospective analysis of 579 patients. J Craniomaxillofac Surg. 2012;40(6):e165–9. https://doi.org/10.1016/j.jcms.2011.08.006.
466. Gandhi S, Ranganathan LK, Solanki M, Mathew GC, Singh I, Bither S. Pattern of maxillofacial fractures at a tertiary hospital in northern India: a 4-year retrospective study of 718 patients. Dent Traumatol. 2011;27(4):257–62. https://doi.org/10.1111/j.1600-9657.2011.00996.x.
467. Hopper RA, Salemy S, Sze RW. Diagnosis of midface fractures with CT: what the surgeon needs to know. Radiographics. 2006;26(3):783–93. https://doi.org/10.1148/rg.263045710.
468. Yokota H, Kurokawa A, Otsuka T, Kobayashi S, Nakazawa S. Significance of magnetic resonance imaging in acute head injury. J Trauma. 1991;31(3):351–7. https://doi.org/10.1097/00005373-199103000-00007.
469. Castilla DM, Dinh CT, Younis R. Pediatric airway management in craniofacial trauma. J Craniofac Surg. 2011;22(4):1175–8. https://doi.org/10.1097/SCS.0b013e31821c00c3.
470. McKellop JA, Bou-Assaly W, Mukherji SK. Emergency head & neck imaging: infections and inflammatory processes. Neuroimaging Clin N Am. 2010;20(4):651–61. https://doi.org/10.1016/j.nic.2010.07.007.
471. Maestre-Vera JR. Treatment options in odontogenic infection. Med Oral Patol Oral Cir Bucal. 2004;9(Suppl 25–31):19–24.
472. Sánchez R, Mirada E, Arias J, Paño JR, Burgueño M. Severe odontogenic infections: epidemiological, microbiological and therapeutic factors. Med Oral Patol Oral Cir Bucal. 2011;16(5):e670–6. https://doi.org/10.4317/medoral.16995.
473. Laskin DM. Contemporary oral and maxillofacial surgery. Alpha Omegan. 2009;102(2):45. https://doi.org/10.1016/j.aodf.2009.04.004.

474. Thomas MV, Puleo DA. Infection, inflammation, and bone regeneration: a paradoxical relationship. J Dent Res. 2011;90(9):1052–61. https://doi.org/10.1177/0022034510393967.
475. Costerton JW, Montanaro L, Arciola CR. Biofilm in implant infections: its production and regulation. Int J Artif Organs. 2005;28(11):1062–8. https://doi.org/10.1177/039139880502801103.
476. Graves DT, Cochran D. The contribution of interleukin-1 and tumor necrosis factor to periodontal tissue destruction. J Periodontol. 2003;74(3):391–401. https://doi.org/10.1902/jop.2003.74.3.391.
477. Offenbacher S, Barros SP, Beck JD. Rethinking periodontal inflammation. J Periodontol. 2008;79(8 Suppl):1577–84. https://doi.org/10.1902/jop.2008.080220.
478. Taub D, Yampolsky A, Diecidue R, Gold L. Controversies in the management of oral and maxillofacial infections. Oral Maxillofac Surg Clin North Am. 2017;29(4):465–73. https://doi.org/10.1016/j.coms.2017.06.004.
479. Rega AJ, Aziz SR, Ziccardi VB. Microbiology and antibiotic sensitivities of head and neck space infections of odontogenic origin. J Oral Maxillofac Surg. 2006;64(9):1377–80. https://doi.org/10.1016/j.joms.2006.05.023.
480. Owtad P, Park JH, Shen G, Potres Z, Darendeliler MA. The biology of TMJ growth modification: a review. J Dent Res. 2013;92(4):315–21. https://doi.org/10.1177/0022034513476302.
481. Liang W, Li X, Gao B, Gan H, Lin X, Liao L, Li C. Observing the development of the temporomandibular joint in embryonic and post-natal mice using various staining methods. Exp Ther Med. 2016;11(2):481–9. https://doi.org/10.3892/etm.2015.2937.
482. Kubosch EJ, Lang G, Furst D, Kubosch D, Izadpanah K, Rolauffs B, Sudkamp NP, Schmal H. The potential for synovium-derived stem cells in cartilage repair. Curr Stem Cell Res Ther. 2018;13(3):174–84. https://doi.org/10.2174/1574888X12666171002111026.
483. Zainal Ariffin SH, Kermani S, Megat Abdul Wahab R, Senafi S, Zainal Ariffin Z, Abdul Razak M. *In vitro* chondrogenesis transformation study of mouse dental pulp stem cells. ScientificWorldJournal. 2012;2012:827149. https://doi.org/10.1100/2012/827149.
484. Zarb GA, Carlsson GE. Temporomandibular disorders: osteoarthritis. J Orofac Pain. 1999;13(4):295–306.
485. Seymour RL, Crouse VL, Irby WB. Temporomandibular ankylosis secondary to rheumatoid arthritis. Report of a case. Oral Surg Oral Med Oral Pathol. 1975;40(5):584–9. https://doi.org/10.1016/0030-4220(75)90367-9.
486. Korioth TW, Romilly DP, Hannam AG. Three-dimensional finite element stress analysis of the dentate human mandible. Am J Phys Anthropol. 1992;88(1):69–96. https://doi.org/10.1002/ajpa.1330880107.
487. Beek M, Koolstra JH, van Ruijven LJ, van Eijden TM. Three-dimensional finite element analysis of the human temporomandibular joint disc. J Biomech. 2000;33(3):307–16. https://doi.org/10.1016/s0021-9290(99)00168-2.
488. Naeije M, Te Veldhuis AH, Te Veldhuis EC, Visscher CM, Lobbezoo F. Disc displacement within the human temporomandibular joint: a systematic review of a 'noisy annoyance'. J Oral Rehabil. 2013;40(2):139–58. https://doi.org/10.1111/joor.12016.
489. Farrar WB, McCarty WL Jr. The TMJ dilemma. J Ala Dent Assoc. 1979;63(1):19–26.
490. Murphy MK, MacBarb RF, Wong ME, Athanasiou KA. Temporomandibular disorders: a review of etiology, clinical management, and tissue engineering strategies. Int J Oral Maxillofac Implants. 2013;28(6):e393–414. https://doi.org/10.11607/jomi.te20.
491. Brooks SL, Westesson PL, Eriksson L, Hansson LG, Barsotti JB. Prevalence of osseous changes in the temporomandibular joint of asymptomatic persons without internal derangement. Oral Surg Oral Med Oral Pathol. 1992;73(1):118–22. https://doi.org/10.1016/0030-4220(92)90168-p.
492. Gynther GW, Holmlund AB, Reinholt FP, Lindblad S. Temporomandibular joint involvement in generalized osteoarthritis and rheumatoid arthritis: a clinical, arthroscopic, histologic, and immunohistochemical study. Int J Oral Maxillofac Surg. 1997;26(1):10–6. https://doi.org/10.1016/s0901-5027(97)80838-7.
493. Galea CJ, Dashow JE, Woerner JE. Congenital abnormalities of the temporomandibular joint. Oral Maxillofac Surg Clin North Am. 2018;30(1):71–82. https://doi.org/10.1016/j.coms.2017.09.003.
494. Kaneyama K, Segami N, Hatta T. Congenital deformities and developmental abnormalities of the mandibular condyle in the temporomandibular joint. Congenit Anom (Kyoto). 2008;48(3):118–25. https://doi.org/10.1111/j.1741-4520.2008.00191.x.
495. Machon V, Hirjak D, Lukas J. Therapy of the osteoarthritis of the temporomandibular joint. J Craniomaxillofac Surg. 2011;39(2):127–30. https://doi.org/10.1016/j.jcms.2010.04.010.
496. Paniagua B, Cevidanes L, Walker D, Zhu H, Guo R, Styner M. Clinical application of SPHARM-PDM to quantify temporomandibular joint osteoarthritis. Comput Med Imaging Graph. 2011;35(5):345–52. https://doi.org/10.1016/j.compmedimag.2010.11.012.
497. Taylan Filinte G, Akan M, Bilgic I, Karaca M, Akoz T. Chondrogenic effect of the perichondrium graft on the internal derangement and osteoarthritis of the temporomandibular joint of the rabbit. J Craniomaxillofac Surg. 2011;39(5):351–8. https://doi.org/10.1016/j.jcms.2010.09.002.

498. Xaymardan M, Cimini M, Weisel RD, Li RK (2008) Bone marrow stem cell: properties and pluripotency. In: Atala A, Lanza R (eds) Principles of regenerative medicine. Burlington, Elsevier, pp. 268–300.
499. Xaymardan M. Orofacial stem cells for cell-based therapies of local and systemic diseases. In: Innovative research on biosis–abiosis intelligent interface. Tokyo: Elsevier; 2017. p. 89–99.
500. Farahani RM, Xaymardan M. Platelet-derived growth factor receptor alpha as a marker of mesenchymal stem cells in development and stem cell biology. Stem Cells Int. 2015;2015:362753. https://doi.org/10.1155/2015/362753.
501. Chong JJ, Xaymardan M, Chandrakanthan V, Asli NS, Li J, Ahmed I, Heffernan C, Menon MK, Scarlett CJ, Rashidianfar A, Biben C, Zoellner H, Colvin EK, Pimanda JE, Biankin AV, Zhou B, Pu WT, Prall OW, Harvey RP. Adult cardiac-resident MSC-like stem cells with a proepicardial origin. Cell Stem Cell. 2011;9(6):527–40. https://doi.org/10.1016/j.stem.2011.10.002.
502. Houlihan DD, Mabuchi Y, Morikawa S, Niibe K, Araki D, Suzuki S, Okano H, Matsuzaki Y. Isolation of mouse mesenchymal stem cells on the basis of expression of Sca-1 and PDGFR-α. Nat Protoc. 2012;7:2103–11.
503. Chen K, Man C, Zhang B, Hu J, Zhu SS. Effect of *in vitro* chondrogenic differentiation of autologous mesenchymal stem cells on cartilage and subchondral cancellous bone repair in osteoarthritis of temporomandibular joint. Int J Oral Maxillofac Surg. 2013;42(2):240–8. https://doi.org/10.1016/j.ijom.2012.05.030.
504. Zhang S, Teo KYW, Chuah SJ, Lai RC, Lim SK, Toh WS. MSC exosomes alleviate temporomandibular joint osteoarthritis by attenuating inflammation and restoring matrix homeostasis. Biomaterials. 2019;200:35–47. https://doi.org/10.1016/j.biomaterials.2019.02.006.
505. Kim H, Yang G, Park J, Choi J, Kang E, Lee BK. Therapeutic effect of mesenchymal stem cells derived from human umbilical cord in rabbit temporomandibular joint model of osteoarthritis. Sci Rep. 2019;9(1):13854. https://doi.org/10.1038/s41598-019-50435-2.
506. Ogasawara N, Kano F, Hashimoto N, Mori H, Liu Y, Xia L, Sakamaki T, Hibi H, Iwamoto T, Tanaka E, Yamamoto A. Factors secreted from dental pulp stem cells show multifaceted benefits for treating experimental temporomandibular joint osteoarthritis. Osteoarthr Cartil. 2020;28(6):831–41. https://doi.org/10.1016/j.joca.2020.03.010.
507. Ishikawa J, Takahashi N, Matsumoto T, Yoshioka Y, Yamamoto N, Nishikawa M, Hibi H, Ishigro N, Ueda M, Furukawa K, Yamamoto A. Factors secreted from dental pulp stem cells show multifaceted benefits for treating experimental rheumatoid arthritis. Bone. 2016;83:210–9. https://doi.org/10.1016/j.bone.2015.11.012.
508. Chen K, Xiong H, Xu N, Shen Y, Huang Y, Liu C. Chondrogenic potential of stem cells from human exfoliated deciduous teeth *in vitro* and *in vivo*. Acta Odontol Scand. 2014;72(8):664–72. https://doi.org/10.3109/00016357.2014.888756.
509. Lee J, Taylor SE, Smeriglio P, Lai J, Maloney WJ, Yang F, Bhutani N. Early induction of a prechondrogenic population allows efficient generation of stable chondrocytes from human induced pluripotent stem cells. FASEB J. 2015;29(8):3399–410. https://doi.org/10.1096/fj.14-269720.
510. Castro-Vinuelas R, Sanjurjo-Rodriguez C, Pineiro-Ramil M, Hermida-Gomez T, Fuentes-Boquete IM, de Toro-Santos FJ, Blanco-Garcia FJ, Diaz-Prado SM. Induced pluripotent stem cells for cartilage repair: current status and future perspectives. Eur Cell Mater. 2018;36:96–109. https://doi.org/10.22203/eCM.v036a08.
511. MacBarb RF, Chen AL, Hu JC, Athanasiou KA. Engineering functional anisotropy in fibrocartilage neotissues. Biomaterials. 2013;34(38):9980–9. https://doi.org/10.1016/j.biomaterials.2013.09.026.
512. Stapleton TW, Ingram J, Fisher J, Ingham E. Investigation of the regenerative capacity of an acellular porcine medial meniscus for tissue engineering applications. Tissue Eng Part A. 2011;17(1–2):231–42. https://doi.org/10.1089/ten.TEA.2009.0807.
513. Dai J, Wang J, Lu J, Zou D, Sun H, Dong Y, Yu H, Zhang L, Yang T, Zhang X, Wang X, Shen G. The effect of co-culturing costal chondrocytes and dental pulp stem cells combined with exogenous FGF9 protein on chondrogenesis and ossification in engineered cartilage. Biomaterials. 2012;33(31):7699–711. https://doi.org/10.1016/j.biomaterials.2012.07.020.
514. Vapniarsky N, Huwe LW, Arzi B, Houghton MK, Wong ME, Wilson JW, Hatcher DC, Hu JC, Athanasiou KA. Tissue engineering toward temporomandibular joint disc regeneration. Sci Transl Med. 2018;10(446):eaaq1802. https://doi.org/10.1126/scitranslmed.aaq1802.
515. Embree MC, Chen M, Pylawka S, Kong D, Iwaoka GM, Kalajzic I, Yao H, Shi C, Sun D, Sheu TJ, Koslovsky DA, Koch A, Mao JJ. Exploiting endogenous fibrocartilage stem cells to regenerate cartilage and repair joint injury. Nat Commun. 2016;7:13073. https://doi.org/10.1038/ncomms13073.
516. Ruscitto A, Scarpa V, Morel M, Pylawka S, Shawber CJ, Embree MC. Notch regulates fibrocartilage stem cell fate and is upregulated in inflammatory TMJ arthritis. J Dent Res. 2020;22034520924656. https://doi.org/10.1177/0022034520924656.
517. Charnley J. Arthroplasty of the hip. A new operation. Lancet. 1961;1(7187):1129–32. https://doi.org/10.1016/s0140-6736(61)92063-3.
518. Healy WL, Iorio R, Lemos MJ. Athletic activity after joint replacement. Am J Sports Med. 2001;29(3):377–88. https://doi.org/10.1177/03635465010290032301.

519. Zwetyenga N, Amroun S, Wajszczak BL, Moris V. Total temporomandibular joint protheses. Rev Stomatol Chir Maxillofac Chir Orale. 2016;117(4):285–93. https://doi.org/10.1016/j.revsto.2016.07.016.
520. Driemel O, Braun S, Muller-Richter UD, Behr M, Reichert TE, Kunkel M, Reich R. Historical development of alloplastic temporomandibular joint replacement after 1945 and state of the art. Int J Oral Maxillofac Surg. 2009;38(9):909–20. https://doi.org/10.1016/j.ijom.2009.01.022.
521. Zheng J, Chen X, Jiang W, Zhang S, Chen M, Yang C. An innovative total temporomandibular joint prosthesis with customized design and 3D printing additive fabrication: a prospective clinical study. J Transl Med. 2019;17(1):4. https://doi.org/10.1186/s12967-018-1759-1.
522. Arakeri G, Brennan PA. Dose-dependent sustained local release of dexamethasone from biodegradable thermosensitive hydrogel of PEG-PLGA-PEG triblock copolymers in the possible prevention of TMJ re-ankylosis (Arakeri's TMJ release technique). Med Hypotheses. 2012;78(5):682–6. https://doi.org/10.1016/j.mehy.2012.02.010.
523. Ueki K, Takazakura D, Marukawa K, Shimada M, Nakagawa K, Takatsuka S, Yamamoto E. The use of polylactic acid/polyglycolic acid copolymer and gelatin sponge complex containing human recombinant bone morphogenetic protein-2 following condylectomy in rabbits. J Craniomaxillofac Surg. 2003;31(2):107–14.
524. Kobayashi M, Oka M. Characterization of a polyvinyl alcohol-hydrogel artificial articular cartilage prepared by injection molding. J Biomater Sci Polym Ed. 2004;15(6):741–51. https://doi.org/10.1163/156856204774196135.
525. Stammen JA, Williams S, Ku DN, Guldberg RE. Mechanical properties of a novel PVA hydrogel in shear and unconfined compression. Biomaterials. 2001;22(8):799–806. https://doi.org/10.1016/s0142-9612(00)00242-8.
526. Dashnyam K, Lee JH, Mandakhbayar N, Jin GZ, Lee HH, Kim HW. Intra-articular biomaterials-assisted delivery to treat temporomandibular joint disorders. J Tissue Eng. 2018;9:2041731418776514. https://doi.org/10.1177/2041731418776514.
527. Spiller KL, Maher SA, Lowman AM. Hydrogels for the repair of articular cartilage defects. Tissue Eng Part B Rev. 2011;17(4):281–99. https://doi.org/10.1089/ten.TEB.2011.0077.
528. Cushing MC, Anseth KS. Materials science. Hydrogel cell cultures. Science. 2007;316(5828):1133–4. https://doi.org/10.1126/science.1140171.
529. Benya PD, Shaffer JD. Dedifferentiated chondrocytes reexpress the differentiated collagen phenotype when cultured in agarose gels. Cell. 1982;30(1):215–24. https://doi.org/10.1016/0092-8674(82)90027-7.
530. Yamaoka H, Asato H, Ogasawara T, Nishizawa S, Takahashi T, Nakatsuka T, Koshima I, Nakamura K, Kawaguchi H, Chung UI, Takato T, Hoshi K. Cartilage tissue engineering using human auricular chondrocytes embedded in different hydrogel materials. J Biomed Mater Res A. 2006;78(1):1–11. https://doi.org/10.1002/jbm.a.30655.
531. Passaretti D, Silverman RP, Huang W, Kirchhoff CH, Ashiku S, Randolph MA, Yaremchuk MJ. Cultured chondrocytes produce injectable tissue-engineered cartilage in hydrogel polymer. Tissue Eng. 2001;7(6):805–15. https://doi.org/10.1089/107632701753337744.
532. Mauck RL, Soltz MA, Wang CC, Wong DD, Chao PH, Valhmu WB, Hung CT, Ateshian GA. Functional tissue engineering of articular cartilage through dynamic loading of chondrocyte-seeded agarose gels. J Biomech Eng. 2000;122(3):252–60. https://doi.org/10.1115/1.429656.
533. Lammi MJ. Current perspectives on cartilage and chondrocyte mechanobiology. Biorheology. 2004;41(3–4):593–6.
534. Najar M, Krayem M, Merimi M, Burny A, Meuleman N, Bron D, Raicevic G, Lagneaux L. Insights into inflammatory priming of mesenchymal stromal cells: functional biological impacts. Inflamm Res. 2018;67(6):467–77. https://doi.org/10.1007/s00011-018-1131-1.
535. Fawzy El-Sayed KM, Hein D, Dorfer CE. Retinol/inflammation affect stemness and differentiation potential of gingival stem/progenitor cells via Wnt/beta-catenin. J Periodontal Res. 2019;54(4):413–23. https://doi.org/10.1111/jre.12643.
536. Badylak SF, Gilbert TW. Immune response to biologic scaffold materials. Semin Immunol. 2008;20(2):109–16. https://doi.org/10.1016/j.smim.2007.11.003.
537. Bayrak A, Tyralla M, Ladhoff J, Schleicher M, Stock UA, Volk HD, Seifert M. Human immune responses to porcine xenogeneic matrices and their extracellular matrix constituents *in vitro*. Biomaterials. 2010;31(14):3793–803. https://doi.org/10.1016/j.biomaterials.2010.01.120.
538. Brown BN, Chung WL, Almarza AJ, Pavlick MD, Reppas SN, Ochs MW, Russell AJ, Badylak SF. Inductive, scaffold-based, regenerative medicine approach to reconstruction of the temporomandibular joint disk. J Oral Maxillofac Surg. 2012;70(11):2656–68. https://doi.org/10.1016/j.joms.2011.12.030.
539. Makris EA, Gomoll AH, Malizos KN, Hu JC, Athanasiou KA. Repair and tissue engineering techniques for articular cartilage. Nat Rev Rheumatol. 2015;11(1):21–34. https://doi.org/10.1038/nrrheum.2014.157.
540. Almarza AJ, Athanasiou KA. Seeding techniques and scaffolding choice for tissue engineering of the temporomandibular joint disk. Tissue Eng. 2004;10(11–12):1787–95. https://doi.org/10.1089/ten.2004.10.1787.
541. Wakitani S, Mitsuoka T, Nakamura N, Toritsuka Y, Nakamura Y, Horibe S. Autologous bone marrow stromal cell transplantation for repair of full-thickness

articular cartilage defects in human patellae: two case reports. Cell Transplant. 2004;13(5):595–600. https://doi.org/10.3727/000000004783983747.
542. Wakitani S, Nawata M, Tensho K, Okabe T, Machida H, Ohgushi H. Repair of articular cartilage defects in the patello-femoral joint with autologous bone marrow mesenchymal cell transplantation: three case reports involving nine defects in five knees. J Tissue Eng Regen Med. 2007;1(1):74–9. https://doi.org/10.1002/term.8.
543. Mata M, Milian L, Oliver M, Zurriaga J, Sancho-Tello M, de Llano JJM, Carda C. *In vivo* articular cartilage regeneration using human dental pulp stem cells cultured in an alginate scaffold: a preliminary study. Stem Cells Int. 2017;2017:8309256. https://doi.org/10.1155/2017/8309256.
544. Fernandes TL, Cortez de SantAnna JP, Frisene I, Gazarini JP, Gomes Pinheiro CC, Gomoll AH, Lattermann C, Hernandez AJ, Franco Bueno D. Systematic review of human dental pulp stem cells for cartilage regeneration. Tissue Eng Part B Rev. 2020;26(1):1–12. https://doi.org/10.1089/ten.TEB.2019.0140.
545. Aryaei A, Vapniarsky N, Hu JC, Athanasiou KA. Recent tissue engineering advances for the treatment of temporomandibular joint disorders. Curr Osteoporos Rep. 2016;14(6):269–79. https://doi.org/10.1007/s11914-016-0327-y.
546. Ripley LS. Estimation of in-vivo miscoding rates. J Mol Biol. 1988;202(1):17–34. https://doi.org/10.1016/0022-2836(88)90514-1.
547. Blunk T, Sieminski AL, Gooch KJ, Courter DL, Hollander AP, Nahir AM, Langer R, Vunjak-Novakovic G, Freed LE. Differential effects of growth factors on tissue-engineered cartilage. Tissue Eng. 2002;8(1):73–84. https://doi.org/10.1089/107632702753503072.
548. Hanaoka K, Tanaka E, Takata T, Miyauchi M, Aoyama J, Kawai N, Dalla-Bona DA, Yamano E, Tanne K. Platelet-derived growth factor enhances proliferation and matrix synthesis of temporomandibular joint disc-derived cells. Angle Orthod. 2006;76(3):486–92. https://doi.org/10.1043/0003-3219(2006)076[0486:PGFEPA]2.0.CO;2.
549. Mumme M, Barbero A, Miot S, Wixmerten A, Feliciano S, Wolf F, Asnaghi AM, Baumhoer D, Bieri O, Kretzschmar M, Pagenstert G, Haug M, Schaefer DJ, Martin I, Jakob M. Nasal chondrocyte-based engineered autologous cartilage tissue for repair of articular cartilage defects: an observational first-in-human trial. Lancet. 2016;388(10055):1985–94. https://doi.org/10.1016/S0140-6736(16)31658-0.
550. de Souza Tesch R, Takamori ER, Menezes K, Carias RBV, Dutra CLM, de Freitas Aguiar M, Torraca TSS, Senegaglia AC, Rebelatto CLK, Daga DR, Brofman PRS, Borojevic R. Temporomandibular joint regeneration: proposal of a novel treatment for condylar resorption after orthognathic surgery using transplantation of autologous nasal septum chondrocytes, and the first human case report. Stem Cell Res Ther. 2018;9(1):94. https://doi.org/10.1186/s13287-018-0806-4.
551. Brown BN, Chung WL, Pavlick M, Reppas S, Ochs MW, Russell AJ, Badylak SF. Extracellular matrix as an inductive template for temporomandibular joint meniscus reconstruction: a pilot study. J Oral Maxillofac Surg. 2011;69(12):e488–505. https://doi.org/10.1016/j.joms.2011.02.130.
552. Almarza AJ, Brown BN, Arzi B, Angelo DF, Chung W, Badylak SF, Detamore M. Preclinical animal models for temporomandibular joint tissue engineering. Tissue Eng Part B Rev. 2018;24(3):171–8. https://doi.org/10.1089/ten.TEB.2017.0341.
553. Williams AC. Facial expression of pain: an evolutionary account. Behav Brain Sci. 2002;25(4):439–55. discussion 455-488
554. Langford DJ, Bailey AL, Chanda ML, Clarke SE, Drummond TE, Echols S, Glick S, Ingrao J, Klassen-Ross T, Lacroix-Fralish ML, Matsumiya L, Sorge RE, Sotocinal SG, Tabaka JM, Wong D, van den Maagdenberg AM, Ferrari MD, Craig KD, Mogil JS. Coding of facial expressions of pain in the laboratory mouse. Nat Methods. 2010;7(6):447–9. https://doi.org/10.1038/nmeth.1455.
555. Iwata K, Takeda M, Oh SB, Shinoda M. Neurophysiology of orofacial pain. In: Farah CS, Balasubramaniam R, McCullough MJ, editors. Contemporary oral medicine. New York: Springer; 2017. p. 1–23.
556. Tsuboi Y, Takeda M, Tanimoto T, Ikeda M, Matsumoto S, Kitagawa J, Teramoto K, Simizu K, Yamazaki Y, Shima A, Ren K, Iwata K. Alteration of the second branch of the trigeminal nerve activity following inferior alveolar nerve transection in rats. Pain. 2004;111(3):323–34. https://doi.org/10.1016/j.pain.2004.07.014.
557. Klasser GD, Goulet J-P, Laat AD, Manfredini D. Classification of orofacial pain. In: Farah CS, Balasubramaniam R, McCullough MJ, editors. Contemporary oral medicine. Cham: Springer; 2017. p. 1–23.
558. Okeson JP. Orofacial pain: guidelines for assessment, classification, and management. Chicago, IL: Quintessence Publishing; 1996.
559. Garlet GP, Giannobile WV. Macrophages: the bridge between inflammation resolution and tissue repair? J Dent Res. 2018;97(10):1079–81. https://doi.org/10.1177/0022034518785857.
560. Goto T, Oh SB, Takeda M, Shinoda M, Sato T, Gunjikake KK, Iwata K. Recent advances in basic research on the trigeminal ganglion. J Physiol Sci. 2016;66(5):381–6. https://doi.org/10.1007/s12576-016-0448-1.
561. Merskey H, Bogduk N (2011) Classification of chronic pain: descriptions of chronic pain syndromes and definitions of pain terms. IASP Taxonomy Working Group.
562. Fortino VR, Pelaez D, Cheung HS. Concise review: stem cell therapies for neuropathic pain. Stem

Cells Transl Med. 2013;2(5):394–9. https://doi.org/10.5966/sctm.2012-0122.
563. Trang T, Beggs S, Salter MW. Brain-derived neurotrophic factor from microglia: a molecular substrate for neuropathic pain. Neuron Glia Biol. 2011;7(1):99–108. https://doi.org/10.1017/S1740925X12000087.
564. Wright EF. Referred craniofacial pain patterns in patients with temporomandibular disorder. J Am Dent Assoc. 2000;131(9):1307–15. https://doi.org/10.14219/jada.archive.2000.0384.
565. Cairns BE, Hu JW, Arendt-Nielsen L, Sessle BJ, Svensson P. Sex-related differences in human pain and rat afferent discharge evoked by injection of glutamate into the masseter muscle. J Neurophysiol. 2001;86(2):782–91. https://doi.org/10.1152/jn.2001.86.2.782.
566. Ossipov MH. Growth factors and neuropathic pain. Curr Pain Headache Rep. 2011;15(3):185–92. https://doi.org/10.1007/s11916-011-0183-5.
567. Cova L, Armentero MT, Zennaro E, Calzarossa C, Bossolasco P, Busca G, Lambertenghi Deliliers G, Polli E, Nappi G, Silani V, Blandini F. Multiple neurogenic and neurorescue effects of human mesenchymal stem cell after transplantation in an experimental model of Parkinson's disease. Brain Res. 2010;1311:12–27. https://doi.org/10.1016/j.brainres.2009.11.041.
568. Wang J, Ding F, Gu Y, Liu J, Gu X. Bone marrow mesenchymal stem cells promote cell proliferation and neurotrophic function of Schwann cells *in vitro* and *in vivo*. Brain Res. 2009;1262:7–15. https://doi.org/10.1016/j.brainres.2009.01.056.
569. Klass M, Gavrikov V, Drury D, Stewart B, Hunter S, Denson DD, Hord A, Csete M. Intravenous mononuclear marrow cells reverse neuropathic pain from experimental mononeuropathy. Anesth Analg. 2007;104(4):944–8. https://doi.org/10.1213/01.ane.0000258021.03211.d0.
570. Siniscalco D, Giordano C, Galderisi U, Luongo L, Alessio N, Di Bernardo G, de Novellis V, Rossi F, Maione S. Intra-brain microinjection of human mesenchymal stem cells decreases allodynia in neuropathic mice. Cell Mol Life Sci. 2010;67(4):655–69. https://doi.org/10.1007/s00018-009-0202-4.
571. Reid AJ, Sun M, Wiberg M, Downes S, Terenghi G, Kingham PJ. Nerve repair with adipose-derived stem cells protects dorsal root ganglia neurons from apoptosis. Neuroscience. 2011;199: 515–22. https://doi.org/10.1016/j.neuroscience.2011.09.064.
572. De Francesco F, Tirino V, Desiderio V, Ferraro G, D'Andrea F, Giuliano M, Libondi G, Pirozzi G, De Rosa A, Papaccio G. Human CD34/CD90 ASCs are capable of growing as sphere clusters, producing high levels of VEGF and forming capillaries. PLoS One. 2009;4(8):e6537. https://doi.org/10.1371/journal.pone.0006537.
573. Vickers ER, Karsten E, Flood J, Lilischkis R. A preliminary report on stem cell therapy for neuropathic pain in humans. J Pain Res. 2014;7:255–63. https://doi.org/10.2147/JPR.S63361.
574. Manion J, Khuong T, Harney D, Littleboy JB, Ruan T, Loo L, Costigan M, Larance M, Caron L, Neely GG. Human induced pluripotent stem cell-derived GABAergic interneuron transplants attenuate neuropathic pain. Pain. 2020;161(2):379–87. https://doi.org/10.1097/j.pain.0000000000001733.
575. Jergova S, Gajavelli S, Varghese MS, Shekane P, Sagen J. Analgesic effect of recombinant GABAergic cells in a model of peripheral neuropathic pain. Cell Transplant. 2016;25(4):629–43. https://doi.org/10.3727/096368916X690782.
576. Meents JE, Bressan E, Sontag S, Foerster A, Hautvast P, Rosseler C, Hampl M, Schuler H, Goetzke R, Le TKC, Kleggetveit IP, Le Cann K, Kerth C, Rush AM, Rogers M, Kohl Z, Schmelz M, Wagner W, Jorum E, Namer B, Winner B, Zenke M, Lampert A. The role of Nav1.7 in human nociceptors: insights from human induced pluripotent stem cell-derived sensory neurons of erythromelalgia patients. Pain. 2019;160(6):1327–41. https://doi.org/10.1097/j.pain.0000000000001511.
577. Namer B, Schmidt D, Eberhardt E, Maroni M, Dorfmeister E, Kleggetveit IP, Kaluza L, Meents J, Gerlach A, Lin Z, Winterpacht A, Dragicevic E, Kohl Z, Schuttler J, Kurth I, Warncke T, Jorum E, Winner B, Lampert A. Pain relief in a neuropathy patient by lacosamide: proof of principle of clinical translation from patient-specific iPS cell-derived nociceptors. EBioMedicine. 2019;39:401–8. https://doi.org/10.1016/j.ebiom.2018.11.042.
578. Pereira T, Gartner A, Amorim I, Almeida A, Caseiro AR, Armada-da-Silva PA, Amado S, Fregnan F, Varejao AS, Santos JD, Bartolo PJ, Geuna S, Luis AL, Mauricio AC. Promoting nerve regeneration in a neurotmesis rat model using poly(DL-lactide-epsilon-caprolactone) membranes and mesenchymal stem cells from the Wharton's jelly: *in vitro* and *in vivo* analysis. Biomed Res Int. 2014;2014:302659. https://doi.org/10.1155/2014/302659.
579. Huang T, He D, Kleiner G, Kuluz J. Neuron-like differentiation of adipose-derived stem cells from infant piglets *in vitro*. J Spinal Cord Med. 2007;30(Suppl 1):S35–40. https://doi.org/10.1080/10790268.2007.11753967.
580. Janebodin K, Horst OV, Ieronimakis N, Balasundaram G, Reesukumal K, Pratumvinit B, Reyes M. Isolation and characterization of neural crest-derived stem cells from dental pulp of neonatal mice. PLoS One. 2011;6(11):e27526. https://doi.org/10.1371/journal.pone.0027526.
581. Gervois P, Struys T, Hilkens P, Bronckaers A, Ratajczak J, Politis C, Brone B, Lambrichts I, Martens W. Neurogenic maturation of human dental pulp stem cells following neurosphere generation induces morphological and electrophysiological characteristics of functional neurons. Stem Cells Dev. 2015;24(3):296–311. https://doi.org/10.1089/scd.2014.0117.

582. Luke AM, Patnaik R, Kuriadom S, Abu-Fanas S, Mathew S, Shetty KP. Human dental pulp stem cells differentiation to neural cells, osteocytes and adipocytes-An *in vitro* study. Heliyon. 2020;6(1):e03054. https://doi.org/10.1016/j.heliyon.2019.e03054.
583. Sasaki R, Aoki S, Yamato M, Uchiyama H, Wada K, Ogiuchi H, Okano T, Ando T. PLGA artificial nerve conduits with dental pulp cells promote facial nerve regeneration. J Tissue Eng Regen Med. 2011;5(10):823–30. https://doi.org/10.1002/term.387.
584. Yang C, Li X, Sun L, Guo W, Tian W. Potential of human dental stem cells in repairing the complete transection of rat spinal cord. J Neural Eng. 2017;14(2):026005. https://doi.org/10.1088/1741-2552/aa596b.
585. Sakai K, Yamamoto A, Matsubara K, Nakamura S, Naruse M, Yamagata M, Sakamoto K, Tauchi R, Wakao N, Imagama S, Hibi H, Kadomatsu K, Ishiguro N, Ueda M. Human dental pulp-derived stem cells promote locomotor recovery after complete transection of the rat spinal cord by multiple neuro-regenerative mechanisms. J Clin Invest. 2012;122(1):80–90. https://doi.org/10.1172/JCI59251.
586. Nakamura M, Okano H. Cell transplantation therapies for spinal cord injury focusing on induced pluripotent stem cells. Cell Res. 2013;23(1):70–80. https://doi.org/10.1038/cr.2012.171.
587. Csobonyeiova M, Polak S, Zamborsky R, Danisovic L. Recent Progress in the regeneration of spinal cord injuries by induced pluripotent stem cells. Int J Mol Sci. 2019;20(15):3838. https://doi.org/10.3390/ijms20153838.
588. Ikeda M, Uemura T, Takamatsu K, Okada M, Kazuki K, Tabata Y, Ikada Y, Nakamura H. Acceleration of peripheral nerve regeneration using nerve conduits in combination with induced pluripotent stem cell technology and a basic fibroblast growth factor drug delivery system. J Biomed Mater Res A. 2014;102(5):1370–8. https://doi.org/10.1002/jbm.a.34816.
589. Wang A, Tang Z, Park IH, Zhu Y, Patel S, Daley GQ, Li S. Induced pluripotent stem cells for neural tissue engineering. Biomaterials. 2011;32(22):5023–32. https://doi.org/10.1016/j.biomaterials.2011.03.070.

Index

© Springer Nature Switzerland AG 2021
S. Hosseinpour et al. (eds.), *Regenerative Approaches in Dentistry*,
https://doi.org/10.1007/978-3-030-59809-9

P

R

S

T

V

X